Methods in Neurosciences

Volume 25

Receptor Molecular Biology

Methods in Neurosciences

Editor-in-Chief

P. Michael Conn

Methods in Neurosciences

Volume 25

Receptor Molecular Biology

Edited by

Stuart C. Sealfon

Dr. Arthur M. Fishberg Research Center in Neurobiology and
Department of Neurology
Mount Sinai School of Medicine
New York, New York

ACADEMIC PRESS
San Diego New York Boston London Sydney Tokyo Toronto

Front cover photograph: Schematic representation of GnRH bound to the transmembrane portion of the GnRH receptor embedded in a membrane environment. The model was assembled by Karel Konvicka in the laboratory of Harel Weinstein, using a phospholipid patch taken from the results of molecular dynamics simulations carried out by David Garmer, a GnRH structure calculated by Frank Guarnieri, and a model of the transmembrane portion of the GnRH receptor constructed by Karel Konvicka.

This book is printed on acid-free paper. ♾

Academic Press, Inc.
A Division of Harcourt Brace & Company
525 B Street, Suite 1900, San Diego, California 92101-4495

United Kingdom Edition published by
Academic Press Limited
24-28 Oval Road, London NW1 7DX

International Standard Serial Number: 1043-9471

International Standard Book Number: 0-12-185295-4

PRINTED IN THE UNITED STATES OF AMERICA
95 96 97 98 99 00 EB 9 8 7 6 5 4 3 2 1

Table of Contents

Contributors to Volume 25

Article numbers are in parentheses following the names of contributors. Affiliations listed are current.

MARJORIE A. ARIANO (21), Department of Neuroscience, The Chicago Medical School, North Chicago, Illinois 60064

JUAN A. BALLESTEROS (19), Departments of Physiology and Biophysics, Mount Sinai School of Medicine, New York, New York 10029

DIANE L. BARBER (12), Department of Stomatology, School of Dentistry, University of California, San Francisco, California 94143

CLAUDIA BOUVIER (5), College of Pharmacy, Oregon State University, Portland, Oregon 97331

JAMES R. BUNZOW (5), Vollum Institute for Advanced Biomedical Research, Oregon Health Sciences University, Portland, Oregon 97201

MARI R. CANDELORE (14), Department of Molecular Pharmacology and Biochemistry, Merck Research Laboratories, Rahway, New Jersey 07065

CYNTHIA J. L. CARRUTHERS (17), Laboratory of Molecular Biology and Biochemistry, Howard Hughes Medical Institute, Rockefeller University, New York, New York 10021

MICHAEL CASCIO (10), Department of Molecular Genetics and Biochemistry, University of Pittsburgh School of Medicine, Pittsburgh, Pennsylvania 15219

BRIAN J. CILIAX (20), Department of Neurology, Emory University School of Medicine, Atlanta, Georgia 30322

J. DAVIDSON (8), Regulatory Peptide Research Unit, Department of Chemical Pathology, University of Cape Town Medical School, Observatory 7925, Cape Town, South Africa

SHARON EDMUNDS (20), Department of Neurology, Emory University School of Medicine, Atlanta, Georgia 30322

C. FLANAGAN (8), Regulatory Peptide Research Unit, Department of Medicine, University of Cape Town Medical School, Observatory 7925, Cape Town, South Africa

TUNG MING FONG (14), Department of Molecular Pharmacology, and Biochemistry, Merck Research Laboratories, Rahway, New Jersey 07065

BOAZ GILLO (6), Department of Physiology, Hadassah Medical School, Jerusalem 91120, Israel

DAVID K. GRANDY (5), Vollum Institute for Advanced Biomedical Research and Department of Cell Biology and Anatomy, Oregon Health Sciences University, Portland, Oregon 97201

JEAN-GÉRARD GUILLET (18), INSERM U152, Institut Cochin de Génétique Moléculaire, F-75074 Paris, France

CRAIG HEILMAN (20), Department of Neurology, Emory University School of Medicine, Atlanta, Georgia 30322

STEVEN M. HERSCH (20), Department of Neurology, Emory University School of Medicine, Atlanta, Georgia 30322

JOHAN HOEBEKE (18), Laboratoire d'Enzymologie et de Chimie des Proteines, Faculte de Medecine, F-37032 Tours Cedex, France

MOSHE JAKUBOWSKI (22), Department of Obstetrics, Gynecology, and Reproductive Biology, Beth Israel Hospital, Harvard Medical School, Boston, Massachusetts 02215

ROLF H. JOHO (4), Department of Cell Biology and Neuroscience, The University of Texas Southwestern Medical Center, Dallas, Texas 75235

EWEN F. KIRKNESS (7), Department of Molecular and Cellular Biology, The Institute for Genomic Research, Gaithersburg, Maryland 20878

BRIAN K. KOBILKA (15), Howard Hughes Medical Institute, Stanford University Medical Center, Stanford, California 94305

ALLAN I. LEVEY (20), Department of Neurology, Emory University School of Medicine, Atlanta, Georgia 30322

CRAIG C. MALBON (23), Department of Molecular Pharmacology, School of Medicine, State University of New York at Stony Brook, Stony Brook, New York 11794

DARIA MALEK (16), Institut fur Pharmakologie und Toxikologie, Universität Ulm, D-89069 Ulm, Germany

R. P. MILLAR (8), Regulatory Peptide Research Unit, Department of Chemical Pathology, University of Cape Town Medical School, Observatory 7925, Cape Town, South Africa

ANITA MISRA-PRESS (13), Vollum Institute for Advanced Biomedical Research, Oregon Health Sciences University, Portland, Oregon 97201

CHRISTOPHER M. MOXHAM (23), Department of Molecular Pharmacology, School of Medicine, State University of New York at Stony Brook, Stony Brook, New York 11794

GERALD MÜNCH (16), The Garvan Institute of Medical Research, St. Vincent's Hospital, Darlinghurst, Sydney, New South Wales 2010, Australia

RANDALL B. MURPHY (2), Department of Chemistry, New York University, New York, New York 10003

KIM A. NEVE (9, 12), Veteran Affairs Medical Center, and Department of Psychiatry and Pharmacology, Oregon Health Sciences University, Portland, Oregon 97201

RACHEL L. NEVE (9), Department of Genetics, Harvard Medical School and McLean Hospital, Belmont, Massachusetts 02178

JOHN C. OWICKI (11), Molecular Devices Corporation, Sunnyvale, California 94089

DIETER PALM (16), Theodor-Boveri-Institut, Physiologische Chemie I, Universität Wurzburg, D-97074 Wurzburg, Germany

MING-GUI PAN (13), Cardiovascular Research Institute, University of California at San Francisco, San Francisco, California 94143

MARY P. ROSSER (12), Department of Biochemical Screening, Bristol-Myers Squibb Company, Wallingford, Connecticut 06492

CARMEN SAEZ (5), Department of Anatomical Pathology, University Hospital Virgen del Rocio, Sevilla, Spain

THOMAS P. SAKMAR (17), Laboratory of Molecular Biology and Biochemistry, Howard Hughes Medical Institute, Rockefeller University, New York, New York 10021

JOHN A. SALON (11), Synaptic Pharmaceutical Corporation, Paramus, New Jersey 07652

DAVID I. SCHUSTER (2), Department of Chemistry, New York University, New York, New York 10003

STUART C. SEALFON (1), Dr. Arthur M. Fishberg Research Center in Neurobiology, and Department of Neurology, Mount Sinai School of Medicine, New York, New York 10029

MEILING SHIH (23), Department of Molecular Pharmacology, School of Medicine, State University of New York at Stony Brook, Stony Brook, New York 11794

DAVID R. SIBLEY (21), Molecular Neuropharmacology Section, National Institutes for Neurological Diseases and Stroke, National Institutes of Health, Bethesda, Maryland 20892

PHILIP J. S. STORK (13), Vollum Institute for Advanced Biomedical Research, Oregon Health Sciences University, Portland, Oregon 97201

CATHERINE D. STRADER (14), Department of Molecular Pharmacology and Biochemistry, Merck Research Laboratories, Rahway, New Jersey 07065

A. DONNY STROSBERG (18), UPR 0415, Immunopharmacologic Moléculaire, Institut Cochin de Génétique Moléculaire, F-75014 Paris, France

SANKURATRI SURYANARAYANA (15), CV Therapeutics, Palo Alto, California 94304

MANAMI TSUTSUMI (6), Dr. Arthur M. Fishberg Research Center in Neurobiology, Mount Sinai School of Medicine, New York, New York 10029

J. CRAIG VENTER (7), Department of Gene Discovery and Physiology, The Institute for Genomic Research, Gaithersburg, Maryland 20878

I. WAKEFIELD (8), Regulatory Peptide Research Unit, Department of Chemical Pathology, University of Cape Town Medical School, Observatory 7925, Cape Town, South Africa

MICHAEL J. WALSH (3), Department of Neurology, Mount Sinai School of Medicine, New York, New York 10029

HAREL WEINSTEIN (19), Departments of Physiology and Biophysics, Mount Sinai School of Medicine, New York, New York 10029

HUI-JUAN ZHANG (4), Department of Cell Biology and Neuroscience, The University of Texas Southwestern Medical Center, Dallas, Texas 75235

QUN-YONG ZHOU (5), Howard Hughes Medical Institute, University of Washington, Seattle, Washington 98195

ROGER D. ZÜHLKE (4), Department of Cell Biology and Neuroscience, The University of Texas Southwestern Medical Center, Dallas, Texas 75235

Preface

With the maturing of DNA technology, we are rapidly arriving at the threshold of a new era in biology. The observational and descriptive foundations of biological science have culminated in the recent elucidation of the amino acid and gene sequences of many proteins of biological significance. With the unfolding of the genome project, we can even foresee the end of cloning: the description of the sequence of every mammalian protein. The challenge for molecular neuroscience is to define research goals in the post-cloning age. Nowhere is this sea change more evident than in the molecular biology of receptors. In little more than a decade the cloning of a receptor has evolved from a nearly inconceivable *tour de force* to become a weekly deluge of new successes. The elucidation of these receptor sequences provides the foundation for the next stage of investigation, a period in which profound insight into receptor structure, regulatory control, and function can be attained.

The contributions of this volume reflect this exciting transitional period in the field. The methodologies of receptor cloning remain critical and several approaches are covered. It is no coincidence that the last word on cloning methods in this volume covers large-scale automated cloning techniques. Most of this volume, however, concentrates on the methodologies that come into play after receptor cloning, including expression systems, structure–function studies, computational modeling, localization, and regulation.

I wish to express my gratitude to the contributors for completing their chapters and for their willingness to share their insights and tricks, to Ms. Betsy Chalfin and to the staff of Academic Press, especially Ms. Shirley Light, for editorial assistance, and to my wife Celia and children Rebecca, Rachel, and Adam for keeping the home front together through yet another deadline.

STUART C. SEALFON

Methods in Neurosciences

Volume 1 Gene Probes
Edited by P. Michael Conn

Volume 2 Cell Culture
Edited by P. Michael Conn

Volume 3 Quantitative and Qualitative Microscopy
Edited by P. Michael Conn

Volume 4 Electrophysiology and Microinjection
Edited by P. Michael Conn

Volume 5 Neuropeptide Technology: Gene Expression and Neuropeptide Receptors
Edited by P. Michael Conn

Volume 6 Neuropeptide Technology: Synthesis, Assay, Purification, and Processing
Edited by P. Michael Conn

Volume 7 Lesions and Transplantation
Edited by P. Michael Conn

Volume 8 Neurotoxins
Edited by P. Michael Conn

Volume 9 Gene Expression in Neural Tissues
Edited by P. Michael Conn

Volume 10 Computers and Computations in the Neurosciences
Edited by P. Michael Conn

Volume 11 Receptors: Model Systems and Specific Receptors
Edited by P. Michael Conn

Volume 12 Receptors: Molecular Biology, Receptor Subclasses, Localization, and Ligand Design
Edited by P. Michael Conn

Volume 13 Neuropeptide Analogs, Conjugates, and Fragments
Edited by P. Michael Conn

Volume 14 Paradigms for the Study of Behavior
Edited by P. Michael Conn

Volume 15 Photoreceptor Cells
Edited by Paul A. Hargrave

Volume 16 Neurobiology of Cytokines (Part A)
Edited by Errol B. De Souza

Volume 17 Neurobiology of Cytokines (Part B)
Edited by Errol B. De Souza

Volume 18 Lipid Metabolism in Signaling Systems
Edited by John N. Fain

Volume 19 Ion Channels of Excitable Membranes
Edited by Toshio Narahashi

Section I

Receptor Cloning

[1] Approaches to Receptor Cloning

Stuart C. Sealfon

In the chapters that follow, specific methodologies which form crucial components of receptor cloning strategies are presented. In this section a brief overview of the relative merits of the available methods is presented. For the sake of clarity and conciseness, no attempt is made to provide comprehensive citations or precise methodologies. Detailed methods specific to receptor cloning can be found in the following chapters and general cloning methodologies of library construction and screening are available in the well-thumbed cloning manuals found in any molecular biology laboratory (1, 2).

It has often been observed that when multiple approaches have developed, they are all likely to have limitations. Successful and efficient receptor cloning requires the development of a reasonable overall strategy and the selection of the most appropriate methodologies for the problem. The writer Gustave Flaubert's comment about the design of a novel, "everything depends on the plan," is equally applicable to cloning. Before a cloning strategy is developed, assessment of the reasons the clone is being isolated and of the receptor class is critical. Representing the two extremes of difficulty is cloning a multi-subunit receptor of unknown structure, such as seminal nicotinic acetylcholine receptor cloning (3), and cloning a fragment of a receptor cDNA across mammalian species for use in nuclease protection assays or *in situ* hybridization, such as a monkey dopamine receptor probe with the sequence already characterized in several other mammalian species (4). In the case of a multiple subunit receptor, expression cloning may be impossible and the arduous task of receptor purification and microsequencing or antibody generation the requisite initial steps. In cloning a cDNA fragment known in another species, an afternoon of PCR with degenerate oligonucleotides will often produce the desired fragment.

The components of receptor cloning can be separated into the intertwined components of screening strategy, candidate testing, and confirmation. The various approaches are summarized in Table I.

Receptor Purification

Receptors represent a low percentage of total cellular protein and their purification is notoriously difficult. Most of the first breakthroughs in receptor cloning of a particular structural class depended on isolation of the receptor,

Methods in Neurosciences, Volume 25

TABLE I Approaches to Screening and Identifying Receptor Clones[a]

Screening strategy

I. Receptor purification*
  A. Microsequencing*
    i. Screen library with degenerate oligonucleotides
    ii. Polymerase chain reaction with degenerate oligonucleotides and screen cDNA or genomic library for full-length clone
  B. Raise antisera to protein
    i. Screen cDNA expression library
II. Expression cloning
  A. *Xenopus* oocyte expression and sib selection*
  B. Cell line expression and screening by binding or functional assay
III. Homology screening
  A. Low-stringency genomic or cDNA library screening with related sequences
  B. Polymerase chain reaction with degenerate oligonucleotides: screen cDNA or genomic library for full-length clone*
IV. Sib selection by hybrid arrest or oocyte expression
V. Sequence entire library and screen by computer using homology analysis*

Testing and confirmation

I. RNA distribution
  A. *In situ* hybridization
  B. Northern blot analysis
II. Functional expression in *Xenopus* oocytes*
  A. Hybrid arrest of expression*
  B. Assay of activation*
  C. Assay of binding*
III. Functional expression in mammalian cells*
  A. Assay of activation*
  B. Assay of binding*

[a] Techniques followed by a star are covered in other chapters in this volume.

including cloning of the *torpedo* nicotinic acetylcholine subunits (3), rhodopsin (5), and the β-adrenergic receptor (6). Because cloning by purification relies on binding for the isolation, this approach has the potential difficulty of leading to the isolation of a protein which binds the ligand but which is not the functional receptor. Examples of this include the cloning of several glutamate binding proteins (7–9). The difficulty of receptor cloning by protein isolation has fueled the development of alternative approaches. With the advent of expression cloning techniques and homology cloning, the role of receptor purification in receptor cloning is diminishing. A discussion of purification techniques can be found in Chapter 2 by Schuster and Murphy. Following purification, the investigator can proceed to microsequencing (see Chapter 3 by Walsh) or to generation of antisera. Partial sequence can be used to isolate a cDNA by library screening or polymerase chain reaction using degenerate oligonucleotide probes. If an antibody to the protein is developed, this reagent can be used to screen a bacterial expression library.

Expression Cloning

Two forms of expression cloning have been developed. One is based on *Xenopus* oocyte expression of RNA transcripts and relies on detection of receptor-mediated activation of signal transduction, most commonly elevation of intracellular calcium leading to opening of a calcium-dependent chloride channel. The first step is to establish expression of the receptor in oocytes and determine the method of detection. For the methods of electrophysiological detection of receptors expressed using heterologous RNA see Chapter 6 by Tsutsumi and Gillo. Oocyte expression cloning is discussed in Chapter 4 Zühlke *et al.* A phage cDNA library is generated from which synthetic capped RNA transcripts can be made. Pools of clones are tested by injection into oocytes, and the positive clone is ultimately identified by progressive division of the positive pool, a process called sib selection. This approach was first used by Masu and co-workers to clone the substance K receptor (10). This approach has usually been restricted to receptors containing a single subunit. Notable exceptions include the successful cloning of a glutamate AMPA receptor subunit (11) and an NMDA receptor subunit (12). Unlike the GABA and nicotinic receptors, single glutamate receptor subunits are capable of aggregating to form functional receptors, thus allowing detection. One limitation of oocyte expression cloning is the high sensitivity of the system which occasionally leads to the cloning of the wrong receptor. This can occur when the test ligand cross-activates other receptors under the screening conditions (13).

An alternative approach to expression cloning relies on mammalian cell expression for screening and selection. A library of cDNAs containing the target of interest is generated in a mammalian expression vector and screened by transforming mammalian cells in microtiter plates. The colonies of transformed cells can be tested by binding to filters or by functional assay. For two examples of this well-utilized approach to cloning, the reader is referred to the cloning of the vasopressin receptor (14) and to the cloning of the activin receptor (15).

Homology Screening

The isolation of increasing numbers of receptor cDNAs has revealed that they belong to structural and evolutionarily related gene families. The high conservation of particular amino acid sequence motifs among these receptors underlies the ability to clone these receptors by homology screening.

Homology cloning has taken several forms. A number of receptors including the G-21 sequence [which later proved to be the serotonin $5HT_{1A}$ receptor (16)] and the dopamine D_2 receptor (17) were isolated by low-stringency library screening, in both cases using the β_2-adrenergic receptor gene as a probe. Many receptors have been cloned using polymerase chain reaction with oligonucleotides designed against two conserved receptor sequences, usually transmembrane domains. Receptors cloned by this approach include the ACTH receptor (18) and the gonadotropin-releasing hormone receptor (19). Detailed methodology for homology cloning can be found in Chapter 5 by Grandy *et al.*

One of the limitations of homology screening is that a receptor sequence may be isolated and classified as a receptor by sequence, but the ligand may be difficult to find. The cannabinoid receptor required extensive ligand screening before its identity was revealed (20) and an increasing number of "orphan receptors," putative receptors sequences of unknown ligand, are being reported (21–25). Orphan receptors can be tested by commercially available ligand screening battery (NOVA screen, Honover, MD) and evaluating mRNA distribution by *in situ* hybridization may provide a clue as to the nature of the receptor. Hybrid arrest studies in oocytes also provide a rapid method of testing the identity of a partial-length receptor cDNA clone (see Chapter 6 by Tsutsumi and Gillo). However, even after these studies, the actual identity of many of these clones remains unknown.

Automated Cloning

The development of large-scale automatic cloning techniques has enabled the partial-length sequencing of entire cDNA libraries. Putative receptor sequences can subsequently be identified by computerized analysis of the sequences generated. This approach, which has the power to identify entire orphanages, is described in detail in Chapter 7 by Kirkness and Venter.

Confirmation

Following isolation of a particular cDNA sequence, the receptor identity, if known, is confirmed by expression in *Xenopus* oocytes or mammalian cells and demonstration of appropriate ligand binding and/or signal transduction coupling. While it may appear self-evident that identification of a receptor cDNA must include functional characterization, this rule has not always been followed. The publication of a long-sought receptor sequence in one species, for example, is commonly followed by the rapid isolation of the receptor cDNA in many mammalian species. Occasionally, the sequence in the new species is published without functional characterization. The existence of expressed receptor pseudogenes (26, 27), however, suggests that it is incorrect to identify a receptor, even one with striking sequence homology to a well-characterized receptor in another species, solely on the basis of sequence. A clone can only be accepted as an authentic cDNA for the receptor after functional characterization of the expressed receptor clone.

Conclusions

The techniques of receptor cloning have advanced dramatically in recent years, as evidenced by the increasing number of receptor cDNA "grails" being attained. While cloning a particular receptor target remains a difficult task, the success of many laboratories is evidence that the pursuit of a flexible and logical strategy is likely to succeed.

References

1. J. Sambrook, E. F. Fritsch, and T. Maniatis, "Molecular Cloning: A Laboratory Manual," 2nd ed. Cold Spring Harbor Lab. Press, Cold Spring Harbor, NY, 1989.
2. S. L. Berger, ed., "Methods in Enzymology," Vol. 152. Academic Press, San Diego, CA, 1987.
3. M. Noda, H. Takahashi, T. Tanabe, M. Toyosato, Y. Furutani, T. Hirose, M. Asai, S. Inayama, T. Miyata, and S. Numa. *Nature* (*London*) **299,** 793 (1982).
4. G. W. Huntley, J. H. Morrison, A. Prikhozhan, and S. C. Sealfon, *Mol. Brain Res.* **15,** 181 (1992).
5. J. Nathans and D. S. Hogness, *Cell* (*Cambridge, Mass.*) **34,** 807 (1983).
6. R. A. F. Dixon, B. K. Kobilka, D. J. Strader, J. L. Benovic, H. G. Dohlman, T. Frielle, M. A. Bolanowski, C. D. Bennett, E. Rands, R. E. Diehl, R. A. Mumford, E. E. Slater, I. S. Sigal, M. G. Caron, R. J. Lefkowitz, and C. D. Strader, *Nature* (*London*) **321,** 75 (1986).
7. K. Wada, C. J. Deschesne, S. Shimaski, R. G. King, K. Kusano, A. Buonanno, D. R. Hampson, C. Banner, R. J. Wenthold, and Y. Nakatani, *Nature* (*London*) **342,** 684 (1989).
8. P. Gregor, I. Mano, I. Maoz, M. McKeown, and V. I. Teichberg, *Nature* (*London*) **342,** 689 (1989).
9. K. N. Kumar, N. Tilakaratne, P. S. Johnson, A. E. Allen, and E. K. Michaelis, *Nature* (*London*) **354,** 70 (1991).

10. Y. Masu, K. Nakayam, H. Tamaki, Y. Harada, M. Kuno, and S. Nakanishi, *Nature* (*London*) **329,** 836 (1987).
11. M. Hollmann, A. O'Shea-Greenfield, S. W. Rogers, and S. Heinemann, *Nature* (*London*) **342,** 643 (1989).
12. K. Moriyoshi, M. Masu, T. Ishii, R. Shigemoto, N. Mizuno, and S. Nakanishi, *Nature* (*London*) **354,** 31 (1991).
13. S. S. Kakar, J. C. Sellers, D. C. Devor, L. C. Musgrove, and J. D. Neill, *Biochem. Biophys. Res. Commun.* **183,** 1090 (1992).
14. M. Birnbaumer, A. Seibold, S. Gilbert, M. Ishido, C. Barberis, A. Antaramian, P. Brabet, and W. Rosenthal, *Nature* (*London*) **357,** 333 (1992).
15. L. S. Mathews and W. W. Vale, *Cell* (*Cambridge, Mass.*) **65,** 973 (1991).
16. B. K. Kobilka, T. Frielle, S. Collins, T. Yang-Feng, T. S. Kobilka, U. Francke, R. J. Lefkowitz, and M. G. Caron, *Nature* (*London*) **329,** 75 (1987).
17. J. R. Bunzow, H. H. M. Van Tol, D. K. Grandy, P. Albert, J. Salon, M. Christie, C. A. Machida, K. A. Neve, and O. Civelli, *Nature* (*London*) **336,** 783 (1988).
18. K. G. Mountjoy, L. S. Robbins, M. T. Mortrud, and R. D. Cone, *Science* **257,** 1248 (1992).
19. M. Tsutsumi, W. Zhou, R. P. Millar, P. L. Mellon, J. L. Roberts, C. A. Flanagan, K. Dong, B. Gillo, and S. C. Sealfon, *Mol. Endocrinol.* **6,** 1163 (1992).
20. L. A. Matsuda, S. J. Lolait, M. J. Brownstein, A. C. Young, and T. I. Bonner, *Nature* (*London*) **346,** 561 (1990).
21. W. Meyerhof, R. Müller-Brechlin, and D. Richter, *FEBS Lett.* **284,** 155 (1991).
22. W. Meyerhof, H. J. Paust, C. Schönrock, and D. Richter, *DNA Cell Biol.* **10,** 689 (1991).
23. H. P. Nothacker and C. Grimmelikhuijzen, *Biochem. Biophys. Res. Commun.* **197,** 1062 (1993).
24. K. A. Eidne, J. Zabavnik, T. Peters, S. Yoshida, L. Anderson, and P. L. Taylor, *FEBS Lett.* **292,** 243 (1991).
25. C. Eva, K. Keinänen, H. Monyer, P. Seeburg, and R. Sprengel, *FEBS Lett.* **271,** 81 (1990).
26. D. K. Grandy, Y. Zhang, C. Bouvier, Q. Y. Zhou, R. A. Johnson, L. Allen, K. Buck, J. R. Bunzow, J. Salon, and O. Civelli, *Proc. Natl. Acad. Sci. U.S.A.* **88,** 9175 (1991).
27. T. Nguyen, J. Bard, H. Jin, D. Taruscio, D. C. Ward, J. L. Kennedy, R. Weinshank, P. Seeman, and B. F. O'Dowd, *Gene* **109,** 211 (1991).

[2] Purification of Receptors

David I. Schuster and Randall B. Murphy

Overview

Purification of receptors for neurotransmitters, neuromodulators, opiates, drugs and other chemical agents that act in the central nervous system as well as the periphery has become an essential step in molecular characterization of such receptors and facilitates elucidation of the biochemical and physiological events which take place on interaction of these agents with their binding sites. Based on pharmacological data, it is now clear that many neurotransmitter receptors exist as multiple subtypes, as in the case of dopamine (D1, D2, D3, D4, and D5, thus far) and serotonin (of which there are at least seven subtypes, e.g., 5HT1A, 5HT1B, and 5HT2), which differ markedly in their affinity for various receptor agonists and antagonists. In such cases, it is of interest to isolate these receptor subtypes as discrete molecular entities in order to determine their structures and their mode of operation. There may also be subtle differences in receptors for the same chemical agent in different tissues and of course in different species. There has been enormous progress in the past few years in the determination of amino acid sequences of receptors using molecular biological techniques, using cDNA probes constructed from parital amino acid sequences of purified receptors or from highly conserved sequences in families of receptors, particularly transmembrane sequences of G protein-linked membrane-bound receptors (1). However, it is frequently the case that the molecular masses of cloned receptors differ from the masses of receptors purified directly from tissue preparations or identified by photoaffinity labeling, so that the native receptor is not absolutely identical to the receptor obtained by cloning. In most cases, such differences have been attributed to glycosylation, although the exact explanation is usually not known. Molecular sizing experiments often show that receptors of interest in neuroscience have molecular masses many times larger than those of cloned receptors, indicating the native receptors exist in some type of complex. Therefore, despite the gains in understanding of receptor structures from molecular biology, there remains active interest in purification of receptors directly from tissue sources.

This chapter focuses on the experimental approaches which have been used successfully for purification of neurotransmitter receptors and other receptors of interest in neuroscience and behavior, and the types of problems

Methods in Neurosciences, Volume 25

which are typically encountered in such studies. Later, specific methodologies which have been used in some particular cases are summarized. For the present purposes, we have not attempted to cover the entire literature of this subject, which is vast, but rather have chosen representative examples which illustrate available methodologies.

Experimental Design Considerations

Solubilization and Prepurification Procedures

Receptor densities in a given tissue are usually determined by standard radioligand binding assays, using ligands which specifically bind to the receptor of interest, usually in a membrane homogenate. A similar radioligand binding assay is generally used at each purification step in the procedure to define the degree of purification. For the purpose of purification, the receptor must first be solubilized from the membrane using a detergent. This has traditionally represented a principal major problem in affinity chromatography of neurotransmitter receptors. Most common detergents, such as the alkyl sulfonates, exemplified by SDS, are highly denaturing due to their charged character. A class of nonionic detergents exemplified by Triton X-100 first began to be used to solubilize membrane proteins in a pharmacologically active state. However, these detergents are still relatively harsh and will denature most G protein-linked receptors. The Triton-type detergents also absorb strongly in the ultraviolet region; hydrogenated Triton derivatives which have been developed to avoid this problem are also reportedly less denaturing. As an alternative, digitonin began to be employed for neurotransmitter receptor solubilization. This naturally occurring glycoside forms a van der Waals complex with cholesterol and is in addition itself a weak detergent. However, its limited solubility in aqueous buffers, particularly at the low temperatures generally employed for purification, severely limits its utility. Alkyl glycosides, typified by octyl glucoside or mannoside, are also commonly employed, as they tend to be relatively nondenaturing. Cholic acid has long been known as a detergent but is a relatively crude material and can be fairly denaturing. A major innovation in detergent methodology has been the development of the zwitterionic detergents, exemplified by 3-[(3-cholamidopropyl)dimethylammonio]-1-propanesulfonate (CHAPS). This and its related sulfoxide derivative CHAPSO are semisynthetic cholic acid derivatives. In summary, at this time the detergents which are used most commonly are CHAPS and the alkyl glycosides, usually separately but occasionally in combination. Solubilization itself often results in a small degree of receptor purification, as measured by specific binding activity [amount of radioligand bound (usually in fmol or pmol) per mg protein in the sample as determined by a standard protein assay].

A second major problem in the affinity purification of neurotransmitter receptors is avoiding protease-induced degradation of the receptor of interest. Since this is a common problem, the solubilization buffer often contains a protease inhibitor [commonly the serine protease inhibitor phenylmethanesulfonyl fluoride (PMSF)] or even a protease inhibitor cocktail. Since there are not especially good inhibitors against many proteases, it is impossible to design a mixture of protease inhibitors which a priori will prevent degradation of a particular receptor. One way to ameliorate this problem is to include a preliminary purification step prior to addition of the solubilized receptor preparation (SRP) to the affinity column. This step itself does not necessarily need to effect a high fold purification; rather, removal of contaminating proteases is the principal goal. This step could for example involve ammonium sulfate precipitation, as in our work on sigma receptors, or passage through an underivatized or "control" column, prepared from a material which does not bind to the receptor of interest. Whatever the methodology, the aim is to nonspecifically remove contaminating proteins and proteases. Some clever ways in which this has been achieved are detailed below.

Affinity Chromatography

Preparation of the Matrix and Adsorption of Receptors

In most cases, the principal procedure for receptor purification involves affinity chromatography. This procedure first involves preparation of an affinity matrix, which consists of a modified receptor ligand attached through a spacer arm to a solid support, in most cases Sepharose, Affigel (an agarose-based gel), or a silica-based gel. This often requires synthesis of a suitable derivative of a potent ligand for the receptor of interest which can be attached either directly to suitably activated support or, preferably, through a spacer arm. It is of course necessary to demonstrate by competitive radioligand binding assays that this derivative retains high affinity for the receptor to be purified. It is desirable to use a receptor ligand which is available in radiolabeled form, so that the synthesis of the affinity matrix can be carried out with receptor ligand "spiked" with a small amount of hot radiolabeled ligand, to determine how much ligand has in fact been successfully linked to the support. The choice of affinity ligand is of course crucial. It is necessary to choose a ligand which selectively binds to the receptor of interest and not to other receptors that might be present in the starting tissue. The ligand must show high affinity for the receptor of interest. On the other hand, if the affinity is too high, it may prove to be difficult to elute the receptor from the column. Following derivitization, the matrix is often treated with a reagent to cap remaining reactive functional groups on the solid support.

It is unclear if a long spacer arm is required for efficient affinity chromatographic purification of solubilized neurotransmitter receptors. If the pharmacologically active ligand is coupled to an affinity gel through a short spacer, it may not be able to access the active binding site of the solubilized receptor. Thus, it is usual practice to use some type of spacer, but few investigators appear to have explicitly demonstrated that a sizable spacer arm actually was requisite for successful affinity purification (2).

Once prepared, the column is generally subjected to a series of washes with the buffer used in the solubilization prior to addition to the SRP. In some cases, these washes specifically include protease inhibitors which presumably are nonspecifically adsorbed onto the solid support and help to prevent degradation of adsorbed receptors. The SRP can be added batchwise to the affinity matrix, and the mixture is then inserted into a column of appropriate dimensions. Alternatively, the matrix can be directly inserted into the column and the SRP is then passed through the column at a relatively slow flow rate; often the SRP is recycled several times through the column to maximize adsorption of receptors onto the column from the SRP. The extent to which the receptor is taken up by the matrix is generally determined by radioligand binding assays on the pass-through solutions. An alternative, which is not frequently used, is to nonspecifically radiolabel the receptor itself (e.g., by radioiodination using Bolton–Hunter reagent) and determine the amount of radioactivity on the column vis a vis the flow-through solutions.

A recent important innovation in affinity chromatography of neurotransmitter receptors is the use of alternative supports to agrose for immobilization of the ligand. These strategies use either a chemically derivatized silica, chemically derivatized microporous membranes, or even chemically derivatized cellulose ester hollow fibers (3). The principal significant advantage that these supports enjoy as compared with agaroses is a very high flow rate, which can be many milliliters per minute in the case of the silica and up to liters per minute in the case of the hollow fiber devices. These fast flow rates obviate many of the concerns as to protease degradation described above.

Elution of Receptors from the Affinity Column

The affinity column is usually washed with several column volumes of elution buffer to remove nonspecifically bound proteins from the matrix. Most directly, one washes the column until the UV absorption of the washes at 280 nm, where aromatic amino acids show absorption, reaches baseline optical density. Elution of the receptor from the column is the most problematic part of the procedure. A generally useful technique is to wash the column with a buffer containing a high concentration of sodium chloride, generally

0.25–1.5 *M*. In some cases, a salt gradient has been utilized successfully. The alternative method, which is most frequently used, is to elute the column with a buffer containing an appropriate concentration of a potent ligand for the receptor being purified. The problem is that such eluting ligands interfere with the radioligand binding assays that are usually performed to determine the recovery of active receptors from the column. In order to remove the ligands and restore receptors to active form for these assays, it is usually necessary to dialyze the eluted fractions (often for an extended time) or to pass them through a desalting column, such as Sephadex G-50. However, in certain cases, typified by the dopamine D2 receptor (see below), active receptors could only be obtained after reconstitution of the purified receptor protein into lipid vesicles. Thus, the failure to detect high binding activity in eluted fractions after dialysis or desalting may be due to the absence of a lipid environment which is required for the receptor protein to assume its biologically active conformation.

It is a common practice for receptors eluted from affinity columns to be further purified by subsequent passage through lectin or hydroxyapatite columns, or both. The lectin procedure is specific for purification of glycosylated proteins, which is generally the case for the receptors of interest in neuroscience. These procedures also serve to concentrate the samples, which simplifies subsequent radioligand binding assays as well as analysis by gel electrophoresis (SDS–PAGE). In some cases, partially purified receptor preparations have been subjected to a second round of purification on an affinity column. The precise nature and number of steps required to achieve the desired level of purification of the receptor of interest, usually complete homogeneity, can only be established by trial and error, using various combinations of the above techniques, as well as special methods that may be applicable in specific circumstances (see examples below).

Analysis of Purified Receptors

The two methods that are almost always used to analyze receptors recovered from purification procedures are radioligand binding assays and gel electrophoresis. The specific activity of the sample, coupled with the protein concentration determined using standard techniques, provides the standard measure of fold purification relative to the crude membrane homogenates or the solubilized preparation. The fold purification needed to achieve homogenity is calculated from the specific activity for binding of the radioligand to the purified vis a vis the unpurified sample (either as a crude homogenate or solubilized preparation) and the presumed molecular mass of the purified receptor, assuming only a single molecule of radioligand binds to the receptor protein. The number and location of bands in gels on SDS–PAGE for the purified receptor compared with membrane homogenates can be used to assess the extent of purification, particularly if the purified receptor migrates on SDS–PAGE as a single sharp band. For example, purified adenosine, dopamine, and adrenergic receptors appear as a single sharp band on SDS–PAGE, while several bands (at least four) are enriched in the purified *N*-methyl-D-aspartate (NMDA) receptor complex (see below). While the protein bands on SDS gels are generally detected using silver staining, nonspecific radioiodination of the protein and detection by autoradiography (on exposure of the gel to X-ray film over several days) are frequently used.

In many cases receptor homogeneity has been confirmed by affinity or photoaffinity labeling utilizing radiolabeled specific receptor ligands. Labeling patterns which demonstrate the pharmacological specificity characteristic of the particular receptor under investigation help to confirm that the purified receptor contains the ligand binding site. The identity of autoradiograms of specifically labeled purified receptors with silver staining of the same receptors following SDS–PAGE is generally taken as evidence that homogeneity has been achieved. Detailed examples for specific systems are discussed later.

If at all possible, it should be demonstrated that the purified receptor shows the characteristic pharmacological profile of the intact membrane-bound receptor, using radioligand binding studies. Other types of studies (e.g., regulatory effects on ion channels) may be appropriate in specific cases. It should be recognized that in some cases (e.g., dopamine D2 and opioid receptors) full biological activity of the purified receptor can only be observed after the receptor protein is reconstituted into a lipid environment, such as artifical vesicles or liposomes. Additionally, specific lipids may be required for full biological activity. Comparatively little is known in this area, and mixtures of phosphatidylcholine, phosphatidylethanolamine, and phosphatidylserine resembling those in crude brain lipid extracts are typically employed.

Another approach which has been used increasingly in the past few years is to raise antibodies against the purified receptor protein and to demonstrate that these antibodies directly affect the biological activity of the membrane-reconstituted receptor protein.

Control Studies

Certain types of control studies are frequently utilized to establish that purification of the receptor protein of interest and not some extraneous protein has been achieved by the experimental protocol utilized. When affinity chromatography is incorporated into the protocol, as it is in the vast majority of cases, one or more of the following types of studies is usually performed: (a) demonstration that when the solubilized receptor is preincubated with a high-affinity ligand for that receptor, as opposed to ligands for other receptors that may be present in the tissue, the receptor of interest is not adsorbed on the affinity matrix; (b) demonstration that affinity matrices that are similar to the one actually utilized, prepared from the same solid support and structurally related organic compounds that do not bind strongly to the receptor of interest, are not effective in purifying the receptor of interest; and (c) demonstration that elution of active receptors from the affinity column can be effected using only specific ligands for the receptor of interest, and not inactive stereoisomeric ligands and/or ligands for other receptor proteins. Specific examples illustrating such studies are described below.

Specific Methodologies

Representative examples from the literature are given below which illustrate techniques which are commonly used for receptor purification, as well as novel methods whose general applications are necessarily more limited.

The β-Adrenergic Receptor

In the classic study by Lefkowitz, Caron, and co-workers (4), β-adrenergic receptors (β-ARs) were purified from frog erythrocytes using a combination of ion-exchange and affinity chromatography. The affinity column consisted of alprenolol immobilized on Sepharose 6B in the presence of potassium persulfate at 25 or 40°C. Unreacted sulfhydryl groups on the support were blocked by treating the gel with iodoacetamide (100 m*M*, 25°C, 2 hr). Frozen purified frog erythrocyte membranes were thawed, washed twice with 25 m*M* Tris–HCl, pH 7.4, containing 2 m*M* $MgCl_2$ at 0–4°C, and treated with 1% digitonin as previously described (*J. Biol. Chem.* 1976, **251,** 2374–2384). Insoluble material was removed by centrifugation at 250,000*g* for 45 min.

Three liters of the 1% digitonin extract (from 700–900 ml of packed erythrocytes) were lyophilized and resuspended in 250–400 ml water prior to desalting on a Sephadex G-50 column (5 × 95 cm) equilibrated with 100 m*M* NaCl, 10 m*M* Tris–HCl, pH 7.4. The desalted extract was reacted batchwise with 50–60 ml of the Sepharose-alprenolol gel and cycled twice through the gel as described earlier. The bound β-AR activity could be eluted either with isoproterenol (agonist) or alprenolol (antagonist). At this point, SDS–PAGE analysis revealed a large number of bands. Therefore the material eluted from the second gel (10–30 ml) was applied to 1 ml of DEAE-Sepharose 6B-Cl ion-exchange column, which was eluted with 20 ml of a linear gradient of NaCl (0–0.5 *M*) in 0.2% digitonin, 10 m*M* Tris–HCl, pH 7.4. Those fractions showing [^{3}H]alprenolol binding activity were then rechromatographed on 3 ml of fresh alprenolol gel. The yield at each step and purification fold relative to crude membranes were as follows: (i) digitonin extraction: 70%, 6.6-fold; (ii) first alprenolol pass: 50–70%, 706-fold; (iii) second alprenolol pass: 50%, 9800-fold; (iv) DEAE-Sepharose: 60–80%, fold not determined; (v) final alprenolol pass: 30–50%, 55,000-fold. The overall yield of purified β-AR was 4–8%.

After two passes through the affinity gel but before DEAE-Sepharose, SDS–PAGE revealed a large number of bands, including polypeptides with mass >90kDa. These contaminants were removed by passage through DEAE-Sepharose. The purified β-AR was labeled nonspecifically using Na^{125}I and chloramine-T before the final passage through the affinity column and subjected to sucrose density gradient centrifugation. Those fractions coincident with [^{3}H]dihydroalprenolol binding activity in cold iodine-labeled preparations were pooled, lyophilized, and desalted, prior to SDS–PAGE. The autoradiogram following a 36-hr exposure showed a prominent band for the β-AR at 58 kDa. Material reactivated after SDS–PAGE revealed binding activity only in the 58-kDa region of the gel. The binding of [^{3}H]alprenolol to purified preparations displayed the affinity, specificity and stereoselectivity characteristic of membrane bound or solubilized β-ARs.

The Dopamine D2 Receptor

Several procedures for purification of the D2 receptor have been published. In the study of Caron and co-workers (5), the affinity matrix was prepared from [(carboxymethylene)oximino]spiperone (CMOS) which was synthesized from spiperone and carboxymethoxylamine hemihydrochloride. Epoxy-Sepharose 4B was converted to a free amino-containing Sepharose as follows: 50 ml of the gel was added to 100 ml of 1 *M* ethylenediamine in 0.1 *M* Na_2CO_3, pH 10.0, 22°C, 16–18 hr. The gel was then washed with 10 vol distilled water, 0.2 *M* acetic acid, and 50 m*M* NaOH and water again until the pH of the effluent was ca. 5. The CMOS was dissolved in dimethyl sulfoxide (DMSO; 100 mg in 50 ml) and amino-Sepharose 4B (50 ml equilibrated in water) was added slowly to the CMOS solution, pH 4.5. 1-Ethyl-3-[3-(dimethylamino)propyl]carbodiimide (EDAC) (1 g/50 ml gel) was added, and the pH was adjusted to 4.5. The mixture reacted for 12–16 hr at 22°C, 1 g of EDAC was added, and reaction continued for 8–10 hr. The derivatized gel was washed with 50% aq DMSO (0.5 liter/50 ml gel) and distilled water

(2.5 liter/50 ml gel) over a scintered glass funnel and stored at 4°C with 0.02% sodium azide.

Bovine pituitary homogenates solubilized with digitonin (DSP) were prepared in the presence of 10 mg/ml leupeptin, 0.1 m*M* PMSF, and 2 m*M* EDTA as protease inhibitors during homogenization and solubilization. After the CMOS-Sepharose was washed with 2–4 bed vol of 50 m*M* Tris–HCl, 100 m*M* NaCl, 2 m*M* $MgCl_2$, and 0.1% digitonin, pH 7.4, the DSP was loaded batchwise by incubation overnight with the gel. Typically, 2 ml of DSP was loaded per milliliter gel. Approximately 70–80% of the [^{3}H]spiperone binding activity was adsorbed on the CMOS-Sepharose gel, while most of the protein in the solubilized preparation (SP) was unretarded. The gel was washed in a column at 4°C with 10–15 bed vol of the above buffer for 2–3 hr at 200 ml/hr. The column was brought to 22°C and eluted with 2 bed vol of 10 m*M* haloperidol in the above buffer at 50 ml/hr. The eluted fractions were collected on ice and desalted by Sephadex G-50 chromatography. D2 receptor activity in the eluant was <5%. Attempts to stabilize the receptor by including stabilizing agents such as DTT, glycerol, BSA, and phospholipids; by replacing digitonin with other detergents; by adding protease inhibitors; or by eluting with other selective agonists and antagonist were unsuccessful. However, a 5- to 10-fold increase in recovered [^{3}H]spiperone binding activity was obtained when the D2 receptor was reinserted into lipid vesicles using a protocol developed earlier for β-adrenergic receptors.

Routinely, the recovery of [^{3}H]spiperone binding activity after reconstitution was 40–50% of that originally adsorbed onto the gel. Binding of ligands to the reconstituted receptor showed the rank order and potency characteristic of binding to dopamine D2 receptors.

Further modifications of the previous methodology and increasing the scale of the procedure resulted (6) in purification of D2 receptor ca. 33,000-fold to apparent homogeneity by sequential use of affinity chromatography on CMOS-Sepharose (see above), *Datura stramonium* lectin agarose, and hydroxyapatite. Preparation of the affinity ligand CMOS on a large scale (> 2 g) required purification as the ethyl ester in order to obtain pure product. Esterified CMOS was purified by silica gel chromatography and characterized by NMR. Pure CMOS was then obtained by base hydrolysis, acidification, and passage through silica gel. The affinity matrix was prepared as above. The solubilization protocol was changed. The membrane preparation was done the same day as solubilization to maximize the yield of solubilized receptor activity. Washed membranes (pellet) were resuspended in buffer A, which consisted of 50 m*M* Tris–HCl, pH 7.2, containing 0.32 *M* sucrose, 100 m*M* NaCl, 10 m*M* EDTA, and 10 m*M* EGTA, as well as a protease inhibitor cocktail of 5 mg/ml each of leupeptin, pepstatin A, α_1-antitrypsin, aprotinin, and soybean trypsin inhibitor; 100 m*M* *N*-ethylmaleimide; 1 m*M*

PMSF; and 0.1 m*M* *p*-APMSF. For solubilization, this buffer also contained digitonin (at a concentration of 2 ml of 1.5% digitonin buffer/g of original tissue wet weight); this was allowed to stir at 4°C for 1 hr. The resultant material was sedimented at 43,000*g* for 1.5 hr at 4°C to give the digitonin-solubilized D2 receptor (DSR).

The DSR was loaded batchwise on CMOS-Sepharose by gently stirring overnight at 4°C. Routinely, a ratio of 5 ml (1.5–2.5 pmol of [^{3}H]spiperone binding activity) of DSR per milliliter gel was employed. This gel–receptor complex was inserted into a 5 × 5-cm column and washed at 4°C with 10 bed vol of buffer A containing 0.1% digitonin at 100 ml/hr, 1 bed vol of a high-digitonin high-salt wash (0.5% digitonin, 0.5 *M* NaCl, 50 m*M* Tris, 10 m*M* EDTA, 10 m*M* EGTA, 1 m*M* PMSF) and then again with 2–4 bed vol of the buffer containing digitonin. The column was brought to ambient temperature (22°C) by rapid equilbration with buffer A–digitonin and then eluted with 2–5 bed vol of the buffer containing 50 m*M* haloperidol and 0.1% digitonin at 20 ml/hr. The eluted receptor was collected on ice and stored at 4°C. A single preparation from ca. 250 g of anterior pituitary containing 175 pmol of solubilized D2 binding activity was loaded onto 100 ml of CMOS-Sepharose and eluted with 200–250 ml of the buffer containing haloperidol and digitonin.

Routinely, eluates from two CMOS affinity columns were pooled and recycled twice at 200 ml/hr over a 0.6-ml column of *D. stramonium* lectin at 4°C. The lectin column was washed with 10–20 bed vol of buffer B (0.05% digitonin, 50 m*M* Tris–HCl, pH 7.2, 100 m*M* NaCl, 10 m*M* EDTA and EGTA, 5 μg/ml leupeptin and pepstatin A, 1 m*M* PMSF, and 1 m*M* haloperidol). The column was eluted batchwise at 4°C for 1–2 hr using 2 ml of buffer B containing 10 m*M* N,N′-diacetylchitobiose. The above eluate (2.2 ml) was applied to a 0.2-ml column of hydroxyapatite preequilibrated with buffer B. This column was washed with 5 ml of buffer B and eluted with 1 ml of 200 m*M* potassium phosphate, pH 7.2, in buffer B. This purified receptor preparation could be stored for 1 week at 4°C with loss of less than 20% of its activity. However, freezing at this point resulted in significant loss in activity.

Reconstitution of D2 receptor activity in the purified materials was performed as follows. The CMOS-purified D2 receptor eluate (0.5 ml) was first desalted by Sephadex G-50 chromatography to reduce the haloperidol concentration. The resulting 1 ml of material was mixed with 2 mg/ml fatty acid-free BSA and 0.9–1.7 mg/ml of sonicated soybean phosphatidylcholine in a total volume of 1.5 ml and incubated on ice for 2 min. Octyl-β-D-glucopyranoside was added to a final concentration of 0.9% (w/v) in total volume of 80 ml, followed immediately by addition of 0.2 g washed SM2 Bio-Beads. The solution was kept on ice with periodic stirring for 20–30 min until turbidity was apparent. Receptor binding activity was assayed on the turbid supernatant. Activity in lectin and hydroxylapatite eluates was assayed by reconstituting 10–50 μl in a total volume of 0.5 ml. Removal of haloperidol was not necessary in these cases because the degree of dilution prevented interference with the binding assay. The addition of the high-digitonin high-salt wash of the affinity column afforded 1500-fold purification compared to the 1000-fold purification reported in the earlier study. Most important, inclusion of the lectin step afforded an additional 10-fold purification with 62% recovery of activity, while the hydroxylapatite step gave yet another 2-fold purification. The overall purification was 33,300-fold in 4.7% overall yield (0.012 mg).

The total protein from the various eluates was labeled by radioiodination using Bolton–Hunter reagent in a buffer of 10 m*M* HEPES, 10 m*M* NaCl, and 0.05% digitonin. After reaction, samples were desalted on Sephadex G-50 and lyophilized before subjection to SDS–PAGE. The lyophilized material containing SDS was incubated with or without competing ligands in a HEPES buffer for 30 min with [^{125}I]bromoacetyl-*N*-(*p*-aminophenethyl) spiperone (^{125}I-labeled Br-Ac-NAPS) at a final concentration of 150–400 p*M* for 15 min. Excess labeling reagent was removed by addition of 1 m*M* cysteine. A single band at M_r 120,000 was seen on SDS–PAGE. Covalent labeling occurred with the pharmacological specificity characteristic of dopamine D2 receptors. The purified D2 receptor showed full activity after reinsertion into phospholipid vesicles. When reinserted into phospholipid vesicles with purified G_i/G_o, the purified receptors were able to mediate the agonist stimulation of ^{35}S-labeled guanosine 5′-*O*-thiotriphosphate binding to brain G proteins with the typical D2 order of potency. The authors therefore concluded that they had purified an intact functional D2 dopamine receptor.

Ramwani and Mishra (7) reported partial (ca. 2000-fold) purification of bovine striatal D2 receptors by affinity chromatography on haloperidol-linked Sepharose, prepared from epoxy-activated Sepharose, haloperidol, and zinc chloride. No adsorption of the receptor on the matrix was observed using Sepharose with the same spacer arm but lacking haloperidol, according to [^{3}H]spiperone binding assays. Preincubation of the cholate-solubilized receptor with 10 μM spiperone, domperidone, or (+)-butaclamol, but not (−)-butaclamol or non-D2 receptor ligands, inhibited adsorption of D2 receptors on the haloperidol-Sepharose column. Spiroperidol at 500 n*M* was approximately four times more effective than 2 m*M* dopamine in eluting receptors from the column; spiroperidol was more effective than haloperidol in eluting bound D2 receptors with ca. 70% efficiency. The partially purified receptors, which were not reconstituted into lipids or subjected to additional purification procedures, showed characteristic D2 pharmacology.

More extensive purification of dopamine D2 receptors from bovine brains was reported by Strange and co-workers (8). In this study, D2 receptors were extracted from bovine brain using sodium cholate and purified 20,000-fold by sequential affinity chromatography on haloperidol-Sepharose and wheat germ agglutinin(WGA)–agarose columns. The purified receptor showed a major diffuse band at M_r 95,000 on SDS–PAGE. The pharmacological specificities of purified and crude solubilized D2 receptors were similar. The authors claimed that this was the first report of purification of brain D2 receptors.

In this study, a mixed mitochondrial–microsomal fraction from bovine caudate nucleus was prepared using a HEPES buffer that additionally contained 0.1 m*M* PMSF, 10 m*M* EDTA, and 1 m*M* EGTA. The membranes were resuspended in 10 m*M* HEPES, pH 7.4, 10 m*M* EDTA, 1 m*M* EGTA (buffer II) containing pepstatin A, leupeptin hemisulfate, aprotinin, chymostatin, and antipain dihydrochloride (all 5 μg/ml). For solubilization, membranes were diluted in buffer to 8 mg protein/ml and mixed with an equal volume of buffer II containing 0.6% w/v sodium cholate, 2*M* NaCl. 0.2 m*M* PMSF, and 10 μg/ml of each of the protease inhibitors listed above for 1 hr at 4°C. The supernatant obtained after centrifugation at 200,000*g* for 1 hr at 4°C was used.

Affinity matrices (type I) were prepared from carboxymethoxylamine derivatives of spiperone and haloperidol which were coupled to AH-Sepharose using EDAC hydrochloride. Another matrix (type II) was prepared from haloperidol hemisuccinate. The extent of coupling of the ligand to the support as determined by UV spectroscopy was three times greater using the second mode of coupling. The D2 affinity of coupled ligands was reduced 55- to 60-fold by type I coupling and 30-fold by type II coupling. The SRP, diluted to 0.225% cholate and 0.75 *M* NaCl (45 ml) and supplemented with 2 m*M* sodium acetate, was incubated at 4°C for 16–20 hr with 10 ml of affinity matrix (generally haloperidol type II) in a column made from a 50-ml syringe equipped with a tap. The matrix was washed with 500 ml of buffer III (buffer II plus 2 m*M* sodium acetate, 0.045% soybean phosphatidylcholine, 0.225% sodium cholate, and 0.75 *M* NaCl). The columns were then eluted over two 24-hr periods with a buffer containing 1 m*M* metoclopamide. Metoclopramide was chosen since it shows high selectivity but only moderate affinity (0.9 μM) for D2 receptors, allowing it to be readily removed from eluates. This procedure afforded 411-fold purification and 25% recovery of the [^{3}H]spiperone binding activity originally bound to column. Portions of each eluate supplemented to 0.3% cholate and 1 *M* NaCl were incubated with WGA-agarose for 90 min at 4°C. After being washed with 30 ml of buffer II, the lectin column was eluted with 10 ml of buffer III containing 10 mM *N*-acetylglucosamine over 90 min at 4°C. In some experiments, metoclopramide eluates were dialyzed against buffer II (five changes of 100 volumes over

72 hr) before [^{3}H]spiperone binding assays. The WGA affinity step gave an additional 50-fold purification with ca. 50% recovery. In this case, it was possible to detect [^{3}H]spiperone binding in eluates without reconstitution into lipids or other forms of manipulation. The pharmacological binding profile of the purified receptor was similar to that in crude membrane and solubilized preparations.

SDS–PAGE showed a major diffuse band at 95 kDa (a closely spaced doublet was often observed) and occasionally faint bands in the 50- to 65-kDa region. The diffuseness of the 95-kDa band was attributed to variations in the extent of glycosylation of the receptor. The major band is not seen if the SRP is preincubated with metoclopramide before application to the affinity column or if the SRP is incubated at 22°C for 24 hr to eliminate [^{3}H]spiperone binding. Specific photoaffinity labeling with [^{3}H]azidomethylspiperone consistently gave a band at 95 kDa, and occasionally one at 32 kDa. Labeling was not observed in the presence of (+)-butaclamol, a potent D2 antagonist.

Bosker *et al.* (9) reported a method for purification of bovine striatal dopamine D2 receptors that showed pharmacological activity without reconstitution into lipid vesicles. A homogenate was prepared from partially thawed bovine striata in the usual manner. The pellets, stored at −30°C, were resuspended in buffer A (0.125 *M* sucrose, 0.5 m*M* EDTA, 0.8 m*M* NaN_3 in 5 m*M* phosphate, pH 7.4) containing 0.1 *M* NaCl and 10 mg/ml digitonin, in a ratio of 0.25 mg protein/mg digitonin. The solubilized receptor was separated from insoluble material by centrifugation, and the supernatant was used immediately. The solubilized preparation contained 320 fmol receptor/mg protein. The activity decreased by 50% over a period of 10 days at 4°C. The affinity gel was made by coupling epoxy-activated Sepharose CL-4B to the *p*-amino derivative of the D2 ligand N-0434. Remaining active groups on the gel were neutralized by hydrolysis to the dihydroxy derivative at pH 10. The degree of substitution, estimated by UV spectroscopy, was 6 μmol N-0434/ml gel. Inactive affinity gels were synthesized in the same way omitting N-0434. Gels prepared from several other agonists and antagonists were found to be unsuitable.

A preliminary step involving passsage of the digitonin-solubilized receptor through a wheat germ agglutinin-Sepharose 6B column containing 5 mg of lectin per ml gel was added to lower the digitonin concentration in order to avoid nonspecific precipitation of protein on the affinity column. In a typical experiment, 20 ml of the solubilized material was applied to a 4-ml column equilibrated with buffer B (buffer A with 100 m*M* NaCl and 1 mg/ml digitonin). The sample was circulated through the column overnight. After the sample was washed with 10 bed vol of buffer B, receptor activity was eluted with 0.3 *M* *N*-acetylglucosamine in buffer B. After application of this solution, the flow was stopped for 30 min and then resumed. The receptor was purified 16-fold at this stage, with 40% recovery relative to membrane homogenate. Four volumes of the eluate from the lectin column were applied to the N-0434 affinity gel. After the gel was washed with 20 vol buffer B, receptor activity was eluted by lowering the pH. This eluate was applied to a second affinity columns and the pH shock treatment was repeated. The overall recovery of active receptor was 3.5%, and the purification was estimated from binding data to be >8000-fold, although the amount of protein was too small to be measured accurately. The purified D2 receptor retained its affinity to spiperone: K_d 1.34 nM for the crude solubilized preparation and 0.67 nM for the purified receptor. No SDS–PAGE gels were reported.

The same group also reported large-scale purification of dopamine D2 bovine striatal receptors using the above methodology in order to obtain an amino acid sequence (10). They started with 1.5 kg striatum and performed digitonin extraction, chromatography on WGA-Sepharose, and N-0434-Sepharose affinity chromatography essentially as above. In the earlier small-scale study, D2 receptors were eluted from the N-0434 column by pH shock, which led to 5000-fold purification. In the large-scale experiments, this turned out to be less efficient than elution with an excess of N-0434 (100 mM). Material was obtained with "only a few bands detectable on a Coomassie blue-stained gel . . . with the dominant component located at 95 kDa," although the gel shown in the paper shows a number of bands. Based on amino acid analysis, the total amount of protein isolated was estimated to be at least 2–3 mg. Nonetheless, all attempts to purify peptides from this band by proteolysis with endoproteinase Lys-C failed. Further attempts to purify larger amounts of D2 receptor for sequencing were therefore abandoned. This experience vividly demonstrates an inherent limitation of this approach to obtaining information about molecular structure of receptor proteins.

Attention was then directed to preparation of antibodies against synthetic peptides derived from the previously determined sequence of the cloned D2 receptor. Three fusion proteins were prepared from β-galactosidase and PCR-synthesized fragments corresponding to amino acids 4–76, 281–377, 389–444 of the cloned D2 receptor. Fusion proteins were expressed in *Escherichia Coli* and, after purification, analyzed on a Western blot using an anti-(β-galactosidase) monoclonal antiserum. Rabbits were immunized with peptide–hemocyanin conjugates, and antisera were tested for recognition of the corresponding fusion protein in Western blots. The most convincing data were obtained with peptide B (281–377). The raised antibody recognizes the D2 receptor in the denatured as well as the native state and recognizes the 95-kDa as well as a 36-kDa protein. The former corresponds to the intact

D2 receptor (see above), but the identity of the latter protein is unclear; it may be a proteolytic fragment of the intact receptor.

The Dopamine D1 Receptor

This receptor from the rat corpus striatum was purified using a straightforward affinity chromatographic protocol by Caron and co-workers (11). The affinity matrix was constructed from a derivative of the selective D1 antagonist SCH 23390 containing an amine moiety at the para position of the 1-phenyl ring. This compound retains high D1 affinity (2.5 n*M*). The ligand was treated with succinic anhydride to introduce a four carbon chain with a terminal carboxyl group. Sepharose 6B was treated with 1,4-butanediol digylcidyl ether to create an epoxy-activated support, and then with 1,6-diaminohexane to create an extended spacer arm with a terminal amino group. This functionalized Sepharose was then coupled with succinylated SCH 39111 using EDAC. Ethanolamine was included to deactivate excess epoxy groups.

Pellets derived from rat brain homogenates were resuspended in 80 vol of solution A (50 m*M* HEPES, pH 7.2, 100 m*M* NaCl, 10 m*M* EDTA, 10 m*M* EGTA) containing a protease inhibitor cocktail (50 m*M* PMSF; 5μg/ml each of pepstatin A, leupeptin, soybean trypsin inhibitor, and aprotinin; and 1 μg/ml α-antitrypsin). This was centrifuged twice at 45,000 *g* for 20 min and the resulting pellet was resuspended in 25 ml/g wet weight striatal tissue of solution B containing 1% w/v digitonin; this was stirred slowly on ice for 60 min. The SRP was obtained by centrifugation at 45,000 *g* for 90 min. Typically 35–40% of [^{3}H]SCH 23390 binding sites were solubilized under these conditions, yielding SRPs containing ca. 1.7–1.9 pmol/ml. Pretreatment with an agonist such as SCH 38393 did not afford improvement in recovery from membranes. The SRP retained characteristic D1 pharmacology.

Typically 50 ml of the SRP was absorbed batchwise by incubation with 5 mol of affinity gel with slow rotation for a 20 hr period at 4°C. The resin was inserted into a 1.5-cm-diameter column and washed sequentially at 4°C with each of the following: (a) 5 bed vol solution C (50 m*M* HEPES, pH 7.2, 100 m*M* NaCl, 5 m*M* EDTA); (b) 5 bed volumes of solution C containing 0.1% digitonin and 250 m*M* NaCl; (c) 5 bed vol of solution C; (d) 5 bed vol of solution C, pH 6.0, at 22°C. The resin was then eluted with solution C, pH 6.0, containing 100 m*M* (+)-butaclamol dissolved in methanol (final MeOH concentration 0.01%) at 22°C at 5 ml/hr, collecting fractions at 30-min intervals. Elutates were collected on ice in tubes containing an equal amount of solution C at pH 7.2 to readjust the final pH to 6.8. The final two eluates were collected at the same flow rate at 3 hr per fraction. The eluates were desalted on a Sephadex G-50 (fine) column (0.6 × 13.5 cm) to separate unbound (+)-butaclamol from receptor. The desalted eluates were incubated with 10 n*M* [^{3}H]SCH 23390 for radioligand binding assays.

Incubation of the SRP with SCH 23390 or (+)-butaclamol prior to exposure to the affinity resin reduced adsorption of [^{3}H]SCH 23390 binding activity by 90 and 70%, respectively. D1 antagonists (SCH 23390, *cis*-flupenthixol, (+)-butaclamol) eluted receptor activity from the affinity matrix, while less-potent stereoisomers (SCH 23388, *trans*-piflupenthixol and (−)-butaclamol) failed. Dopaminergic agonists and ligands for other receptors were also ineffective as eluents. The resin adsorbed 75–85% of [^{3}H]SCH 23390 binding activity; ca. 90% of total protein was not retained. (+)-Butaclamol was chosen for biospecific elution because of its ease of removal from the purified receptor by Sephadex G-50 gel filtration. The best recovery was obtained when elution was carried out at room temperature with the pH of the elutant at 6.0, which maximized dissociation of the receptor from the affinity matrix. By this procedure, 35–55% of the adsorbed receptor activity was recovered, resulting in 200- to 250-fold purification relative to the SRP. This was estimated to be a factor of 40 less than the theoretical specific activity estimated for pure receptor. The purified D1 receptor was characterized by its specific pharmacology.

A quite novel and interesting approach for purification of the D1 receptor was published by Sidhu (12). This procedure took advantage of an earlier demonstration (13) that sulfhydryl groups associated with the ligand binding site were susceptible to alkylation by *N*-ethylmaleimide (NEM). NEM inactivation was 100% blocked in the presence of an agonist such as SKF R-38393, perhaps due to an agonist-mediated conformational change. Antagonists were unable to fully protect the binding sites, as only 50% of binding activity was lost on NEM treatment. These findings led to the proposal that if membranes were pretreated with NEM in the present of SKF R-38393 prior to solubilization, one might be able to covalently alkylate all or most of the SH groups in membrane receptors except those specifically associated with the D1 receptors. If the solubilized receptors were then exposed to a mercury column, the D1 receptors would be immobilized on the mercury column and could then be recovered by reduction with a suitable agent.

Striatal membranes prepared as previously described were resuspended at a protein concentration of 0.8 mg/ml in buffer A (50 m*M* Tris—HCl, pH 7.4, 120 m*M* NaCl, 5 m*M* KCl, 2 m*M* $CaCl_2$, 1 m*M* $MgCl_2$). The membranes were pretreated with 30 m*M* SKF R-38393 for 25 min at 37°C, and then NEM was added to a final concentration of 5 m*M*. After incubation for an additional 25 min at 37°C, the NEM-treated membranes were then washed and resuspended in buffer A containing 1 m*M* PMSF at a protein concentration of 1 mg/ml. After dilution with an equal volume of 50 m*M* Tris-HCl, pH 7.4, and centrifugation at 18,000*g* for 20 min, treated membranes were resuspended at a protein concentration of 2 mg/ml in buffer S (50 m*M* Tris–HCl, pH 7.4, 5 m*M* KCl, 1 m*M* EDTA, 1 m*M* $MgCl_2$, 2 m*M* $CaCl_2$, 250 m*M* sucrose, 1.5 m*M* PMSF, 1 m*M* DTT, and 1 *M* NaCl). Sonicated phospholipids were added to a final concentration of 1.2 mg/ml, and sodium cholate (20% solution in water) was added to a final concentration of 1%. After 15–20 min on ice, the mixture was centrifuged at 31,300*g* for 45 min. The clear supernatant was removed and stored frozen at −80°C for up to 1 year without appreciable loss in ligand binding activity.

Five milliliters of mercury agarose (14) was packed in a 1 × 15-cm column and sequentially washed at room temperature with 5 bed vol each of 50 m*M* sodium acetate, pH 5.0; 50 m*M* sodium acetate, pH 5.0, containing 4 m*M* mercuric acetate; buffer C (50 m*M* Tris–HCl, pH 7.4, 5 m*M* KCl, 1 m*M* $MgCl_2$, 2 m*M* $CaCl_2$, 0.2 mg/ml sonicated phospholipids, 0.4% sodium cholate, 250 m*M* sucrose, 0.5 m*M* PMSF, and 5 μg/ml each of leupeptin and pepstatin). The sodium cholate-solubilized receptors were thawed on ice, and protease inhibitors were added to give final concentrations of 5 μg/ml (leupeptin and pepstatin) and 0.5 m*M* (PMSF). The cholate-solubilized mixture was applied to the column at room temperature in 2.5-ml batches. After each 2.5 ml application, proteins were allowed to bind to the column for 15–20 min at room temperature prior to additional sample application. The binding capacity of each column was approximately 5 column volumes of soluble extract, corresponding to a protein concentration of ca. 1.5 mg/ml. After sample application, the column was brought to 4°C and allowed to equilibrate for 20—30 min. All subsequent steps were performed at 4°C. The column was washed with 50 bed vol of ice-cold buffer C from which phospholipids were omitted. Bound receptors were specifically eluted with 10 m*M* β-merceptoethanol (ME) in buffer D (buffer C containing 1 m*M* EDTA and sonicated lipids at a final concentration of 0.6 mg/ml), and 5 ml fractions were collected at a flow rate of 0.3 ml/min. The ME was removed from eluted fractions by desalting on Ultrogel AcA 202 columns. The desalting gel in 1.5 × 20-cm columns was washed with 50 m*M* Tris-HCl, pH 7.4, and then equilibrated with buffer D. Then, ME-eluted fractions were applied to the column and desalted receptors were collected in the void volume. Desalted fractions were either stored at −80°C or used directly in binding assays. Samples frozen in the presence of ME did not display any specific binding activity, while samples frozen in the absence of ME were stable for several months. Purified ME-free samples had to be reconstituted into vesicles prior to measurements of binding activity using buffer E (50 m*M* Tris–HCl, pH 7.4, 5 m*M* KCl, 2 m*M* $CaCl_2$, 1 m*M* $MgCl_2$, 250 m*M* sucrose, 1 m*M* EDTA, and protease inhibitors as above) containing 120 m*M* NaCl and 0.4% sodium cholate. Sonicated phospholipids (maintained on ice for 5 min) were added at twice the desired final concentration. An equal volume of buffer E containing phospholipids was added, and the detergent was removed using SM-2 Bio-Beads. This particular procedure vastly improved the binding activity of the purified receptor, as measured with [^{125}I]SCH 23982. The receptor-binding activity in fractions eluted from mercury–agarose showed up in a broad peak representing routinely 80—90% of applied activity.

Conditions for each of the above steps were optimized. In particular, removal of ME using dialysis gave fairly labile receptor samples, and results were inconsistent. The overall recovery of D1 receptors relative to the crude membrane preparation was 33%, and 8000-fold purification was achieved by this method, somewhat short of the 10,400-fold purification required for complete purification. SDS–PAGE showed two major bands at 74 and 54 kDa, which were not present in fractions not displaying [^{125}I]SCH 23982-binding activity or from membranes which were treated with NEM in the absence of D1-specific agonist. Photoaffinity labeling of membrane-bound receptors with [^{125}I]MAB ([^{125}I]8-hydroxy-3-methyl-1-(4-azidophenyl)-2,3,4,5-tetrahydro-1*H*-3-benzazepine) specifically labeled three major polypeptides at 74, 51, and 25 kDa with characteristic D1 pharmacospecificity. It was suggested that the purified polypeptides may represent glycosylated and deglycosylated forms of the D1 dopamine receptor.

The Adenosine A1 Receptor

Two reports have appeared by Nakata (15, 16) on purification of A1 adenosine receptors from rat and human brain membranes by affinity chromatography using agarose immobilized with xanthine amine congener (XAC). In the first study (15), receptors from rat brain membranes were solubilized using 1% digitonin and 0.1% sodium cholate with a yield of 30%. When solubilized membranes were applied to a 2.5 × 14-cm agarose column at a flow rate of 50 ml/hr, more than 95% of proteins passed through, although 80% of [^{3}H]DPCPX ([^{3}H]8-cyclopentyl-1,3-dipropylxanthine) binding activity was retained on the column. Little loss of activity was observed on washing the column with 5 bed vol of 50 m*M* Tris acetate buffer, pH 7.2, containing 100 m*M* NaCl, 1 m*M* EDTA, and 0.1% digitonin (bugger A). The binding activity was specifically eluted with 3 vol of buffer A containing 100 m*M* CPT (8-cyclopentyltheophylline), a potent A1 antagonist, at a flow rate of 15 ml/hr. More than 90% of the eluted activity was present in the column volume (70 ml). This resulted in 2500-fold purification over the solubilized preparation and a yield of 40%. The CPT-eluted fractions were then applied to a 0.5-ml column of hydroxyapatite at a flow rate of 20 ml/hr. This column was washed with 5 ml of buffer A and eluted successively with 10 ml of 10, 110, and 500 mM potassium phosphate buffer, pH 7.0, each containing 100 mM NaCl and 0.1% digitonin. This step gave additional 10-fold purification as well as 14-fold concentration. The 3-ml eluate using the highest salt concentration (3 ml) was diluted 2-fold with buffer A and applied to a second XAN-agarose column (1 × 5 cm) at a flow rate of 10 ml/hr. The column was washed with 6 vol of buffer A, and receptors were then eluted with 1.5–2 vol of buffer A containing 100 mM CPT. More than 50% of the applied activity was eluted in the first 5 ml of the eluting buffer. Ligand was separated from the receptor by desalting on a Sephadex G-50 column (0.6 × 13.5 cm). Based on the [^{3}H]DPCPX binding, the overall purification of A1 receptors achieved was ca. 50,000-fold with an overall yield of 4% relative to intact membranes.

After lyophilization of the material following the second affinity chromatography, SDS–PAGE showed a single broad band at 34 ± 1 kDa (silver staining), suggesting microheterogeneity. Minor bands were observed inconsistently at 97 and 29 kDa. The same 34-kDa band was seen under nonreducing conditions. Affinity labeling of the purified receptor preparation using *p*-[^{3}H]DITC-XAC, a high-affinity acylating antagonist, resulted in a band at 34 kDa after analysis by SDS–PAGE using fluorography. The same band was seen on autoradiography of nonspecifically ^{125}I-labeled purified receptors.

On gel permeation chromatography on a TSK-3000 SW steric exclusion column (7.5 × 300 mm) at 0.35 ml/min at 4°C, [^{3}H]DPCPX-binding activity eluted as a single peak with M_r = 150,000, coincident with appearance of the 34-kDa band on SDS–PAGE. The latter is smaller than that (63 kDa) determined by radiation inactivation in intact membranes, suggesting that the 34-kDa protein may be a subunit of the intact A1 receptor.

The same affinity matrix, XAN–agarose, was used to purify and characterize human brain A1 adenosine receptors (17). Human cerebral cortices were homogenized and crude membranes were solubilized with 1% digitonin/0.1% cholate. The solubilized preparation (400 ml) was applied to the XAN–agarose affinity column (5 × 8.5 cm) and the column was washed with 450 ml of 50 mM Tris acetate, pH 7.2, containing 100 mM NaCl, 1 mM EDTA, and 0.1% digitonin. The receptor was eluted with 300 ml of 100 μM CPT in the same buffer. Active fractions were pooled (120 ml) and applied to a small (0.5 ml) hydroxyapatite column. After the column was washed with 5 ml 200 mM potassium phosphate buffer, the receptor was eluted with 2 ml of 500 mM phosphate buffer, pH 7.0, containing 100 mM NaCl and 0.1% digitonin. The eluate (2 ml) was diluted twice with buffer and applied to a small (1 × 7-cm) XAN–agarose column. After being washed with 20 ml buffer A, the receptor was eluted with 10 ml buffer A containing 100 μM CPT. Eluted receptor fractions were concentrated to ca. 200 ml in a Centricon 30 (Amicon) and injected in 100-ml aliquotes to tandem-linked TSK-3000SW columns. Active eluted fractions were saved and analyzed. The overall recovery of active receptor was 1.6% with 13,000-fold purification.

Aliquots of purified preparations were radioiodinated using chloramine-T for analysis by SDS–PAGE. The final purified receptor showed a broad band on autoradiography with apparent mass of 35 ± 1 kDa, similar to that of purified rat brain adenosine A1 receptors (see above). The purified receptor was irreversibly labeled by DITC-[^{3}H]Xan, a specific affinity label for the A1 receptor, demonstrating that the purified protein contains the ligand binding site of the A1 adenosine receptor.

Bovine Striatal Opioid-Binding Protein

Simon and co-workers have described an improved procedure for purification of an active opioid-binding protein from digitonin-solubilized bovine striatal membranes by a rapid two-step procedure: affinity chromatography on β-naltrexylethylenediamine (NED)-CH-Sepharose 4B followed by lectin affinity chromatography on WGA-agarose (18). The first step yields a protein fraction that binds opiates stereospecifically in a saturable manner, with specific activity enriched 4000- to 7000-fold over membrane-bound or soluble receptors. High specific activity in recovered receptors was achieved by washing with 0.05% digitonin 40–50 times the gel volume, followed by washes with high salt (0.5 *M*) and high digitonin (0.25%) and finally with 0.1% digitonin, and performing at least two elutions using 2.5 μM naloxone. The protein assay on eluates was also improved using a radioiodination procedure, by quantitative evaluation of Coomassie blue staining on SDS–PAGE gels, and by scaling up the amount of material processed.

β-Naltrexylethylenediamine was coupled to CH-Sepharose 4B. Gel and digitonin-solubilized material (1 ml of gel/10–12 ml of solubilized preparation) were incubated with frequent shaking for 45 min at 25°C. The flow-through was collected by pouring the incubated mixture into a 2.5 × 30-cm column and then rinsing the gel with 40-50 bed vol of buffer A (50 m*M* Tris–HCl, 1 m*M* K_2EDTA, 0.5 *M* NaCl, pH 7.4) containing 0.05% digitonin, 1 gel volume of buffer containing 0.25% digitonin, and then 1 gel volume of buffer with 0.1% digitonin. The retained binding sites were eluted with 2.5 μM naloxone in buffer containing 0.1% digitonin over a 40-min period at 25°C. Before binding assays with [^{3}H]bremazocine were performed, eluates were treated with Bio-Beads SM-2 that had been previously incubated for at least 2 hr with 0.1% digitonin.

Wheat germ agglutinin–agarose was preequilibrated with buffer containing

0.05% digitonin or 0.05% digitonin in 50 mM K_2HPO4, 1 mM EDTA, and 100 mM NaCl. The eluate from the NED column was incubated with the WGA–agarose gel for 45 min at 4°C with frequent shaking. The flow-through was collected and the gel was then rinsed with either of the above buffers at 5–10 times the applied sample volume. The gel was then eluted by incubation with 0.3 M N-acetylglucosamine in either buffer for 45 min at 4°C. Specific binding assays employed 1.5–2.0 nM [^{3}H]bremazocine in the presence and absence of 2 μM unlabeled naloxone. The additional enrichment in binding activity after lectin chromatography was 10- to 20-fold, corresponding to an overall purification factor of 65,000-fold. The overall yield of purified receptors was 5.8%.

The amount of affinity-purified protein from striata of five cow brains (ca. 50 g of tissue) was determined to be 60–80 μg. This material showed five bands on SDS–PAGE. After the lectin chromatography, the yield of binding activity was 61 ± 4% (n = 3). One clean band was seen on SDS–PAGE at 65 kDa under reducing conditions (100 mM DTT), and at 54 kDa under nonreducing conditions, suggesting there are intramolecular S-S bonds. A variety of evidence, including naloxone-inhibited cross-linking to human β-[^{125}I]endorphin, led the authors to conclude that the purified 65-kDa protein has the characteristics of a μ opioid receptor.

While this purified opioid binding protein (OBP) bound μ-antagonists with the potency of the membrane-bound receptor, its affinity toward μ-agonists was several orders of magnitude less than that of native receptors. Since coupling of the μ-OBP to G proteins is required for high-affinity binding to opioid agonists but not to antagonists, the lack of agonist-binding of the purified receptor was attributed to the absence of the requisite G proteins. In order to establish that the protein they had purified was indeed the μ-OBP, the purified OBP was reconstituted into liposomes as follows (19). A CHAPS extract of bovine striatal membranes lacking opioid binding activity was prepared by extracting the membranes with 5 mM CHAPS in 50 mM Tris buffer, pH 7.4, containing 1 mM EDTA but lacking NaCl, and heating the supernatant after centrifugration at 37°C for 30–60 min to destroy any remaining opioid receptor activity. NaCl was added to this solution at a final concentration of 0.5 M. The purified OBP (10 μl, 200–500 fmol) was added and the mixture was then treated with an equal volume of 40% polyethyleneglycol in Tris/EDTA. After centrifugation at 10,000g for 10 min at 4°C, the pellet containing the liposomes was rinsed and resuspended in Tris/EDTA containing 10 mM $MgCl_2$. This preparation now bound μ-selective opioid agonists [morphine and [D-Ala^2, N-methyl-Phe^4, Gly-ol^5]-enkephalin (DAGO)] with appropriately high affinity, whereas δ- and κ-selective ligands showed two to three orders of magnitude lower affinities. The characteristic stereospecificity associated with naloxone binding was also observed, i.e., (−)-naloxone competed against [^{3}H]DAGO with high affinity (K_i = 2 nM) while (+)-naloxone was inactive. Binding of [^{3}H]DAGO was also abolished in the presence of GTPγS, as in crude membrane preparations. These results confirm that the purified OBP is indeed an opioid binding site of the μ type.

The Ah Receptor for Dioxin and Related Substances

The *Ah* receptor is a soluble protein found in a variety of vertebrates and mediates the biological responses produced by 2,3,7,8-tetrachlorodibenzo-*p*-dioxin and other halogenated aromatics, including induction of P-450 isozymes, wasting syndrome, and tumor production. This receptor is presumed to be a member of the erb-A superfamily of receptors which are DNA binding proteins. This family includes steroid hormone receptors, thyroid hormone receptors, and retinoic acid receptors. A major obstacle to its study has been the inability to purify it to a degree useful in antibody generation and amino acid sequencing. Data indicate the *Ah* receptor is present in heptatic cytosol of C57BL/6J mice at a concentration of 100 fmol/mg and that during homogenization as much as 40% of the parent 95-kDa protein is proteolyzed to yield a 70-kDa fragment. In order to purify these two species to homogeneity, enrichments of 170,000- and 360,000-fold are needed, respectively. The novel procedure used for purification of this receptor involved photoaffinity labeling prior to ion-exchange chromatography and final purification using high-performance liquid chromatography (HPLC) (20).

Livers of C57BL/6J mice were removed and homogenized in 9 vol of MβENG buffer [25 mM MOPS, 0.02% sodium azide (w/v), pH 7.5, containing 10 mM β-mercaptoethanol (ME), 1 mM EDTA, 10% v/v glycerol] plus 5 mM EGTA. The supernatant from the cytosolic fraction was used in these studies. The receptor was first photoaffinity labeled using 2-azido-3-[^{125}I]iodo-7,8-dibromodibenzo-*p*-dioxin. About 5% (100 ml) of the cytosolic preparation from 200 g liver (total vol 2 liter, 8–9 mg protein/ml) was taken and diluted with MENG buffer to 2 mg protein/ml. The photoaffinity label was added to a final concentration of 3×10^6 dpm/ml and the sample was incubated for 30 min at 20°C. Unbound radioligand was removed by addition of 10 ml of charcoal/gelatin (final concentration 1:0.1% w/v) in the MOPS–azide buffer, mixing with a vortex mixer for 5 sec, and incubation at 20°C for 10 min. The charcoal was removed by centrifugation at 4°C. The supernatant, after a second centrifugation, was irradiated with an 80-W lamp at 310 nm for 1 min. β-Mercaptoethanol was added to a final concentration of 10 mM to quench any remaining free radicals. The photolabeled material was then pooled with the bulk of the cytosolic preparation, which had a specific activity corresponding to approximately 2 fmol of photoaffinity ligand/mg of cytosolic protein.

The ion-exchange chromatography was carried out in a room maintained at 4°C. The NaCl in the pooled cytosolic preparation was raised to 80 m*M*, and this material was loaded onto a phosphocellulose column (10 cm i.d. × 14 cm; column volume, 1 liter) at a flow rate of 15 cm/hr. The column was washed with MENG buffer containing 80 m*M* NaCl until the UV absorbance returned to baseline. The receptor was then eluted with MENG buffer containing 225 m*M* NaCl at 30 cm/hr. The enriched fraction had a volume of 500 ml. This eluate was diluted with MENG buffer containing 165 m*M* NaCl and loaded onto a DEAE-cellulose column (5 cm i.d. × 13 cm; volume, 250 ml). The column was washed with MENG buffer containing 165 m*M* NaCl until absorbance at 280 nm returned to baseline. The receptor was then eluted with MENG buffer containing 300 m*M* NaCl. The total volume collected was ca. 100 ml which was stored until further use at −80°. Specific activity increased ca. 100-fold with a 46% receptor recovery by this sequence of steps.

Further purification was done using denaturing conditions by chromatography on a preparative-scale RP-HPLC C4 silica-based column (2.2 × 25 cm) with a large particle size (15–20 μm) and elution using a linear gradient of acetonitrile in aqueous trifluoroacetic acid (TFA). Lithium dodecyl sulfate and ME were added, each to 2%, to the fraction collected above containing 1.2 mg protein, and the solution was heated at 56°C for 30 min. The sample was then precipitated by heating at 56°C for 2 min after addition of *n*-propanol/trichloroacetic acid (final concentrations 20:0.1%). The precipitate was collected by centrifugation, solubilized in formic acid (0.5 ml/mg protein), diluted with 10 vol of equilibration solvent (water/acetonitrile/trifluoroacetic acid 60.4:39.5:0.1), and filtered through a 0.45-mm membrane. Sample was then loaded onto the preparative HPLC column at a flow rate of 2 ml/min and eluted using a gradient of acetonitrile in water (with 0.1% TFA as a ion-pairing solvent) at 56°C. The radioactivity in each 2-ml fraction was quantified by scintillation counting. The 95-kDa protein eluted at 51.2% acetonitrile and the 70-kDa protein at 52% acetonitrile. Pooled fractions of the 95-kDa protein from two runs on the prep HPLC columns (as identified by SDS–PAGE) were diluted with 0.5 vol of equilibration solvent (70:21.2:8.8 water/propanol/formic acid) and loaded onto a semipreparative C4 column (1 × 25 cm; 5-μm particle size). After being washed with equilibration solvent until UV absorbance again returned to baseline, the receptor was eluted with a gradient of *n*-propanol in water, using 8.8% formic acid as the ion-pairing agent. The 95-kDa receptor eluted as a sharp peak at 26.3% *n*-propanol.

The final purification step was performed on a prewashed analytical C4

HPLC column (4.6 mm × 25 cm, 5-μm particle size). The column was washed with equilibration solvent consisting of 57.6 : 42.3 : 0.1 water/acetonitrile/TFA and receptor was then eluted using a shallow linear gradient of acetonitrile in water with 0.1% TFA as the ion-pairing solvent. The elution was monitored by radioactivity and SDS–PAGE. The peak of maximum radioactivity (fraction 19) came out after the peak intensity of the 95-kDa band (fraction 16), indicating that the unlabeled and photoaffinity-labeled receptor separated under these HPLC conditions. The HPLC fractions containing the peak amount of 95-kDa protein were pooled, subjected to SDS–PAGE, and electrotransferred onto a PVDF membrane. The band was visualized using Coomassie blue R250, and the quantity estimated by staining intensity using laser densitometry. In a typical experiment, 3–5 μg of the 95-kDa receptor was obtained from 10 g of cytosolic protein, corresponding to an overall recovery of 5%. The overall purification factor was 180,000-fold. A consensus N-terminal amino acid sequence of the purified *Ah* receptor was obtained by Edman sequencing of three separate samples of the purified material.

The NMDA/PCP Receptor Complex

The *N*-methyl-D-aspartate (NMDA) receptor is one of the best-characterized excitatory glutamate receptors in the mammalian cental nervous system and has been shown to be involved in a host of physiological responses. It is also well-established that phencyclidine (PCP) and related substances modulate the activity of NMDA receptors, and that the binding of [^{3}H]TCP, an analog of PCP, is modulated by NMDA receptor ligands. The purification of the NMDA/PCP receptor complex by affinity chromatography has been described by Ikin *et al.* (21). A critical feature of the successful protocol was the utilization of a protease inhibitor cocktail at each stage of the process, i.e., homogenization, solubilization, and affinity purification.

Rat forebrains were homogenized in 20 vol of ice-cold 50 m*M* Tris–HCl, pH 7.4, containing 0.32 *M* sucrose, 1 m*M* EGTA, 3 m*M* EDTA, and a mixture of protease inhibitors consisting of 0.1 m*M* PMSF, 5 units/ml aprotinin, and 5 μg/ml pepstatin A. The pellet after centrifugation was resuspended in 20 m*M* Tris–HCL, pH 7.4, containing 2 m*M* EDTA and the protease inhibitor cocktail. The final protein concentration was 8 mg/ml. This suspension was mixed with an equal volume of 3% sodium cholate in 5 m*M* Tris–HCl, pH 7.4, to give final detergent and protein concentrations of 1.5% and 4 mg/ml, respectively. The mixture was shaken for 1 hr and then centrifuged at 100,000 *g* for 1 hr. The supernatant was dialyzed for 4 hr against 500 vol of 20 m*M* Tris–HCl, pH 7.4, 2 m*M* EDTA, and 0.1 m*M* PMSF to remove the detergent, which inhibits binding of PCP-like ligands. The dialysate was used as the SRP.

The affinity column was prepared by coupling of amino-PCP to an agarose gel prepared from (*p*-nitrophenyl)agarose using DCC, with the amino-PCP present in 5–10 times the concentration of active ester on the agarose matrix. The extent of coupling, ca. 8 mmol/ml gel, was estimated from the concentration of *p*-nitrophenolate released in the coupling reaction. The PCP-agarose column (5.5 × 1.5 cm) was preequilibrated with 20 m*M* Tris–HCl, pH 7.4, 0.05% sodium cholate, 2 m*M* EDTA, and the mixture of protease inhibitors (buffer A). The SRP was applied at 60 ml/hr using a peristaltic pump. The column was then washed with 40 ml of buffer A containing 10^{-4} *M* of the competitive glutamate antagonist DL-AP-5. The column was then washed overnight with 800 ml of buffer containing 10^{-5} *M* of this antagonist. The receptor was eluted with 36–40 ml of buffer A containing 10 μM glutamate, 1 μM glycine, and 10 μM PCP. The eluate was dialyzed for 4 hr at 4°C against 200–500 vol of buffer A containing only 0.1 m*M* PMSF as protease inhibitor to remove free PCP. The dialysis buffer was exchanged every hour. Two columns were run simultaneously to increase percentage yields and amounts of purified receptors.

The recovery of [^{3}H]TCP binding sites after solubilization was 24%, with a 60% decrease in specific activity. Between 40 and 50% of the binding sites in the SRP were retained on the affinity column; the remainder were in the pass-through of the column. About 28% of the [^{3}H]TCP binding sites were specifically recovered on elution with the mixture of glutamate, glycine, and PCP. The yield of purified receptor from 10 rat forebrains after one pass through the amino-PCP-agarose affinity column was 7.6 pmol of [^{3}H]TCP binding sites, corresponding to a total yield of 7% and 3700-fold purification relative to membrane homogenate. After silver staining, SDS–PAGE revealed four major bands at 67, 57, 46, and 33 kDa. The pharmacological profile of the purified NMDA/PCP complex was similar to that of the corresponding membrane bound and solubilized receptors. Photoaffinity labeling using [^{3}H]azido-PCP showed specific and irreversible labeling of proteins at 67–68, 52–57, and 42 kDa.

The Neurotensin Receptor

Neurotensin (NT) is a putative neurotransmitter/neuromodulator in the CNS, and in the spinal cord it is potentially important in pain pathways. Photoaffinity labeling with NT derivatives suggests that the NT receptor contains two proteins of 49 and 51 kDa. Neurotensin receptors were purified (22) using membranes from bovine brain cortex prepared in *N*-tris[hydroxymethyl]-

methyl-2-aminoethanesulfonic acid (TES) buffer in the presence of protease inhibitors [1 mM benzamidine HCl, 0.02% (w/v) bacitracin, and 0.002% (w/v) soybean trypsin inhibitor (STI)]. The membrane pellets were resuspended in TES buffer to which digitonin and asolectin were added to give final concentrations of 2% (w/v) and 0.06%, respectively. After sedimentation at 120,000g, the supernatant was stored at -20°C before use.

Affi-Gel 10 (2 ml) prewashed with 3 vol of ice-cold water was gently agitated for 3 hr at 40°C in 2 ml of 20 mM HEPES (pH 7.4) containing 12 mmol NT and a trace amount of [3,11-tyrosyl-3,5-^{3}H]NT, followed by 20 min incubation at 20°C. Unbound ligand was removed by washing and remaining active sites were blocked by reaction with 1 M ethanolamine HCl, pH 8.0, for 1 hr at 20°C. The gel was subsequently washed with 10 vol of 20 mM HEPES (pH 7.4) followed by 5 vol of 10 mM TES buffer (pH 7.5), and stored at 4°C. Based on the radiolabel, the amount of bound NT was estimated to be 4–5 mmol/ml packed gel. The coupling yield was 30%. The NT-Affi-Gel 10 was preequilibrated with 5 vol TES buffer containing 10 mM 1,10-phenanthroline, 0.1% (w/v) digitonin, and 0.003% asolectin. The crude SRP in the same medium was loaded at 20–30 ml/hr onto a 2-ml column of the gel, which was washed with 40 vol of equilibration buffer from which STI and bacitracin were omitted, and asolectin increased to 0.06%. Bound receptors were eluted with 2.5 vol of the latter buffer containing 250 mM NaCl and dialyzed against 5 liters of 10 mM TES (pH 7.5), 1 mM EGTA, 2 mM $MgSO_4$, and 1 mM benzamidine-HCl. Fractions were analyzed immediately for [^{3}H]NT binding activity. The overall recovery of active NT receptors was 14.8% with 18,000- to 36,000-fold purification relative to crude membranes. Poor recovery was found when the affinity column was eluted with 10^{-5} M NT.

A single band for purified NT receptor was seen on SDS–PAGE after denaturation with 100 mM dithiothreitol at M_r 72,000 after either Coomassie blue or silver staining. Under nonreducing conditions, a single band was seen at 50 kDa, suggesting the presence of intramolecular disulfide bonds in the native receptor. When excess free NT was added to the solubilized preparation prior to binding to the NT-Affi-Gel column, no protein bands were detected on final SDS–PAGE analysis, even on overstaining of the gel. This preblocking effect confirms that the 72-kDa polypeptide is specific to the NT receptor. The 49- and 51-kDa subunits found previously in rat brain membranes could arise from proteolytic degradation or deglycosylation of the 72-kDa protein.

Another group (23) reported purification of digitonin-solubilized extracts of NT receptors from newborn rat cerebral cortices using the same affinity matrix and a similar protocol. In this case, the affinity gel was equilibrated with 10 vol of TES buffer [100 mM TES (pH 7.5), 1 mM EGTA, 2 mM

$MgSO_4$ 1 m*M* benzamidine HCl, 10 m*M* 1,10-phenanthroline, 200 m*M* KCl, and 0.1% digitonin] containing protease inhibitors (0.02% bacitracin, 0.002% trypsin inhibitor). The column was washed with at least 10 vol of binding buffer from which bacitracin and trypsin inhibitor were omitted. Bound NT receptors were eluted with 2 ml of 10 m*M* TES·KOH (pH 7.5), 1 m*M* EGTA·K, 2 m*M* MgSO4, 1 m*M* benzmidine HCl, 10 m*M* 1,10-phenanthroline, 0.1% digitonin, and 500 m*M* NaCl. Samples diluted with soluble receptor binding assay buffer were assayed immediately. This procedure afforded 14,000-fold purification and 5.2% recovery of active receptors.

Purified receptors were concentrated by centrifugation and ultrafiltration using Ultrafree-C3 (MW fraction 10,000, Millipore) and dissolved in standard SDS buffer for PAGE, which showed a single protein band at 55 kDa under reducing conditions (in presence of 2-mercaptoethanol) and 54 kDa under nonreducing conditions. The SDS–PAGE autoradiogram of membrane receptors and of ^{125}I-labeled NT photocross-linked with SANAH also showed a band at 55 kDa, which was protected from photolabeling by the presence of 1 μM unlabeled NT. Nonspecific labeling of a band at 76 kDa under these conditions was also observed. The difference between the mass of the protein in this study (55 kDa) and that of the cloned 424 amino acid NT receptor (M_r 47,052) is attributed to *N*-glycosylation.

Imidazoline Receptors

Reis and co-workers (24) recently reported the purification of imidazoline receptors from bovine adrenal chromaffin cells. Clonidine, idazoxan, and related agents have been generally believed to act exclusive at α_2-adrenergic receptors, but evidence has been accumulated to show they also bind to a novel nonadrenergic receptor, the imidazoline receptor (IR). The IR and α-$_2$-adrenergic receptors appear to be molecularly distinct: they are independently expressed in different tissues and brain areas and may utilize different signal transduction mechanisms. The authors recently discovered that adrenal chromaffin cells express only IRs and not α_2-adrenergic receptors, and therefore sought to isolate IRs from these cells using imidazoline agents as ligands. Two affinity matrices are described, PAC-ReactiGel in which *p*-aminoclonidine is linked to Trisacryl GF-200, and IDA-agarose in which idazoxan is coupled to PharmaLink agarose. The coupling reaction for preparation of the PAC matrix is straightforward, involving the *p*-amino group of PAC and the commercially derivatized ReactiGel. For the IDA-agarose, a Mannich-type reaction involving the terminal amino group of the DADPA-derivatized PharmaLink gel, formaldehyde, and idazoxan was utilized. In both cases, trace amounts of radiolabeled ligands were used to determine the extent of coupling, which was 0.5–1 mg PAC/ml gel and 0.2–0.4 mg idazoxan/ml gel, respectively.

Chromaffin cells from bovine adrenal medulla were isolated, homogenized by sonication, and solubilized in 50 m*M* Tris–HCl buffer, pH 7.4, containing a protease inhibitor cocktail consisting of 0.3 m*M* PMSF, 0.1 m*M* EGTA, 10 m*M* ε-aminocaproic acid, 0.8 *M* pepstatin A, 0.1 m*M* benzamide, and 0.1 m*M* benzamidine hydrochloride. The protease inhibitors were added at each step of the procedure. SRP was equally divided (approximately 30 mg) and loaded in 25 ml vol onto the IDA-agarose (2 ml bed vol) and PAC-ReactGel (5 ml bed volume) columns which had previously been washed with solubilization buffer containing 0.5% CHAPS. The columns were washed with solubilization buffer and then running buffer (with 0.05% CHAPS) until UV absorbance at 280 nm returned to baseline. The columns were sequentially eluted at 4°C with (a) 0.05% CHAPS/Tris–HCl buffer, pH 7.4, containing either 30 m*M* KCl, 30 m*M* NaCl, 100 μ*M* rauwolscine, or 1 m*M* epinephrine (to remove contamination by small amounts of α_2-adrenergic receptors), (b) 100 μ*M* idazoxan or 100 μ*M* cirazoline, and (c) 1 *M* KCl. Fractions from each elution were pooled (total volume 25–30 ml), dialyzed (4 liters, three changes) against 50 m*M* Tris–HCl, pH 7.4, containing 0.01% CHAPS, and concentrated in the dialysis membrane using PVP-360 (Sigma) to a final volume of 300–500 μl for radioligand binding assays and analysis by SDS–PAGE.

The eluted proteins retain the ligand-binding properties of receptor in intact membranes. Proteolytic degradation and receptor inactivation were minimized by the relatively short purification time (10–16 hr) and the use of protease inhibitors throughout the isolation process. Use of the IDA-agarose column and elution with idazoxan yielded two distinct protein bands, a major one at 70 kDa and a minor one at 55 kDa. Many additional proteins were observed in the final KCl eluate. With the PAC-ReactiGel, elution with idazoxan following prewashing usually yielded a 70-kDa protein, but in some cases afforded a mixture of 62, 55, and 20-kDa proteins. Nonspecific radioiodination of the purified receptor isolated from both affinity matrices afforded only a protein band at 70 kDa. Proteins which bound [^{3}H]idazoxan were not retained on control matrices that were devoid of the two affinity ligands. Protein with M_r 70,000 was seen on electrophoresis under either reducing or nonreducing conditions, indicating this protein is not linked by disulfide bonds. The 55-kDa band seen on double silver staining in some runs is assumed to be a degradation product of the 70-kDa protein. By this single-step procedure, 700-fold purification of a [^{3}H]idazoxan-binding protein was achieved, with a yield of 5%.

Rabbit antisera raised against the isolated 70-kDa protein were found by Western blot analysis to specifically label a 70-kDa protein out of numerous membrane proteins and to inhibit binding of [^{3}H]idazoxan to chromaffin cell membranes. Antisera also immunoextracted [^{3}H]idazoxan-binding activity from a solubilized chromaffin cell membrane prep. These observations support the suggestion that the purified 70-kDa protein is the ligand-binding entity of the imidazoline receptor.

Approaches Using Avidin and Biotin

The interaction between biotin and the naturally occurring proteins avidin (derived from egg white) or streptavidin (produced bacterially) is of extremely high affinity, even compared with neurotransmitter receptor binding. Many receptor ligands can be chemically derivatized with biotin in a convenient manner, since numerous reactive biotin analogues are commerically available which will allow coupling with free amino or sulfhydryl groups. The attractive feature of this system with respect to receptor purification is that a biotinylated ligand complexed to a neurotransmitter receptor will almost always bind to an affinity gel which contains covalently coupled avidin or streptavidin. For example, Howl and his colleagues (25) prepared a biotinylated derivative of vasopressin [1-phenylacetyl, 2-*O*-methyl-D-Try,6-Arg, 8-Arg, 9-lysinamide]vasopressin. This peptide, prepared by solid-phase synthesis, was capable of simultaneously binding to the rat liver V1a vasopressin receptor and to avidin. This approach could conveniently be adapted for affinity purification.

In a similar manner, Akiyama and colleagues (26) performed an affinity purification of the endothelin receptor from human placenta with the aid of [9-Lys]-biotinylated endothelin-1 and avidin agarose. The purified endothelin receptor using this method was relatively homogenous on SDS–PAGE and was pharmacologically fully active in endothelin-binding assays.

The avidin–biotin interaction is of such high affinity that the receptor can on occasion be purified as a complex with other associated proteins, if conditions used for the purification are sufficiently mild to avoid dissociation of the complex. Alternatively, such complexes, for example with G proteins, can be chemically stabilized (e.g., by addition of agonist) during the purification process. This provides a novel approach to examining receptor–protein interaction. With the aid of this approach Brown and Schonbrunn (27) purified a somatostatin receptor–G protein complex. This was done by solubilizing washed pancreatic membranes with dodecyl-β-D-maltoside. Requirements for detergents were investigated and it was found that only a small number of alkyl glycosides could be used successfully to obtain a reasonable degree of recovery. The solubilized receptor preparation was treated with a biotinylated somatostatin analogue, *N*-biotinyl-[8-Leu, 22-D-Trp, 25-Tyr] somatostatin-28. The receptor–biotinylated ligand complex was passed through a streptavidin–agarose column, which retained a large amount of the biotinylated activity. When this column was washed with 100 μM GDP, the interaction released the entire G protein–receptor complex, which could be analyzed on SDS gels. Western blots of the gel bands using antibodies selective for Gi/Gs indicated that the bands observed corresponded to the G protein component, and presumably also contained the putative receptor component. This strategy is potentially of quite general utility for isolation of G protein–receptor complexes in an intact state without the need for covalent cross-linking. Since in principle this allows the receptor to be regenerated in a pharmacologically intact form, binding studies could be performed if denaturing conditions are not used in the gel analysis.

A continuing problem with the avidin–biotin system has been the strength of the avidin–biotin interaction; sometimes even a large excess of biotin applied to wash the column will not completely remove the bound material. This may in part be a kinetic problem. It would be of interest to examine in detail the kinetics of avidin–biotin association on a matrix under conditions where a large mass excess of biotin was present. One way to circumvent this problem is to use a cleavable spacer linking biotin to the desired ligand. Such spacers are commerically available in a form where they can be readily biotinylated and typically contain a disulfide bond, so that brief treatment with 2-mercaptoethanol or DTT will break the cross-link. The freed receptor in principle can then be eluted with ordinary buffer, as it will not be retained on the column. The only problem is that if the native receptor itself contains disulfide linkages necessary for pharmacological activity, this activity will be destroyed by this procedure and probably cannot be readily regenerated by removal of the reducing agent using dialysis.

An interesting strategy has been employed by Ozyhar and colleagues (28) to purify the ecdysteroid receptor (EcdR) from a nuclear extract of *Drosophila*. This soluble receptor, which is an insect steroid receptor, is known to function by binding to DNA. The DNA sequences required for specific binding of the EcdR are also accurately known from previous work. Thus, it was possible to construct a double-stranded 28-mer oligonucleotide which bound to the EcdR with high affinity.This oligonucleotide was labeled on its 5′-end with biotin. Magnetic beads, which are commerically available in a variety of chemically derivatized forms, were coupled with streptavidin to yield a magnetic affinity support. This magnetic affinity support coupled to streptavidin was then added to a soluble nuclear extract from *Drosophila*, and the beads after being shaken were then magnetically separated and washed with buffer. The receptor was liberated by washing with excess biotin. In this single separation step, which took a total of about 1.5 hr, a 29,000-fold purification to homogeniety was achieved. This simple approach could be applied to other receptors, such as the *Ah* receptor system (see above), which are known to have sequence specificities for DNA binding.

Sigma Receptors

A putative opiate receptor subtype termed sigma (σ) was proposed to be the site of action of *N*-allylnormetazocine (SKF-10,047), which produced characteristic dysphoric effects as well as autonomic stimulation in animals which were distinct from the analgesic and sedative effects of typical opiates such as morphine. Suggestions that the σ site had co-identity with a PCP receptor were later found to be unsubstantiated when the PCP site was identified as a component of the NMDA receptor. While a number of biological functions have been attributed to sigma receptors in the brain and the periphery (29), their precise function remains unknown, although it is clear that they have physiological roles in modulation of hippocampal NMDA responses. For example, there is evidence that σ receptors may represent a link between the central nervous system and the endocrine and immune systems.

Using an affinity ligand structurally related to 3-(3-hydroxyphenyl)-*N*-propylpiperidine (3-PPP), Arnold (30) succeeded in purifying two proteins of molecular mass 63 and 65 kDa from bovine and rat cerebellar preparations which exhibited pharmacology characteristic of the σ receptors/binding site. Ehrlich, Schuster, and Murphy (31, 32) have purified a component of the rat liver σ receptor, where σ receptor densities are much higher than those in the CNS, and obtained an N-terminal amino acid sequence of this material.

Rat liver homogenates were prepared in 50 m*M* Tris–HCl, pH 8.00, containing 250 μM PMSF (buffer A). Following centrifugation, resuspension, and residementation, the pellet was suspended in 10 vol (w/v) buffer B (buffer A containing 5 m*M* CHAPS). The clear supernatant obtained after centrifugation was filtered through a 0.22-μm cellulose acetate filter and then diluted with buffer B to yield a final protein concentration of ca. 1.4 mg/ml. The resultant homogenate was fractionally precipitated on treatment with ammonium sulfate (60% saturation) for 30 min at 4°C. The precipitate, which contained essentially all the σ binding activity (as measured using [^{3}H]haloperidol or ^{3}H-labeled (+)-pentazocine) was resuspended in 150–200 ml of buffer B. This solution was rapidly passed through a 5-g portion of underivatized NuGel P-AP aminosilica gel in a 60-ml sintered glass funnel of medium porosity at 4°C in order to remove proteins that might be nonspecifically adsorbed on the affinity gel. The resulting SRP was stored at −80°C until used.

An oximino derivative of haloperidol (OB-101) was used as the affinity ligand. The succinimide ester of OB-101 was coupled to NuGel P-AP silica-based gel. Using trace amounts of [^{3}H]OB-101, this procedure was found to produce greater than 1.25 μmol ligand/ml gel. The SRP diluted fourfold with phosphate-buffered saline (PBS) was applied to a 20-g portion of the affinity gel packed into a 1.6 $\times$ 25-cm column which had been preequilibrated with PBS. The solution which passed through the column showed substantial [^{3}H]haloperidol binding activity and was reserved. The column was pre-washed with ca. 15 bed vol of 500 mM NaCl in PBS to remove nonspecifically absorbed proteins until the absorbance of the effluent solution at 280 nm was at a background level. Elution using 50 ml of 1 mM dextrallorphan or 50 ml of 1 nM (+)-pentazocine in PBS containing 250 μM PMSF at 4°C gave material which showed characteristic σ binding activity after dialysis. The SDS—PAGE of this material showed prominent bands at 55 and 65 kDa. Since the amount of purified protein was too small to measure accurately (<1 μg/ml), the fold purification achieved based on the specific binding of [^{3}H]haloperidol was estimated to be at least 560 assuming the σ receptor has M_r 65,000; this is about a factor of 6 less than that required for complete purification.

The procedure was repeated on a preparative scale in order to obtain enough material for N-terminal amino acid analysis. In this procedure, the pass-through from addition of the SRP to the affinity column was applied a second time to deposit additional σ receptors, and the column was eluted twice with 1 mM dextrallorphan. The dextrallorphan eluates were dialyzed against 4–6 liters of PBS in 12,000–14,000 MWCO dialysis tubing with at least two changes of dialysate over a 24-hr period at 4°C. Dialysates were concentrated by ultrafiltration against YM10 membrane (Amicon). On this material, SDS–PAGE showed a substantial increase in the relative amounts of the protein bands at 28 and 40 kDa compared to the 65-kDa band. The pharmacological activity of the concentrated dextrallorphan eluate decreased dramatically on concentration, which could be due to loss of a critical binding component of the receptor complex. Transfer of the concentrated denatured material from polyacrylamide gels onto PVDF membranes and N-terminal analysis of the 28-kDa protein yielded a 15 amino acid sequence which was a perfect match with that of rat cyclophilin A (17 kDa), which is the major site of action of the immunosuppressant agent cyclosporin A. This result was reproduced using a freshly purified sample of the protein. Assuming that the cyclophilin was not adventitiously carried along as a contaminant during the multistep purification process, which has yet to be firmly established, the results suggest that a 28-kDa cyclophilin analog may be a component of the rat liver σ receptor. The mass of this protein is similar to that of proteins identified by σ-specific photoaffinity labeling using

[^{3}H]azido-di-*o*-tolylguanidine (30, 33, 34) and [^{125}I]azidococaine (35) in a number of mammalian tissues.

Immunoaffinity Approaches

The approach of immunoaffinity chromatography is useful in cases where the primary sequence of a receptor is known, and it is desired to purify the receptor from different tissues or, for example, to study the effects of posttranslational modification or interaction with G proteins. In this approach, monoclonal antibodies are used against a portion of the receptor sequence which is known to be exposed to the cytosol and which therefore is presumed to be relatively antigenic. Frequently, it is desired to compare the affinity purification procedure using several monoclonal antibodies with specificities directed against different domains or even subunits of the receptor.

Obviously, this approach is not useful if the receptor sequence is unknown. It is possible that immunoaffinity approaches could be used to purify unknown receptors which are highly homologous to known receptors, if domains of the receptor which were expected to be relatively invariant were selected. For example, in the case of G protein-linked receptors, certain transmembrane segments are highly conserved across subclasses of these receptors (1), so this approach might work if the peptides were sufficiently antigenic. Since there is no particular requirement for monoclonal antibodies in this schema, it is possible that polyclonal antibodies might be preferred for such an experiment. Various investigators have used antibody fragments; antibodies rather than IgG molecules could probably also be employed.

As examples of this approach, two groups have used antibodies directed against different regions of the $GABA_A$ receptor to affinity purify it from detergent-solubilized brain membrane preparations. Park and colleagues (36) prelabeled the benzodiazepine binding site of the GABA receptor with [^{3}H]flunitrazepam, in order to provide a radioactive "signal." In separate experiments, they passed the solubilized, photoaffinity-labeled receptor preparation over an Affi-Gel 10 column, which had been derivatized with several different monoclonal antibodies directed at presumed exposed domains of the receptor. Pollard and colleagues (37) used a different set of monoclonal antibodies directed against the same receptor to purify it from cortex and cerebellum. Kuriyama and colleagues (38) used a similar approach to purify the $GABA_B$ receptor from CHAPS-solubilized bovine cerebral cortical membranes.

The main advantage of this approach is that purification to homogeneity can usually be achieved in a single step. The other hallmark of this approach is that high recoveries are usually achieved, frequently in excess of 40%. One potential difficulty with immunoaffinity chromatography is that sometimes it is difficult to completely remove the bound material from the column. Frequently this has to be accomplished under denaturing conditions or by the use of high salt concentrations, which can seriously affect receptor structure and activity.

References

1. W. C. Probst, L. A. Synder, D. I. Schuster, J. Brösius, and S. C. Sealfon, *DNA Cell Biol.* **11,** 1 (1992).
2. C. D. Ebert and S. W. Kim, *Thromb. Res.* **26,** 43 (1982).
3. M. Nachman, A. R. Azad, and P. Bailon, *J. Chromatogr.* **597,** 155 (1992).
4. R. G. L. Shorr, R. J. Lefkowitz, and M. G. Caron, *J. Biol. Chem.* **256,** 5820 (1981).
5. S. E. Senogles, N. Amlaiky, A. L. Johnson, and M. G. Caron, *Biochemistry* **25,** 749 (1986).
6. S. E. Senogles, N. Amlaiky, P. Falardeau, and M. G. Caron, *J. Biol. Chem.* **263,** 18996 (1988).
7. J. Ramwani and R. K. Mishra, *J. Biol. Chem.* **261,** 8894 (1986).
8. R. A. Williamson, S. Worrall, P. L. Chazot, and P. G. Strange, *EMBO J.* **7,** 4129 (1988).
9. F. J. Bosker, F. J. Van Bussel, A. P. Thielen, Y. L. Soei, G. T. Sieswerda, J. Dijk, P. G. Tepper A. S. Horn, and W. Moller, *Eur. J. Pharmacol.* **163,** 319 (1989).
10. M. J. Plug, J. Dijk, J. A. Maassen, and W. Mollwe, *Eur. J. Biochem.* **206,** 123 (1992).
11. J. A. Gingrich, N. Amlaiky, S. E. Senogles, W. K. Chang, R. D. McQuade, J. G. Berger, and M. G. Caron, *Biochemistry* **27,** 3907 (1988).
12. A. Sidhu, *J. Biol. Chem.* **265,** 10065 (1990).
13. A. Sidhu, *Biochemistry* **27,** 3907 (1988).
14. L. S. Aglio, J. M. Maturo, and M. D. Hollenberg, *J. Cell. Biochem.* **28,** 143 (1985).
15. H. Nakata, *Mol. Pharmacol.* **35,** 780 (1989).
16. H. Nakata, *J. Biol. Chem.* **264,** 16645 (1989).
17. H. Nakata, *Eur. J. Biochem.* **296,** 171 (1992).
18. T. L. Gioannini, A. D. Howard, J. M. Hiller, and E. J. Simon, *J. Biol. Chem.* **260,** 15117 (1985).
19. T. L. Gioannini, L.-Q. Fan, L. Hyde, D. Ofri, Y.-H. Yao, J. M. Hiller, and E. J. Simon, *Biochem. Biophys. Res. Commun.* **194,** 901 (1993).
20. C. A. Bradfield, E. Glover, and A. Poland, *Mol. Pharmacol.* **39,** 13 (1991).
21. A. F. Ikin, Y. Kloog, and M. Sokolovsky, *Biochemistry* **29,** 2290 (1990).
22. A. Mills, C. D. Demoliou-Mason, and E. A. Barnard, *J. Biol. Chem.* **263,** 13 (1988).

23. Y. Miyamoto-Lee, S. Shiosaka, and M. Tohyama, *Peptides* (*N.Y.*) **12,** 1001 (1991).
24. H. Wang, S. Regunathan, M. P. Meeley, and D. J. Reis, *Mol. Pharmacol.* **42,** 792 (1992).
25. J. Howl, X. Wang, C. J. Kirk, and M. Wheatley, *Eur. J. Biochem.* **213,** 711 (1993).
26. N. Akiyama, O. Hiraoka, Y. Fujii, H. Terashima, M. Satoh, K. Wada, and Y. Furuichi, *Protein Express. Purif.* **3,** 427 (1992).
27. P. J. Brown and A. Schonbrunn, *J. Biol. Chem.* **268,** 6668 (1993).
28. A. Ozyhar, M. Gries, H. H. Kiltz, and O. Pongs, *J. Steroid Biochem. Mol. Biol.* **43,** 629 (1992).
29. J. M. Walker, W. D. Bowen, F. O. Walker, R. R. Matsumoto, B. De Costa, and K. C. Rice, *Pharmacol. Rev.* **42,** 355 (1990).
30. F. Arnold, Ph.D. Thesis, New York University (1988); see also D. I. Schuster, F. J. Arnold, and R. B. Murphy, *Brain Res.,* in press.
31. G. K. Ehrlich, Ph.D. Thesis, New York University (1993).
32. D. I. Schuster, G. K. Ehrlich, and R. B. Murphy, *Pharmacol. Lett. Life Sci.* **55,** PL151 (1994).
33. M. Kavanaugh, B. Tester, M. Scherz, and J. F. W. Keana, *Proc. Natl. Acad. Sci. U.S.A.* **85,** 2844 (1988).
34. S. B. Hellewell and W. D. Bowen, *Brain Res.* **527,** 244 (1990).
35. J. R. Kahoun and A. E. Ruoho, *Proc. Natl. Acad. Sci. U.S.A.* **89,** 1393 (1992).
36. D. Park, J. Vitorica, G. Tous, and A. L. de Blas, *J. Neurochem.* **56,** 1962 (1991).
37. S. Pollard, M. J. Duggan, and F. A. Stephenson, *J. Biol. Chem.* **268,** 3753 (1993).
38. K. Kuriyama, H. Nakayasu, H. Mizutani, R. Narihara, and T. Ichida, *Neurochem. Res.* **18,** 377 (1993).

[3] Protein and Peptide Microsequencing: Applications in Neuroscience and Receptor Research

Michael J. Walsh

Introduction

Almost 30 years have gone by since Edman and Begg (1) described a machine that automated the sequential chemistry to determine the linear arrangement of amino acids along the backbone of peptides and proteins. Solid-phase immobilization chemistries were subsequently introduced and allowed complete amino acid sequences of many proteins to be determined including the entire complement of about 52 eubacterial ribosomal proteins, the greatest peptide sequence enterprise ever undertaken (2). An omnibus instrument which incorporated almost every important development of previous years in machine design, material science, component miniaturization, and chemical purification, as well as detection of phenylthiohydantoin-amino acids (PTH-AA) was presented in 1981 (3). The timing was fortuitous. In fact, the prompt conjunction of recombinant DNA and Edman sequence chemistries led astonishingly soon to the determination of the primary structures of the pentameric nicotinic acetylcholine receptor (4) and high M_r sodium channel subunit (5), the sine qua non requirement for correlation of structure with function which is now a central aim of molecular neurobiology. In fact, no sector of cell biology has been greater impacted by peptide microchemistry than the neurosciences. The implication of the βAPP in Alzheimer's disease, the structural characterization of the major peptide component of the infectious prion particle, the molecular characterization of the synaptic vesicle, the postsynaptic density, the cloning of many myelin membrane proteins, and neuropeptide transporters are just a few of the areas of neuroscience where peptide microchemistry has impacted.

The first receptor-cloning projects relied on screening cDNA libraries by hybridization with oligonucleotide mixtures synthesized on the basis of partial amino acid sequence data obtained from receptor polypeptides eluted from one-dimensional electrophoresis (1DE) gels and chemically or enzymatically fragmented. The primary structures of the nicotinic acetylcholine receptor (4), sodium channels (5), subunits of the major neurotransmitter receptors for glycine (6), and γ-aminobutyric acid (GABA) (7) were obtained

Methods in Neurosciences, Volume 25

this way. Later, a higher resolution technique for complex protein mixtures, two-dimensional electrophoresis (2DE) (8), was married to these two technologies, i.e., peptide microsequencing and cDNA cloning, with the demonstration that N-terminal sequences of proteins blotted to sequencer stable membrane supports could be routinely obtained (9–11). Finally, since some proteins are naturally N-terminally blocked as well as inadvertantly blocked during purification and electrophoresis, the development of methods to chemically or enzymatically fragment proteins immobilized in the gel matrix (9, 11) or on nitrocellulose membrane (12) then separation of the peptides generated by high-performance liquid chromatography (HPLC) or electrophoresis and blotting to suitable supports for sequencing gave every lab with even modest biochemical expertise an approach for obtaining peptide sequences for protein identification, gene cloning, or production of synthetic peptides.

For optimal results with these microchemical methods and with minute amounts of protein meticulous technique is necessary at every step. This includes sample preparation, gel electrophoresis, electroblotting, and choice of enzyme and digestion conditions to generate peptides (9–14). The subsequent HPLC and peptide sequence procedure must also be first class. There is a price to be paid throughout for suboptimal technique. Much is written about what is possible with choice low M_r soluble proteins (e.g., myoglobins, lactalbumins); the sequences of these proteins were determined decades ago, and they are useless methodological models for the laboratory which has to operate with minute quantities of relatively insoluble and usually high M_r protein. The test of methods in peptide microchemistry is the ability to discover and sequence proteins, have confidence in the use of the sequences obtained to make antibodies using synthesized peptides, and propose the right oligonucleotide mixtures to direct cloning projects.

Basic Methodology

Protein Purification

The variety of ways receptors are purified are reviewed in Chapter 2 by Schuster and Murphy. In general, methods that assure functional reconstitution of isolated membrane receptors and high specific activities measured by ligand binding are also good ways to prepare proteins for peptide sequencing. The recovery of free N-termini and ability to identify tryptophan, methionine, and cysteine (after alkylation) by chemical sequencing is a good measure that the appropriate precautions are taken during protein/peptide purification. To obtain intact proteins severe precautions against proteolysis may also be necessary. Failure to observe this injunction led to insistent claims that the AChR was an oligomer of a single polypeptide and that the mammalian sodium channel had an apparent M_r of about 65,000.

If a protein is already damaged during purification then all the precautions recommended for the subsequent electrophoresis steps are superfluous. There are a few general guidelines which are common to all purification procedures whether involving fractionation by centrifugation, chromatography, or other techniques. These include using water with ≥18 megohm/cm resistivity, using the highest quality reagents and buffers, avoiding pH extremes, limiting exposure to detergents with unacceptable peroxide and aldehyde contamination (15), using amine-containing buffers when possible especially if urea is used, and incorporating a thiol reagent generally fresh 1 m*M* dithiothreitol in all buffers including gradient solutions and column elution buffers. Aliquots of 1 *M* dithiothreitol which are stable stored at −20°C for months are thawed immediately before use for addition to buffers. The sequences of almost all of the more than 52 eubacterial ribosomal proteins by Edman degradation were determined using urea chromatography-purified proteins (2) indicating that with the right precautions a wide range of options for protein solubilization is available to the researcher. Some important approaches and guidelines for protein purification are detailed in recent monographs (16); approaches to receptor purification of particular interest for neuroscientists are reviewed by Schuster and Murphy (Chapter 2).

Procedure for Gel Electrophoresis

Since gel electrophoresis is performed in most laboratories that require peptide sequences a description of basic gel procedures is omitted. Gel systems based on the 1DE Laemmli system (17) and the O'Farrell system of two-dimensional electrophoresis (8) are most in use but the basic methodology is entirely compatible with numerous different gel systems. Some features of the Laemmli and O'Farrell gel systems make them attractive however for peptide sequence applications:

1. The chemicals for these procedures meet highest quality specifications so purification in the laboratory is no longer necessary.
2. Preelectrophoresis with scavengers which rids gels of a variety of species with the potential to damage amino acids does not compromise the high resolving power of the 1DE or 2DE gel systems (11).

3. Two-dimensional electrophoresis library catalogs are now reported for several tissues as well as several cell lines of importance in cell biology, oncology, and toxicology. The relevant studies were recently reviewed (18). By using the most generally standardized gel methods investigators have the option of correlating the 1DE and 2DE coordinates of proteins of interest with such libraries.

Routine Precautions for One- and Two-Dimensional Electrophoresis Gels

Source of Chemicals

Suppliers whose reagents routinely perform satisfactorily are reported from several laboratories (10–14).

Gel Polymerization

Sodium Dodecyl Sulfate–Polyacrylamide Gel Electrophoresis (SDS–PAGE)

Some publications enlarge by a factor of 4 or 5 the amount of ammonium persulfate (APS) required to obtain a good-quality polymerization. Polymerization is a free radical initiated reaction and excess APS generates molecular species which modify amino acids including N-terminal blockage. Ammonium persulfate generates a crackling sound on solvation and should be prepared daily when gels are to be used to prepare proteins for sequencing. For 30 ml 10% acrylamide gel solution, 45 μl 10% APS and 18 μl *N,N,N′,-N′*-tetramethylethylenediamine (TEMED) gives satisfactory polymerization in 30 min. If the gel solution is degassed about 10% less APS will suffice. After pouring, the resolving gel is overlaid gently from the edge with water or a gentle spray of 0.1% SDS rather than organic solvent of uncertain purity. For gradient gels the amount of APS and TEMED is reduced by 10% in case polymerization occurs in the pump tubing while the gel is poured. The lower percentage (3–4% acrylamide) stacking gel is polymerized using 20 μl APS and 8 μl TEMED per 10 ml gel solution. Gels may be stored at 4°C for days to weeks. Gels are routinely polymerized at least 24 hr before use to permit decay of free radicals and short-lived species.

Isoelectric Focusing/Nonequilibrium pH Gradient Electrophoresis (IEF/NEpHGE)

There are some special considerations in preparation of the first-dimension gels for the 2DE procedure. Ultrapure urea is required. For 10 ml gel solution, 10 μl 10% APS and 6 μl TEMED result in polymerization within 30 min.

When the urea concentration is ≥9 M solvation occurs by swirling the gel solution for a few minutes under warm water rather than by prolonged stirring. Usually, IEF and NEpHGE gels are 4% acrylamide; this can be reduced to as low as 3% if especially large proteins are being studied but 3% gels are fragile and require a lot of care in manipulation.

There is a wide range of detergents including nonionic (NP-40, Triton X-100), zwitterionic (CHAPS), and ionic (sodium deoxycholate, SDS) available. For brain-derived subcellular fractions including fractions enriched in membrane proteins (19), for ribosomal proteins (11), serum, and cerebrospinal fluid (20) addition of any detergent has not been noticeably effective as measured by the amount of protein entering the IEF gel or quality of resolution obtained. Numerous techniques involving detergent manipulation to enhance the entry and resolution of membrane proteins have not been found to be very useful. A number of detergents have such high levels of peroxides and aldehydes (15) as to cause concern about amino acid modifications of denatured and thiol-reduced protein. Therefore, if detergent does not enhance the separation, omit it. If detergent is omitted from the gel omit it also in the lysis buffer. When detergent is found to enhance separation (e.g., desmosomal glycoproteins) NP-40 is used because of its low level of impurities (15).

The IEF/NEpHGE gels are run several hours after polymerization. The chief consideration is concern about spontaneous formation of cyanate from the urea. Reaction with cyanate blocks the N-terminal amino group for Edman chemistry and modifies side chains of several other amino acids (21). After being poured, IEF gels are overlaid with water until polymerized. It is important to remove the water overlay immediately on polymerization and protect the gel surface with an overlay solution containing 10 *M* urea; if this is not done, urea in the gel matrix diffuses out and the concentration of urea at the gel application surface may be lowered critically so that protein with borderline solubility in the sample precipitates and then impedes solubilized proteins from entering the gel. The IEF/NEpHGE gels are loaded directly onto the SDS–PAGE gel. Equilibration gives a nicer 2DE gel from a cosmetic viewpoint but there are dramatic losses of lower M_r proteins through diffusion from the low-porosity first dimension gel.

Preelectrophoresis of One- and Two-Dimensional Electrophoresis Gels

Experience indicates that when precautions including preelectrophoresis are omitted then the N-terminus is regularly blocked. When precautions are taken including preelectrophoresis with charged thiol reagents then a free

N-terminus (i.e., for proteins that are not N-terminally blocked in the native state) is routinely obtained and amino acids especially sensitive to oxidative modification or destruction are recovered quantitatively during peptide sequencing (11, 19). We have been able to obtain N-terminal sequences of a number of proteins which were reported to be blocked. The quantitative recovery of methionine, cysteine, and tryptophan is highly desirable when peptide sequences are used to guide oligonucleotide synthesis to clone proteins from cDNA libraries since these amino acids use one or two codons. Since half of all mammalian proteins may be N-terminally blocked and many proteins become blocked if the purification steps prior to 1DE and 2DE are extended most investigators elect to omit direct N-terminal sequencing and pursue internal peptide sequences by chemical or enzyme digestion after electrophoresis. If the latter approach is used, numerous new peptide N-termini are available for sequencing whereas if the protein is N-terminally sequenced and blocked it is with some exceptions (22) lost for amino acid sequence determination. Moreover, if peptide sequences are sought chiefly for protein identification as in construction of 2DE data bases (18), then the gaps so familiar in large-scale sequencing projects representing damaged or modified amino acids are a marginal problem. These observations should suffice to direct an investigator as to the most suitable approach.

The methods which follow *routinely* assure N-terminal protection and quantitative recovery of methionine, tryptophan, and cysteine (after modification).

IEF Gels

Gels are polymerized 3–6 hr before use in the case of IEF/NEpHGE tube gels. IEF gels are prefocused for at least 2 hr at 250 V overlaid with 10 *M* urea solution adjusted to 10 m*M* with thioglycolic acid. Thioglycolic acid is an efficient scavenger and protects tryptophan even during high-temperature acid hydrolysis for amino acid composition determination (23). This charged thiol moves through the IEF gel as scavenger. Other harmful molecular species, cyanates, and acrylic acid are also removed during this step.

Sodium Dodecyl Sulfate–Polyacrylamide Gel Electrophoresis Gels

Gels are polymerized days or weeks and stored at 4°C until used. Gels are preelectrophoresed after addition of 200 μl thioglycolic acid per liter to the running buffer. The duration of electrophoresis is optional from several hours or overnight. Two hours at 60 V for 1-mm thick and 16-cm-long gels is appropriate. Times are modified depending on gel dimensions. The upper buffer must be replaced with new running buffer but omitting thioglycolic acid before restarting electrophoresis after sample application. A concern is sometimes raised about loss of protein resolution by preelectrophoresis due to loss of the pH discontinuity at the interface of the stacking and resolving gel. However, in practice resolution is not routinely compromised. In the rare case a gradient gel is employed. As a general rule, the lowest percentage gel which permits the protein of interest to be resolved should be used as the efficiency of subsequent electroelution or electroblotting is enhanced. It is not necessary if it is not convenient to immediately load a sample (for 1DE). Instead, the prerun gel is stored at 4°C until ready for use.

Other scavengers and gel techniques have been reported to protect proteins for sequence analysis. The results of meticulous attention to protein protection as reported here are well documented by the pioneering microchemical achievements of the 1980s which contributed to the determination of the primary structures of most of the major channel and receptor protein families [see, especially, Hunkapiller *et al.* (24)].

Sample Preparation for One- and Two-Dimensional Electrophoresis

Fractions enriched in proteins of interest are often obtained in solutions or concentrations not suitable for direct application to either 1DE or 2DE gels. A few examples suffice: SDS precipitates in the presence of guanidine salts, excess of detergent, phosphate buffers, and high salt concentration may severely degrade the resolution of SDS–PAGE, high concentrations of some detergents such as SDS cause problems with IEF, and even small amounts of SDS are incompatible with NEpHGE. Standard approaches of preparative biochemistry such as dialysis and column exchange are not discussed here; with small amounts of material sample losses are unacceptable. Methods involving organic solvent precipitation have been most widely used for this purpose. The method of Wessel and Flügge (25) is used routinely to provide salt and detergent free proteins for N-terminal sequence analysis and is described here.

The Procedure of Wessel and Flügge

The procedure is performed in polypropylene microcentrifuge tubes; the smallest volume tube with enough capacity is used to reduce the amount of surface for nonspecific binding of protein. Loss of protein through binding to surfaces is a key concern when small quantities of protein are processed.

For 100-μl sample add 400 μl Methanol, vortex vigorously, centrifuge at 10,000*g* for 2 min, add 100 or 200 μl chloroform (200 μl is added if there is a high phospholipid content as in brain white matter). Vortex again vigorously, and spin at 10,000*g* for 2 min. Add 300 μl water, vigorously vortex, and spin at 10,000*g* for 1 min to obtain phase separation. The upper phase is removed observing carefully to avoid protein that may line the side of the tube and the protein at the interface (a 1- or 2-ml syringe with a bent needle and the bevelled end cut off is used). Four to eight volumes of methanol are added and the precipitated protein is recovered at the tube bottom by centrifugation at 10,000*g* for 5 min. The following are convenient modifications:

a. If the amount of protein is considerable, e.g., a protein pellet is easily visible after addition of methanol and centrifugation, the methanol, chloroform, and water may be added at once, the tube centrifuged for 1 min for phase separation, the upper phase discarded, and the protein pellet recovered by methanol addition and centrifugation as described above.

b. If the protein solution is very dilute the efficiency of protein precipitation is less; therefore, the volume is first reduced by rotary evaporation. This is especially useful to reduce the volume of electroeluted protein samples containing SDS. If the sample does not have SDS, the tube may be frozen in liquid nitrogen and the volume reduced by the same means, or aliquots may be added to the tube stepwise and precipitated, and further aliquots added sequentially until the entire sample is processed. Sequential processing in one tube is used to mitigate somewhat the problem of losses by irreversible peptide binding to surfaces.

This method is easily scaled upward, e.g., for precipitation of several milliliters of sample in 15 or 50 ml polypropylene tubes. A variation of this technique, i.e., precipitation with at least 4 volumes of methanol or at least 8 volumes of acetone overnight at -20°C is also routinely used. We have used this method for the preparation of hundreds of proteins for Edman chemical sequencing. A very clean first cycle chromatogram is routinely obtained. The method also provides a clean, salt-free sample for solubilization in Laemmli sample or O'Farrell lysis buffer.

Precipitation with trichloroacetic acid is also widely used in preparation of samples for SDS–PAGE and for preparation of proteins for peptide sequencing. There are disadvantages—this corrosive acid may modify amino acids and affect acid-labile linkages to proteins and it is rather difficult to subsequently obtain quantitative resolubilization of the sample even if SDS is used.

Whichever method is used the precipitated pellet is conveniently dried by application of a gentle vacuum in a benchtop dessicator after the tube is capped with needle-punctured Parafilm or with a gentle stream of high-quality nitrogen or argon.

Sample Solubilization for One- and Two-Dimensional Electrophoresis

Precipitated samples may be notoriously difficult to resolubilize for 1DE and 2DE. Quantitative solubilization is of key importance when the amount of sample is limiting. Fortunately, when sample is prepared by organic solvent precipitation as described above, solubilization is often readily achieved by the addition of Laemmli SDS sample buffer or O'Farrell urea lysis buffer. Minor useful modifications of the standard SDS sample buffer include supplementing the Laemmli buffer to 4–5% SDS with 50 m*M* dithiothreitol (this is stored as 0.5-ml aliquots at −20°C), the sample is heated at 56°C for 20 min, and then the tube is vigorously agitated on a device such as the Eppendorf shaker for several hours or overnight. Heating at 100°C for 5 min which was once routine is usually unnecessary and may be harmful in terms of solubilization of some membrane proteins as well as producing amino acid modifications. Samples for 2DE are solubilized in 20 m*M* Tris–HCl or other amine containing buffer with 10 *M* urea; a buffer with pH <6.5 may be preferable because of lower reactivity of the cyanate with protein under acidic conditions. Samples dissolved in urea for not more than 4 hr and samples dissolved in SDS for as long as 24 hr are spun for 10 min and supernatants applied to the 1D or IEF/NEpHGE gels. It is important to centrifuge well (10,000*g* for 10 min is usually adequate); protein that is not in solution does not enter a gel but impedes the electrophoresis of solubilized protein.

Several factors determine the dimensions, specifically width and length, and percentage acrylamide used for the 1DE gel and the diameter and length of the IEF/NEpHGE tube gel. When proteins are intended for electroblotting for peptide digestion and sequence determination it is particularly desirable to achieve saturation of the electroblotting membrane with protein and achieve the immobilization of the largest amount of protein possible on the smallest area of membrane. The importance of this will be emphasized later. The IEF/NEpHGE is often performed in disposable 0.5- to 1-mm disposable microcapillaries (11). In this instance sample volumes to a maximum of about 25 μl dependent on salt concentration are consistent with highly resolved spots, but 0.4-cm i.d. tube gels permit applications of as much as 400 μl of sample, 1–1.5 mg of total protein (the amount and resultant resolution of the sample are determined by the protein complexity of the sample, the relative solubilities of the sample components, and other factors). For the wider diameter tube gels one or two additional crystals of urea are dropped into the tube gel with the aid of a spatula to assure saturating amounts of urea (>10 *M*); it is not necessary that the urea be in solution. This is effectively semipreparative 2DE.

Choice of Electroblotting Membrane

The first electroblotting supports introduced for direct N-terminal sequence analysis were based on glass fiber supports "activated" and derivatized in a variety of ways (9–11) and were generally characterized by high initial yield on chemical sequencing and high repetitive yield. Nitrocellulose as the established protein blotting membrane was not employed as it is labile to the chemicals used in Edman chemical sequencing (10). The introduction of the organic polymer blotting membrane, Immobilon, based on polyvinylidene difluoride (PVDF) by Millipore (Bedford, MA), led to a move away from glass fiber-based modified membranes because of commercial availability, convenience of use especially ease of detection of blotted proteins, and acceptable performance during the sequencing procedure. Initial evaluation of Immobilon (11, 26) was positive because of these considerations. A superior detection method which requires no staining at all is as follows (11): as the PVDF membrane dries, spots become detectable as areas much grayer than the surrounding membrane; second, with further drying, areas containing protein are easily seen as zones more intensely white than the surrounding PVDF when transilluminated with white light. As little as 100 ng individual protein was detectable without use of dyes. For some lots of PVDF stained with amido black or Coomassie R-250, spot visualization is often optimal while the membrane dries, and sometimes it is observed that spots or bands not clearly visible after staining become clear if the blot is left in water overnight. There is a large literature now on which membranes are best and which detection method is optimal. Part of the confusion seems to relate to changes in the manufacturing process. Eckerskorn and Lottspeich (27) have evaluated all the most frequently used blotting membranes for efficiency of protein binding during electroblotting procedures and performance during Edman sequencing. Comparisons and quantitation as presented in this study are extraordinarily demanding in operator and instrument time; the conclusions of this study should be carefully assessed by researchers using such membranes. Conclusions from our less-comprehensive studies over the past several years are very congruent. There are other comparisons of membranes which provide useful information (28). Eckerskorn and Lottspeich (27) have not addressed the performance of nitrocellulose, Immobilon PSQ (Millipore), or ProBlott (Applied Biosystems), the three most used membranes, as supports for enzymatic digestion of membrane-immobilized protein. In this laboratory, for several proteins, more or less identical recoveries and HPLC peptide patterns were obtained from digests done on the three supports (not shown).

Electroblotting of One- and Two-Dimensional Electrophoresis Resolved Proteins

Tank Electrophoretic Transfer

A variety of buffers have been reported in addition to the standard blot buffers. These include acetic acid (1%) for acetic acid-urea gels (10, 11), sodium borate (9), and pH 11 buffers without and with methanol (26). For wet blots, 25 m*M* Tris–HCl, pH 8.4, with 0.5 m*M* dithiothreitol (10) is used. Methanol is omitted. There is much written about the optimal buffer for electroblotting. When a blot is used for immunologic detection of protein quantitative transfer of protein is not necessary. For blots intended for sequence studies quantitative transfer of protein is imperative. In our experience, quantitative transfer to many different membranes is reliably obtained by the following procedure. The buffer is 25 m*M* Tris–HCl, pH 8.4, with 0.5 m*M* dithiothreitol (10). Electroblotting is carried out using prechilled and degassed buffer at 4°C with stirring. After 1–2 hr at 150–250 mA, blotting is interrupted, 1 ml of 20% SDS per liter of blot buffer is added, the solution is stirred for 20 min, and blotting is recommenced at 500 mA for 4 hr or 150 mA overnight. In the first stage of the procedure, most of the protein in the M_r region less than 60,000 is efficiently transferred. There is some swelling of the gel facilitating protein transfer but the SDS is electrophoretically stripped from the larger proteins still embedded in the gel matrix. The interruption of electrophoresis and SDS supplementation allows the SDS–protein complex to reform and is then efficiently transferred. This simple procedure has without fail allowed the near quantitative transfer of a very wide range of soluble, membrane and cytoskeletal proteins from many different sources, as well as brain sodium channels, spectrin dimers, desmoplakins, microtubule-associated proteins including MAP II, i.e., all proteins with apparent M_r from >200,000 to 400,000 range from 1DE and 2DE gels including gels heavily overloaded with protein. Peptide sequence determination on some of these proteins has been reported (19, 29). With this approach methanol, which may strip SDS from protein, but enhance protein binding to the blotting membrane, is never necessary. Investigators can vary somewhat the conditions reported here depending on any unique features of the protein of interest and gel dimension and porosity. The use of gels with the lowest acrylamide concentration and in some instances gradient gels has been alluded to above.

Semi-Dry Electrophoretic Transfer

This blotting method is about a popular as wet blotting. A single buffer gives excellent electrophoretic transfer and is prepared as follows (30):

	Concentration	For 1000 ml
Tris base	48 m*M*	5.8 g
Glycine	39 m*M*	2.9 g
SDS	0.037% (vol/vol)	0.37 g
Methanol	20%	200 ml
Distilled water		Make up to 1000 ml

The SDS may be replenished as for the wet blotting procedure by adding fresh absorbent paper soaked in transfer buffer on the cathodic side of the gel. Although both methods of blotting are popular the tank blot procedure offers by far the widest choice of conditions and flexibility and is also less expensive.

Electroelution of One- and Two-Dimensional Electrophoresis Resolved Proteins

Electroelution was the approach of choice for isolation and sequencing of proteins before methods for digestion of proteins on membranes were introduced. Without doubt the greatest accomplishments in sequencing of receptor, ion channel, and other important proteins were achieved using proteins purified in this manner (24). For N-terminal sequence determination this approach retains advantages:

i. It may be mechanically difficult to load sufficient membrane immobilized electroblotted protein in the reactor chamber of the sequencer to obtain a clear sequence especially for larger proteins.

ii. Overloading of the reactor cartridge may result in nonhomogenous delivery of the sequencing solvents and reagents resulting in incomplete coupling and cleavage and rapid development of overlap rendering identification of a clear and extended N-terminal sequence difficult.

iii. Some chemistries such as complete reduction and alkylation are more efficiently performed on the pure protein in solution rather than immobilized on membrane supports. N-terminal sequence analysis of the epidermal membrane glycoproteins, desmoglein, and desmocollins I and II, after alkylation, permitted extended sequence analysis as well as identification of conserved cysteines and tryptophans which permitted placement of the desmosomal glycoproteins as members of the cadherin family of cell adhesion proteins even before the complete primary structures were deduced by molecular cloning (29, 31).

An excellent and complete description of the methodology most often used is provided by Hunkapiller *et al.* (24) using an electroelution device constructed in the workshop and we have used it without any modification except for addition of 1 m*M* dithiothreitol to the elution buffer, omission of the electrodialysis step, and recovery of the protein by methanol precipitation or phase separation as described above after reduction of the volume of the eluted protein by vacuum centrifugation. The electroelution approach has been chosen by most investigators involved in purification of receptor and ion channel proteins and proteins enriched in brain subcellular fractions (4–7, 19). Other elution devices have also been reviewed (32).

Digestion of Electroeluted and Electroblotted Proteins

There are three main requirements for efficient digestion of protein immobilized on membranes. First, as much protein as possible is immobilized on the least surface area of membrane (12, 14). This may be accomplished by several approaches. Gel loading must be maximized while resolution of individual proteins as discrete spots or bands is preserved. Prior subcellular fractionation to reduce the total protein complexity of the sample and semipreparative 2DE accomplishes this such as reported for the neuronal postsynaptic and synaptosomal fraction (19). A procedure for reelectrophoresis of gel fragments of a single protein in one well after excision from multiple gels and then electroblotting has been widely used (14). Second, quantitative electrotransfer is desirable and the excision of the protein spot should be conservative and not include membrane without any bound protein. Third, membrane sites not blocked entirely by blotted protein must be efficiently saturated with nonproteinaceous material for efficient enzyme digestion to occur (12–14). If this condition is not met unsatisfactory digests or no digestion at all is the result.

The procedure of Aebersold *et al.* (12) for blocking the nitrocellulose (NC) membrane and enzyme digestion is described. Wet membrane with immobilized protein is pooled in a microcentrifuge tube and incubated for 30 min at 37°C in 1 ml of 0.5% PVP-40 in 100 m*M* acetic acid. The tube may be agitated if membrane pieces are adherent. After being blocked, the pieces are washed at least five times with 1.2 ml water; for each wash the tube is capped and vortexed, and the water removed with a needle and syringe. Nitrocellulose pieces are then chopped in pieces about 1 × 1 mm while still wet and transferred to a 0.5-ml tube. A volume of buffer, 100 m*M* NH_4HCO_3, or 100 m*M* Tris–HCl/acetonitrile (95:5 v/v), sufficient to submerge the NC, is added. Protease of choice is then added and the tube is incubated at 37°C overnight. After digestion, the whole reaction mixture is frozen at −20°C or loaded onto the HPLC system after acidification. Tempst and colleagues have analyzed each of these steps, including choice and amount of enzyme, duration of digestion, and cysteine derivatitization, as well as separation and detection of peptides obtained (14) and this report and that of Aebersold *et al.* (11) should be carefully reviewed to obtain peptides suitable for sequence determination by this approach.

Much of the quantitative data on yield of this procedure has been obtained using standard, generally soluble, and for the most part lower M_r proteins. The method generally results in the release of hydrophilic relatively short (<15 amino acids) peptides (13). This may be a unique if not anticipated advantage of this technique since a less-complex peptide chromatogram is more easily obtained, the chances that unique HPLC peaks represent single peptides is enhanced, rendering rechromatography and the inherent losses avoidable. In instances where the standard method yields a poor quality chromatogram, the following procedure has been used (29, 31). Blot immobilized protein is suspended in 200 m*M* Tris–HCl, pH 8, containing 0.2% Tween-20, 2 m*M* EDTA, and 5 m*M* dithiothreitol to block the membrane (the blocking procedure of Aebersold *et al.* may also be used). The nitrocellulose pieces are washed six times with water with vortexing each time, cut as described above, then suspended in a volume of 200 m*M* Tris–HCl, pH 8, containing 10 *M* urea and 2 m*M* dithiothreitol just sufficient to submerge the membrane pieces. After agitation for 1 hr (e.g., using an Eppendorf shaker), the urea concentration is adjusted to 5 *M* by water dilution and enzyme added (an enzyme to substrate ratio of 1:40 is used). After overnight shaking at room temperature, an additional aliquot of enzyme is added and digestion continued for 2 hr. The digest is recovered with a Hamilton syringe, an equal volume of 200 m*M* Tris–HCl and 10 *M* urea is added to the tube, the suspension is vortexed and agitated for 30 minutes, and the recovered supernatants are combined. After acidification and centrifugation for 10 min the combined digest is loaded onto the HPLC column. As an alternative, the membrane may be suspended in 200 m*M* Tris–HCl and 7 *M* guanidine hydrochloride, pH 8, and appropriately diluted subsequently by addition of water to 3.5 *M*, and then 2.75 *M* guanidine HCl, with addition of further enzyme aliquots after each dilution. This method achieves a quantitative digestion of the immobilized protein and total release of peptides including many peptides between 20 to 30 amino acids or longer (29, 31). The benefits of a complete digestion and release of peptide must be contrasted with the hugely complex HPLC chromatogram resulting from near complete digestion of large proteins. Most peaks require rechromatography, and enzyme autodigestion becomes significant because of the low enzyme substrate ratios required to obtain digestion under significantly denaturing conditions (e.g., 5 *M* urea, 2.75 *M* guanidine). Even with careful control digestion of blank membrane, identification of peaks due to enzyme is difficult and unnecessary sequencing is undertaken. If a facility is available for mass spectrometry, 1/10th to 1/20th peak aliquots can be screened in a few minutes for characteristic protease-derived peptide masses and peptide purity. If mass spectrometry is not available, this method is now reserved for instances where the original procedure (12) is unsatisfactory.

High-Performance Liquid Chromatography

The handling of minute quantities of peptide derived from digests of membrane-bound protein is the next problem to be confronted. Losses of peptide become rather dramatic even under optimal storage, e.g., frozen at −80°C. Peptide peaks in the range of 2–10 pmol may vanish entirely even if stored only briefly (33). This is assumed to be due to peptide decomposition and irreversible binding to surfaces. Also, as a rule, the more dilute the peptide solution the greater is the likelihood of loss. With traditional HPLC equipment it was common to use larger (4.6 × 250 mm) columns and flow rates of 1 ml per min to separate peptide digests. More recently narrowbore (2.1 mm i.d.) and microbore (1 mm or less i.d.) columns are generally used (e.g., 11, 29, 31, 33–35). The eluent peak volume from a column is inversely proportional to the square of the column diameter and directly proportional to column length. A 20-fold reduction in flow rate (1 to 0.05 ml per min) by shifting from a 4.6- to 1-mm i.d. results in an equivalent reduction in peak volume (i.e., 1200 to 60 μl) and equivalent sensitivity of detection with 1/20th the sample load; this aspect has been reviewed several times (34, 35). The recovery of peptide from columns in a small volume and in concentrated form is the main reason for the use of narrowbore and microbore columns.

The instrumental requirements for the separation and recovery of low picomole and subpicomole quantities of peptides have been described (33–35). Several laboratories have been able to do major replumbing of their analytical HPLC systems for operation with narrowbore (≤2.1 mm i.d.) and microbore (≤1 mm i.d.) columns and there are now several manufacturers of dedicated narrowbore and microbore chromatography systems. We have used the Model 172 Micropurification System from Applied Biosystems (Foster City, CA). Earlier versions of this instrumentation (Models 120 and 130) were used for several years in this laboratory for peptide purification for sequencing and have performed satisfactorily. The significant features of this instrumentation are extremely low dead volumes, precise flow rates, sensitive detection due to optimal flow cell geometry, recovery of peptide peak in a small volume (typically, 30–60 μl), and the requirement for manual recovery of fractions. The solvent delivery system is the Model 140B with a 75-μl dynamic mixer. A Rheodyne 8125 valve is substituted for the older Rheodyne 7125 and has smaller flow passages optimized for microbore chromatography. All plumbing is now done with polyetheretherketone (PEEK) tubing; connections from the dynamic mixer to the injection valve are 0.007 in i.d. The column outlet is connected directly to a 2.4-μl flow cell of a 785 variable wavelength detector (AB) with 0.005 in i.d. PEEK tubing. Post-flow cell tubing is fused silica with 0.004 in i.d. Solvent A is 0.1% trifluoroacetic acid (TFA) in fresh Milli Q water and solvent B is 0.085% TFA in 80% acetonitrile.

Individual HPLC peak fractions of a complex chromatogram such as those obtained from the complete enzymatic digestion of large proteins usually require rechromatography to obtain pure peptides for sequencing. Peptide complexity may be determined by mass spectrometry if available using 1/10th to 1/20th aliquots of individual peaks. For rechromatography, we use 20 m*M* sodium acetate, pH 5, as solvent A and solvent B is 15 m*M* sodium acetate in 80% acetonitrile. This gives resolution equal or superior to a number of more basic buffers we have assayed. To reduce the number of rechromatography procedures and minimize peptide losses by dilution, several peaks are combined from the TFA separation (each rechromatographed peak is separated by at least 5% solvent B from the neighbor with which it is combined); the combined peaks are acidified, diluted 1:1 with 7 *M* guanidine hydrochloride, and loaded onto the same or another HPLC column of smaller configuration (e.g., from C_4 to C_{18} 1 × 150 mm column) using a shallower gradient. A similar approach is to add TFA but to a final concentration of at least 2–20% after at least 1:1 dilution of the peptide peak (33). Vydac microbore columns (C_4 and C_{18}; 1 × 150 mm) and Aquapore narrowbore and microbore columns have been most generally used but many other supports are now available.

Peptide Mapping in Gels

This has become one widely used procedure in biochemistry and cell biology laboratories. Its chief utility is in determining the relatedness of proteins by size comparison among SDS–PAGE of the fragments produced after chemical and enzymatic fragmentation. Proteins of interest are excised from gels after detection by methods such as overlay of gel autoradiograms or by gel staining; the gel fragment is compressed into the well of a higher percentage SDS–PAGE gel, overlaid with buffer containing a protease solution, and *in situ* digested after the electroeluted protein and protease are stacked at the interface of the stacking and resolving gel. The peptide mapping method is applicable to proteins in solution as well as proteins resolved by 1DE and 2DE. An excellent account of the methodology and variables of the technique is available (36). The gel containing the protease-generated fragments may be blotted to an appropriate support and directly N-terminally sequenced or blotted to nitrocellulose and probed with specific antibodies to localize specific fragments containing antibody epitopes. The ability to obtain peptide sequences of N-terminally blocked proteins and long peptide sequences (9, 11) has been demonstrated. For high M_r proteins where complete digestion is desired, longer gradient gels are preferable; doubling the concentration of Tris in the resolving gel and Tris and glycine in the running buffer (37) also enhances peptide resolution and such gels have been found to give high resolution of peptides in the M_r range of 3000 (11) albeit with low sensitivity in detection. In summary, peptide digestion in gels followed by blotting is useful where proteins are refractory to digestion on blots, where longer sequences are desired, where it is desirable to correlate an immunoreactive band with its protein sequence, or where state of the art HPLC facilities are not available. It is expected to be useful also in detecting phosphorylated or glycosylated regions of proteins—after gel digestion region bands containing radiolabel or reactive with lectin may be blotted and sequenced at least to loosely define the location of these posttranslational protein modifications.

Four methods are commonly used to detect protein before excision without chemical fixation of the protein in the gel. The first and quite sensitive method is staining for 30 min to 1 hr with gentle agitation with 0.2% Coomassie R-250 in water (36). Destaining is hastened by frequent water changes and addition of Kimwipes to mop up excess dye. Proteins may be visualized also by soaking in ice-cold 1 *M* KCl, or 4 *M* NaOAc at room temperature for 10–30 min (38). With KCl, the protein bands are visible as opaque zones on a clear background and visualization is facilitated by placing the gel on a glass plate and viewing against a dark background. With NaOAc, protein appears as clear zones on an opaque background. Metal based stains for proteins have been described and are the most sensitive—staining is reversible with the use of chelators (39). There is some risk of metal-related thiol oxidation with this method; nevertheless, it has been useful in identifying proteins for excision for sequence analysis and peptide digestion. Overall, the Coomassie in water method is now the most reliable, consistent, and generally useful method of protein detection in this laboratory.

Special Topics

Cysteine Modification

There are four reasons to break disulfide bonds and stably modify cysteine before peptide sequencing:

i. Proteins treated with high concentrations of denaturants may still have a large amount of residual higher order structure because of disulfide bonds. This results in resistance to proteolytic or chemical cleavage. Albumin with 35 cysteines involved in 17 disulfide bonds is markedly resistant to a variety of proteases unless the disulfide linkages are broken (not shown; Ref. 14).

ii. Suboptimal Edman degradation occurs because of steric hindrance if protein is not fully unfolded.

iii. Cysteine is unstable during Edman chemistry and the PTH-cysteine is not readily identifiable.

iv. If there is a cysteine identified it simplifies oligonucleotide probe construction to screen libraries. Reduction and alkylation with 4-vinylpyridine (4-VP) is most commonly used now (40). The reaction with 4-VP is high-yield, and very specific for cysteine, the S-vinylpyridylethylated-cysteine is stable during Edman chemistry, and the PTH derivative is readily identified by HPLC. Moreover, the valuable reaction is very straightforward to perform.

Reduction-Alkylation Procedure in Solution

Protein is suspended in 100 μl 0.5 *M* Tris–HCl, pH 8, containing 2% SDS, 20 m*M* dithiothreitol, 5 m*M* EDTA, at room temperature, and under argon. As an alternative to SDS, use 7 *M* guanidine HCl or 10 *M* urea. After 2 hr at room temperature with agitation and 30 min at 56°C (heating is omitted if urea is used as denaturant), 10 μl 4-VP (freshly diluted 1:5 in ethanol) is added and the tube agitated for an additional 30 min at room temperature in the dark. The modified protein is recovered by any of a variety of precipitation methods, e.g., 4–8 volumes of methanol or methanol/acetone 50:50 or by the methanol chloroform procedure with phase separation as described above. This procedure is routinely carried out (19, 29, 31) and permits extended N-terminal sequencing of membrane and other proteins with clear identification of cysteines in protein and peptide sequencing. We have not observed any deterioration in quality of sequence results if the PE protein is precipitated and stored frozen for several months. Pre-1DE gel alkylation is also done (13, 14) as follows: reduction is carried out in Laemmli sample buffer with 0.5% 2-mercaptoethanol as reductant for 10 min at 60°C followed by 20 min at 37°C. A 20% v/v solution of 4-VP in ethanol is added to yield a final concentration of 1.5%. The alkylation reaction is done in the dark for 30 min at room temperature and then immediately loaded on a gel for SDS–PAGE.

Postdigest Alkylation of *in Situ* Digestions

It is not advisable to do an alkylation of total cell extracts prior to 2DE; if all cysteines are not uniformly modified heterogeneities in 2DE coordinates may cause confusion in spot identification. Alkylation of *in situ* digested peptide may be done instead (14) as follows: 2-mercaptoethanol is added to the supernatant after enzyme digestion at a final concentration of 0.1% and reduction carried out for 30 min at 37°C. Next, 4-VP (20% solution in ethanol) is added to a final concentration of 0.3% and the reaction is carried out for 30 min at room temperature in the dark. The mixture is then immediately injected for HPLC separation after acidification. The 2-mercaptoethanol and 4-VP are eluted by a long isocratic wash in 5% solvent B before the gradient is initiated.

On-Sequencer Protein Alkylation

Protein already immobilized on a membrane such as PVDF can also be alkylated. The procedure of Kao and Chung (41) is given. The PVDF membrane is prewetted in methanol, rinsed with water, and dipped in 250 m*M* *N*-ethylmorpholine buffer, pH 8.5. Excess buffer is removed by blotting from the edge with a piece of filter paper. It is positioned on the lower cartridge block of the sequencer. A glass fiber disc is fitted onto the upper cartridge block recess and wetted with 20 μl of freshly prepared reaction mixture (100 μl acetonitrile, 2 μl tributylphosphine, and 5 μl 4-VP). This mixture is prepared in a well-ventilated fume hood as tributylphosphine is toxic and malodorus. The reaction cartridge is assembled immediately without the cartridge seal and incubated for 10 min at 48°C. At the end of the incubation period, the glass fiber disc is discarded and the PVDF membrane rinsed 10 times with water before reinstallation in the ethanol-rinsed cartridge for Edman sequencing.

Identification of Phosphorylated Amino Acids

Phosphorylation may be the most important mechanism by which extracellular signaling produces biological pleiotropic responses in target neurons (42). Most early localizations of phosphate groups in protein primary structures depended on identification of [^{32}P]phosphate radiolabels introduced *in vitro* using purified protein kinases. If the primary structure of the phosphorylated substrate protein is known the putative phosphoamino acid may be identified after isolation of the phosphopeptide, peptide hydrolysis, and 1D chromatography or 2DE chromatography-electrophoresis. The combination of 1DE and 2DE with blotting to PVDF membranes (43, 44) has greatly simplified the analysis of phosphopeptides and identification of the specific amino acids modified by phosphorylation.

Identification of Phosphoserine by Chemical Sequencing

It is also possible to determine sites that are phosphorylated *in vivo*. Meyer and colleagues (45) have described a generally useful procedure for the identification of phosphoserine and the method is described here. The solution mixture is prepared by combining reagents as follows: 60 μl ethanethiol, 200 μl dimethyl sulfoxide, 80 μl ethanol, and 65 μl 5N NaOH; to this is added, last, 200 μl water. Ethanethiol is toxic and malodorous and the preparation is carried out in a well-ventilated fume hood. Add 50 μl of this mixture to a microcentrifuge tube containing the dried phosphopeptide. The tube is capped under nitrogen or argon, sealed with Parafilm, and incubated for 1 hr at 50°C. The tube is then cooled to room temperature and 10 μl acetic acid added. The *S*-ethlycysteine-modified peptide is applied as soon as possible to the sequencer. The PTH-*S*-ethylcysteine is readily identifiable during sequencing. The method is only applicable to the detection of phosphoserine which, however, accounts for >90% of phosphorylated residues in proteins. There are methods available for the identification of phosphothreonine and phosphotyrosine during Edman sequencing but they require more specialized techniques and are not presented here.

Protein Glycosylation

The majority of membrane and especially receptor proteins are modified by glycosylation especially N-linked glycosylation of asparagine residues within the consensus sequon Asn-Xaa-Ser/Thr where "X" is any amino acid except proline. Although glycosylated amino acids are cleaved normally during chemical sequencing, the PTC-derivatized amino acid with attached carbohydrate is not extracted from the reactor and no PTH derivative is seen for the corresponding cycle. Protein may also be digested with enzyme, the enzyme inactivated, and glycopeptides in the protein digest isolated by lectin affinity chromatography. Peptides eluted from the lectin column are then separated by HPLC and sequenced. The application of this approach to determination of N-linked glycosylation sites of desmosomal membrane glycoproteins has been reported (46). Alternatively, HPLC fractions of digested proteins may be screened by dot-blotting with specific lectins to identify glycopeptide fractions which are then sequenced.

Mapping Monoclonal Antibody (mAb) Epitopes

This is a valuable application of chemical sequencing technology. Protein is digested enzymatically, the enzyme is inactivated, and the peptide digest is incubated with specific monoclonal antibodies. The mAb–peptide immune complex is precipitated with protein A or protein G-agarose and the specific mAb-bound peptide(s) eluted with acetic acid and resolved by HPLC for N-terminal sequencing. The peptide if necessary may first be separated from the mAb by centrifugal ultrafiltration through Centricon-10 (Amicon) membrane as has been done to separate peptides from MHC molecules by Jardetzky and colleagues (47) before HPLC. A specific example of this application is given (29): *S*-vinylpyridyl ethylated desmoglein was digested with *staphylococcal* V8 protease as described by Cleveland (36) using 0.25% SDS; after 2 hr at room temperature, enzyme was inactivated by heating at 100°C for 2 min and the SDS concentration adjusted to 0.1% by dilution with water and then added to protein G–Sepharose to which mAb was already affinity bound. After rotation at room temperature for 1 hr, the protein G–Sepharose immune complex was recovered by centrifugation and washed five times with 50 m*M* Tris–HCl, pH 7.4, containing 150 m*M* NaCl, 1% Triton X-100, 0.1% SDS, and 5 m*M* EDTA, and subsequently washed five times with water alone to remove traces of detergent which interfere with HPLC. The pellet was then suspended in 2 *M* acetic acid for 10 min at room temperature to elute the antigen–antibody complex, the protein G–Sepharose was pelleted, and the supernatant was directly loaded onto the HPLC system.

References

1. P. Edman and G. Begg, *Eur. J. Biochem.* **1,** 80 (1967).
2. L. Giri, W. E. Hill, H. G. Wittmann, and B. Wittmann-Liebold, *Adv. Protein Chem.* **36,** 1 (1984).
3. R. M. Hewick, M. W. Hunkapiller, L. E. Hood, and W. J. Dreyer, *J. Biol. Chem.* **256,** 7990 (1981).
4. M. Noda, H. Takahashi, T. Tanabe, M. Toyosato, Y. Furutani, T. Hirose, M. Asai, S. Inayama, T. Miyata, and S. Numa, *Nature (London)* **299,** 793 (1982).
5. M. Noda, S. Shimizu, T. Tanabe, T. Takai, T. Kayano, T. Ikeda, H. Takahashi, H. Nakayama, Y. Kanaoka, N. Minamino, K. Kangawa, H. Matsuo, M. A. Raftery, T. Hirose, S. Inayama, H. Hayashida, T. Miyata, and S. Numa, *Nature (London)* **312,** 121 (1984).
6. G. Grenningloh, A. Rienetz, B. Schmitt, C. Methfessel, M. Zensen, K. Beyreuther, E. D. Gundelfinger, and H. Betz, *Nature (London)* **328,** 215 (1987).
7. P. R. Schofield, M. G. Darlison, N. Fugita, D. R. Burt, F. A. Stephenson, H. Rodriguez, L. M. Rhee, J. Ramachandran, V. Reale, T. A. Glencorse, P. H. Seeburg, and E. A. Bernard, *Nature (London)* **328,** 221 (1987).
8. P. H. O'Farrell, *J. Biol. Chem.* **250,** 4007 (1975).
9. J. Vandekerckhove, G. Bauw, M. Puype, J. van Damme, and M. van Montagu, *Eur. J. Biochem.* **152,** 9 (1985).
10. R. H. Aebersold, D. R. Teplow, L. E. Hood, and S. B. H. Kent, *J. Biol. Chem.* **261,** 4229 (1986).

11. M. J. Walsh, J. McDougall, and B. Wittmann-Liebold, *Biochemistry* **27,** 6867 (1988).
12. R. H. Aebersold, J. Leavitt, R. A. Saavedra, L. E. Hood, and S. B. H. Kent, *Proc. Natl. Acad. Sci. U.S.A.* **84,** 6970 (1987).
13. G. Bauw, J. Van Damme, M. Puype, J. Vandekerckhove, B. Gesser, G. P. Ratz, J. B. Lauridsen, and J. E. Celis, *Proc. Natl. Acad. Sci. U.S.A.* **86,** 7701 (1989).
14. P. Tempst, A. J. Link, L. R. Riviere, M. Fleming, and C. Elicone, *Electrophoresis (Weinheim, Fed. Repub. Ger.)* **11,** 537 (1990).
15. Y. Ashani and G. N. Catravas, *Anal. Biochem.* **109,** 55 (1980).
16. M. P. Deutscher, ed., "Methods in Enzymology," Vol. 182. Academic Press, San Diego, CA, 1990.
17. U. K. Laemmli, *Nature (London)* **227,** 680 (1970).
18. J. E. Celis, ed., *Electrophoresis (Weinheim, Fed. Repub. Ger.)* **11,** 1091 (1993).
19. M. J. Walsh and N. Kuruc, *J. Neurochem.* **59,** 667 (1992).
20. M. J. Walsh, W. W. Tortellotte, J. Roman, and W. Dreyer, *Clin. Immunol. Immunopathol.* **35,** 313 (1985).
21. G. R. Stark, *Biochemistry* **4,** 1030 (1965).
22. W. Jahnen, L. D. Ward, G. E. Reid, R. L. Moritz, and R. J. Simpson, *Biochem. Biophys. Res. Commun.* **166,** 139 (1990).
23. H. Matsubara and R. M. Sasaki, *Biochem. Biophys. Res. Commun.* **35,** 175 (1969).
24. M. W. Hunkapiller, E. Lujan, F. Ostrander, and L. E. Hood, *in* "Methods in Enzymology" (C. His and S. Timasheff, eds.), Vol. 91, p. 277. Academic Press, New York, 1983.
25. D. Wessel and U. I. Flügge, *Anal. Biochem.* **138,** 141 (1984).
26. P. Matsadaira, *J. Biol. Chem.* **262,** 10035 (1987).
27. C. Eckerskorn and F. Lottspeich, *Electrophoresis (Weinheim, Fed. Repub. Ger.)* **14,** 831 (1993).
28. J. Mozdzanowski, P. Hembach, and D. W. Speicher, *Electrophoresis (Weinheim, Fed. Repub. Ger.)* **13,** 59 (1992).
29. P. J. Koch, M. J. Walsh, M. Schmelz, M. D. Goldschmidt, R. Zimbelmann, and W. W. Franke, *Eur. J. Cell Biol.* **53,** 1 (1990).
30. E. Harlow and D. Lane, "Antibodies: A Laboratory Manual," p. 488. Cold Spring Harbor Lab., Cold Spring Harbor, NY, 1988.
31. P. J. Koch, M. D. Goldschmidt, M. J. Walsh, R. Zimbelmann, M. Schmelz, and W. W. Franke, *Differentiation (Berlin)* **47,** 29 (1991).
32. M. G. Harrington, *in* "Methods in Enzymology" (M. Deutscher, ed.), Vol. 182, p. 488. Academic Press, San Diego, CA, 1990.
33. C. Elicone, M. Lui, S. Geromanos, H. Erdjument-Bromage, and P. Tempst, *J. Chromatogr.* **676,** 121 (1994).
34. B. Greggo, I. R. Van Driel, P. A. Stearne, J. W. Goding, E. C. Nice, and R. J. Simpson, *Eur. J. Biochem.* **148,** 485 (1985).
35. R. J. Simpson, R. L. Moritz, G. S. Begg, M. R. Rubira, and E. C. Nice, *Anal. Biochem.* **177,** 221 (1989).
36. D. W. Cleveland, *in* "Methods in Enzymology" (S. Fleischer and B. Fleischer, eds.), Vol. 96, p. 222. Academic Press, New York, 1983.

37. S. P. Fling and D. S. Gregerson, *Anal. Biochem.* **155,** 83 (1986).
38. R. C. Higgins and M. E. Dahmus, *Anal. Biochem.* **93,** 257 (1979).
39. J. K. Dzandu, J. F. Johnson, and G. E. Wise, *Anal. Biochem.* **174,** 157 (1988).
40. M. L. Friedman, L. H. Krull, and J. F. Cavins, *J. Biol. Chem.* **245,** 3868 (1970).
41. M. C. C. Kao and M. C. M. Chung, *Anal. Biochem.* **215,** 82 (1993).
42. S. I. Walaas and P. Greengard, *Pharmacol. Rev.* **43,** 299 (1991).
43. M. P. Kamps and B. M. Sefton, *Anal. Biochem.* **176,** 22 (1989).
44. E. Hildebrandt and V. A. Fried, *Anal. Biochem.* **177,** 407 (1989).
45. H. E. Meyer, E. Hoffmann-Posorske, and L. M. G. Heilmeyer, *in* "Methods in Enzymology" (T. Hunter and B. Sefton, eds.), Vol. 201, p. 69. Academic Press, San Diego, CA, 1991.
46. P. J. Koch, M. D. Goldschmidt, M. J. Walsh, R. Zimbelmann, and W. W. Franke, *Eur. J. Cell Biol.* **55,** 200 (1991).
47. T. S. Jardetzky, W. S. Lane, R. A. Robinson, D. R. Madden, and D. C. Wiley, *Nature (London)* **353,** 326 (1991).

[4] *Xenopus* Oocytes: A System for Expression Cloning and Structure–Function Studies of Ion Channels and Receptors

Roger D. Zühlke, Hui-Juan Zhang, and Rolf H. Joho

Introduction

In 1971, Gurdon and colleagues published the first report demonstrating expression of foreign genes in oocytes of *Xenopus laevis,* an African frog species (1). The authors were, at the time, specifically interested in studying the mechanism of translation after microinjection of mRNA in *Xenopus* oocytes. Soon attention shifted toward the characterization of functional properties of oocyte-expressed proteins, in particular plasma membrane proteins. In 1984, functional expression was reported of the cloned acetylcholine receptor after microinjection of cDNA-derived mRNA (cRNA) (2). Subsequently, the oocyte expression system was successfully used to clone interleukin 4 by functional expression (3). These early events eventually triggered the immense popularity of the *Xenopus* oocyte as a heterologous expression system for the functional characterization of cloned ion channels, receptors, and transporters (4–9).

The advantages of the *Xenopus* oocyte compared to other expression systems are severalfold. An individual frog egg is fairly large (0.5–1.0 mm in diameter); therefore, it can be handled with relative ease. Endogenous receptors and ion channels are mainly expressed in the layers of follicular cells surrounding the vitellin membrane of the oocyte. Because the follicular layers are removed before or after microinjection of cRNA, it is possible to express a channel or receptor of interest in "denuded" oocytes without signal interference of endogenously expressed gene products. The *Xenopus* oocyte not only possesses the necessary machinery for correct translation of foreign mRNAs but, in many instances, is capable also of carrying out posttranslational modifications which may be needed for the proper function of an oocyte-expressed protein.

The unique characteristics of the oocyte expression system made it possible to express legions of foreign genes. It enabled the cloning-by-function of many cDNAs for which no peptide or nucleotide sequence information was available (see Table I for a list of cDNAs isolated by the expression cloning approach). In addition, many insightful structure–function studies

Methods in Neurosciences, Volume 25

were successfully conducted on ion channels and receptors using *Xenopus* oocytes. In this article, we outline both approaches. We first review and assess the expression cloning strategy by discussing reports by others as well as our own work. Furthermore, we describe basic procedures and techniques to maintain and use the *Xenopus* oocyte expression system, including the handling of frogs and oocytes and the preparation of high-quality cRNA ready for microinjection. Eventually, a procedure is described for the production of many distinct mutant cDNA molecules of a particular gene of interest with high frequency.

Expression Cloning in *Xenopus* Oocytes

With the advent of recombinant DNA techniques, a flurry of reports appeared over the past decade documenting thousands of cloned genes from many different species. Nevertheless, there are still many proteins whose primary structures are not yet known although they have been functionally characterized, often both *in vivo* and *in vitro*. The *Xenopus* oocyte expression system may, therefore, provide a helpful alternative for the cloning of genes (cDNAs) whose functions can be detected after expression of mRNA from a suitable source. Expression cloning strategies using the *Xenopus* oocyte system have been successfully applied to isolate a fairly large number of cDNAs (see Table I). Our laboratory has used this approach to clone a voltage-gated K^+ channel (10) and the thyrotropin-releasing hormone receptor (11). Detailed descriptions for the construction of high-quality cDNA libraries in transcription-competent vectors and for the cloning procedure have been published earlier (12, 13). Here, we discuss and evaluate the requirements, advantages, and drawbacks of the cloning-by-expression strategy using *Xenopus* oocytes.

1. The first important step is to explore the possibility that the oocyte provides a functional assay for a particular gene of interest. Electrophysiological procedures have allowed the direct measurement of ionic currents through the oocyte membrane. Voltage- and ligand-gated ion channels, and receptors coupled to a transduction system, all of which eventually lead to an electrophysiologically measurable signal, are good candidates for expression cloning in *Xenopus* oocytes (Table I). Numerous G protein-coupled receptors have been cloned due to the fact that the receptor under investigation activated phospholipase C (PLC) through an endogenous G protein. Inositol 3-phosphate(IP_3)-dependent intracellular Ca^{2+} release could be measured electrophysiologically as a current through Ca^{2+}-activated chloride channels, or as Ca^{2+}-triggered light emission of microinjected aequorin. Other successfully applied assays were the detection of uptake or secretion of radioactively

labeled substrates for particular transport systems or the direct measurement of enzyme activity in the medium or in oocyte extracts. In Table I, we have attempted to list most of the cDNAs cloned over the past few years, and we have included useful information such as library sizes (number of cloning events), tissue sources, and functional assays used to screen the library.

2. The selection of a suitable source of RNA is very important. Large tissues or cultured cells provide an almost unlimited supply of mRNA. However, the selection of the RNA source should primarily be guided by the abundance of the particular mRNA species to be detected after functional expression in oocytes. The success rate for expression cloning may be increased (a) by enriching for the mRNA to be cloned using size fractionation of poly$(A)^+$ RNA by sucrose-gradient centrifugation (13), (b) by the use of highest quality reverse transcriptase for the synthesis of as many full-length cDNA molecules as possible, (c) by the directional cloning of cDNA, and (d) by appropriately subdividing the original unamplified cDNA library. A detailed description may be found in Refs. 12 and 13. In most cases when the signal of the expressed gene was detected through the PLC/IP_3 pathway, the sizes of the sublibraries used for the initial screen were larger than 20,000 cloning events, reaching up to 300,000 in one instance (Table I).

Isolation of a clone from a large "positive" sublibrary can be very time consuming. It is worth while to take some precautions that decrease the probability of losing the clone of interest during the lengthy procedure of sib selection. If cRNA from a large pool of phages has led to the desired signal of interest in oocytes, we recommend subdividing this pool into as many small sublibraries as convenient. If there were e.g., a "positive" pool comprised of 10,000 cloning events, we would generate sublibraries by growing 20 to 30 agar plates, each with a density of 1000 phages per plate. Assuming a Poisson distribution, the probability that the clone of interest will be present at least once in 20 sublibraries is 86.5%, and it is 95% for 30 sublibraries. Before eluting phages from the agar plates, which would amount to creating sublibraries of 1000 cloning events, we cut each agar plate into eight pieces corresponding to an area of approximately 125 phage clones. Phages are then separately eluted from each agar piece to create 160 to 240 sublibraries, each derived from 125 cloning events. Small aliquots of eight sublibraries are combined, and this pool is analyzed for the presence of the clone of interest. If a response is detected from such a pool, the "positive" clone must be present in one of the eight sublibraries. The next round of selection, therefore, will be confined only to the analysis of the eight sublibraries used. Once such a sublibrary derived from 125 cloning events has been confirmed, we pick a few hundred individual phage clones. We repeat the above procedure by pooling aliquots of 10 individual clones for subsequent

TABLE I Proteins Cloned by Functional Expression in *Xenopus* Oocytes

Cloned protein	Tissue source	Initial library size screened[a]	Functional screening assay	cDNA size (kb)	mRNA size (kb)[b]	Refs.[c]
IgG1 induction factor (Interleukin 4)	Cultured murine T cells	250 (16)	Activity of secreted IgG1 induction factor	0.75	0.65	1
Na^+/glucose cotransporter	Rabbit intestinal brush border	300–500 (4-6)	Na^+-dependent methyl-glucopyranosid-uptake	2.2	2.3	2
Substance K receptor	Bovine stomach	300,000	Agonist-induced, PLC/IP_3-mediated, Ca^{2+}-dependent Cl^- current	2.0	nr	3
Serotonin 5HT1C receptor	Murine choroid plexus	35,000	Agonist-induced PLC/IP_3-mediated, Ca^{2+}-dependent Cl^- current	1.9[d]	5.0	4
Serotonin 5HT1C receptor	Murine choroid plexus	100,000 (5)	Agonist-induced, PLC/IP_4-mediated, Ca^{2+}-dependent Cl^- current	3.0	5.2	5
Slowly activating voltage-gated K^+ channel	Rat kidney	90,000	Voltage-dependent K^+ current	0.6	0.7	6
Delayed rectifier K^+ channel	Rat brain	100,000 (10)	Voltage-dependent K^+ current	3.4	4.3	7
Glutamate receptor	Rat forebrain	44,000 (18)	Agonist-induced depolarization	3.0	5.2	8
Substance P receptor	Rat brain	80,000	Agonist-induced, PLC/IP_3-mediated, Ca^{2+}-dependent Cl^- current	3.4	nr	9
Steroid 5α-reductase	Female rat liver	150–200 (20)	Activity of secreted 5α-reductase	2.5	2.5	10
Voltage-gated Cl^- channel	*Torpedo marmorata* electric organ	3,000	Voltage-dependent Cl^- current	0.3 (2.7)[d]	9–10	11
Bombesin/gastrin-releasing hormone receptor	Murine Swiss 3T3 fibroblasts	50,000 (16)	PLC/IP_3-mediated Cl^- current/Ca^{2+}-dependent aequorin luminescence	2.5	9.5	12
Neurotensin receptor	Rat brain	25,000	Agonist-induced, PLC/IP_3-mediated, Ca^{2+}-dependent Cl^- current	3.6	3.8	13

Endothelin receptor	Bovine lung	30,000	Agonist-induced, PLC/IP_3-mediated, Ca^{2+}-dependent Cl^- current	3.2	3.3; 4.2	14
Thyrotropin-releasing hormone receptor	Mouse pituitary thyrotropic tumor	20,000 (20)	Agonist-induced, PLC/IP_3-mediated, Ca^{2+}-dependent Cl^- current	3.8	nr	15
Na^+/bile acid cotransporter	Rat liver	500–250,000	Na^+-dependent taurocholate uptake	1.7	1.7; 7.5; 9.5	16
Histamine H1 receptor	Bovine adrenal medulla	20,000 (65)	Agonist-induced, PLC/IP_3-mediated, Ca^{2+}-dependent Cl^- current	3.0	3.0	17
Na^+/phosphate cotransporter	Rabbit renal cortex	500–1,500 (7)	Na^+-dependent phosphate uptake	1.8	~2.0	18
Na^+/phosphate cotransporter	Human kidney cortex	1,000	Na^+-dependent phosphate uptake	2.6	2.7; 4.8	44
Na^+/phosphate cotransporter	Rat kidney cortex	1,000	Na^+-dependent phosphate uptake	2.5	2.7	44
B_2 bradykinin receptor	Rat uterus	20,000 (5)	Agonist-induced, PLC/IP_3-mediated, Ca^{2+}-dependent Cl^- current	4.1	4.0; 5.7; 6.5	19
Thrombin receptor	Human megakaryocyte-like cell lines	20,000 (50)	Agonist-induced, PLC/IP_3-mediated, Ca^{2+}-dependent Cl^- current	3.5	3.5	20
Platelet-activating factor receptor	Guinea pig lung	60,000	Agonist-induced; PLC/IP_3-mediated, Ca^{2+}-dependent Cl^- current	3.0	~3.5	21
Metabotropic glutamate receptor	Rat cerebellum	28,000	Agonist-induced, PLC/IP_3-mediated, Ca^{2+}-dependent Cl^- current	4.3	4.4; 7.4	22
NMDA receptor	Rat forebrain	3,000	NMDA-induced inward current	4.2	4.2; 4.4	23
Presynaptic Ca^{2+} channel subunit	*Torpedo californica* electric lobe	10–20 (~20)	Voltage-dependent Ca^{2+} current	1.55 (7.4)[d]	7.5	24,25
Na^+/Cl^--dependent betaine transporter	MDCK cells	200	Na^+-dependent betaine and GABA uptake	2.9	3.0	26
Na^+/*myo*-inositol cotransporter	MDCK cells	1,000 (6)	Na^+-dependent *myo*-inositol uptake	2.9	10.5	27

(continued)

TABLE I (*continued*)

Cloned protein	Tissue source	Initial library size screened[a]	Functional screening assay	cDNA size (kb)	mRNA size (kb)[b]	Refs.[c]
Na^+-independent neutral amino acid transporter	Rat kidney	8,000	L-Leu and L-Phe uptake	2.3	2.4	28
Transporter for cystine and dibasic amino acids	Rabbit kidney	1,000	L-Arg and L-Ala uptake	2.2	2.3; 3.9	29
Transporter for cystine and dibasic amino acids	Rat kidney	600–700	Cystine uptake	2.3	2.4; 4.4	30
Transporter for cystine and dibasic amino acids	Human kidney	1,000	L-Arg uptake	2.3	2.5	45
Vasopressin receptor	Rat liver	50,000 (24)	Agonist-induced, PLC/IP_3-mediated, Ca^{2+}-dependent aequorin luminescence	1.4	2.1	31
Oxytocin receptor	Human myometrium	6,000 (16)	Agonist-induced, PLC/IP_3-mediated, Ca^{2+}-dependent Cl^- current	4.1	4.4	32
Gonadotropin-releasing hormone receptor	Murine gonadotropic cell line	50,000 (5)	Agonist-induced, PLC/IP_3-mediated, Ca^{2+}-dependent aequorin luminescence	1.2	1.6; 3.5	33
Ciliary neurotrophic factor (CNTF)	Rat C6 glioma cells	20,000 (25)	CNTF activity of oocyte extracts on cultured primary neurons	0.6	nr	34
High-affinity glutamate transporter	Rabbit small intestine	300–400	Glutamate uptake	3.4	3.5	35
Amiloride-sensitive Na^+ channel	Rat colon	6,000 (10)	Amiloride-sensitive Na^+ current	3.2	3.5	36
Amiloride-sensitive Na^+ channel	Rat colon	5,000 (20)	Amiloride-sensitive $^{22}Na^+$-flux and Na^+ current	3.1	nr	37
ATP receptor	Mouse neuroblastoma NG108-15	200,000 (5)	ATP-induced inward currents	2.3	2.4	38
Thiazide-sensitive Na^+/Cl^- cotransporter	Flounder bladder	700	Thiazide-sensitive, Cl^--dependent Na^+ uptake	3.7	3.7; 3.0	39

ATP-regulated inward rectifier K^+ channel	Rat kidney medulla	600	Voltage-dependent, Ba^{2+}-sensitive inward K^+ current	2.1	2.2; 3.2; 6; 9; 12	40
Voltage-gated inward rectifier K^+ channel	Mouse macrophage cell line (J774)	5,000 (16)	Voltage-dependent, Ba^{2+}-sensitive inward K^+ current	5.5	5.5	41
Apolipoprotein B mRNA editing protein	Rat intestine	1,000,000	RNA editing activity in oocyte extracts	0.9	1.2	42
Vasopressin-regulated urea transporter	Rabbit kidney	nr	Radiolabeled urea uptake	3.1	3.0; 4.0	43
Na^+/sulfate cotransporter	Rat kidney cortex	1,000	Na^+-dependent sulfate uptake	2.2	2.3; 2.9	46
G protein-activated K^+ channel	Rat atrium	40,000	5-HT-evoked K^+ current	2.1	2.3; 2.9; 3.9; 6.2	47
Ca^{2+}-sensing receptor	Bovine prathyroid	nr	Agonist (gd^{3+})-induced inward currents	5.3	5.3	48

[a] Numbers in parentheses indicate how many sublibraries of this size were screened (if mentioned in the original paper).

[b] As determined by Northern blots; nr, not reported.

[c] References: (1) Y. Noma, P. Sideras, T. Naito, S. Bergstedt-Lindquist, C. Azuma, E. Severinson, T. Tanabe, T. Kinashi, F. Matsuda, Y. Yaoita, and T. Honjo, *Nature* (*London*) **319,** 640 (1986); (2) M. A. Hediger, M. J. Coady, T. S. Ikeda, and E. M. Wright, *Nature* (*London*) **330,** 379 (1987); (3) Y. Masu, K. Nakayama, H. Tamaki, Y. Harada, M. Kuno, and S. Nakanishi, *Nature* (*London*) **329,** 836 (1987); (4) H. Lübbert, B. J. Hoffman, T. P. Snutch, T. Van-Dyke, A. J. Levine, P. R. Hartig, H. A. Lester, and N. Davidson, *Proc. Natl. Acad. Sci. U.S.A.* **84,** 4332 (1987); (5) D. Julius, A. B. MacDermott, R. Axel, and T. M. Jessell, *Science* **241,** 558 (1988); (6) T. Takumi, H. Ohkubo, and S. Nakanishi, *Science* **242,** 1042 (1988); (7) G. C. Frech, A. M. J. VanDongen, G. Schuster, A. M. Brown, and R. H. Joho, *Nature* (*London*) **340,** 642 (1989); (8) M. Hollmann, A. O'Shea-Greenfield, S. W. Rogers, and S. Heinemann, *Nature* (*London*) **342,** 643 (1989); (9) Y. Yokota, Y. Sasai, K. Tanaka, T. Fujiwaa, K. Tsuchida, R. Shigemoto, A. Kakizuka, H. Ohkubo, and S. Nakanishi, *J. Biol. Chem.* **264,** 17649 (1989); (10) S. Andersson, R. W. Bishop, and D. W. Russell, *J. Biol. Chem.* **264,** 16249 (1989); (11) T. J. Jentsch, K. Steinmeyer, and G. Schwarz, *Nature* (*London*) **348,** 510 (1990); (12) E. R. Spindel, Giladi, P. Brehm, R. H. Goodman, and T. P. Segerson, *Mol. Endocrinol.* **4,** 1956 (1990); (13) K. Tanaka, M. Masu, and S. Nakanishi, *Neuron* **4,** 847 (1990); (14) H. Arai, S. Hori, I. Aramori, H. Ohkubo, and S. Nakanishi, *Nature* (*London*) **348,** 730 (1990); (15) R. E. Straub, G. C. Frech, R. H. Joho, and M. C. Gershengorn, *Proc. Natl. Acad. Sci. U.S.A.* **87,** 9514 (1990); (16) B. Hagenbuch, B. Stieger, M. Foguet, H. Lübbert, and P. J. Meier, *Proc. Natl. Acad. Sci. U.S.A.* **88,** 10629 (1991); (17) M. Yamashita, H. Fukui, K. Sugama, Y. Horio, S. Ito, H. Mizuguchi, and H. Wada, *Proc. Natl. Acad. Sci. U.S.A.* **88,** 11515 (1991); (18) A. Werner, M. L. Moore, N. Mantei, J. Biber, G. Semenza, and H. Murer, *Proc. Natl. Acad. Sci. U.S.A.* **88,** 9608 (1991); (19) A. E. McEachern, E. R. Shelton, S. Bhakta, R. Obernolte, C. Bach, P. Zuppan, J. Fujisaki, R. W. Aldrich, and K. Jarnagin, *Proc. Natl. Acad. Sci. U.S.A.* **88,** 7724 (1991); (20) T.-K. H. Vu, D. T. Hung, V. I. Wheaton, and S. R. Coughlin, *Cell* (*Cambridge, Mass.*) **64,** 1057 (1991); (21) Z.-I. Honda, M. Nakamura, I. Miki, M. Minami, T. Watanabe, Y. Seyama, H. Okada, H. Toh, K. Ito, T. Miyamoto, and T. Shimizu, *Nature* (*London*) **349,** 342 (1991); (22) M. Masu, Y. Tanabe, K. Tsuchida, R. Shigemoto, and S. Nakanishi, *Nature* (*London*) **349,** 760 (1991); (23) K. Moriyoshi, M. Masu, T. Ishii, R. Shigemoto, N. Mizuno, and S. Nakanishi, *Nature* (*London*) **354,** 31 (1991); (24) J. A. Umbach and C. B. Gundersen, *Ann. N.Y. Acad. Sci.* **635,** 443 (1991); (25) C. B. Gundersen and J. A. Umbach, *Neuron* **9,** 527 (1992); (26) A. Yamauchi, S. Uchida, H. Moo-Kwon, A. S. Preston, R. Brooks-Robey, A. Garcia-Perez, M. B. Burg, and J. S. Handler, *J. Biol. Chem.* **267,** 649 (1992); (27) H. Moo-Kwon, A. Yamauchi, S. Uchida, A. S. Preston, A. Garcia-Perez, M. B.

(*continued*)

TABLE I (*continued*)

Burg, and J. S. Handler, *J. Biol. Chem.* **267,** 6297 (1992); (28) S. S. Tate, N. Yan, and S. Udenfriend, *Proc. Natl. Acad. Sci. U.S.A.* **89,** 1 (1992); (29) J. Bertran, A. Werner, M. L. Moore, G. Stange, D. Markovich, J. Biber, X. Testar, A. Zorzano, M. Palacin, and H. Murer, *Proc. Natl. Acad. Sci. U.S.A.* **89,** 5601 (1992); (30) R. G. Wells and M. A. Hediger, *Proc. Natl. Acad. Sci. U.S.A.* **89,** 5596 (1992); (31) A. Morel, A.-M. O'Carroll, M. J. Brownstein, and S. J. Lolait, *Nature* (*London*) **356,** 523 (1992); (32) T. Kimura, O. Tanizawa, K. Mori, M. J. Brownstein, and H. Okayama, *Nature* (*London*) **356,** 526 (1992); (33) J. Reinhart, L. M. Mertz, and K. J. Catt, *J. Biol. Chem.* **267,** 21281 (1992); (34) A. Lam, J. Kloss, F. Fuller, B. Cordell, and P. A. Ponte, *J. Neurosci. Res.* **32,** 43 (1992); (35) Y. Kanai and M. A. Hediger, *Nature* (*London*) **360,** 467 (1992); (36) C. M. Canessa, J.-D. Horisberger, and B. C. Rossier, *Nature* (*London*) **361,** 467 (1993); (37) E. Lingueglia, N. Voilley, R. Waldmann, M. Lazdunski, and P. Barbry, *FEBS Lett.* **318,** 95 (1993); (38) K. D. Lustig, A. K. Shiau, A. J. Brake, and D. Julius, *Proc. Natl. Acad. Sci. U.S.A.* **90,** 5113 (1993); (39) G. Gamba, S. N. Saltzberg, M. Lombardi, A. Miyanoshita, J. Lytton, M. A. Hediger, B. M. Brenner, and S. C. Hebert, *Proc. Natl. Acad. Sci. U.S.A.* **90,** 2749 (1993); (40) K. Ho, C. G. Nichols, W. Lederer, J. Lytton, P. M. Vassilev, M. V. Kanazirska, and S. C. Hebert, *Nature* (*London*) **362,** 31 (1993); (41) Y. Kubo, T. J. Baldwin, Y. N. Jan, and L. Y. Jan, *Nature* (*London*) **362,** 127 (1993); (42) B. Teng, C. F. Burant, and N. O. Davidson, *Science* **260,** 1816 (1993). (43) G. You, C. P. Smith, Y. Kanai, W.-S. Lee, M. Stelzner, and M. A. Hediger, *Nature* (*London*) **365,** 844 (1993); (44) S. Magagnin, A. Werner, D. Markovich, V. Sorribas, G. Stange, J. Biber, and H. Murer, *Proc. Natl. Acad. Sci. U.S.A.* **90,** 5979 (1993); (45) J. Bertran, A. Werner, J. Chillaron, V. Nunes, J. Biber, X. Testar, A. Zorzano, X. Estivill, H. Murer, and M. Palacin, *J. Biol. Chem.* **268,** 14842 (1993); (46) D. Markovich, J. Forgo, G. Stange, J. Biber, and H. Murer, *Proc. Natl. Acad. Sci. U.S.A.* **90,** 8073 (1993); (47) N. Dascal, W. Schreibmayer, N. F. Lim, W. Wang, C. Chavkin, L. DiMagno, C. Labarca, B. L. Kieffer, C. Gaveriaux-Ruff, D. Trollinger, H. A. Lester, and N. Davidson, *Proc. Natl. Acad. Sci. U.S.A.* **90,** 10235 (1993); (48) E. M. Brown, G. Gamba, D. Riccardi, M. Lombardi, R. Butters, O. Kifor, A. Sun, M. A. Hediger, J. Lytton, and S. C. Hebert, *Nature* (*London*) **366,** 575 (1993).

[d] Only partial length cDNA clones were isolated by suppression cloning (hybrid arrest technique). Numbers in parentheses correspond to length in kilobases of full-length clones, isolated from libraries using the partial-length cDNAs as hybridization probes.

analysis. A response allows us to isolate the clone of interest from the remaining 10 phage clones.

Expression cloning usually results in isolation of cDNA clones carrying the entire coding region; however, there are some exceptions. The cloning-by-function strategy selects for a highly expressible cDNA which often but not always is a full-length clone. Indeed, although the voltage-gated K^+ channel cloned by Frech *et al.* (10) was incomplete at its 5′ end, it led to very large currents when expressed in oocytes. A subsequently isolated full-length clone encoded an open reading frame with an extra four amino acids at the N-terminus. Although translation of the incomplete clone started with the methionine corresponding to position 5 in the full-length open reading frame, the incomplete cDNA expressed higher levels of currents, and presumably channel protein, than the longer cDNA isolate (14). This might have been due to negative effects on the transcription by *cis*-acting sequences in the 5′ untranslated region of several hundred base pairs in length of the full-length cDNA clone.

Table I shows that the longest cDNA clone isolated by expression cloning has a length of 5.5 kb. The size of the mRNA of interest represents an obvious limiting factor, mainly due to the fact that full-length cDNA larger than 6–7 kb is difficult to synthesize. An alternative approach would be "suppression cloning" using hybrid arrest techniques. This procedure has been thoroughly described by Lotan (15). Suppression cloning offers two major advantages. First, it can be utilized to clone proteins encoded by large mRNAs and, second, it may detect an mRNA encoding only one of several subunits required to form a functional heteromultimer. In the latter instance, application of the regular expression cloning design may eventually fail if the mRNAs encoding the subunits forming a multimer were no longer present in the same sublibrary.

Expression of Cloned Ion Channels and Receptors in *Xenopus* Oocytes

Initially, we used the *Xenopus* oocyte system for expression cloning of the voltage-gated K^+ channel Kv2.1 (DRK1) and for the thyrotropin-releasing hormone receptor (10, 11). More recently, we have expressed many mutant K^+ channels to identify important structural domains and to elucidate their roles for particular ion channel functions (16–19). Here, we outline the essential strategies, and we describe in detail the procedures to perform structure–function studies of cloned ion channels in *Xenopus* oocytes. In particular, we give a simple description to generate and handle large numbers of different mutant cDNAs and to express and study their function in *Xenopus* oocytes.

Preparation of Xenopus Oocytes

In our typical procedure for *Xenopus* oocyte preparation we pay special attention to several aspects: the temperature for frog maintenance and oocyte incubation, the procedure for defolliculation of oocytes, and the selection of healthy oocytes for RNA injection. So far, we have never experienced seasonal variations of oocyte quality. The frogs can be subjected to partial ovariectomy many times without affecting oocyte quality. The number of operations is primarily limited by the institutional regulations.

Maintenance and Surgery of Xenopus laevis

Xenopus laevis females were purchased from Nasco (Fort Atkinson, WI). Frogs are kept in tanks of dechlorinated water at 17–19°C, and the frog room is under a fixed 12-hr light–dark cycle. Frogs are fed every 2 days, and the water is changed a few hours after feeding. To dissect ovarian tissue and prepare enough oocytes for microinjection, we basically follow the procedures described by Colman (20).

All dissection instruments used are treated with 75% ethanol for about 30 min before surgery. A single frog is immersed in approximately 500 ml of tap water containing 0.2% of tricaine (3-aminobenzoic acid ethyl ester, methane sulfonate salt, Sigma). After about 20 min, the frog should be checked for its responsiveness every 5 min until it is fully anesthetized. To remove ovarian tissue, place the frog on its back on a flat moist pad and make a small incision (1 cm) with scissors through the loose skin and body wall on the posterior ventral side of the animal. Using blunt forceps, gently tease out several lobes of the ovary. Use scissors to remove as few lobes as necessary, and place them in culture medium (see Table II for composition). Gently push the remaining ovarian tissue back into the abdomen. The incisions in the body wall and the skin are sutured using plain gut and silk suture, respectively (Ethicon, Inc., Somerville, NJ). To recover, the frog is placed on a slope partially immersed in shallow water. Keep the head moist, and make sure that the head is above the water surface to prevent the frog from drowning.

Defolliculation

Gently separate the oocytes with forceps into clumps of about 20–30 cells. The clumps are incubated in Ca^{2+}-free OR-2 solution (21) (Table II) containing 2 mg/ml collagenase (Sigma, Type IA) for 2 hr at room temperature

TABLE II Composition of Solutions

Component	Ca^{2+}-free OR-2 (m*M*)	ND-96 (m*M*)	Culture medium (m*M*)	Culture medium with horse serum (m*M*)
NaCl	82.5	96.0	96.0	96.0
KCl	2.0	2.0	2.0	2.0
$MgCl_2$	1.0	1.0	1.0	1.0
$CaCl_2$	—	1.8	1.8	1.8
HEPES	5.0	5.0	5.0	5.0
Gentamycin	—	—	50 μg/ml	50 μg/ml
Sodium pyruvate	—	—	2.5	2.5
Horse serum	—	—	—	5%
pH	7.4	7.4	7.4	7.4

(22 ± 1°C) without agitation. Oocytes are washed several times with culture medium (see Table II). At this point, oocytes are either in clumps or separated but still covered with the follicular cell layer. They are transferred to a petri dish with culture medium containing 5% horse serum (22). The enzyme-treated oocytes are kept overnight at 17°C, and healthy cells are sorted out for manual defolliculation. We found it easier to remove the follicular cell layer manually after 2 hr collagenase treatment without agitation than with no enzyme treatment. Oocytes generally appear healthier if agitation is omitted during enzyme treatment. A sharp pair of forceps (Dumont No. 55) is used to grab the stalk of the clump or a small part of the follicular cell layer of an individual oocyte, while a second pair of forceps is used to pull on the follicular layer, much like removing the skin of a grape. This part of the preparation is done under a stereomicroscope. We find a magnification of 20–40× suitable. Defolliculation is successful when no blood vessels are associated with the isolated oocyte. The defolliculated stage V and VI oocytes (23) are transferred to culture medium containing 5% horse serum for at least 2 hr at 17°C before injection. We have kept oocytes up to 10 days before RNA injection and subsequent successful K^+ channel expression. It is important to sort out the healthy cells and transfer them to fresh culture medium with horse serum every day.

Injection

Pipettes for microinjection are prepared by pulling 3.5-in. replacement tubes (Drummond Cat. No. 3-00-203-G/X) on a horizontal pipette puller (Flaming/Brown micropipette puller, Sutter Instrument Co., Model P-87). Under visual inspection, the tip of the pipette is broken against the base of filament of a microfuge (Narishige Scientific Instrument lab, MF-83). The external diameter of the pipette is about 15 μm and the tip is bevelled. This makes the pipette less damaging to the oocyte. We always prepare a set of injection pipettes, under nuclease-free conditions, just before microinjection. Additionally, the pipettes can be heat-treated (200°C for 6 hr) and stored in a sterile box prior to use. Pipettes are filled from the back with light mineral oil using a 1-ml syringe with a 27-gauge hypodermic needle. Take care that there are no bubbles and that the pipette is completely filled. Slowly slide the pipette over the metal plunger of the microinjector (Nanoject variable automatic injector, Drummond Scientific Company) mounted on a micromanipulator. Carefully but firmly tighten the nut of the injector. Before filling the pipette with solution, test for easy ejection of oil. The pipette is acceptable when mineral oil flows smoothly through the tip. Then deposit a volume as small as necessary of the solution to be injected on a piece of Parafilm. Be careful not to let the solution sit for too long on the Parafilm to prevent concentration of the precious cRNA sample by evaporation of water. Following your movements under the stereomicroscope, carefully move the pipette tip into the solution. To avoid suction of air, slowly fill the pipette making sure that the tip is always in the solution during loading. We transfer 10 healthy-looking oocytes to a 35-mm petri dish containing ND-96 solution (Table II) and a nylon mesh (0.5 mm mesh size) at the bottom. We inject in the vegetal pole or near the equator. Avoid the animal pole to minimize damage to the nucleus of the egg. With some experience, the injection pipette can be used to roll the oocyte carefully into the proper orientation. We arrange the oocytes in a line; this makes it easier to keep track of injection. Injecting 50 nl will inflate the oocyte slightly. This may conveniently serve as a visual indication of a successful injection. After injection, oocytes are transferred to a multiwell cell culture dish with culture medium containing 5% horse serum, and they are incubated at 17°C for up to 10 days.

Synthesis of cRNA for Microinjection into Xenopus Oocytes

Preparation of DNA Template

Ever since the early eighties when cloning vectors bearing promoter regions recognized by bacteriophage RNA polymerases were developed (24), it has been routine to use such vectors for the expression of ion channels and receptors. Despite the fact that all three bacteriophage RNA polymerases are now available from recombinant sources, SP6 RNA polymerase is still considerably more expensive than T3 or T7 RNA polymerase. Therefore, it is advisable to use transcription-competent vectors containing either T3 or

T7 promoters, or both. The use of cloning vectors which carry two different RNA polymerase promoters flanking the polylinker site is highly recommended. It will allow for easy transcription of sense or antisense RNA from the same template DNA after linearizing with the appropriate restriction enzymes. Antisense RNA, although not essential for direct functional expression in oocytes, will facilitate studies of gene expression using *in situ* hybridization or hybrid-arrest techniques. Often, it is beneficial to position the coding region to be expressed close to the promoter site by removing long 5′ untranslated sequences and as much as practical of the polylinker sequences. These precautions may decrease the chances that negatively *cis*-acting elements upstream of the initiation site or palindromic polylinker sequences with the potential to form stable secondary structures reduce translational efficiency (25). Some laboratories have improved the translation of foreign genes in *Xenopus* oocytes by cloning into vectors containing polylinkers flanked by 5′ and 3′ untranslated sequences derived from *X. laevis* β-globin mRNA (26). It may be especially helpful for cRNAs encoding otherwise poorly translated proteins. This strategy has been used to express multimeric K^+ channels which were encoded by concatemerized cDNA constructs (27, 28). The influence of a poly(A) tail at the 3′ end of the cDNA template on the translation efficiency in *Xenopus* oocytes is somewhat controversial. It has been shown that the presence of a poly(A) tail may improve the stability and increase the efficiency of translation of microinjected cRNA up to 20-fold (29). It, however, did not apply to all tested cRNAs. Using cloned genes, derived from an oligo(dT)-primed cDNA library, will in many cases provide at least part of the original poly(A) segment. If necessary, a poly(A) tail may be introduced at the appropriate site of the polylinker by subcloning (30). Alternatively, however, less desirably, up to several hundred adenylyl residues may be added posttranscriptionally using *Escherichia coli* poly(A) polymerase (25).

To optimize the synthesis of full-length cRNA, corresponding to the size of the cloned insert, two additional prerequisites regarding the DNA templates need to be mentioned: (i) the presence of an appropriate unique restriction site 3′ to the coding region or the poly(A) tail, and (ii) the quality of the DNA template. The restriction enzyme used to linearize the DNA template should, if possible, generate 5′ protruding or blunt ends. If 3′ overhanging ends are generated, they may serve as "false" transcription-initiation sites for bacteriophage RNA polymerases (25). This, in turn, may lead to additional synthesis of the "wrong" RNA strand. If necessary, 3′ protruding termini may be removed by treatment with the Klenow fragment of *E. coli* DNA polymerase or with T4 DNA polymerase. Linearization of the template needs to be carried to completion to prevent transcription of vector DNA. We therefore always analyze the size and integrity of the linearized template by analytical agarose gel electrophoresis. The quality of the DNA preparation used as template for cRNA synthesis is critical for a high yield of full-length cRNA. For phage DNA we recommend banding of the insert-bearing vector by centrifugation in a CsCl gradient. This will result in DNA template of the highest possible purity. For plasmid DNA, it may be sufficient to use a combination of alkaline lysis and precipitation with polyethylenglycol (PEG). This is convenient when large numbers of different DNA templates need to be prepared, as may be the case after site-directed mutagenesis using degenerate oligonucleotide mixtures encoding many distinct mutations. If a particular template is used for several transcriptions, we recommend linearizing and purifying a large amount of DNA template and storing appropriate aliquots at −20° to −80°C. When many different templates need to be handled at once, we prepare just enough linearized DNA to perform one transcription reaction. After complete digestion of the DNA, it is highly recommended that the linearized template in the digestion buffer be treated with proteinase K (50 μg/ml) in 0.5% SDS at 37°C for 30–60 min prior to extraction with phenol-chloroform and precipitation with ethanol. At this moment, all reagents, tubes, tips, and pipettes have to be RNase-free.

In Vitro RNA Transcription

The following protocol describes the synthesis of capped and radioactively labeled cRNA using a small-scale procedure that limits expenditures for expensive reagents but provides enough cRNA for initial studies. Capping the 5′ end of the cRNA using the commercially available RNA cap analog $m^7G(5')ppp(5')G$ is essential for stability of the cRNA and efficient translation after injection into *Xenopus* oocytes (29). The percentage of capped cRNA corresponds linearly to the molar ratio of $m^7G(5')ppp(5')G$ to GTP in the transcription reaction (G. Frech and R. H. Joho, unpublished results). Usually, we use a $m^7G(5')ppp(5')G$ to GTP ratio of 4:1. This leads to approximately 80% capped cRNA molecules. In the protocol described below, this ratio and the concentrations of the nucleotides represent a compromise to obtain efficient capping and acceptable yields while spending a minimum of money for the quite expensive cap analog. The yield of cRNA depends primarily on the limiting GTP concentration. Our standard reaction volume is 25 μl with a limiting GTP concentration of 0.25 m*M*. Under these conditions, approximately 4 μg of cRNA may be synthesized before the GTP concentration drops below half its initial value. If one intends to microinject several mutant cRNAs for direct comparison of expression levels and functional properties, or for coexpression of different protein subunits, it is highly recommended that the cRNA concentrations are determined as accurately as possible. We routinely use [α-^{32}P]UTP of low specific activity to monitor synthesis and integrity of the cRNA.

Reagents

5× transcription buffer: 200 m*M* Tris–HCl (pH 8.0), 40 m*M* $MgCl_2$, 250 m*M* NaCl, 10 m*M* spermidine. The buffer is autoclaved and stored in small aliquots at −20°C. Alternatively, the buffer provided by the manufacturer with the RNA polymerase may be used.

10 m*M* ATP, 10 m*M* CTP, 6.25 m*M* GTP, 10 m*M* [α-^{32}]UTP (20,000–50,000 cpm/nmol): Small aliquots are diluted from 100 m*M* NTP stock solutions in sterile 10 m*M* Tris–HCl (pH 7.5) and stored at −20°C. We strongly recommend that the concentrations of at least the GTP and the radiolabeled [α-^{32}P]UTP be determined photometrically. A small aliquot of the radiolabeled and diluted UTP can be used as a standard for cRNA yield determinations.

10 m*M* cap analog: $m^7G(5')ppp(5')G$ is available only from few manufacturers, and the prices differ considerably. We have used the cheapest obtainable cap analog (currently New England Biolab, Inc.). This compound still remains the most expensive chemical in the transcription reaction. The exact concentration should be determined photometrically.

1 *M* DDT: Made up in sterile water and stored in small aliquots at −20°C.

RNasin 40 units/μl: Obtainable from Promega Corp.

Bacteriophage RNA polymerases (22–50 units/μl): Provided by most enzyme suppliers.

RNase-free DNase I (10 units/μl): Provided by most enzyme suppliers.

Procedure

1. To 10 μl of a template solution containing approximately 200 ng/kb insert DNA, add 5 μl 5× transcription buffer, 1 μl of each of the four NTP solutions, 2.5 μl of 10 m*M* cap analog, 0.75 μl 1 *M* DTT, 40 units RNasin, sterile RNase-free water, and 25–50 units RNA polymerase to a final volume of 15 μl. Mix at room temperature to prevent DNA precipitation by spermidine, and incubate subsequently for 30–60 min at 37°C.

2. If the DNA template is precious, monitor the progress of the reaction by determining the incorporation of radiolabeled UMP into trichloroacetic acid (TCA)-precipitable RNA from a small aliquot of the reaction mixture (1 nmol incorporated UMP corresponds to 1.3 μg RNA synthesized).

3. Add 5 units RNase-free DNase I and incubate for 30 min.

4. Chill the reaction on ice and add 10 m*M* Tris–HCl (pH 7.5) up to 200–250 μl. Extract twice with phenol–chloroform–isoamyl alcohol (25:24:1) and once with chloroform–isoamyl alcohol (24:1). Adjust to 2 *M* NH_4acetate, add 2.5 volumes of ethanol, and precipitate the cRNA overnight at −20°C.

5. Recover the cRNA by centrifugation at 10,000 rpm for 30 min at 4°C. Wash the pellet twice with 80% ethanol, and air-dry the cRNA.

6. Dissolve the cRNA in 10–15 μl 0.1 *M* KCl or sterile water. Use 0.5–1 μl of the solution to determine TCA-precipitable radioactivity, and use a similar amount to evaluate size and integrity by electrophoresis through either a denaturing glyoxal–DMSO agarose gel or a denaturing formaldehyde agarose gel.

Using this procedure, we routinely obtain 1–3 μg cRNA which is approximately 80% capped and more than 90% of the radiolabel is located in one major band of expected size.

All reagents and materials used in the protocol need to be RNase-free. Addition of RNasin will reduce RNase activity during enzymatic treatment. During or after extraction, we do not add RNase inhibitor to cRNA. We dissolve cRNA either in sterile water or 0.1 *M* KCl. To achieve higher yields of RNA, it is not advisable to simply increase the NTP concentrations. Concentrations above 4 m*M* may inhibit the efficiency of RNA synthesis (26). Instead, we recommend increasing the reaction volume.

Electrophysiological Recording

This part depends on the particular application. Here, we describe the procedure currently in use in the laboratory for the detection of voltage-gated K^+ channels. We have successfully recorded from oocytes kept at 17°C for up to 10 days after injection. For measurements using a two-electrode voltage-clamp protocol, a single oocyte is placed at the bottom of a recording chamber (RC 13, Warner Instrument Corp.). The oocyte is superfused with ND-96 at a flow rate of 1 ml/min at room temperature (22 ± 1°C). Electrodes are pulled from borosilicate glass capillaries with an inner filament (Warner Instrument Corp. Cat. No. GT12) on a Flaming/Brown micropipette puller. Electrodes are filled with 3 *M* KCl in 10 m*M* HEPES (pH 7.4). The resistances of the electrodes are between 0.2 and 1 MΩ. Whole-cell currents are recorded using a two-microelectrode voltage-clamp amplifier (OC-725A, Warner Instrument Corp.). The pClamp system (version 6.0) and TL-1 DMA interface (Axon Instruments, Inc.) are used to generate the voltage pulse protocol and for data acquisition. Linear capacitive and leakage currents are subtracted on line using a P/4 protocol. Many useful reviews have been published elsewhere describing in detail protocols for the electrophysiological recording from *Xenopus* oocytes (21, 31).

Structure–Function Studies using *in Vitro* Mutagenesis

Today, several different methods for *in vitro* mutagenesis are widely used [for an overview, see McPherson (32)]. In addition to site-directed mutagenesis, the use of chimeric gene constructs has provided a wealth of interesting data in the recent past (17). Over the past few years, we had great success with a commercially available mutagenesis kit for the targeted introduction of single and multiple base changes. We describe some modifications that have been useful to generate a large number of mutants with a high mutagenesis frequency for a reasonable cost in a fairly short time period, and we also present a quick and efficient procedure to screen many mutants. The DNA targeted for mutagenesis needs to be transferred into the single-stranded phage vector M13 or into a double-stranded phagemid such as pBluescript. The latter may also be used to generate a single-stranded DNA template.

Construction of Mutagenesis Templates and Design of Oligonucleotides

Before starting an extensive mutagenesis project, one should carefully consider the template DNA. First, unique restriction sites should be present close to the sequence to be mutated. If this is not the case, the procedure described below can be easily applied to create useful silent restriction sites (19). Second, the size of the insert subcloned into M13 should be as small as convenient to prevent spontaneous deletions during infection and propagation in *E. coli*. Nevertheless, it is advantageous to subclone a fragment large enough so it may be used for mutagenesis reactions at different positions. Most of our mutagenesis work was performed on a 1.7-kb-long *Eco*RI-*Sph*I fragment of the cloned K^+ channel Kv2.1 (DRK1) (10). This enabled us to use the same DNA template to introduce mutations at different positions in the ion channel core region. Unique restriction sites were used to subclone the smallest possible mutant DNA fragment into the full-length parent cDNA (17, 19).

The use of a phagemid vector as source of the single-stranded DNA may be suitable when the size of the cDNA is small (e.g., less than 1 kb). Additional subcloning steps after the mutagenesis reaction may be avoided; however, the entire cDNA insert ought to be resequenced. Often the extra work of subcloning relatively small fragments is compensated by much less DNA sequencing.

Over the past several years, we have followed a simple rule to design oligonucleotides for site-directed mutagenesis. We have always been successful introducing up to three basic changes using oligonucleotides with at least 10 nucleotides or as many as necessary to reach a melting temperature of at least 36°C (in 0.9 *M* NaCl) on either side of the nearest mismatch. The melting temperature has been estimated using the formula K_m (in °C) = $2(A + T) + 4(G + C)$; where A, T, G, and C are the numbers of the respective bases present in the oligonucleotide. Whenever possible, we try to include a C or G at the 3′ end of the oligonucleotide primer.

Site-Directed Mutagenesis Reaction

The reaction is based on the method described by Eckstein and co-workers (33, 34). We use a commercially available kit (oligonucleotide-directed *in vitro* mutagenesis sytem, version 2.1, Amersham Corp.), and we basically follow the manufacturer's recommendations and protocols with the modifications described below. The kit includes enough reagents for 10 full-scale reactions. The limiting supplies are the nucleotide mix containing the dCTPαS and the number of filter units provided in the kit. These filter units can be purchased separately from Schleicher & Schuell, Inc. (Centrex Microfilter, Order No. 02320). We downscale the reaction volumes fourfold with some adjustments: First, the total amount of DNA template and oligonucleotide used in the reaction is downscaled only twofold, i.e., the DNA concentration in the reaction mix is increased by a factor of 2. Second, we do not decrease the sample volumes for the filtration step during the procedure. Third, we do not withdraw the indicated aliquots for gel electrophoresis, since this does not really allow a sound assessment of the reaction. All the adjustments lead to an increased yield of mutated M13 DNA and increases simultaneously the number of possible mutagenesis reactions per kit up to 40. The protocol is particularly suitable for the simultaneous mutation of many templates. We have successfully included up to 20 reactions at the same time to create amino acid substitutions in as many different positions. We strongly recommend including the supplied control template and oligonucleotide in each mutagenesis reaction. This enables one to evaluate and predict the efficiency of the reaction (see Concluding Remarks).

Generation of Mutant Libraries

While picking single mutant plaques after the mutagenesis reaction, isolating single-stranded DNA, and screening them by sequencing may be the simplest way to obtain a few mutant clones, this strategy is not very convenient if large numbers of mutants are desired. We prefer a simple, alternative procedure that avoids sequencing of single-stranded phage DNA. Using degenerate oligonucleotides as primers may lead to several different mutations. Many sequencing reactions would have be to be carried out on DNA derived from single plaques, and, from this point on, every mutant would have to be processed individually. We, instead, pool a number of plaques (5–10 times more than the number of different mutants expected). An aliquot of the pool is used to grow a mutant phage mixture for the preparation of single-stranded DNA. This mixture of single-stranded DNA should contain all the mutant DNA molecules obtained in the original mutagenesis reaction, and it will serve as a mixed template for the synthesis of the second strand. An easy and fast procedure to prepare double-stranded mutant DNA fragments is described below. This double-stranded DNA is then subcloned into the parent cDNA in an expression vector from which cRNA may be transcribed. Plasmid DNA is prepared from individual colonies, and the DNA is only sequenced at this point.

1. To 5–10 μg mutant single-stranded DNA (derived from pooled plaques) add 5–10 pmol primer, 4 μl 5× reaction buffer [200 m*M* Tris–HCl (pH 7.5), 100 m*M* $MgCl_2$, 250 m*M* NaCl]. Adjust volume to 20 μl with sterile water. Heat to 65°C for 5 min and cool slowly to room temperature.
2. Add 1.2 μl 1 *M* Tris–HCl (pH 7.5), 1 μl dNTP-mix [10 m*M* of dATP, dCTP, dGTP, and dTTP in 10 m*M* Tris–HCl, (pH 7.5)], sterile water, and 5–10 units Klenow fragment to a final volume of 40 μl. Incubate 30–45 min at 37°C. Inactivate the enzyme for 10 min at 75°C.
3. Add 5–10 μl 10× digestion buffer, sterile water, and 3–5 units restriction enzyme per μg M13 ssDNA to a final volume of 100 μl. Incubate 2–3 hr at 37°C. With two different restriction enzymes the reaction conditions may have to be adjusted accordingly.
4. Separate the restriction fragment by preparative gel electrophoresis using low-melting agarose (e.g., SeaPlaque GTG agarose from FMC BioProducts). Excise the fragment in the thinnest possible gel slice and recover the DNA using gelase (Epicentre Technologies) or another β-agarase following the manufacturer's recommendations.
5. Directly use the liquified gel solution for ligation of the isolated DNA fragment into the appropriately digested and dephosphorylated parent cDNA (however, some manufacturers of β-agarases recommend precipitating the DNA first). Transform a suitable strain of *E. coli* with an aliquot of the ligation reaction and eventually spread the cells on a selective agar plate to create a mutant library.

For each individual mutation event expected, we pick 3–5 colonies to inoculate small (2–3 ml) overnight cultures. We keep 500 μl of each culture as 7% DMSO stock at −80°C. The remaining culture is used to purify plasmid DNA (any fast way of obtaining plasmid DNA suitable for sequencing is fine). Before sequencing, we verify that the subcloned inserts can be excised and are of the correct sizes. It allows us to select for clones with intact restriction sites at the cloning boundaries, and, therefore, with a correct reading frame. Double-stranded DNA sequencing is then performed to ensure that each clone contains the appropriate mutation. We use a primer just outside the inserted DNA, and we always sequence the entire subcloned fragment. In cases of small fragments (e.g., 150–300 bp), one sequencing reaction will be sufficient to screen for mutants as well as to read through the entire length of the insert. This ensures selection of mutants which do not carry any undesired mutations.

Concluding Remarks

The usefulness of the *Xenopus* oocyte as a tool for the characterization of cloned ion channels, receptors, transporters, and other membrane-bound proteins is well documented by numerous publications. It is apparent from the references in Table I that the number of genes cloned using a cloning-by-expression strategy did not exhibit the same exploding development as the conventional cloning procedures. There are sometimes many obstacles to overcome and many requirements to fulfill before a successful expression cloning project using *Xenopus* oocytes can seriously be considered. Even if there is only limited amino acid sequence information available for a gene of interest, it may still be easier and more promising to pursue traditional cloning-by-hybridization strategies. Table I lists 46 DNA clones isolated by expression cloning in *Xenopus* oocytes. The fact that most of the genes encoding these proteins were the first of their kind cloned at the time underlines the importance, the usefulness, and the impact of the *Xenopus* oocyte expression system.

The use of a commercially available mutagenesis kit has advantages (no preparation of reagents, effectiveness of enzymes tested and guaranteed, easy-to-use design and protocol) and most drawbacks (price, small number of reactions per kit) can be minimized as described above. By circumventing single-strand DNA sequencing and creating pools of mutants, our protocol provides a fast and easy way of applying targeted mutagenesis to receptors, ion channels, or transporters. Over the past year, we have generated, sequenced, and expressed approximately 100 mutants of a voltage-gated K^+ channel. Table III summarizes the results of 10 mutagenesis reactions. We used degenerate oligodeoxynucleotides and the mutant pool protocol. Muta-

TABLE III Mutagenesis Reactions Using Degenerate Oligodeoxynucleotides

Original codon	Degenerate primer sequence[a]	Number of clones analyzed[b]	Expected mutated codons (number of corresponding clones in parentheses)	Number of wild-type codons	Mutation frequency[c]
TGC	(ACG)(ACT)C	38[d]	AAC (9), AAC (3), ATC (1), CAC (0), CCC (0), CTC (2), GAC (10), GCC (6), GTC (2)	2	87%
TGC	T(ACT)C	9	TAC (3), TCC (2), TTC (2)	2	78%
TGC	(AGT)GG	10	AGG (2), GGG (4), TGG (2)	2	80%
TGC	GA(CG)	9	GAC (8), GAG (1)	0	100%
TGC	(AC)AG	3	AAG (2), CAG (1)	0	100%
TGT	(ACG)(ACT)T	37	AAT (1), CAT (2), GAT (10), ACT (1), CCT (1), GCT (9), ATT (1), CTT (2), GTT (4)	6	84%
TGT	(ACG)GT	7[e]	AGT (1), CGT (1), GGT (5)	0	100%
TGT	AA(CG)	2	AAC (1), AAG (1)	0	100%
TGT	T(TA)T	8	TTT (4), TAT (3)	1	88%
TGT	(CG)AG	6	CAG (1), GAG (2)	3	50%

[a] Degenerate positions of the mutagenic oligonucleotides are given in parentheses.

[b] Unless otherwise stated the nature of the mutation was determined by sequencing of double-stranded DNA from individual clones after subcloning pooled mutant DNA fragments into parent cDNA.

[c] Expressed as percentage of the number of mutants obtained compared to the number of clones analyzed.

[d] Three mutants carried a codon (TCC) at the mutated position that could not be derived from the degenerate oligodeoxynucleotide.

[e] Clones were screened by sequencing of single-stranded DNA from individual M13 plaques.

tion frequencies were consistently high (>80%) and independent of the degree of degeneration of the oligonucleotides. Table IV recapitulates our recent mutagenesis reactions, separated into four groups, depending on mutagenesis and screening strategies, and it depicts the corresponding mutagenesis frequencies. Considering the uneven distribution of mutations obtained using highly degenerate oligonucleotides (see Table III), it is apparent that the frequency with which all expected mutations were obtained compared to the total number of clones screened, is higher with less degenerate oligonucleotides.

Finally, the described mutagenesis protocol may also be adapted for random mutagenesis, either by using highly degenerate "spiked" oligonucleotides (35) and the same mutagenesis protocol, or by applying chemical mutagenesis strategies to single-stranded DNA (36).

TABLE IV Combinations of Different Mutagenesis and Screening Strategies

Method[a]	Mutations attempted	Correct mutations obtained	Mutation frequency[b]
Single mutation, ssDNA screening	15	14	20/29 (69%)
Single mutation, dsDNA screening	20	20	37/49 (75%)
Multiple mutations, ssDNA screening	3	3	7/7 (100%)
Multiple mutations, dsDNA screening	35	33	107/122 (88%)

[a] Single mutation, using mutagenic oligonucleotides for a single amino acid replacement; multiple mutations, using degenerate oligonucleotides to target multiple changes; ssDNA screening, sequencing single-stranded DNA derived from individual mutant M13 plaques; dsDNA screening, sequencing double-stranded DNA of individual clones after subcloning pooled mutant DNA fragments into parent cDNA.

[b] Expressed as number of mutants (first number) compared to the number of clones screened (second number).

Acknowledgments

This work was supported in part by NIH Grant NS28407 (RHJ), by a grant of the Muscular Dystrophy Association (RHJ), and by the Swiss National Science Foundation (RDZ).

References

1. J. B. Gurdon, C. D. Lane, H. R. Woodland, and G. Marbaix, *Nature* (*London*) **233,** 177 (1971).
2. M. Mishina, T. Kurosaki, T. Tobimatsu, Y. Morimoto, M. Noda, T. Yamamoto, M. Terao, J. Lindstrom, T. Takahashi, M. Kuno, and S. Numa, *Nature* (*London*) **304,** 604 (1984).
3. Y. Noma, P. Sideras, T. Naito, S. Bergstedt-Lindquist, C. Azuma, E. Severinson, T. Tanabe, T. Kinashi, F. Matsuda, Y. Yaoita, and T. Honjo, *Nature* (*London*) **319,** 640 (1986).
4. N. Dascal, *CRC Crit. Rev. Biochem.* **22,** 317 (1987).
5. T. P. Snutch, *Trends NeuroSci.* **11,** 250 (1988).
6. L. Kushner, J. Lerma, M. V. Bennett, and R. S. Zukin, *Methods Neurosci.* **1,** 3 (1990).
7. J. J. Heikkila, *Int. J. Biochem.* **22,** 1223 (1990).
8. E. Sigel, *J. Membr. Biol.* **117,** 201 (1990).
9. H.-C. Wang, B. Beer, D. Sassano, A. J. Blume, and M. Reza Ziai, *Int. J. Biochem.* **23,** 271 (1991).

10. G. C. Frech, A. M. J. VanDongen, G. Schuster, A. M. Brown, and R. H. Joho, *Nature (London)* **340,** 642 (1989).
11. R. E. Straub, G. C. Frech, R. H. Joho, and M. C. Gershengorn, *Proc. Natl. Acad. Sci. U.S.A.* **87,** 9514 (1990).
12. G. C. Frech and R. H. Joho, *Gene Anal. Technol.* **6,** 33 (1989).
13. G. C. Frech and R. H. Joho, *in* "Methods in Enzymology" (B. Rudy and L. E. Iverson, eds.), Vol. 207, p. 592. Academic Press, San Diego, CA, 1992.
14. G. C. Frech, Ph.D. Dissertation, Baylor College of Medicine, Houston (1991).
15. I. Lotan, *in* "Methods in Enzymology" (B. Rudy and L. E. Iverson, eds.), Vol. 207, p. 605. Academic Press, San Diego, CA, 1992.
16. A. M. J. VanDongen, G. C. Frech, J. A. Drewe, R. H. Joho, and A. M. Brown, *Neuron* **5,** 433 (1990).
17. H. A. Hartmann, G. E. Kirsch, J. A. Drewe, M. Taglialatela, R. H. Joho, and A. M. Brown, *Science* **251,** 942 (1991).
18. G. E. Kirsch, J. A. Drewe, H. A. Hartmann, M. Taglialatela, M. deBiasi, A. M. Brown, and R. H. Joho, *Neuron* **8,** 499 (1992).
19. R. D. Zühlke, H.-J. Zhang, and R. H. Joho, *Receptors and Channels* **2,** 237 (1994).
20. A. Colman, *in* "Transcription and Translation: A Practical Approach" (B. D. Hames and S. J. Higgins, eds.), p. 271. IRL Press, Oxford, 1984.
21. J. P. Leonard and T. P. Snutch, *in* "Molecular Neurobiology: A Practical Approach" (J. Chad and H. Wheal, eds.), p. 161. IRL Press, Oxford, 1991.
22. M. W. Quick, J. Naeve, N. Davidson, and H. A. Lester, *BioTechniques* **13,** (1992).
23. J. N. Dumont, *J. Morphol.* **136,** 153 (1972).
24. D. A. Melton, P. A. Krieg, M. R. Rebagliati, T. Maniatis, and K. Zinn, and M. R. Green, *Nucleic Acids Res.* **12,** 7035 (1984).
25. M. Wormington, *Methods Cell Biol.* **36,** 167 (1991).
26. P. A. Krieg and D. A. Melton, *Nucleic Acids Res.* **12,** 7057 (1984).
27. E. R. Liman, J. Tytgat, and P. Hess, *Neuron* **9,** 861 (1992).
28. T. E. Lee, L. H. Philipson, A. Kuznetsov, and D. J. Nelson, *Biophys. J.* **66,** 667 (1994).
29. D. R. Drummond, J. Armstrong, and A. Colman, *Nucleic Acids Res.* **13,** 7375 (1985).
30. W. Stühmer, M. Stocker, B. Sakmann, P. Seeburg, A. Baumann, A. Grupe, and O. Pongs, *FEBS Lett.* **242,** 199 (1988).
31. W. Stühmer, *in* "Methods in Enzymology" (B. Rudy and L. E. Iverson, eds.), Vol. 207, p. 319. Academic Press, San Diego, CA, 1992.
32. M. J. McPherson, "Directed Mutagenesis: A Practical Approach." IRL Press, Oxford, 1991.
33. J. R. Sayers, W. Schmidt, and F. Eckstein, *Nucleic Acids Res.* **16,** 791 (1988).
34. J. R. Sayers, W. Schmidt, A. Wendler, and F. Eckstein, *Nucleic Acids Res.* **16,** 803 (1988).
35. J. D. Hermes, S. M. Parekh, S. M. Blacklow, H. Köster, and J. R. Knowles, *Gene* **84,** 143 (1989).
36. R. M. Myers, L. S. Lerman, and T. Maniatis, *Science* **229,** 242 (1985).

[5] Cloning of G Protein-Coupled Opioid Receptors Using Degenerate PCR and Low-Stringency Homology Screening

David K. Grandy, Qun-Yong Zhou, Claudia Bouvier, Carmen Saez, and James R. Bunzow

Introduction

Stereoselective and saturable binding of opioid ligands to specific brain receptors was first described in 1973 (1–3). In 1976 Martin *et al.* proposed the existence of three opioid receptor types based on the distinct responses that morphine, ketocyclazocine, and SKF10,047 produced in the nondependent, chronic spinal dog model (4). These receptors came to be known as mu (μ, morphine), kappa (κ, ketocyclazocine), and sigma (σ, SKF10,047). Subsequently delta (δ, vas deferens) and epsilon (ε, β-endorphin) opioid receptor types were also described (5, 6).

Immediately following their initial pharmacological description the biochemical and physiological characterization of the opioid receptors began. These efforts revealed that opioid receptors are glycosylated proteins (7) with approximate M_r's of 60,000–95,000 (8) that contain disulfide bonds that are important for binding activity (9). Furthermore, agonist binding to opioid receptors is modulated by GTP and its analogs (10) and the μ, δ, and κ opioid receptors all couple to the inhibition of adenylyl cyclase activity, an action that is blocked by pretreatment with pertussis toxin (10). Activation of the opioid receptors characteristically results in the hyperpolarization of the neuron as a consequence of activating potassium channels (11) and the inhibition of neurotransmitter release (10). Taken together these results are consistent with the opioid receptors belonging to the G protein-coupled receptor (GPCR) superfamily.

All members of the GPCR family share a characteristic structure consisting of seven putative transmembrane domains. Interestingly, when the amino acid sequences of related receptors are aligned the maximum identity is usually localized within these seven hydrophobic domains. This extensive sequence conservation within the putative transmembrane domains has been exploited to clone members of several GPCR families.

In the past our laboratory has employed low-stringency homology screening and degenerate primers in the polymerase chain reaction (degenerate

Methods in Neurosciences, Volume 25

PCR) to clone members of the dopamine receptor family (12). More recently we have applied similar strategies to the cloning of members of the opioid receptor family.

Experimental Procedures

Homology Cloning of G Protein-Coupled Receptors: An Overview of the Strategy

The two stratagies that we have used to clone members of the opioid receptor families are summarized in Table I. The details for each step are presented in the following sections.

Synthesis of First-Strand cDNA

The successful application of any cloning strategy involving cDNA depends on the preparation of high-quality RNA from a source known to express the receptor(s) of interest. For the opioid receptors we chose several sources of RNA including the neuroblastoma × glioma cell line NG108-15 which expresses about 1 pmol of δ opioid receptors/mg soluble protein (13) and the human neuroblastoma cell line SH-SY5Y which expresses both μ and δ opioid receptors (14). We have also prepared RNA from regions freshly dissected from rat brain that contain opioid receptors based on ligand autoradiographic surveys (15). Among the brain regions that we have examined are the locus coeruleus which expresses μ receptors; the thalamus which expresses μ, δ, and κ receptors; and the caudate putamen which also expresses μ, δ, and κ opioid receptors (15).

To prepare total RNA from tissue culture cell lines we have used the guanidinium thiocyante/cesium chloride centrifugation method (16). However, when working with smaller amounts of material, such as dissected brain regions, we routinely use the RNaid kit (Bio101). The RNA that is prepared from both of these methods is suitable for synthesizing first-strand cDNA using the SuperScript preamplification system (Gibco,BRL). Both random and oligo(dT) primers are supplied with this kit allowing the investigator maximum flexibility. It is also possible to synthesize cDNA that is enriched in one particular known receptor sequence. In this case the synthesis of cDNA is primed with an antisense oligonucleotide that is complementary to sequences specific to the receptor of interest. Typically 1 μg of total RNA is sufficient to synthesize enough first-strand cDNA for several PCR reactions.

TABLE I Summary of Degenerate PCR and Homology Cloning by Low-Stringency Hybridization

Cloning by degenerate PCR
1. Select G protein-coupled receptors of interest
2. Align the amino acid sequences of the selected receptors to identify regions of conservation; usually, but not exclusively located in the putative transmembrane domains
3. Design two pairs of degenerate primers (outer and nested pairs) for use in the PCR
4. Synthesize first-strand cDNA from target tissue or cell line using oligo(dT) and random hexamers
5. Perform the first PCR on the cDNA[a] template using the outer primers; use an aliquot of this first PCR reaction as template and the inside, nested primers in the second round of PCR
6. Analyze the PCR products by gel electrophoresis and Southern blotting (choose probe such that primer sequences are not included)
7. Subclone the PCR products
8. Determine the sequence of selected inserts
9. Use both nucleotide and deduced amino sequences to query databases
10. Isolate full-length cDNAs corresponding to the interesting PCR products; use the PCR product to probe libraries prepared from the appropriate tissue or cell line
11. Clone the complete coding region of the receptor cDNA into an expression vector
12. Express the new receptor in tissue culture cell lines for use in pharmacological and/or physiological analyses
Homology cloning by low-stringency hybridization
1. Identify representative receptor
2. Obtain clone and prepare radiolabeled probe
3. Screen cDNA[a] libraries [random hexamer and oligo(dT) primed] synthesized from RNA prepared from appropriate source (determined by Northern blotting, for example) under conditions of low stringency
4. Identitify positive clones, plaque purify and map inserts using restriction enzymes and Southern blotting
5. Subclone inserts into sequencing vectors and determine open reading frame
6. Use both nucleotide and amino acid sequences to query databases
7. Subclone the coding regions of interesting receptor(s) into eukaryotic expression vectors
8. Express the new receptor(s) in tissue culture cell lines for pharmacological and/or physiological analyses

[a] cDNA is recommended here because all G protein-coupled receptors to date are encoded by genes containing introns. However, not all coding regions are interrupted by introns. Therefore, a thorough analysis will include genomic DNA.

Design of Oligonucleotide Primers

When the amino acid sequences of GPCRs that bind the same or similar ligands are aligned the regions of greatest evolutionary conservation (homology) correspond to the putative transmembrane domains. We, and others (17), have compiled alignments of GPCR sequences that have helped us to design degenerate oligonucleotide primers for use in the PCR. With respect to the opioid receptors of particular interest to us have been the somatostatin receptors because CTOP (DPhe-Cys-Tyr-DTrp-Orn-Thr-Pen-Thr-NH_2), a cyclic somatostatin analog, has high affinity for the μ opioid receptor (18). After aligning consensus sequences present in several GPCRs we designed several pairs of degenerate oligonucleotide primers. One of these pairs consisted of a degenerate sense oligonucleotide (oligo 1) that corresponds to sequences in the C-terminal portion of putative transmembrane domain III with the addition of the *Sal*I recognition sequence at its 5′ end (Fig. 1). The other primer (oligo 2) is in the antisense orientation and corresponds to sequences in the amino terminal end of transmembrane domain VI with the *Eco*RI recognition sequence added to its 5′ end. The different restriction enzyme sequences were added to the 5′ end of each primer to facilitate the directional subcloning of the resulting PCR products. These restriction enzyme recognition sequences are preceded by two additional nucleotides because in our experience their presence enhances the enzyme's cutting efficiency. The degenerate positions in the two primers are indicated by multiple bases in parentheses. The "I" in the sequence stands for deoxyinosine, a neutral base that pairs equally well with all four nucleotides. The use of deoxyinosine reduces the need to make guesses when highly degenerate codons are encountered. For a more thorough discussion of the uses of deoxyinosine the reader is referred to Sambrook, Fritsch, and Maniatis (19).

With the publication of the mouse δ opioid receptor cDNA (20, 21) three oligonucleotide primers specific for the mouse receptor sequence were also designed (Fig. 1, oligos 3, 4, and 5). Primer 3 is complementary to the noncoding sequence that is ~200 bp 3′ of the receptor's termination codon while primer 4 is also in the antisense orientation but its cognate sequence

oligo #1: 5'-GAGTCGACCTGTG(C/T)G(C/T)(C/G)AT(C/T)(A/G)CIIT(G/T)GAC(C/A)G(C/G)TAC-3'

oligo #2: 5'-CAGAATTCAG(T/A)AGGGCAICCAGCAGAI(G/C)(G/A)(T/C)GAA-3'

oligo #3: 5'-AGGAATCACCAGCTCTGC-3'

oligo #4: 5'-CCTCTAGAAGCCTTTGATCCCTCAGG-3'

oligo #5: 5'-CCAAGCTTTAAAGGCTGGGTCCCTGC-3'

oligo #6: 5'-ATGAATTCAC(GATC)(AG)T(GC)ATGAG(CT)GT(GC)GAC(CA)G(CA)TA-3'

oligo #7: 5'-TTGTCGAC(GA)TA(GA)AG(GA)A(CT)(GATC)GG(GA)TT-3'

oligo #8: 5'-CCAAGCTTCCGCAGCAGGCCTTCAGCACC-3'

oligo #9: 5'-CCTCTAGAAGTACCAAGCTTAGCGAGGGT-3'

FIG. 1 Sequences of oligonucleotides used to clone the opioid receptors.

(plus the *Xba*I recognition sequence) is ~150 bp 3′ of the receptor's termination codon. Primer 5 was designed to be identical to sequences ~50 bp 5′ of the initiation codon of the mouse δ opioid receptor with the addition of the *Hin*dIII recognition sequence. Oligonucleotide 3 was designed to serve as a δ opioid receptor-specific primer in the synthesis of first-strand cDNA from NG108-15 RNA (see above). Primers 4 and 5 were designed to amplify the complete coding region of the δ opioid receptor from the NG108-15 cDNA. In our experience the use of a pair of nested primers, such as 3 and 4, greatly increases the probability that the target sequence (in this case the mouse δ opioid receptor cDNA) will be amplified by the PCR.

Based on the mouse δ opioid receptor sequence two degenerate primers were designed (Fig. 1, oligos 6 and 7). The sequence of the sense primer (6) was selected based on the putative transmembrane domain III of the δ opioid receptor sequence and includes an *Eco*RI recognition sequence. The sequence of the antisense primer (7) was based on residues found in transmembrane domain VII of the mouse δ receptor sequence with the addition of a *Sal*I recognition sequence.

Polymerase Chain Reaction (PCR)

We routinely use approximately 1 μg of first-strand cDNA, or genomic DNA, as the template in a typical polymerase chain reaction containing the thermostable polymerase buffer, 80 pmol of each degenerate primer, and 200 μ*M* of each deoxynucleotide in a total volume of 50 μl. Once an equal volume of mineral oil is layered over the reaction the sample is placed in a thermocycler programmed to hold at 100°C for 10 min to denature all duplexes in the template. Following the denaturation step the thermocycler ramps down to 80°C as quickly as possible and holds at this temperature for 5 min during which time the enzyme is then added. It has been our experience that if we combine the "hot start" with 1 unit of the enzyme *Thermus flavus* (Epicenter Technologies, Madison, WI) in the presence of 5% DMSO we obtain more specific priming with our degenerate oligonucleotides. Following enzyme addition the temperature is held at 94°C for 1 min (denaturation step), then dropped to 50°C for 1.5 min (annealing step), and then held at 72°C for 1.5 min (extension step). This cycle is repeated 35 times. It should be noted that except for the extension step, the denaturation and annealing temperatures were estabished empirically for each new deoxyoligonucleotide primer and template DNA combination.

To evaluate the success of the amplification typically 1/10th of the reaction mix is electrophoresed through a 1% agarose gel along with appropriate size standards (e.g., the i-kb ladder, Gibco,BRL), stained with ethidium bromide, and photographed along with a fluorescent ruler. At this point the gel can be processed for Southern blotting and probed if desired. The types of probes that we have hybridized to blots of this type include radiolabeled cDNA fragments prepared from the δ opioid receptor as well as kinased oligonucleotides whose sequences correspond to conserved regions present in other members of the same receptor family. To avoid false-positive signals it should be remembered that the best probes of PCR products will not contain either primer sequence. For low-stringency hybridization we have used 35% formamide, 5 × SSC, 5 × Denhardt's, and 1% SDS at 37°C followed by several washes, first at room temperature and then at 50–55°C in 2 × SSC plus 0.1% SDS (see below). If products in the size range of 400–750 bp are visible we usually treat the remainder of the polymerase chain reaction with 100 μg/ml proteinase K for 30 min at 37°C followed by phenol chloroform extraction and ethanol precipitation. The DNA is then resuspended in buffer and digested overnight with *Eco*RI (Gibco,BRL or New England Biolabs) and *Sal*I (Boehringer-Mannheim) at 37°C. The digested DNA is electrophoresed in a 1% agarose gel, stained with ethidium bromide and the 400- to 750-bp products are cut from the gel and extracted from the agarose using either the GeneClean (Bio101) or the Prep-A-Gene (Bio-Rad) DNA isolation kits. These DNA fragments are ligated to a plasmid vector, such as pBluescript (Stratagene), that has also been digested with *Eco*RI and *Sal*I and the ligation products are used to transform competent bacteria (19), such as the strain XL-1 (Stratagene), to ampicillin resistance. The bacteria that received the plasmid will grow on ampicillin-containing nutrient agar plates, and in the presence of X-Gal and IPTG those bacteria harboring a plasmid with an insert will appear white. The number of bacterial transformants that are obtained vary greatly from ligation to ligation even though the same primers were used. If 12 or fewer white, amp^r clones grow we will pick them all and grow them up in nutrient broth. Double-stranded DNA is then prepared by the alkaline lysis method (19) and usually enough DNA is obtained from a 2-ml culture to perform five sets of double-stranded, dideoxy chain termination sequencing reactions using the Sequenase version 2 kit (USB). Alternatively, if the number of clones is high rather than sequencing a large number of uninteresting inserts the amp^r bacterial colonies are picked and transferred to both a numbered master ampicillin plate and a nylon filter (Colony/Plaque Screen, NEN) that is placed on top of a second ampicillin plate. Following an overnight incubation the master plate is stored at 4°C and the bacterial colonies on the nylon filter are denatured, neutralized, and baked as described (19). The filter is then prehybridized in a low-percentage formamide solution followed by hybridization with the appropriate ^{32}P-labeled probe (see below). The hybridization is terminated by low-stringency washes as described below.

Once a PCR product has been sequenced and determined to be of further interest (see Database Searching below) it is necessary to isolate a full-length clone for expression studies. The most straightforward method to isolate a full-length cDNA is to screen a cDNA library [whose synthesis was primed with both oligo(dT) and random hexamers] prepared from the tissue or cell line of interest with the radiolabeled PCR product. At this stage of the analysis we have found that high-stringency hybridization (50% formamide, 37°C) and washing conditions (0.5 × SSC, 0.1% SDS, 55°C) are appropriate.

Low-Stringency Hybridization Screening

The publication of the mouse δ opioid receptor cDNA sequence (20, 21) provided new probes that could be used to screen libraries for other members of the opioid receptor gene family by low-stringency homology screening. The probes we chose to use were derived from the mouse δ opioid receptor. To obtain the full-length mouse δ opioid receptor we prepared first-strand cDNA from NG108-15 RNA using a primer complementary to 3′ noncoding sequence of the mouse receptor (oligo 3). The resulting specifically primed cDNA was then used as the template in a PCR experiment. Primers 4 and 5 are exactly complementary to sequences in the 5′ and 3′ untranslated regions flanking the δ receptor's coding region. As with our other PCR primers, these δ-specific primers included restriction enzyme recognition sites, in this instance *Sal*I and *Eco*RI, to facilitate the cloning of the receptor cDNA into both shuttle (pBluescript, Stratagene) and expression (pRc/RSV, In Vitrogen) vectors. The resulting PCR product was sequenced to confirm its identity and to also confirm that no sequence artifacts had been introduced as a result of the PCR. To generate radiolabeled probes from double-stranded cDNA longer than 500 bp we use either random priming or nick translation (19). For DNA probes less than 500 bp we prefer to label by random priming. In the case of the full-length δ receptor, the cDNA was isolated from an agarose gel following restriction digestion with *Hin*dIII and *Xba*I and labeled using a random priming kit (Boehringer-Mannheim Biochemicals). The resulting probe is denatured and then added to the prehybridization solution ($\sim 5 \times 10^6$ cpm/ml of hybridization buffer).

The library that we screened for additional opioid receptors was generously provided by Dr. Terry Snutch (University of British Columbia). The cDNA was synthesized from rat brain mRNA. First-strand synthesis was primed with both oligo(dT) and random primers and the resulting cDNA was size-selected prior to cloning in λZAP (Stratagene). Once titered the library was plated onto twelve, 130-mm plates at a density of about 60,000 plaques per plate. Replica nylon filters (Colony/Plaque Screen, NEN) were pulled, processed according to the manufacturer's recommendations, baked, and prehybridized in 35% formamide, 5× SSC, 5× Denhardt's, and 1% SDS overnight at 37°C. The hybridization was initiated by the addition of the probe and allowed to proceed for 24–36 hr at which time it was terminated by two room temperature washes in 2× SSC, 0.1% SDS followed by two washes at 55°C for approximately 15 min each. The filters were then exposed to Kodak XAR film with an intensifying screen overnight at −70°C. Clones that hybridized to the probe on duplicate filters were plaque-purified and analyzed by PCR, Southern blotting, restriction enzyme digestion, and nucleotide sequencing.

Database Searching

The importance of having access to the most current databases cannot be overemphasized. When this project was begun it was routine to screen sequence databases that were 6 months or more out of date. These databases are still of value and with the development of very powerful algorithms such as FASTA they can be quickly scanned for a sequence match. However, for the most up-to-date database information we recommend using the services offered by the National Center for Biotechnology Information at the National Library of Medicine. The NCBI is now responsible for updating the nucleic acid database GenBank and they also support the EMBL, Swiss prot, and PIR databases, among several others. In our experience the most valuable database is GenBank which is updated daily and can be rapidly queried using the sequence similarity search program BLAST. To perform a BLAST search use e-mail and address the message to blast@ncbi.nlm.nih.gov. For more information on BLAST send a "help" message. The response that comes back will contain all of the most recent information regarding BLAST searches, the databases that can be searched, and the necessary commands, as well as a contact person and telephone number for more information or on-line help. To retrieve sequences address your inquiry to retrieve@-ncbi.nlm.nih.gov and for more information type "help" followed by a return. The NCBI is in the process of compiling over 20 specialized databases and these can be downloaded by connecting to ftpncbi.nlm.nih.gov or ftp130.14.25.1. For more information on any of these services contact *NCBI News,* a publication of the National Center for Biotechnology Information and the National Library of Medicine (e-mail: info@ncbi.nlm.nih.gov; phone: 301-496-2475).

Opiate Receptors Cloned by the Homology Approach

Cloning of the Rat κ Opioid Receptor (R21) Using Degenerate Primers

Using the first set of degenerate primers (oligos 1 and 2) in a polymerase chain reaction an orphan G protein-coupled receptor, R21, was amplified from rat striatum cDNA. When the sequence of R21 was used to query the GenBank, EMBL, and the protein database PIR in 1991 it failed to make a highly significant match with any published receptor. This situation changed dramatically with the publication of the mouse δ opioid receptor cDNA sequence (20, 21). When the PCR fragment of R21 was aligned with the published mouse δ opioid receptor cDNA sequence the two were found to share approximately 70% of their amino acid sequences in putative transmembrane domains III through VI. For us this extensive homology strongly suggested that R21 is a member of the opioid receptor gene family.

To obtain a clone that contained the complete coding region of R21 both genomic and striatal cDNA libraries were screened simultaneously with R21 as the radiolabeled probe. The genomic screen revealed that the coding region of R21 is spread over several exons. Eventually the complete coding region of R21 was isolated from total rat brain cDNA. The results of our *in situ* hybridization studies in mouse brain using R21 as probe were consistent with a κ receptor distribution (22a). As this work was in progress both the mouse (22) and the rat (23) κ opioid receptors were published and the alignment of all three sequences revealed that R21 is identical to the rat κ opioid receptor (Fig. 2). A compilation of the average identity between the rat κ, μ, and δ opioid receptors is presented in Table II.

Cloning of the Rat μ Opioid Receptor (TS11) by Low-Stringency Homology Screening

The low-stringency screening of the rat brain cDNA library with the mouse δ opioid receptor probe resulted in the identification of another receptor sequence that belongs in the opioid receptor gene family. This clone, referred to as TS11, contains a 1.3-kb open reading frame encoding a G protein-coupled receptor 398 amino acid residues long (Fig. 2) that is 49% identical to the rat κ and 51% identical to the mouse δ opioid receptors (Table II).

To determine what type of opioid receptor TS11 encodes we cloned its open reading frame into the expression vector pRc/RSV (InVitrogen). This step was made easier by using the PCR and a 5′ noncoding primer (Fig. 1, oligo 8) that contained a *Hin*dIII site and a 3′ noncoding primer (Fig. 1,

```
Amino acid alignment

LC132                                                  MESLFPAPYWEVL
Rat μ-Opioid Receptor        MDSSTGPGNTSDCSDPLAQASCSPAPGSWLNLS
Mouse δ-Opioid Receptor                             MELVPSARAELQSS
Rat κ-Opioid Receptor                     MESPIQIFRGDPGPTCSPSACLLP

                                                          I
LC132 HGSHFQGNLSLLNETVPHHLLLNASHSAFLPLGLKVTIVGLILAVCIGGLLGNCL
μ-OR  HVDGNQSDPCGLNRTGLGGNDSLCPQTGSPSMVTAITIMALYSIVCVVGLFGNFL
δ-OR  PLVNLSDAFPSAFPSAGANASGSPGARSASSLALAIAITALYSAVCAVGLLGNVL
κ-OR  NSSSWFPNWAESDSNGSVGSEDQQLESAHISPAIPVIITAVYSVVFVVGLVGNSL

                           II
LC132 VMYVILRTPKMKTATNIYIFNLALADTLVLLTLPFQGTDILLGFWPFGKALCKTV
μ-OR  VMYVIVRYTKMKTATNIYIFNLALADALATSTLPFQSVNYLMGTWPFGTILCKIV
δ-OR  VMFGIVRYTKLKTATNIYIFNLALADALATSTLPFQSAKYLMETWPFGELLCKAV
κ-OR  VMFVIIRYTKMKTATNIYIFNLALADALVTTTMPFQSAVYLMNSWPFGDVLCKIV

          III                                        IV
LC132 IAIDYYNMFTSTFTLTAMSVDRYVAICHPIRALDVRTSSKAQAVNVAIWALASVV
μ-OR  ISIDYYNMFTSIFTLCTMSVDRYIAVCHPVKALDFRTPRNAKIVNVCNWILSSAI
δ-OR  LSIDYYNMFTSIFTLTMMSVDRYIAVCHPVKALDFRTPAKAKLINICIWVLASGV
κ-OR  ISIDYYNMFTSIFTLTMMSVDRYIAVCHPVKALDFRTPLKAKIINICIWLLASSV

                                                        V
LC132 GVPVAIMGSAQ   VEDEEIECLVEIPAP QDYWGPVFAICIFLFSFIIPVLIISV
μ-OR  GLPVMFMATTK   YRQGSIDCTLTFSHP TWYWENLLKICVFIFAFIMPILIITV
δ-OR  GVPIMVMAVTQ   PRDFAVVCMLQFPSP SWYWDTVTKICVFLFAFVVPILIITV
κ-OR  GISAIVLGGTKVREDVDVIECSLQFPDDEYSWWDLFMKICVFVFAFVIPVLIIIV

                                                  VI
LC132 CYSLMIRRLRGVRLLSGSREKDRNLRRITRLVLVVVAVFVGCWTPVAVFVLVQGL
μ-OR  CYGLMILRLKSVRMLSGSKEKDRNLRRITRMVLVVVAVFIVCWTPIHIYVIIKAL
δ-OR  CYGLMLLRLRSVRLLSGSKEKDRSLRRITRMVLVVVGAFVVCWAPIHIFVIVWTL
κ-OR  CYTLMILRLKSVRLLSGSREKDRNLRRITKLVLVVVAVFIICWTPIHIFILVEAL

                          VII
LC132 GVQPGSETAVAIL RFCTALGYVNSCLNPILYAFLDENFKACFRKFCCASSLHRE
μ-OR  ITIPETTFQTVSW HFCIALGYTNSCLNPVLYAFLDENFKRCFREFCIPTSSTIE
δ-OR  VDINRRDPLVVAALHLCIALGYANSSLNPVLYAFLDENFKRCFRQLCRTPCGRQE
κ-OR  GSTSHSTAALSSY YFCIALGYTNSSLNPVLYAFLDENFKRCFRDFCFPIKMRME

LC132 MQVSDRVRTIAKDVGLGCKTSETVPRPA 367
μ-OR  QQNSTRVRQNTREHPSTANTVDRTNHQLENLEAETAPLP 398
δ-OR  PGSLRRPRQATTRERVTACTPSDGPGGGAAA 372
κ-OR  RQSTNRVRNTVQDPASMRDVGGMNKPV 380
```

FIG. 2 Alignment of the amino acid sequences of the putative opioid receptor LC132 with the rat μ, mouse δ, and rat κ opioid receptors. To achieve maximum alignment, gaps were introduced into the sequences. Residues that are conserved among all four receptors are indicated by boldface print. Reproduced from ref. 30.

TABLE II Amino Acid Identity between Rat LC132 and the Rat μ and κ and Mouse δ Opioid Receptors

	TM[a]I	TMII	TMIII	TMIV	TMV	TMVI	TMVII	TM avg	Avg/all
LC132[b] vs rat μ[b]	58[c]	67	77	48	67	59	85	66	48
LC132 vs rat κ[d]	35	67	82	43	71	73	80	64	47
LC132 vs mouse δ[e]	46	67	77	52	63	59	75	63	46
rat κ vs mouse δ	62	83	91	57	75	64	90	75	52
rat μ vs mouse δ	69	90	86	48	83	77	85	77	51
rat μ vs rat κ	54	80	91	33	75	73	95	72	49

[a] Transmembrane domain.
[b] Bunzow *et al.* (28).
[c] Percent.
[d] Nishi *et al.* (23).
[e] Evans *et al.* (20) and Kieffer et al. (21).

oligo 9) that contained an *Xba*I site. COS-7 cells were then transfected with this construction using the $CaPO_4$ precipitation method (24) and 36 to 48 hr later the monolayers were mechanically scraped, washed four times with ice-cold 50 m*M* Tris buffer (pH 7.7), homogenized in a glass dounce homogenizer, and centrifuged at 44,000 *g* for 15 min at 4°C. The protein concentration in the membrane preparation was determined by the method of Lowry *et al.* (25). The equilibrium saturation and binding assays were performed at room temperature for 1 hr using 18–50 μg of protein per assay tube, at pH 7.7, with [^{3}H]diprenorphine. To reduce nonspecific binding the GF/B filters were presoaked in 0.5% polyethyleneimine. The rank order of ligand affinities for the TS11 receptor fit the pharmacological profile expected for the μ receptor with high affinity for DAMGO (K_i = 1.6 n*M*), levorphanol (0.6 n*M*), β-endorphin (0.7 n*M*), morphine (0.8 n*M*), U50488 (910 n*M*), DPDPE (3170 n*M*), and dextrorphan (4100 n*M*). Functional coupling of the receptor to G proteins was demonstrated by GTP shift experiments and the inhibition of forskolin-stimulated cAMP production. At the time that this work was in progress the cloning of the rat μ opioid receptor was published (26) and a comparison of this sequence to that of TS11 revealed that they encode identical receptors.

To examine the distribution of TS11 mRNA in the brain and spinal cord we prepared an ^{35}S-labeled antisense riboprobe complementary to its 5′ end. This probe was hybridized *in situ* to coronal rat brain sections as previously described (27). When the *in situ* hybridization results with the TS11 probe were compared to the distribution of μ opioid receptors as determined by ligand autoradiography studies (15) there was very good agreement (28).

It is worth noting that for studies of opioid receptor abundance and distribution Northern blot analysis may be more difficult to interpret than results from either *in situ* or PCR-based experiments. The reason for this is that the mRNAs for the rat μ and δ opioid receptors are large, on the order of 11 and 16 kb, respectively (29). In retrospect the large size of the opioid receptor mRNAs is probably the reason why most oligo(dT)-primed cDNA libraries are deficient in opioid receptor sequences (unpublished observations).

Cloning of LC132, a Novel Member of the Opioid Receptor Family, by Degenerate PCR

The δ opioid receptor-based degenerate primers (oligo 6 and 7) were used to amplify fragments of sequence out of rat brain cDNA. These amplification products were subcloned and screened with a radiolabeled δ opioid receptor probe using the conditions described under Experimental Procedures for low-stringency hybridization. The most abundant PCR product that was obtained by this approach, LC132, was a 600-bp fragment of a new GPCR. LC132 shares significant sequence identity with the mouse δ opioid receptor and when the complete coding region was obtained (Fig. 2) it was found to be 46, 47, and 48% identical overall to the mouse δ, rat κ, and the rat μ opioid receptors (Table I). When only the putative transmembrane domains are compared the average identity approaches 65%. Somewhat surprisingly, however, was the observation that when individual cytoplasmic and transmembrane domains of LC132 are compared with the mouse δ, rat κ, and rat μ opioid receptors the greatest sequence conservation is found in the third cytoplasmic loop. This domain is 83% identical to the same domain in the μ opioid receptor and 87% identical to the δ and κ. This result is interesting because it suggests that the LC132 receptor couples to the same second messengers as the δ, κ, and μ opioid receptors. Therefore, solely on the basis of structural considerations, we propose that LC132 is a GPCR and that it is a member of the opioid receptor gene family.

To test this hypothesis we cloned LC132 into the expression vector Rc/RSV and expressed the receptor transiently in COS-7 cells and stably in Ltk^- cells. Membranes were prepared from cells expressing the LC132 transcript (as determined by Northern blots, data not shown) and binding assays were performed using several standard radiolabeled opioid ligands including DAMGO, DPDPE, U69593 diprenorphine, and ethylketocyclazocine. None of these ligands bound the LC132 membranes with high affinity. However, the LC132-containing membranes did bind low amounts of labeled naloxone and bremazocine. Our interpretation of these results is that LC132 is closely related to the other known opioid receptors but it does not bind the standard drugs used to discriminate opioid receptor subtypes. This result is not surprising given the fact that only a small number of the more than 20 known opioid peptides are available as radiolabeled ligands. It is also possible that the endogenous ligand for LC132 remains to be identified.

In an attempt to clarify the function of LC132 *in situ* hybridization was performed to determine the distribution of its mRNA. Although the *in situ* hybridization pattern that we obtained overlaps with the opioid ligand autoradiography results (15), the distribution of LC132 mRNA is clearly distinct from them as well. Now that we know the brain regions that express high levels of LC132 mRNA we predict that one function of this receptor is to participate in analgesia and the perception of pain.

Summary

In summary we have described methods that we have used to clone the classical μ, δ, and κ opioid receptors and a new receptor, LC132, that based on structural arguments we predict is also a member of the opioid receptor gene family. The combination of low-stringency homology screening and degenerate PCR provides a powerful strategy for the cloning of members of this G protein-coupled receptor family. When these cloning techniques are combined with transient and/or stable expression and classical pharmacological analyses it is often possible to determine which receptor subtype is encoded by the cloned cDNA. Given the sensitivity of the molecular approaches, in particular PCR, and what has been observed for other G protein-coupled receptors that bind the same, or similar ligands, we expect that more receptor genes will be identified than were orginally predicted from earlier pharmacological data. The cloning of LC132 is a perfect example of the situation that molecular neuropharmacologists now face; that is the ability to clone and express a new receptor type whose sequence allows its inclusion within an established gene family yet whose endogenous ligand is unknown. We are confident, however, that in time the endogenous ligand for orphan receptors such as LC132 will be determined by combining expression in tissue culture with pharmacological and physiological analyses.

Acknowledgments

The authors thank Drs. M. Kelly, O. Ronnekleiv, G. Zhang, J. T. Williams, T. F. Murray, A. Mansour, M. Low, and O. Civelli for their assistance and comments concerning various aspects of this work. We also thank M. Mortrud, S. Hagen, Z. W. Zhu, A. Unteutsch, M. Bosch, and B. Naylor for their technical assistance.

This work was supported in part by grants to DKG from NIDA (DA08562), NIDDK (DK47093), and NIMH (MH48991). The sequences of LC132 and the μ opioid receptor have been assigned GenBank accession numbers U01913 and U02083, respectively.

References

1. E. J. Simon, J. M. Hiller, and I. Edelman, *Proc. Natl. Acad. Sci. U.S.A.* **70,** 1947 (1973).
2. L. Terenius, *Acta Pharmacol. Toxicol.* **32,** 317 (1973).
3. C. B. Pert and S. H. Snyder, *Science* **179** 1011 (1973).
4. W. R. Martin, C. G. Eades, J. A. Thompson, R. E. Huppler, and P. E. Gilbert, *J. Pharmacol. Exp. Ther.* **197,** 517 (1976).
5. J. A. H. Lord, A. A. Waterfield, J. Hughes, and H. W. Kosterlitz, *Nature* (*London*) **267,** 495 (1977).
6. M. Wüster, R. Schultz, and A. Herz, *Neurosci. Lett.* **15,** 193 (1979).
7. T. Gioannini, B. Foucaud, J. M. Hiller, M. E. Hattren, and E. J. Simon, *Biochem. Biophys. Res. Commun.* **105,** 1128 (1982).
8. A. D. Howard, Y. Sarne, T. L. Gioannini, J. M. Hiller, and E. J. Simon, *Biochemistry* **25,** 357 (1986).
9. J. M. Bidlack and L. G. Abood, *Adv. Biochem. Pyschopharmacol.* **33,** 415 (1982).
10. H. Ueda, M. Nozaki, and M. Satoh, *Comp. Biochem. Physiol. C* **98C,** 157 (1991).
11. A. North, J. T. Williams, A. Suprenant, and M. Christie, *Proc. Natl. Acad. Sci. U.S.A.* **84,** 5487 (1987).
12. O. Civelli, J. R. Bunzow, and D. K. Grandy, *Annu. Rev. Pharmacol. Toxicol.* **32,** 281 (1993).
13. W. F. Simonds, G. Koski, R. A. Streay, L. M. Hjelmeland, and W. A. Klee, *Proc. Natl. Acad. Sci. U.S.A.* **77,** 4623 (1980).
14. V. C. Yu, S. Eiger, D.-S. Duan, J. Lameh, and W. Sadée, *J. Neurochem.* **55,** 1390 (1990).
15. A. Mansour, H. Khachaturian, M. E. Lewis, H. Akil, and S. J. Watson, *J. Neurosci.* **7,** 2445 (1987).
16. J. M. Chirgwin, A. E. Przbyla, R. J. MacDonald, and W. J. Rutter, *Biochemistry* **18,** 5924 (1979).
17. W. C. Probst, L. A. Synder, D. I. Schuster, J. Brosius, and S. C. Sealfon, *DNA Cell Biol.* **11,** 1 (1992).
18. K. Gulya, J. T. Pelton, V. J. Hruby, and H. I. Yamamura, *Life Sci.* **38,** 2221 (1986).
19. J. Sambrook, E. F. Fritsch, and T. Maniatis, "Molecular Cloning: A Laboratory Manual," 2nd ed. Cold Spring Harbor Lab. Press, Cold Spring Harbor, NY, 1989.
20. C. J. Evans, D. E. Keith, Jr., H. Morrison, K. Magendzo, and R. H. Edwards, *Science* **258,** 1952 (1992).
21. B. L. Kieffer, K. Befort, C. Gaveriaux-Ruff, and C. G. Hirth, *Proc. Natl. Acad. Sci. U.S.A.* **89,** 12048 (1992).

22. K. Yasuda, K. Raynor, H. Kong, C. D. Breder, J. Takeda, T. Reisine, and G. I. Bell, *Proc. Natl. Acad. Sci. U.S.A.* **90,** 6736 (1993).
22a. C. Saez, Q.-Y. Zhou, O. Civelli, J. R. Bunzow, and D. K. Grandy, *Regulatory Peptides* **Suppl. 1,** 13 (1994).
23. M. Nishi, H. Takeshima, K. Fukuda, S. Kato, and K. Mori, *FEBS Lett.* **330,** 77 (1993).
24. C. Chen and H. Okayama, *Mol. Cell. Biol.* **7,** 2745 (1987).
25. O. H. Lowry, N. J. Rosebrough, A. L. Farr, and R. J. Randall, *J. Biol. Chem.* **193,** 265 (1951).
26. Y. Chen, A. Mestek, J. Liu, J. A. Hurley, and L. Yu, *Mol. Pharmacol.* **44,** 8 (1993).
27. O. K. Ronnekleiv, B. R. Naylor, C. T. Bond, and J. P. Adelman, *Mol. Endocrinol.* **3,** 363 (1989).
28. J. R. Bunzow, G. Zhang, C. Bouvier, C. Saez, O. Ronnekleiv, M. Kelly, and D. K. Grandy, *J. Neurochem.* **64,** in press (1995).
29. K. Fukuda, S. Kato, K. Mori, M. Nishi, and H. Takeshima, *FEBS Lett.* **327,** 311 (1993).
30. J. R. Bunzow, C. Saez, M. Mortrue, C. Bouvier, J. T. Williams, M. Low, and D. K. Grandy, *FEBS Lett.* **347,** 284–288 (1994).

[6] Hybrid Arrest Screening in Oocytes

Manami Tsutsumi and Boaz Gillo

Overview

Oocyte Bioassay

Xenopus laevis oocyte possesses the basic machinery to translate exogenous eukaryotic mRNA into a fully functional protein. The unusually large size (over 1 mm in diameter), the ease of manipulation, and the high sensitivity for functional expression has made the oocyte an ideal expression system for many mRNAs [see Snutch *et al.* (1, 2) for possible limitations]. Following injection of RNA from various sources (tissues, cells, synthesized RNA) into the oocyte, the expression of mRNAs can be quantitatively measured on appropriate stimulation. Since the first demonstration of mRNA expression in *Xenopus* oocytes in 1971 (3), the oocyte expression system has emerged as a popular bioassay for various mRNA expression and regulation. Using this bioassay, the identities of various cloned receptors (4–13), channels (14–16), transporters (17), enzymes, and secretory proteins (18–20) have been confirmed.

Cloning Strategies Using the Oocyte Bioassay

There are two cloning strategies in which the oocyte bioassay has been used: (i) expression cloning and (ii) cloning by hybrid arrest screening. For expression cloning, which is described in Chapter 4 of this volume, mRNAs are transcribed *in vitro* from a cDNA library containing the receptor of interest. A receptor cDNA clone is identified by its expression in the oocyte following injection of the transcribed mRNA. The original pool of cDNA clones is purified by stepwise fractionations of the response-evoking cDNA mixture until a single clone is obtained. To achieve proper expression of a protein in the oocyte, injection of the entire coding sequence of the mRNA is required. In cases where multiple subunits are required for activity, such as in the case of the nicotinic Ach receptor (11), mRNA for all subunits must be present for the expression of a functional receptor but this may not always be feasible from a given cDNA library.

The second cloning strategy in which the oocyte bioassay is used is the technique of hybrid arrest screening. As early as 1977, hybrid arrest technique

Methods in Neurosciences, Volume 25

was used to study gene regulation and structural analysis of regulatory sequences in adenovirus 2 DNA (21), regulation of Rous sarcoma virus replication (22), and regulation of tumor necrosis factor expression (23) in cell culture and cell-free systems. Hybrid arrest in the oocyte was first described in 1985 by Kawasaki in which the translation of the lymphokines interleukin 2 and interleukin 3 was inhibited (18).

This chapter describes the method of hybrid arrest screening in the oocyte which we used in cloning of the gonadotropin-releasing hormone (GnRH) receptor (12). Using this method, the identity of a partial receptor sequence is quickly and reliably determined by its ability (as an antisense DNA or RNA) to block the translation of a specific receptor mRNA derived from tissue or cellular RNA. Although one must have access to some sequence information on the cDNA to be cloned before this technique can be employed, hybrid arrest technique eliminates the necessity of having the entire coding sequence in order to confirm the identity of the cDNA. The confirmed partial sequence information is then used to screen a cDNA library to isolate a cDNA encoding the entire receptor sequence. Hybrid arrest technique can be used for both screening and confirmation of cDNA clones. In addition to the GnRH receptor, this method has been used in the cloning of the strychnine-sensitive glycine receptor (24), serotonin 5HT1c receptor (7), the type I iodothyronine 5′-deiodinase (25), and the delayed-rectifier-type K^+ (Isk) channel (26).

Hybrid arrest in the oocyte has also been used to elucidate functional roles and regulatory domains of already cloned cDNA species (9, 20, 27–34). Hybrid arrest technique is particularly valuable when the receptor is multisubunit—requiring more than one subunit for function—because this technique can be used to test a putative subunit for contribution to the functional receptor (27).

Theoretical and Experimental Design Considerations

Mechanism of Hybrid Arrest by Antisense DNA

As illustrated in Fig. 1, when RNA isolated from cells or tissue containing a receptor mRNA is injected into the oocyte, mRNA is translated into functional protein within 48 to 72 hr. The functional expression of this receptor can be measured using a standard two-electrode voltage-clamp recording technique. Other assay methods depend on the nature of the expressed proteins and may include adenylate cyclase activation, transporter flux, and other biochemical or electrophysiological measurements. The concentration

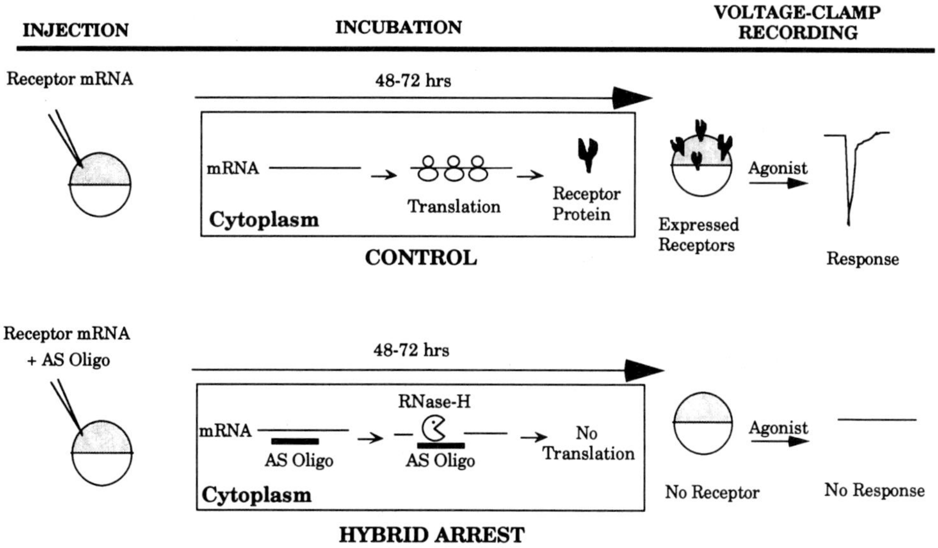

FIG. 1 Schematic illustration of hybrid arrest in oocytes. In control oocytes, RNA isolated from cells or tissue containing receptor mRNA is injected. Oocytes are incubated for 48–72 hr to allow for translation and insertion of receptor protein into the membrane. The level of receptor response is recorded by a standard two-electrode voltage-clamp recording. When an AS oligo is co-injected with receptor mRNA, the mRNA moiety of the mRNA·AS oligo hybrid is degraded by endogenous RNase H in the oocyte and causes irreversible arrest of the receptor mRNA expression.

of injected mRNA is usually linearly correlated with the amplitude of agonist-induced response (8, 35) allowing for quantitative analysis of functionally expressed receptors.

Co-injection of a receptor-specific antisense DNA or RNA with the receptor mRNA inhibits mRNA translation. The antisense fragment can be either DNA or RNA, but we focus on the use of antisense DNA (oligonucleotide) in this chapter. The use of antisense RNA is limited in some respects as discussed in the following section. Figure 1 illustrates the predominant mechanism of mRNA translation inhibition by antisense oligonucleotide (AS Oligo) in the *Xenopus* oocyte. When a specific complementary AS oligo is co-injected with receptor mRNA, the RNA moiety of the RNA·DNA hybrid is selectively and irreversibly degraded by the endogenous RNase-H (28, 36–41). Thus, translation of receptor mRNA is irreversibly blocked. The AS oligo is also degraded within 30 min and the remaining target mRNA fragments are eliminated over the next several hours, presumably by cellular nucleases (39, 42).

About Antisense RNA

Antisense RNA has been used successfully to arrest mRNA translation by several investigators [43, 44, and see Green *et al.* (45)]. But the use of antisense RNA may be limited in some respects. First, the arrest of translation by antisense RNA is partial and/or reversible. RNA·RNA hybrid is believed to inhibit translation by blocking ribosome binding and/or translation initiation (46, 47). However, the injected antisense RNA apparently does not form stable hybrids *in vivo* with its complementary mRNA due to the presence of an RNA duplex unwinding activity by an endogenous enzyme. RNA·RNA hybrids are transient in systems such as fertilized oocytes and developing *Xenopus* embryos (although minimal in *Xenopus* oocytes) (42, 48, 49) leading to unreliable hybrid arrest results. Second, hybrid arrest by antisense RNA is effective only when the antisense RNA includes complementary sequences to the 5′ end or the translation initiation site region of the target mRNA. Hybrid arrest in oocytes by antisense RNA has been found to result in inconsistent inhibitions by some investigators (29).

Experimental Design

For cloning of the GnRH receptor, we utilized the ability of the GnRH receptor to couple to phospholipase C in the oocyte. Injection of RNA isolated from the rat anterior pituitary or the mouse gonadotrope cell line αT3-1 into the oocyte will result in the expression of the GnRH receptor (50). Activation of the receptor by GnRH stimulates phosphatidylinositol 4,5-bisphosphate hydrolysis, resulting in the production of second messengers diacylglycerol and inositol 1,4,5-trisphosphate (IP_3). In the oocyte, IP_3 mobilizes intracellular Ca^{2+}. This is turn leads to the activation of the Ca^{2+}-sensitive Cl^- channel causing a characteristic depolarizing Cl^- current (50–52). It is this Cl^- current that is measured as an index of receptor expression in the oocyte in the case of phospholipase C-coupled receptors. But the electrophysiology that is described in this chapter can also be utilized to measure the activation of other receptors and channels such as the acetylcholine nicotinic receptor/channel complex and voltage-sensitive channels.

The major liability of oocyte hybrid arrest screening is potential variability of the responses in oocytes. Thus, to clone the GnRH receptor (12), we developed the oocyte hybrid arrest technique into a reliable screening assay.

To increase reliability of screening, we have utilized the following controls as illustrated in Fig. 2. First, RNA containing a known mRNA sequence was co-injected as an internal control. In order to ascertain the specificity of hybrid arrest and the identity of a clone, a proper internal control must be included in the experiment. For this, we used the 5HT1C receptor, another member of the G protein-coupled receptor superfamily which also couples to phospholipase C in the oocyte (6, 7). In addition to RNA from αT3-1 cell line containing the GnRH receptor mRNA (53), oocyte was co-injected with whole-brain rat RNA which contained the 5HT1C receptor mRNA. Second, 5HT1C receptor AS oligo effect and test AS oligo effect on targeted response (i.e., GnRH) versus 5HT1C response were compared. Expression of irrelevant mRNAs should not be affected by a receptor-specific AS oligo. As seen in Fig 2A, the internal control 5HT1C receptor AS oligo abolished the 5HT response without affecting the GnRH response. The test oligo, on the other hand, completely abolished the GnRH response while leaving the 5HT response unaffected (Fig. 2B). These results demonstrated that this putative GnRH receptor AS oligo (WZ7) specifically and completely inhibits the GnRH receptor mRNA expression. This oligo was later confirmed to encode a partial sequence of the GnRH receptor.

Obtaining a Partial Sequence Representing a Putative Receptor

In order to utilize the technique of hybrid arrest, it is first necessary to obtain a partial sequence representing a putative receptor from which AS oligos can be designed. There are several ways to obtain partial sequences of a putative receptor. These include polymerase chain reaction (PCR)-based homology cloning, cDNA library sequencing of a source enriched in the desired sequence (see Chapter 7 by Kirkness and Venter), low stringency cDNA library screening, and data from the amino acid sequence.

For cloning of the GnRH receptor, PCR-based homology cloning was used to synthesize putative GnRH receptor fragments using degenerate oligonucleotide primers. The GnRH receptor was predicted to be a G protein-coupled receptor; therefore, putative receptor cDNA fragments were amplified by PCR cloning using primers directed against homologous regions of other G protein-coupled receptor sequences (i.e., the third and sixth transmembrane regions). Polymerase chain reaction products of the predicted size were sequenced. A unique AS oligo was designed for each PCR-amplified clone containing sequence characteristic of a G protein-coupled receptor (such as other hydrophobic transmembrane regions IV and V).

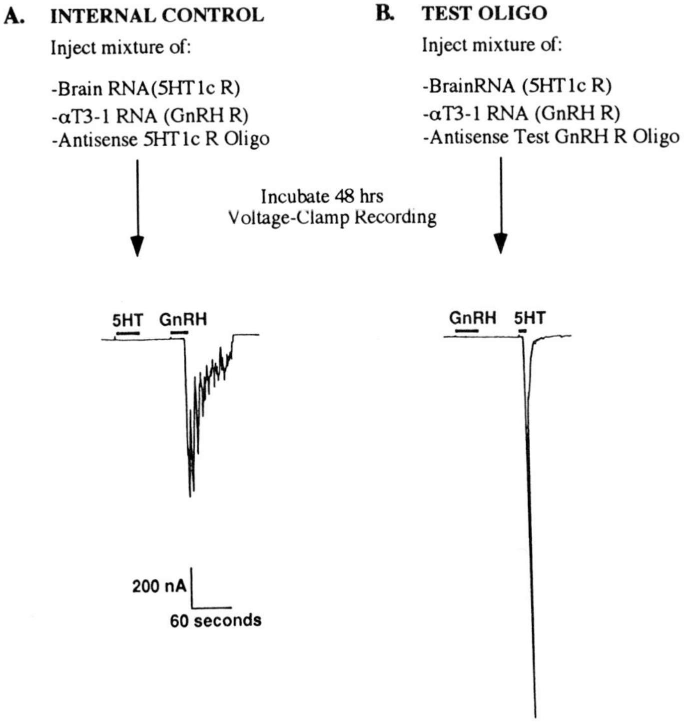

FIG. 2 Experimental design in hybrid arrest screening. Oocytes are injected with RNA mixture containing 1 mg/ml each of 5HT1C (internal control) and GnRH receptor (test) mRNA. One set of oocytes is co-injected with 5 ng of 5HT1C receptor AS oligo (internal control) and the other set is injected with 5 ng of test oligo containing a putative GnRH receptor sequence. Both oligos were 15 nucleotides in length. The specificity of hybrid arrest is determined by comparing the two responses as illustrated. The electrophysiological tracings illustrate hybrid arrest of 5HT1C receptor of GnRH receptor expression by respective AS oligo. One hundred nanomolars 5HT or 200 n*M* GnRH were introduced into the bath at the horizontal lines. (A) Response to 5HT and GnRH in oocytes previously injected with a mixture of rat brain RNA (for the 5HT response), αT3-1 RNA (for the GnRH response), and the antisense 5HT1C receptor oligo. Sixteen cells showed identical responses. (B) Response to GnRH and 5HT in oocytes previously injected with a mixture of rat brain RNA, αT3-1 RNA, and antisense WZ7 (GnRH R) oligo. Twenty-four cells had identical response. [Reproduced with permission from Tsutsumi *et al.* (12); © The Endocrine Society.]

Oligonucleotide Design

AS oligos can be directed against any part of the coding sequence of target mRNA. In designing an AS oligo, its specificity and its ability to stably hybridize with the target mRNA should be considered. An AS oligo should contain a unique sequence of the putative receptor clone (e.g., PCR product) about 11–30 nucleotides in length. Ideally, it should be long enough to ensure stable hybridization and it should exclude homologous regions of receptors within that receptor family.

We have found that 5 ng of 11-mer AS oligo can selectively and completely abolish the expression of 5HT1C receptor in oocytes injected with 50 ng of total rat brain RNA (unpublished results). Another study reported that a stretch of a complementary sequence of 10 consecutive bases is sufficient to produce significant (> 60% at 50 ng) and specific antisense degradation of a moderately abundant endogenous mRNA in oocytes (54). Oligos shorter than 10 nucleotides may work as well. However, in addition to decreased hybridization stability, the chance occurrence of a shorter sequence in a given RNA pool is greater. This can increase nonspecific hybridization.

Theoretically, increased length minimizes nonspecific hybridization by increasing the affinity of the oligonucleotide to the target mRNA while reducing the likelihood of its hybridization to irrelevant mRNAs. For this reason, oligonucleotides as long as 40–90 nucleotides have been used by some investigators (55). However, when screening for a specific receptor sequence among homologous receptors such as in the case of G protein-coupled receptor superfamily, increased oligo length could increase the probability of its running into homologous sequences.

Longer oligos also tolerate mismatches better than shorter oligos. For example, approximately 60% sequence homology with AS oligos of 25–30 nucleotides in length, but not AS oligo of 15 nucleotides, was reported to cause some degradation of compelmentary mRNA by RNase H-like activity (39, 54).

Oligonucleotide Concentration

The concentration of oligo used to effectively hybrid arrest a target mRNA may vary according to the source of mRNA. In one study, *in vitro*-transcribed amylase mRNA was mixed with AS oligo prior to injection with 1:1 stoichiometry (oligo:mRNA). This mixture, when injected into oocytes, hybrid-arrested greater than 50% of amylase synthesis compared to control (19). This particular study, however, represents an ideal case of hybrid arrest in which only the pure amylase mRNA transcript was hybridized to the AS oligo prior to injection. In the absence of a cDNA clone, RNA isolated from cells or a mixture of mRNA transcripts is used instead of a single mRNA transcript. Under these conditions, the level of target mRNA in that heterogeneous RNA population is not always known. Therefore, the concentration of AS oligo to effectively inhibit mRNA translation in your system needs to be determined empirically by using a control mRNA and control AS oligo in the same oocyte using the same RNA (such as 5HT1C R discussed under Experimental Design).

When using cellular mRNA for hybrid arrest in oocytes, a molar ratio of ~200:1 (oligo:target mRNA) appears to be required to cause complete degradation of target mRNA (39). A molar ratio of 1000 to 10,000 should yield complete or near complete arrest of target mRNA translation in most cases when injecting heterogeneous RNA. The lowest concentration of the internal control AS oligo producing maximal inhibition of internal control mRNA expression should be used as a guide to a starting point for screening test receptor clones since too high a concentration of AS oligo could result in nonspecific hybridization.

When screening several test clones, a mixture of several AS oligo sequences can be used in one hybridization reaction by either co-injecting several unique oligomers or by using degenerate oligos. Using degenerate oligomers may be less expensive since it may take screening of several dozen putative receptor clones before a positive clone is identified. When using mixed or degenerate oligos, the desired molar ratio of each oligo sequence to target mRNA should be maintained. That is, determine how much total oligo concentration has to be injected to meet the concentration of each unique sequence needed to hybrid arrest the target mRNA. For example, if 5 ng of any one oligo sequence is expected to inhibit expression (based on the results of the internal control), then 50 ng of mixed test oligos containing 10 unique sequences should be co-injected with the mRNA. Once a pool of degenerate oligos is found which can effectively hybrid arrest the target mRNA expression, then unique oligos from that particular pool of mixed oligos can be synthesized and tested individually for the identification of the positive clone.

Modified Oligonucleotides

Oligonucleotides are highly susceptible to endogenous nucleases *in vivo*. To increase the efficiency of RNase H digestion and hybrid arrest, several modified oligos have been developed and tested for their resistance to nucleases. Since unmodified oligos have worked well for our purposes, we have no experience with the modified oligos. If problems with endogenous nucleases are encountered with unmodified oligos, one may consider using modified oligos. For reviews on modified oligos, readers may refer to Tolume and Helene (56) and Stein and Cohen (57). Of the modified oligos, only two are RNase H substrates: (i) a phosphorothioate derivative which substitutes one of the phosphodiester oxygen atoms for sulfur (49, 58) and (ii) oligos linked to an intercalating agent (acridine derivative) terminally, which increases binding affinity of oligos (37). These modified oligos are much more stable (> 3 hr) in oocytes than the unmodified counterparts (< 30 min) and are more efficient in inhibiting mRNA translation. Caution should be used, however, since modified oligos are reported to be more toxic to the oocyte and length- and concentration-dependent nonspecific inhibition has been reported within the concentration range safely used for unmodified oligos. For an example, the acridine derivative-linked oligo inhibits [α-^{32}P]CTP incorporation in oocytes suggesting inhibition of cellular transcription (37). Modified acridine-linked α-oligo (but not α-oligos), although it is not an RNase H substrate, has been shown to inhibit translation by an RNase H-independent mechanism when directed at the cap site of the target mRNA, presumably by blocking ribosome binding or the formation of translation complexes (59).

Hybridization

The hybridization/injection buffer used for *in vitro* hybridization of RNA and oligos contains NaCl (necessary for hybridization) and Tris, pH 7.4. The proper pH of the injection buffer is important since solutions of improper pH are correlated with a greater percentage of cell death (B. Gillo, unpublished results). Concentrations of 200 mM NaCl (of 50–200 mM range) and 10 mM Tris·HCl (of 5–10 mM range) are typical and appear to be optimal for prehybridization of AS oligo and receptor mRNA (12, 18, 34). Whether the mRNA and the AS oligo are prehybridized or not, heat denaturation of mRNA prior to injection has been shown to enhance hybrid arrest by AS oligo, presumably by elimination of secondary structures of mRNA that can interfere with hybridization (34, 37). Prehybridization is not absolutely necessary since hybridization can also occur in the oocyte cytoplasm. Even in the absence of NaCl (no hybridization can occur), near-complete hybrid arrest of interleukin 2 mRNA translation in the oocyte was observed when mRNA and AS oligo were mixed prior to injection (18). Injecting mRNA and oligos separately, however, more notably reduced the efficiency of hybrid arrest. Prehybridization probably yields greater efficiency of hybrid arrest since it allows the oligos to hybridize to the target mRNA prior to injection and decreases the chance of oligo degradation *in vivo*.

Injection and Recording

When planning to inject the oocyte for recording session it is important to estimate how many cells one is able to assay on a given day. However, keep in mind that some oocyte deaths are expected and that for each experimental condition, at least one dozen oocytes should be injected to ensure adequate number of oocyte responses for data analysis ($n > 5$). Usually we prefer to dissect the frog and to enzymatically defolliculate the oocyte one day before the injection. We found that in most cases, the largest number of oocyte deaths occur within the first 24 hr. If there is more than 60% cell death, we discard the entire preparation since their chance of survival in the next 2 days is small. Healthy oocytes should be transferred to fresh enriched ND96 medium (see Specific Methodologies) to continue incubation until recording. In general, oocytes should be checked every day for dead oocytes and good oocytes transferred to fresh medium. Both injection and electrophysiological recording are performed at room temperature; however, remaining oocytes should be kept in a 20°C incubator to protect them from heat. Temperature beyond 25°C can modify the oocyte response and its passive electrophysiological properties. Sometimes, we observe a slight, overall decrease in the oocyte response amplitude over the course of a few hours. It is therefore desirable to record different experimental groups of oocytes alternately so that if occyte responses begin to diminish throughout the day, you have proper controls. This is particularly important if oocyte recording takes several hours to complete.

Analysis of Results

Five to fifty nanograms of a 15-mer AS oligo should inhibit translation of target receptor mRNA significantly (> 80%) if there is 100% sequence complementarity. If no significant hybrid arrest is observed at 50 ng of AS oligo, then the AS oligo is not significantly complementary to the receptor of interest. When significant hybrid arrest is observed, make sure that the arrest is specific to the target receptor. To confirm the identity of the clone from which the AS oligo was designed, synthesize another AS oligo directed against a different region of the same clone and test for its ability to arrest receptor mRNA translation before going on to screen a cDNA library.

Isolation and Characterization of the Full-Length Receptor cDNA

The confirmed partial sequence of the receptor is used as a probe to screen a cDNA library for isolation of the full-length receptor cDNA. Once a putative full-length receptor cDNA is obtained, mRNA is transcribed from the cDNA and is injected into oocytes. The identity of the full-length receptor cDNA can be confirmed by oocyte expression. Details of characterization and expression of cloned receptors in the oocyte can be found elsewhere in this volume (see Chapter 4 by Zühlke *et al.*)

Specific Methodologies

Note: Keep everything RNase-free (60).

RNA Preparation for Oocytes (61)

Solutions

Lysis Buffer

- 6 *M* Urea
- 3 *M* LiCl
- 10 m*M* NaOAc, pH 5.2

Mix in glass of distilled untreated water. Filter through Nalgene (0.45 μm) and store at 4°C.

Add 0.05% SDS just before use.

Solubilization Buffer

- 10 m*M* NaCl
- 10 m*M* Tris·HCl, pH 7.5
- 5 m*M* EDTA, pH 8.0
- 1.0% SDS

Mix from stocks and keep at room temperature. Stock solutions should be made with RNase-free H_2O (diethylpyrocarbonate-treated autoclaved H_2O) and filtered through Nalgene filter (0.45-μm pore size).

Phenol:Chloroform:Isoamylalcohol

TE-saturated phenol that has been recently distilled (within 30 days) kept at 4°C. Make 24:1 (vol:vol) mixture of $CHCl_3$:isoamylalcohol (IAA) and store at room temperature. Mix $CHCl_3$/IAA 1:1 (vol:vol) with phenol just prior to use.

Procedure

1. Homogenize cells or tissue in lysis buffer at 4°C. Use approximately 25 ml buffer/g tissue or 50 ml buffer/wt g cells. Use homogenizer or a 22-gauge needle and a syringe to homogenize (20–30 times).

2. Leave overnight at 4°C.

3. Spin at 10,000 *g* for 90 min at 4°C.

4. Discard supernatant. Invert on Kimwipe to drain thoroughly. Wipe the top of the tube with Kimwipe.

5. Suspend pellet in solubilization buffer by pipetting a small volume up and down. Use between 0.5 and 3 ml/g cells as final volume of solubilization buffer.

6. Extract with phenol:$CHCl_3$:IAA three times. Spin for 5 min at 10,000 *g* at room temperature each time.

7. Extract with $CHCl_3$:IAA two to three times. Remove all traces of phenol since phenol is toxic to oocytes. During the course of extractions, if you leave a reasonable interface, you will lose about one-third of your original volume.

8. Add 1/10 volume of 3 *M* NaOAc and 2.5 volumes of 100% ethanol. Precipitate overnight at −20°C.

9. Spin at 10,000 *g* for 10 min at 4°C. Wash the pellet with 70% ethanol. Spin again for 5 min at 4°C.

10. Air-dry the pellet by inverting on Kimwipe for 30 min or more. You can also dry the pellet under vacuum, but be careful not to overdry (check every 2 min or so).

11. Resuspend the pellet in a small volume of DEPC-treated water and reprecipitate with 1/10 volume NaOAc and 2.5 volumes of ethanol.

12. Resuspend in DEPC water at desired concentration of RNA for injection and/or poly$(A)^+$ isolation (60). [It is not necessary to select for poly$(A)^+$ RNA given that there is high enough concentration of the receptor mRNA present in your heterogeneous RNA preparation.] When resuspending for injection, keep in mind that RNA will be diluted by the addition of oligos and the hybridization/injection buffer; therefore, your RNA solution should be at least 3× concentrated. An RNA concentration of 1–5 mg/ml range is a good starting point. Oocytes can tolerate at least 50 ng of oligos and 250 ng of total RNA (concentration of 1 mg/ml oligo and 5 mg/ml RNA when injection 50 nl).

Note: The amount of RNA to be injected should be determined empirically by first performing a concentration–response curve between about 0.1 and 5 mg/ml. If your target mRNA is extremely rare, then a higher concentration range should be used. In our case, due to the abundance of GnRH receptor mRNA in αT3-1 cells (as much as 0.5–3% of total RNA), 1mg/ml total RNA is sufficient to induce a 1000-nA signal in oocytes when exposed to GnRH. A total of 1.25 ng of GnRH receptor mRNA transcript produces enough receptors to generate greater than 5000 nA of signal.

Oligonucleotide Preparation

Custom designed oligonucleotides can be purchased from various sources or synthesized using an automatic DNA synthesizer (60). Determine oligo concentration by spectrophotometry reading at OD 254. If they are degenerate oligos, determine what concentration of each oligo is present in the mixture (see discussion on oligo concentration). Make sure that the oligos are RNase-free by phenol:$CHCl_3$ extraction, followed by two chloroform extractions. Resuspend oligos in H_2O to the desired concentration (~1 mg/ml), aliquot, and store at −20°C. Dilute down to the appropriate concentration when mixing with RNA and buffer prior to injection.

Frog Purchase and Maintenance

Purchase adult *X. laevis* females (Xenopus One Co., Ann Arbor, MI: Nasco, Fort Atkinson, WI). Mature females are about 9.0–10.5 cm in length and weigh approximately 90–150 g. Frogs are kept in dark tanks or, alternatively, the bottom is covered by a black mat. Commercial, inexpensive storage boxes with tight perforated lids will adequately fit this purpose. The tank is filled with 15–20 cm of dechlorinated water that is passed through a filter or set water in open reservoir for 24 hr before use. The number of frogs per tank depends on the tank size, at density of 1–2 frogs per gallon. The water temperature should be 18–20°C and never over 23°C. Daily light/dark cycle is set to 15/9 hr, respectively. The frogs are fed twice a week with frog pellets (Nasco, Fort Atkinson, WI) or 10 g of chicken or beef liver per frog per feeding. Tank should be cleaned a few hours after each feeding (~5 hr).

Oocyte Preparation

Solutions

Ca^{2+}-free ND96, pH 7.5
- 96 m*M* NaCl
- 2 m*M* KCl
- 1 m*M* $MgCl_2$
- 5 m*M* HEPES·NaOH

Collagenase

Ca^{2+}-free ND96
2 mg/ml collagenase (Sigma type Ia)

ND96

Ca^{2+}-free ND96
1.8 mM $CaCl_2$

ND96E, pH 7.6

ND 96
2.5 mM pyruvate
100 units/ml penicillin
100μg/ml streptomycin

Instruments

Two pairs of fine forceps
One pair of small blunt forceps
Large scissors
Small scissors
Sutures
Clamp
Scalpel
Sterile Pasteur pipettes with a fire-polished large broken tip

Operation

1. Put a frog into a shallow bath of 0.2% tricaine solution (Sigma A-5040, 1 g/500 ml tap water) for 20–30 min, or until the frog is immobile.
2. Sterilize instruments by soaking in 70% ethanol. Air-dry on sterile pad.
3. Swab frog abdomen with ND96E and make a small incision with a scalpel (1 cm or less) in abdominal right or left lower quadrant. Make sure to cut both skin and peritoneum.
4. Pull a section of ovary through the incision with small blunt forceps.
5. Cut off a clump of ovary and put it in a dish with Ca^{2+}-free ND96 for collagenase treatment.
6. Suture the incision. Make sure to suture both muscle layers and skin.
7. Let the frog recover for 2 hr in a shallow bath and then return it to the "recovery tank." Remove suture after approximately 1 week. The frog should not be operated on again for at least 1 month. Perform next operation on this frog on the opposite side of the abdomen.

Defolliculation

1. Cut a clump of ovarian tissue into clusters of 15–20 oocytes each using fine forceps and small scissors.
2. Wash twice in Ca^{2+}-free ND.
3. Transfer oocytes to 15-ml conical tubes with screw cap with 5 ml freshly prepared collagenase solution (see Solutions).
4. Incubate oocytes in collagenase at room temperature on a rotating platform for 1.5 to 3 hr until oocytes are separated and defolliculated. (Defolliculated oocytes will have a glow at the perimeter with incidential illumination.) Change the collagenase solution after about 45–60 min. Time required for complete defolliculation may vary between different batches of collagenase and from different oocyte donors.
5. Wash the oocytes three times with 5 ml ND96 and transfer them into a petri dish using a fire-polished Pasteur pipette.
6. Select the oocytes carefully. Take only the largest (stage 5 and 6) and undamaged oocytes.
7. Let oocytes recover overnight before injection. They can be injected after 1 hr or so, but a small percentage of oocytes always die overnight, so increase the number to compensate for possible loss (see discussion on injection and recording).

Hybridization

1. An experiment should contain three sets of RNA mixture to inject: (i) RNA containing both target mRNA and internal control mRNA; (ii) RNA mixture plus internal control AS oligo; and (iii) RNA mixture plus test AS oligo. RNA, AS oligos, and hybridization buffer are three times the final concentration prior to mixing. Mix on ice 2 μl of 3× hybridization buffer (600 m*M* NaCl, 15 m*M* Tris·HCl, pH 7.4), 2 μl 3× RNA (~3–30 mg/ml), and 2 μl 3× AS oligo (~3–300 μg/ml) or H_2O in an RNase-free 500 μl Eppendorf tube.
2. Denature RNA mixture at 65°C for 2 min. Incubate the mixture at 37°C for 10 min for hybridization (I have found that this is not necessary, but some investigators incubate the mixture for as long as 3 hr).
3. Cool to room temperature then put on ice.
4. Proceed to Injection.

Injection

Materials

Microinjector or digital microdispenser (10 μl Drummond microinjector, Drummond Scientific Co., Broomall, PA)

10-μl pipette capillaries (VWR, Cat. No. 53508-400; baked at 250°C for 2 hr) pulled to a relatively long tip (Narishige Puller, Model PP-83, Narishige Scientific Instrument Lab., Tokyo, Japan) and tip broken down to 20 μm under microscope against a fire-cracker glass slide. Select the needles with tip broken with slanted end similar to a syringe needle for best results (see below).
Injection chamber: Petri dish with mesh (polypropylene mesh monofilament mesh, opening 500 μm (Small Parts, Inc., Miami, FL. Cat. No. CMP-500) attached to the bottom by wax and sterilized by EtOH or under UV.
RNA/oligonucleotide mix
Mineral oil

1. Backfill the top of the capillary with 1 cm or less mineral oil.
2. Pull the plunger all the way back and slide the capillary over the plunger until it rests against the stop and screw down the collet. The oil drop should be pushed to the tip leaving room for 2–3 μl of RNA mixture. If no space is left, the pipette is too short.
3. Turn the plunger forward (clockwise) until a drop of mineral oil comes out of the tip of the capillary.
4. Mount the injector on the micromanipulator.
5. On a piece of Parafilm, place 1–3 μl of RNA mixture on the stage of a dissection microscope.
6. Pull up RNA mixture into the capillary. Calculate the RNA mixture volume in the capillary and the number of oocytes that can be injected. Approximately one dozen oocytes can be injected with 1 μl of the reaction mixture. Do not inject more than you are able to assay in a given day (not more than 72 oocytes).
7. Fill the injection chamber with ND96 and spread 10–15 oocytes on the mesh with a fire-polished round tip capillary. Oocytes will rest securely on the mesh. All oocytes should be facing the same direction so that they can all be injected in the same location. We usually inject at the animal hemisphere, about one-third the distance between the perimeter and the pole to avoid penetration into the nucleus.
8. Impale the oocyte to depth of 200–300 μm and inject 50 nl of the RNA mixture.
9. Pull the pipette out slowly and move to the next oocyte. After completing the injection of a group of occytes, make sure that the pipette is not clogged by pushing out one drop of RNA mix in open air before impaling the next group of oocytes.
10. Set the oocyte to recover for 30–60 min. Discard the damaged oocytes. Transfer the rest into a petri dish in ND96E incubation medium.
11. Incubate for 2–3 days at 20°C.

Electrophysiological Recording

A standard two-microelectrode voltage-clamp technique is used to record oocyte membrane current with high voltage-clamp amplifier (such as Dagan 8500 amplifier, Dagan Corp. Minneapolis, MN). The membrane current and voltage are monitored by a chart recorder and by a dual-beam storage oscilloscope. More detailed descriptions of the electrophysiological recording system can be found elsewhere (62, 63). Microelectrodes are made with capillaries without filament, 1.6 mm in diameter (Jencons Glass, Jencons Scientific, England). It is pulled at low heat and weak pull to produce a tip with very low resistance (0.5 to 2 megohm). Oocyte is recorded at room temperatrue in a 1-ml perfusing bath (the maximum rate of perfusion is approximately 10 ml/min). The bathing solution is grounded with a 3 *M* KCl agar bridge. The response is recorded on a chart recorder.

1. Fill the electrodes with 3 *M* KCl. Test electrode impedence. A resistance of 0.3–1 megohm is recommended.
2. Impale oocyte with both current and voltage electrodes. A healthy oocyte has a resting membrane potential of −40 mV or higher. Oocytes with lower membrane potential may still respond well.
3. Voltage-clamp at −70 mV. The holding current clamp should not exceed 200 nA (R_m > 0.3 magaohm). If the oocyte is unable to hold the clamp, discard.
4. First,check the response for oocytes injected with RNA only. Administer agonist (30-sec perfusion) for the internal control receptor and determine the response amplitude. The amplitude of this response is set as 100% for that receptor. Discard oocyte.
5. Before recording the next oocyte, make sure that there is no residual drug in the bathing solution by washing carefully with perfusing ND96 buffer.
6. Take the next oocyte and voltage-clamp as before. Repeat with agonist for the test receptor (target mRNA). This is 100% response for the target receptor.
7. Test oocytes for hybrid arrest:

A. For oocytes co-injected with the internal control oligo: Administer appropriate agonist by perfusion for 30 sec. If hybrid arrest occurred, you will observe a decreased response. Lack of response 2 min following agonist administration suggests complete hybrid arrest. To make certain that the arrest is specific, administer the agonist for the target receptor to the same oocyte (this response should be the same as the response observed in step 6).

B. For oocytes injected with the test oligo: Repeat above using an agonist to activate the target receptor. If there is hybrid arrest, the amplitude of the response should be reduced. Response to the control agonist should be unaffected (same as the response observed in step 4).

Potential Problems

1. Toxic RNA: Oocyte death may occur if RNA is contaminated with phenol. When extracting RNA with phenol, make sure to follow it up with at least two extractions with chloroform/IAA. Another possible cause of oocyte death is improper pH of the RNA mixture. Make sure that the pH of the mixture is near pH 7.4 prior to injection.

2. RNA degradation and/or extremely low-abundance target mRNA: These problems can lead to oocytes that do not express the protein of interest. When handling RNA, make sure to wear gloves and use RNase-free supplies and solutions.

3. Viscous RNA: Viscous or sticky RNA will cause the syringe to clog up during injection. Sticky RNA is caused by DNA contamination. This is usually caused by not shearing the DNA properly during cell lysis. After homogenizing cells or tissue in lysis buffer, if the solution is still viscous, it should be shaken vigorously before being left overnight for precipitation.

4. Injection problems: The best injection result is obtained with a tip broken to 20 μm in diameter with a slanted end similar to a syringe needle. If the diameter is too large, it will damage the oocyte by leaving a large hole on the surface and cause the oocyte yolk to spill out. If the diameter is too small, the syringe will clog up easily and will not eject RNA. Always make sure that the injector tip is not clogged by pushing out 10–20 nl of RNA into air before penetrating the next oocyte. Also, an injection that is too shallow can cause RNA leakage backward from the injection hole after the withdrawal of the pipette. On the other hand, going in too deep into oocyte with a syringe will damage the oocyte. The best way to get started is to inject some concentrated dye into the oocyte. You will be able to monitor the drop location and leakage and train yourself easily.

Acknowledgments

We gratefully acknowledge Dr. Stuart Sealfon for his input and suggestions throughout the course of this work. We thank Drs. Andrea Gore, Lenore Synder, and Niva Almaula for critical reading of the manuscript. M. Tsutsumi is an Aaron Diamond Foundation Fellow and this work was supported in part by the Aaron Diamond Foundation.

References

1. T. P. Snutch, and G. Mandel, *In* "Methods in Enzymology" (B. Rudy and L. Iverson, eds.), Vol. 207, p. 297. Academic Press, San Diego, CA, 1992.
2. T. P. Snutch, J. P. Leonard, M. M. Gilbert, H. A. Lester, and N. Davidson, *Proc. Natl. Acad. Sci. U.S.A.* **87,** 3391 (1990).
3. J. B. Gurdon, C. D. Lane, H. R. Woodland, and G. Marbaix, *Nature* (*London*) **233,** 177 (1971).
4. Y. Masu, K. Nakayam, H. Tamaki, Y. Harada, M. Kuno, and S. Nakanishi, *Nature* (*London*) **329,** 836 (1987).
5. R. E. Straub, G. C. Frech, R. H. Joho, and M. C. Gershengorn, *Proc. Natl. Acad. Sci. U.S.A.* **87,** 9514 (1990).
6. D. Julius, A. B. MacDermott, R. Axel, and T. M. Jessell, *Science* 241, 558 (1988).
7. H. Lubbert, B. J. Hoffman, T. P. Snutch, T. van Dyke, A. J. Levine, P. R. Hartig, H. A. Lestr, and N. Davidson, *Proc. Natl. Acad. Sci. U.S.A.* **84,** 4322 (1987).
8. Y. Oron, R. E. Straub, P. Traktman, and M. C. Gershengorn, *Science* **238,** 1406 (1987).
9. B. K. Kobilka, C. MacGregor, K. Daniel, T. S. Kobilka, M. G. Caron, and R. J. Lefkowitz, *J. Biol. Chem.* **262,** 15796 (1987).
10. M. Tsutsumi, S. C. Laws, and S. C. Sealfon, *Mol. Endocrinol.* **7,** 1625 (1993).
11. M. Mishina, T. Kurosaki, T. Tobimatsu, Y. Morimoto, M. Noda, T. Yamamoto, M. Terao, J. Lindstrom, T. Takahashi, M. Kuno, and S. Numa, *Nature* (*London*) **307,** 604 (1984).
12. M. Tsutsumi, W. Zhou, R. P. Millar, P. L. Mellon, J. L. Roberts, C. A. Flanagan, K. Dong, B. Gillo, and S. C. Sealfon, *Mol. Endocrinol.* **6,** 1163 (1992).
13. P. R. Schofield, M. G. Darlinson, N. Fujita, D. R. Burt, F. A. Stephenson, H. Rodriguez, L. M. Rhee, J. Ramachandran, V. Reale, T. A. Lencorse, P. H. Seeburg, and E. A. Barnard, *Nature* (*London*) **328,** 221 (1987).
14. L. C. Timpe, T. L. Schwarz, B. L. Tempel, D. M. Papazian, Y. N. Jan, and L. Y. Jan, *Nature* (*London*) **331,** 143 (1988).
15. G. C. Frech, and R. H. Joho *in* "Methods in Enzymology" (B. Rudy and L. Iverson, eds), Vol. 207, p. 592. Academic Press, San Diego, CA. 1992.
16. M. Noda, T. Ikeda, H. Suzuki, H. Takeshima, T. Takahashi, M. Kuno, and S. Numa, *Nature* (*London*) **322,** 826 (1986).
17. M. A. Hediger, M. J. Coady, T. S. Ikeda, and E. M. Wright, *Nature* (*London*) **330,** 379 (1987).
18. E. S. Kawasaki, *Nucleic Acids Res.* **13,** 4991 (1985).
19. M. S. Urnes, and D. Carroll, *Gene* **95,** 267 (1990).
20. J. Sharifi, and D. L. St. Germain *J. Biol. Chem.* **267,** 12539 (1992).
21. B. M. Paterson, B. E. Roberts, and E. L. Kuff, *Proc. Natl. Acad. Sci. U.S.A.* **74,** 4370 (1977).
22. P. C. Zamecnik, and M. L. Stephenson, *Proc. Natl. Acad. Sci. U.S.A.* **75,** 280 (1978).
23. A. M. Wang, A. A. Creasey, M. B. Ladner, L. S. Lin, J. Strickler, J. N. Van Arsdell, R. Yamamoto, and D. F. Mark *Science* **228,** 149 (1985).

24. G. Grenningloh, A. Rienitz, B. Schmitt, C. Methfessel, M. Zensen, K. Beyreuther, E. D. Gundelfinger, and H. Betz, *Nature* (*London*) **328,** 215 (1987).
25. D. L. St. Germain, W. Dittrich, C. M. Morganelli, and V. Cryns, *J. Biol. Chem.* **265,** 20087 (1990).
26. K. Folander, J. Smith, J. Antanavage, C. Bennett, R. B. Stein, and R. Swanson, *Proc. Natl. Acad. Sci. U.S.A.* **87,** 2975 (1990).
27. I. Lotan, P. Goelet, A. Gigi, and N. Dascal, *Science* **243,** 666 (1989).
28. W. Meyerhof, and D. Richter, *FEBS Lett.* **266,** 192 (1990).
29. I. Lotan *in* "Methods in Enzymology" (B. Rudy and L. Iverson, eds.), Vol. 207, p. 605. Academic Press, San Diego, CA. 1992.
30. D. F. Slish, D. B. Engle, G. Varadi, I. Lotan, D. Singer, N. Dascal, and A. Schwartz, *FEBS Lett.* **250,** 509 (1989).
31. A. Davidson, G. Mengod, N. Matus-Leibovitch, and Y. Oron *FEBS Lett.* **284,** 252 (1991).
32. D. W. DeSimone, P. A. Norton, and H. R. O, *Dev. Biol.* **149,** 357 (1992).
33. F. Fournier, P. Charnet, E. Bourinet, C. Vilbert, F. Matifat, G. Charpentier, P. Navarre, G. Brule, and D. Marlot, *FEBS Lett.* **317,** 118 (1993).
34. I. Lotan, A. Volterra, P. Dash, S. A. Siegelbaum, and P. Goelet, *Neuron* **1,** 963 (1988).
35. S. C. Sealfon, S. C. Laws, J. C. Wu, B. Gillo, and W. L. Miller, *Mol. Endocrinol.* **4,** 1980 (1990).
36. J. Minshull, and T. Hunt, *Nucleic Acids Res.* **14,** 6433 (1986).
37. C. Cazenave, N. Loreau, N. T. Thuong, J. J. Toulme, and C. Helene, *Nucleic Acids Res.* **15,** 4717 (1987).
38. J. Shuttleworth, and A. Colman, *EMBO J.* **7,** 427 (1988).
39. P. Dash, I. Lotan, M. Knapp, E. Kandel, and P. Goelet, *Proc. Natl. Acad. Sci. U.S.A.* **84,** 7896 (1987).
40. R. C. Smith, M. B. Dworkin, and E. Dworkin-Rastl, *Genes Dev.* **2,** 1296 (1988).
41. C. Jessus, C. Cazenave, R. Ozon, and C. Helene, *Nucleic Acids Res.* **16,** 2225 (1988).
42. M. R. Rebagliati and D. A. Melton, *Cell* (*Cambridge, Mass.*) **48,** 599 (1987).
43. K. Sumikawa, and R. Miledi, *Proc. Natl. Acad. Sci. U.S.A.* **85,** 1302 (1988).
44. G. Dahl, T. Miller, D. Paul, R. Voellmy, and R. Werner, *Science* **236,** 1290 (1987).
45. P. J. Green, O. Pines, and M. Inouye, *Annu. Rev. Biochem.* **55,** 569 (1986).
46. D. A. Melton, *Proc. Natl. Acad. Sci. U.S.A.* **82,** 144 (1985).
47. R. Harland, and H. Weintraub, *J. Cell Biol.* **101,** 1094 (1985).
48. B. Bass, and H. Weintraub, *Cell* (*Cambridge, Mass.*) **48,** 607 (1987).
49. T. M. Woolf, C. G. B. Jennings, M. Rebagliati, and D. A. Melton, *Nucleic Acids Res.* **18,** 1763 (1990).
50. S. C. Sealfon, B. Gillo, S. Mundamattom, P. L. Mellon, J. J. Windle, E. Landau, and J. L. Roberts, *Mol. Endocrinol.* **4,** 119 (1990).
51. N. Dascal, B. Gillo, and Y. Lass, *J. Physiol.* (*London*) **366,** 299 (1985).
52. B. Gillo, E. M. Landau, T. M. Moriarty, J. L. Roberts, and S. C. Sealfon, *J. Physiol.* (*London*) **417,** 47 (1989).
53. J. J. Windle, R. I. Weiner, and P. L. Mellon, *Mol. Endocrinol.* **4,** 597 (1990).

54. T. M. Woolf, D. A. Melton, C. G. Jennings, *Proc. Natl. Acad. Sci. U.S.A.* **89,** 7305 (1992).
55. N. Dascal, I. Lotan, E. Karni, and A. Gigi, *J. Physiol.* (*London*) **450,** 469 (1992).
56. J.-J. Tolume, and C. Helene, *Gene* **72,** 51 (1988).
57. C. A. Stein, and J. S. Cohen, *Cancer Res.* **48,** 2659 (1988).
58. C. Cazenave, C. A. Stein, N. Loreau, N. T. Thuong, L. M. Neckers, C. Subasinghe, C. Helene, J. S. Cohen, J.-J. Toulme, *Nucleic Acids Res.* **17,** 4255 (1989).
59. C. Boiziau, R. Kurfurst, C. Cazenave, V. Roig, N. T. Thuong, and J.-J. Toulme *Nucleic Acids Res.* **19,** 1113 (1991).
60. J. Sambrook, E. F. Fritsch, and T. Maniatis, "Molecular Cloning: A Laboratory Manual," 2nd ed. Cold Spring Harbor Lab. Press, Cold Spring Harbor, NY, 1989.
61. P. Dierks, A. Van Ooyen, N. Mantei, and C. Weissman, *Proc. Natl. Acad. Sci. U.S.A.* **78,** 1411 (1981).
62. B. Gillo, Y. Lass, E. Nadler, and Y. Oron *J. Physiol.* (*London*) **342,** 349 (1987).
63. N. Dascal, *CRC Crit. Rev. Biochem.* **22,** 317 (1987).

[7] Receptor Cloning: High-Throughput Sequencing of cDNA Tags for Identification of Novel Genes

Ewen F. Kirkness and J. Craig Venter

Introduction

Historically, the cloning of receptor genes has often benefited from an extensive prior knowledge of the pharmacological and physiological properties of the receptor proteins. These properties have been utilized indirectly, for receptor purification, protein sequencing, and oligonucleotide probe design (see Chapter 2, 3), or directly, for various forms of expression cloning (e.g., see Chapter 4). These approaches are largely responsible for our present classification of receptor gene families, and they remain the most reliable means for identifying new structural families. However, the methods employed by these approaches are relatively time consuming and expensive and must therefore be confined to only a few examples of interest in any one laboratory.

During the past decade, the cloning of numerous receptor genes has revealed two general principles. First, for almost all families of receptors (e.g., adrenergic, $GABA_A$), the diversity of receptor subtypes has been found to far exceed that predicted by previous pharmacological studies. Second, many families of receptors that appeared to be pharmacologically and functionally distinct (e.g., muscarinic receptors and rhodopsins), were found to be related by common structural features. These findings have been exploited to clone additional members of receptor gene families by using conserved motifs to design suitable DNA probes. Cloning strategies that are based on the cross-hybridization of gene sequences have provided a fast and direct route to novel receptor genes. This approach has been particularly useful for the identification of new receptor subtypes, although it is generally unable to detect more distantly related homologs of known receptor genes.

The majority of cloned receptor genes, now numbering several hundred, can be classified within a few superfamilies that each exhibit characteristic structural motifs (e.g., G protein-coupled, tyrosine kinases, and ligand-gated ion channels). These features of diversity and conservation make the identification of novel receptor genes a highly productive area of research within large-scale cDNA sequencing projects. Unlike the cross-hybridization ap-

Methods in Neurosciences, Volume 25

proaches described above, these projects can uncover new receptor genes that share only a minimal homology with known receptor gene families. In addition, the comparison of randomly sampled cDNA populations from distinct tissues or cell types can provide a measure of gene expression patterns for both known and newly identified genes.

In this chapter we describe the approaches used at The Institute for Genomic Research (TIGR) for high-throughput sequencing and analysis of cDNA tags. This approach involves extensive sampling of clones from cDNA libraries. Each clone is subjected to a single sequencing reaction that yields a short segment of the cDNA sequence (300–500 bp), termed an expressed sequence tag (EST). After computer analyses of the EST data, each clone is classified in terms of its putative identification, tissue source, abundance, etc. The principal aim of this EST project is to obtain partial sequence information from the majority of expressed human genes. Pilot projects have already demonstrated the utility of this approach for the identification of novel genes that are expressed in the human brain (1, 2). At TIGR, this analysis has been expanded to include a wide range of human tissues. A daily throughput of approximately 1000 ESTs has permitted the identification of many more new genes. As would be expected, a significant proportion of these represent new members of the receptor gene superfamilies.

This chapter, describing methodology used in large-scale EST projects, is divided into three sections, each containing one or more subsections (Fig. 1).

Construction of cDNA Libraries

For cDNA libraries used in EST projects, the methods of library construction are of crucial importance for obtaining the maximum amount of useful sequence data from each sequencing reaction. Some aspects of the construction are obligatory, owing to special requirements of the automated sequencing process. Others are choices that reflect the present priorities of each project. It would be expected that, as EST projects evolve, the methods of library construction will also change to fulfill more specialized roles (e.g., various types of subtracted libraries). Before describing the methodologies used for library construction, features that have been incorporated into current library designs are discussed briefly.

With the aim of obtaining ESTs from the majority of human genes, we, and collaborating laboratories, have constructed over 100 libraries from a variety of human tissues, tumor samples, cultured primary cells, and immortalized cell lines. In addition, multiple libraries have been prepared from specific cell lines to monitor changes in gene expression patterns during processes such as differentiation, aging, and cell-cycle progression. We use

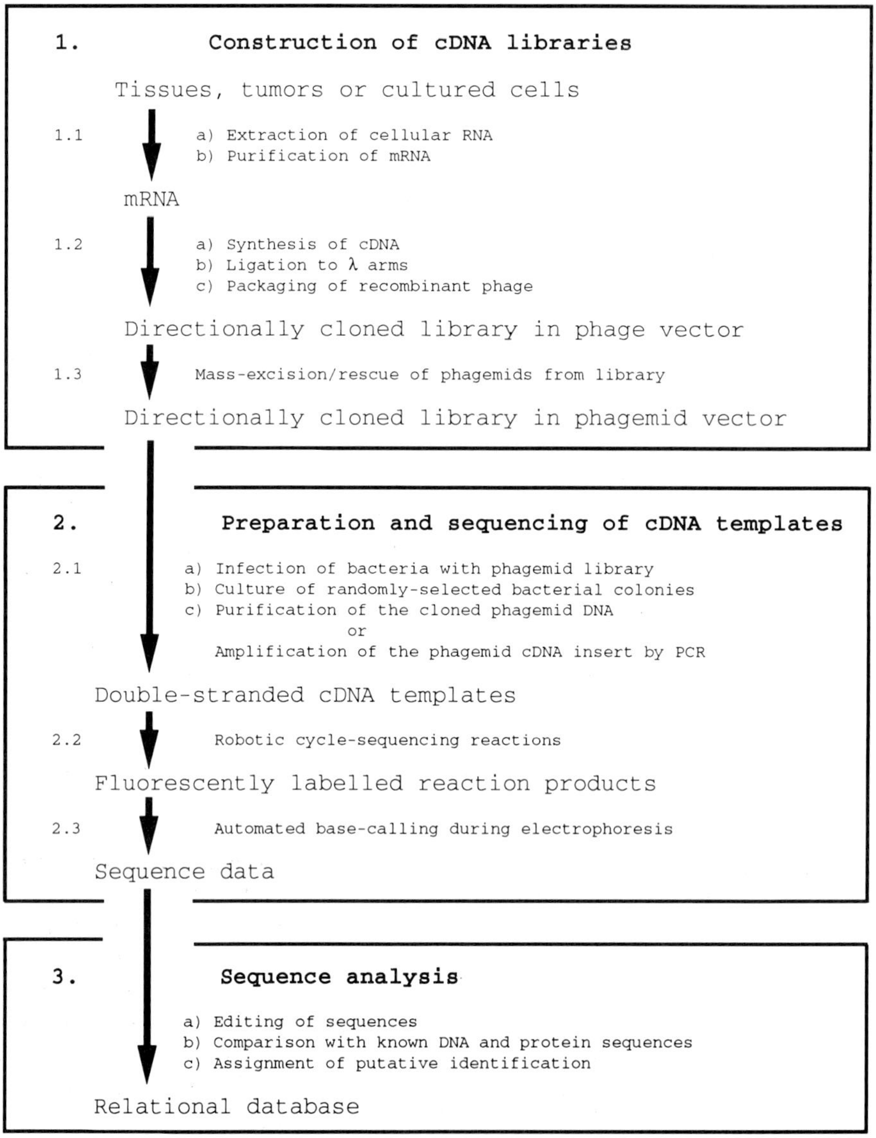

FIG. 1 Flow diagram for a typical EST project. Messenger RNA is purified from a selected tissue or cell line and reverse-transcribed to cDNA. The cDNAs are cloned directionally into a λ phage vector and converted to a phagemid derivative. The cDNA inserts of individual, randomly selected clones are purified and used as templates for DNA sequencing. Sequencing reactions and the generation of sequence data are largely automated processes.

only the poly(A)$^+$ fraction of RNA as template for cDNA synthesis. Although cytoplasmic mRNA transcripts of nuclear genes have been detected in a non-polyadenylated form, their physiological significance is questionable and there is presently no convenient method for purifying these species from the bulk of nonpolyadenylated RNA (rRNA, tRNA, etc.).

Ultimately, the cloned cDNA is sequenced on automated ABI 370A sequencers (Applied Biosystems, Inc., Foster City, CA) using fluorescently labeled primers. Currently, four different primers are available for this approach (M13-forward, M13-reverse, SP6- and T7-promoter sequences). It is therefore necessary to clone the cDNA within a vector that contains at least one of these sequences in close proximity to the cloning site. We have found the phagemid pBluescript (Stratagene, La Jolla, CA) to be a suitable vector, permitting reliable sequencing of inserted cDNA from either end, using the labeled M13 primers.

For synthesis and cloning of cDNAs we have generally used oligo(dT)-primers and a directional cloning strategy. The resulting libraries have several advantages over random hexamer- or oligo(dT)-primed nondirectional libraries. Sequencing the 5′ end of each cDNA clone provides the greatest likelihood of obtaining protein-coding sequence and, hence, the best chance of assigning a putative identification to an unknown cDNA. However, owing to premature terminations during cDNA synthesis, multiple cDNA copies of a single mRNA species may differ in sequence at their 5′ ends. Sequencing from the 3′ end of the clones can normally reveal their common identity and can therefore provide a measure of the redundancy of the library. For most libraries, redundancy has not yet become a significant problem, and cDNA clones are routinely sequenced only from the 5′ ends.

The cDNAs are cloned into a λ phage vector before conversion to phagemid form. Although it would appear more efficient to clone the newly synthesized cDNA directly into plasmid or phagemid vector, it has been our experience that such an approach leads to an overrepresentation of short cDNAs in the library. Ideally, we would like the libraries to be representative, containing all sequences of the original mRNA population in the same relative frequencies. The use of λ phage ensures that a suitable proportion of long cDNA inserts are represented in the library.

Preparation of mRNA

Samples of human tissues, snap-frozen in liquid nitrogen 4–8 hr after death, have been generously provided by the National Disease Research Interchange (Philadelphia, PA). Primary cultures of human cells have been obtained from Clonetics Corporation (San Diego, CA) and immortalized cell lines from American Type Culture Collection (Rockville, MD). At least

1 mg of total cellular RNA is routinely prepared from tissues (1–5 g) or cultured cells (1×10^8) by the method of Chomczyniski and Sacchi (3). This protocol utilizes guanidinium salts which effectively inhibit the activity of endogenous ribonucleases. However, they also destroy subcellular organelles and so preclude the isolation of subcellular RNA fractions. Occasionally we have also employed methods that permit the fractionation of nuclear, cytoplasmic, or polysomal RNAs (4–6). Following extraction, the RNA is quantified by measurement of optical density (260 nm) and its integrity checked by electrophoresis in a standard denaturing (e.g., formaldehyde) agarose gel (7, 8).

Methods for the purification of poly(A)$^+$ mRNA rely on the base-pairing between oligo(dT) sequences that are coupled to a solid support and the poly(A)$^+$ tail of the mRNAs. We have obtained consistently good results using either oligo(dT)-linked latex particles (9) or oligo(dT)-coated magnetic beads (10). These can be obtained commerically in kit form, as Oligotex (Qiagen, Chatsworth, CA) and Dynabeads (Dynal, Lake Success, NY), respectively. The purification process generally yields 5–15 μg of poly(A)$^+$ RNA per 500 μg total cellular RNA. Ideally, 5μg of poly(A)$^+$ RNA would be used for library construction, although it is feasible to use smaller amounts if necessary.

Construction of Directionally Cloned Libraries in λ Phage

In view of the requirements discussed above, the λ ZAP vectors (Stratagene) have proved to be particularly suitable for construction of cDNA libraries used in EST projects. Insert cDNAs, up to 10 kb in length, can be cloned directionally, and with high efficiency, within the phage vector. The phage clones can then be converted to phagemid form, thereby permitting convenient manipulation of the cDNA insert. Considering the expense and effort required for each library construction, it is essential that all components of the synthesis are well-optimized. For this reason, most investigators now resort to commercially available kits. We routinely use the ZAP-cDNA synthesis kit, containing the Uni-ZAP XR vector, and Gigapack II Gold packaging extract (Stratagene). Radiolabeled [α-^{32}P]dATP (800 Ci per mmol; Dupont, Wilmington, DE) is also required to monitor the cDNA synthesis.

In this system, first-strand cDNA synthesis is primed with a hybrid oligonucleotide, comprised of poly(dT) coupled to a 32-base linker sequence. The poly(dT) binds to the poly(A) tail of the mRNA, while the linker provides a *Xho*I restriction site that permits directional cloning at a later stage. The mRNA is transcribed to cDNA using Moloney murine leukemia virus reverse transcriptase, with a mixture of dATP, dGTP, dTTP, and 5-methyl dCTP. The use of 5-methyl dCTP instead of dCTP results in the synthesis of a hemimethylated cDNA that is protected from digestion by *Xho*I. In consequence, only the unmethylated restriction site within the terminal linker sequence can be cleaved by *Xho*I. The quality and quantity of first-strand synthesis can be assessed by performing a side reaction (10% of the main reaction) in the presence of ^{32}P-labeled dATP. The labeled reaction products are analyzed alongside a sample of the second-strand reaction (see below).

Before synthesis of second-strand cDNA, the bound RNA is nicked with RNase H. The resulting RNA fragments serve as primers for DNA polymerase I, which catalyzes the second-strand synthesis. In this reaction, dCTP replaces 5-methyl dCTP so as to avoid hemi-methylation of the linker *Xho*I site. The second-strand synthesis is performed in the presence of ^{32}P-labeled dATP. At this point it is advisable to examine a sample of the reaction products, together with the labeled first-strand reaction, by alkaline agarose gel electrophoresis (8, 11). These denaturing gels demonstrate the size range of the first- and second-strand cDNA and can reveal major hairpinning that may have occurred during the second-strand reaction. Generally, the size range of cDNAs should extend from <1 to >10 kb, with greatest abundance in the range of 1.5–3 kb. Poor yields, or an overabundance of short products (<1 kb) is suggestive of insufficient or degraded mRNA, contamination with rRNA or DNA, or the presence of reaction inhibitors (e.g., SDS) in the mRNA preparation.

If the gel analysis indicates an adequate yield and size range, the ends of the cDNA are blunted with Klenow and ligated to *Eco*RI adapters. These adapters are short, double-stranded oligonucleotides that have one blunt end and one *Eco*RI-cohesive end. Only the blunt end of the adapter is phosphorylated. This ensures that only the blunt end can participate in a ligation reaction, thereby avoiding the concatamerization of multiple adapters and cDNAs. Following the ligation, the cDNA possesses *Eco*RI-cohesive ends which must be phosphorylated in preparation for their ligation with the λ vector arms. Digestion of the constructs with *Xho*I releases the *Eco*RI adapter and residual linker from the 3′ end of the cDNA. This small fragment is separated from the cDNA constructs during size fractionation on a gel filtration column. The cDNA now has *Eco*RI and *Xho*I cohesive ends which can be ligated directionally into the corresponding ends of the λ phage arms. For size fractionation, we routinely use gravity-flow Sephacryl S-500 columns (BRL, Gaithersburg, MD) rather than Sephacryl S-400 spin columns. Gravity-flow columns permit the collection of a greater number of size fractions. These fractions can be ligated to the λ arms in independent reactions, thereby ensuring an adequate representation of longer cDNAs in the final, combined library. The size range of cDNAs in each fraction can be determined by electrophoresis of ~10% of the sample on a nondenaturing acrylamide gel. Fractions that contain cDNAs of less than 500 bp are normally discarded. The efficient ligation of cDNAs to λ arms is highly dependent on the relative molar ratios. It is therefore advisable to accurately quantitate the DNA content of the relevant cDNA fractions, e.g., by fluorescence of ethidium bromide in an agarose plate assay (8).

Following the ligation of cDNA and λ arms, the reaction products are packaged and plated on the *Escherichia coli* strain, XL-1Blue MRF′. This strain is $mcrA^-$, $mcrB^-$ and therefore fails to restrict methylated DNA. After passing the library through these cells, the cDNAs are no longer hemimethylated and can be grown on the more common $mcrA^+$, $mcrB^+$ strains. When titering, the recombination frequency can be checked by color selection, or by comparing plating efficiency of a library with that of λ arms that have been ligated in the absence of cDNA. For color selection, dilutions of the packaged reaction are plated in the presence of IPTG and X-Gal (8). Recombinant plaques are clear and should be at least 50-fold more abundant than the background, blue plaques. The primary library can be unstable. It is therefore advisable to prepare a stable, high-titer stock for long-term storage by amplifying a portion of the primary preparation. However, amplification can alter the composition of the library by enhancing the representation of the faster growing clones. For this reason, we generally use the primary phage library for the mass excision of recombinant phagemids that will be used for EST analysis.

Mass Excision of Phagemids from λ Phage Library

The Uni-ZAP XR vector was designed to permit *in vivo* excision and recircularization to form a phagemid containing the cloned insert. When the λ phage and a helper phage co-infect a susceptible strain of *E. coli,* the region of λ containing the phagemid (plus insert) is copied as a single-stranded DNA molecule. This DNA is circularized, packaged as a filamentous phage particle, and secreted from the bacteria. Once harvested, these particles can be used to infect a fresh strain of *E. coli* in which the phagemid now replicates as a regular double-stranded plasmid.

For mass excision of libraries, we use a modification of the protocol that was designed for excision of individual λ clones. Cultures of XL1-Blue MRF′ cells are harvested while growing in log phase and suspended in 10 m*M* $MgSO_4$ at ~10^8 cells per ml. The bacterial cells (1 ml) and primary packaged library (~10^6 pfu) are incubated in 50-ml conical tubes (37°C, 15 min). ExAssist helper phage (~10^{10} pfu; Stratagene) are added and allowed to absorb (37°C, 15 min). After addition of Luria broth (LB) medium (10 ml), the mixture is agitated slowly (<200 rpm, 37°C, 2 hr). The *E. coli* are then killed by heating (70°C, 20 min) and removed by centrifugation (2500*g*, 10 min). The supernatant containing single-stranded phage particles, is relatively stable at 4°C for several weeks.

Preparation and Sequencing of cDNA Templates

The preparation and sequencing of templates lie at the heart of any large-scale sequencing project and have formally been the rate-limiting step in the overall process. However, many of the procedures involved are simple and repetitive and therefore lend themselves to some degree of automation. Automated DNA sequencers, which can detect fluorescently labeled DNA during electrophoretic separation, have been available commercially since 1987. Workstations that can automate sample preparation and sequencing reactions have been introduced more recently. This technology has increased both the amount and consistency of sequence data that can be generated by individual laboratories. In consequence, the rate-limiting step of many sequencing projects has now shifted toward analysis of the acquired data.

Template Preparation

The cloning and culture of recombinant phagemids is a relatively standard procedure (see below). However, there are several approaches to the subsequent step of template purification. Use of entire phagemid DNA as template permits good sequence data to be obtained consistently from either end of the cDNA insert. In addition, the yield of template DNA is relatively independent of the length of the cDNA insert. A major disadvantage of this approach is the difficulty of adapting suitable protocols to a 96-well format that would increase the throughput and consistency of template production. The most suitable methods for large-scale phagemid purification are also relatively expensive (~\$3 per preparation). In contrast, the preparation of cDNA templates by PCR is readily adaptable to a 96-well format and requires less-costly reagents (~\$0.50/per preparation). However, templates that have been generated by PCR do not provide reliable sequence data at their 3′ ends (perhaps because of the addition or deletion of bases within the poly(A-T) tail of the cDNA during the amplification process). A second disadvantage of this approach is a reduced representation of very long cDNA inserts that are not amplified efficiently by PCR. We have used both purified phagemid and PCR-amplified templates to generate EST sequence data. The choice of method is clearly dependent on the requirements of each particular project.

Cloning and Culture of Phagemids

After mass excision of the primary phage library, the packaged phagemid particles are used to infect a fresh strain of *E. coli*. This strain permits replication of the phagemid as double-stranded DNA, but does not support infection or replication of contaminating λ phage and ExAssist helper phage. Colonies of infected bacteria can be selected by their acquired resistance to ampicillin. Typically, we use SOLR cels (Stratagene) that are harvested while growing in log phase and suspended in 10 m*M* $MgSO_4$ at $\sim 10^8$ cells per ml. The supernatants of harvested phagemid particles (1 μl) and SOLR cells (200 μl) are combined and incubated (37°C, 15 min), before plating on LB/ampicillin. The following step of picking isolated colonies for liquid culture generally requires a colony density of less than 200 per 10-cm plate. It is therefore often necessary to optimize the infection/plating process prior to a large-scale preparation. Typically, 200 colonies are sampled from a new library. If the library is subsequently judged to contain an adequate diversity of cDNA inserts, the sampling is increased to 2000–5000 colonies. The colonies are cultured overnight in Terrific broth/ampicillin (50 μg per ml), using 96-well, square-sided plates (1.3 ml per well) or regular 15-ml culture tubes (3 ml per tube). Samples of the overnight culture are frozen in glycerol for long-term storage.

Template Preparation

For preparation of phagemid DNA, we have used the Qiawell8 Ultimate system (Qiagen), essentially in accordance with the manufacturer's directions. The system utilizes a series of filter strips to bind and purify the phagemid DNA. Eight preparations are purified simultaneously by using a custom vacuum manifold. The preparation regularly yields 3–10 μg of DNA from 3 ml of overnight culture.

The preparation of cDNA templates by PCR has been developed from several existing protocols, with emphasis on conditions that promote the amplification of long DNA fragments (12). The PCR reactions and purification of reaction products are performed in 96-well plates. For each template, the 50-μl reaction mix contains 30 m*M* Tricine (pH 8.4), 2 m*M* $MgCl_2$, 5 m*M* β-mercaptoethanol, 0.1% Thesit, 0.01% gelatin, 0.2 m*M* dNTPs, 10 ng of primers A and B, 0.5 U *Taq* polymerase, and 1.0 μl overnight bacterial culture. Primers A and B are 21-mers, located 150 bases outside the M13 forward and reverse primer sequences of pBluescript. The reaction mixture is incubated at 94°C for 5 min and then cycled at 94°C (1 min), 55°C (1 min), and 72°C (2 min) for 30 cycles (GenAmp System 9600, Perkin—Elmer). After a final incubation at 72°C for 5 min, the reaction is held at 4°C. The reaction products are transferred to a 96-well polystyrene plate and mixed with an equal volume of PEG-8000 (13%)/sodium acetate (1.6 *M*). The plates are cooled on ice (15 min) before centrifugation (Sorvall, 3000 rpm, 45 min). Pellets are washed three times by addition of 70% ethanol (150 μl) and centrifugation (Beckman GS6R, 3000 rpm, 15 min). Pellets are then air-dried and resuspended in H_2O (35 μl). The yield of template DNA can be assessed by optical density, or by Cytoflor analysis (Millipore, Milford, MA).

Sequencing Reactions

Sequencing reactions for the ABI 373A automated sequencer (see below) involve the synthesis of dye-labeled DNA in a dideoxy-termination reaction (13, 14). When labeled primers are used (as in most EST projects), a single primer carries a different dye in each of the four termination reactions. The primers are extended by *Taq* polymerase in a cycle-sequencing protocol (15) and then combined prior to analysis. As the major requirements of the protocol involve the pipetting of reagents and temperature cycling, it can be performed easily by a robotic workstation. We have had most experience using the CATALYST-800 LabStation (Applied Biosystems). Other robots (e.g., Biomek-1000; Beckman, Fullerton, CA) have also been developed for this purpose.

The CATALYST consists of an enclosed work surface in which automated reagent transfer and thermal cycling are performed. It is controlled by Macintosh software and optimized for DNA sequencing with fluorescence-labeled primers or terminators. The robot requires approximately 4 hr to process 24 sequencing reactions, but can be loaded with 48 templates for consecutive runs overnight. A cocktail is prepared from the Prism cycle sequencing kit (Applied Biosystems), consisting of buffer, deoxy- and dideoxynucleotides, dye-labeled primer, and *Taq* polymerase. After this reaction mix and the DNA templates (0.2 μg per μl) are loaded into a cold storage area, the reactions proceed automatically. On completion, the reacted samples are centrifuged (10,000*g*, 20 min, 4°C) to pellet the DNA. Pellets are washed with 70% ethanol and vacuum-dried (5 min). The reaction products are stored in the dark at −20°C until required for electrophoresis.

Collection of Sequence Data

Automated sequencing technology, as developed by Applied Biosystems, monitors the relative mobility of dye-labeled DNA molecules during electrophoresis through a laser beam (14). The fluorescence patterns that are generated by the activated dyes are recorded digitally. Automatic analysis of this data provides an output of DNA sequence, together with a graphic representation of the fluorescence patterns. In comparison to manual sequencing methods, the process is considerably faster and much less labor-intensive. Currently, we use ABI 373A DNA sequencers, each of which can process 36 sequences simultaneously. The sequencers are operated by Macintosh software, generally as recommended by the manufacturer.

The acrylamide gels used for electrophoresis are prepared essentially as for manual sequencing methods. However, the region of the plates that is scanned by the laser (about 25 cm from the top) must be cleaned thoroughly. The gel is placed in an electrophoresis chamber and the run parameters are selected (run-length, comb-size, etc.). A run length of 12 hr will permit the collection of approximately 600 bases (normally the maximum length of unambiguous sequence data). However, shorter run times (8–9 hr) can provide sufficient data (300–500 bases) for EST analysis and permit the machines to be used for two runs per day. After the comb is inserted and the buffer chambers are filled, the gel is prerun for about 5 min. During this time, the sequencing reaction samples are denatured in formamide/EDTA at 90°C for 2 min. Odd-numbered wells are flushed, loaded with samples, and electrophoresed. After 5 min, even-numbered wells are treated similarly. During the subsequent electrophoresis, laser scanning and data collection occur automatically.

After the run is completed, the quality of the analysis is assessed by viewing sequence chromatograms. Samples that display no signal at the start, or display a sudden drop-off, often require retracking and/or reanalysis. It is critical to ensure that the tracking assigns the correct file name to each sequence. Ambiguous data, which are often generated at the end of long sequencing runs, are deleted manually. When tracking and editing are satisfactory, the sample sheets on the Macintosh hard disc are copied to the EST sequence database.

Computer Analysis of EST Sequence Data

Computer analysis is essential for effective manipulation and interpretation of DNA sequence data. For large-scale sequencing projects, the management and maintenance of the sequence data present additional computing requirements. This is particularly true for EST projects, in which all of the partial cDNA sequences are initially of unknown identity. For some ESTs, comparison with previously characterized sequences can permit the assignment of a putative identification. This process has resulted in the identification of many new members of known gene families (e.g., G protein-coupled receptors).

However, a major proportion of the ESTs do not exhibit homology with the current pool of characterized genes. It is therefore useful to assemble all of the information that can be associated with these anonymous sequences (tissue distribution, relative abundance, homology to other ESTs, etc.). This is achieved most conveniently by using a relational data base, in which selected information from many different tables can be compared simultaneously. An outline of the information flow that we have developed for EST analysis is illustrated in Fig. 2. All data that relate to the acquisition or analysis of the ESTs are stored in Sybase tables.

Data that are generated by the automated sequencers are first edited to remove vector sequence from the start of the EST. They are also examined for sequence quality (extent of unambiguous sequence) to determine if they are suitable for data base searching. It is now possible to design programs that permit these tasks to be performed automatically. Generally, sequences of less than 100 bases, or which contain more than 5% ambiguous sequence, are not used for subsequent database homology searches. Suitable sequences are prescreened before being used to search public sequence databases. The prescreen aims to detect ESTs that exhibit exact alignment with a selection of DNA sequences that have been assembled in custom data bases. The BLAST algorithm (16) permits fast comparative searches, particularly suitable for detecting exact alignments. Prescreening identifies many of the ESTs that contain repetitive elements. The presence of these elements in many characterized genes can obscure any weak, but more meaningful, homologies with other regions of the EST. The identification and subsequent masking of these elements permit the remaining EST sequence to be used for more productive homology searches. Prescreening also permits the immediate identification of ESTs that contain λ, *E. coli,* or rRNA sequences. An abundance of these "trash" sequences signifies some problem with the construction of the source library. The final prescreen is a search against a nonredundant collection of all known human cDNA sequences. The ESTs that are identified at this stage can be cataloged automatically, leaving only novel cDNA sequences to be evaluated for their homology to other known genes.

Prescreened ESTs are first searched against the GenBank database of nucleotide sequences using the BLAST algorithm. As the BLAST alignment does not permit gaps, and gene families are rarely well-conserved at the DNA level, only very closely related sequences are identified by this approach. For more productive searches, the EST is first translated in all six frames. Each of the peptides is compared with known protein sequences in the GenPept, PIR and SWISS-PROT data bases. For these searches, we use the BLAZE algorithm (Intelligenetics), which aligns identical or similar amino acid residues and can accommodate gaps. Unfortunately, BLAZE requires substantial computational resources and

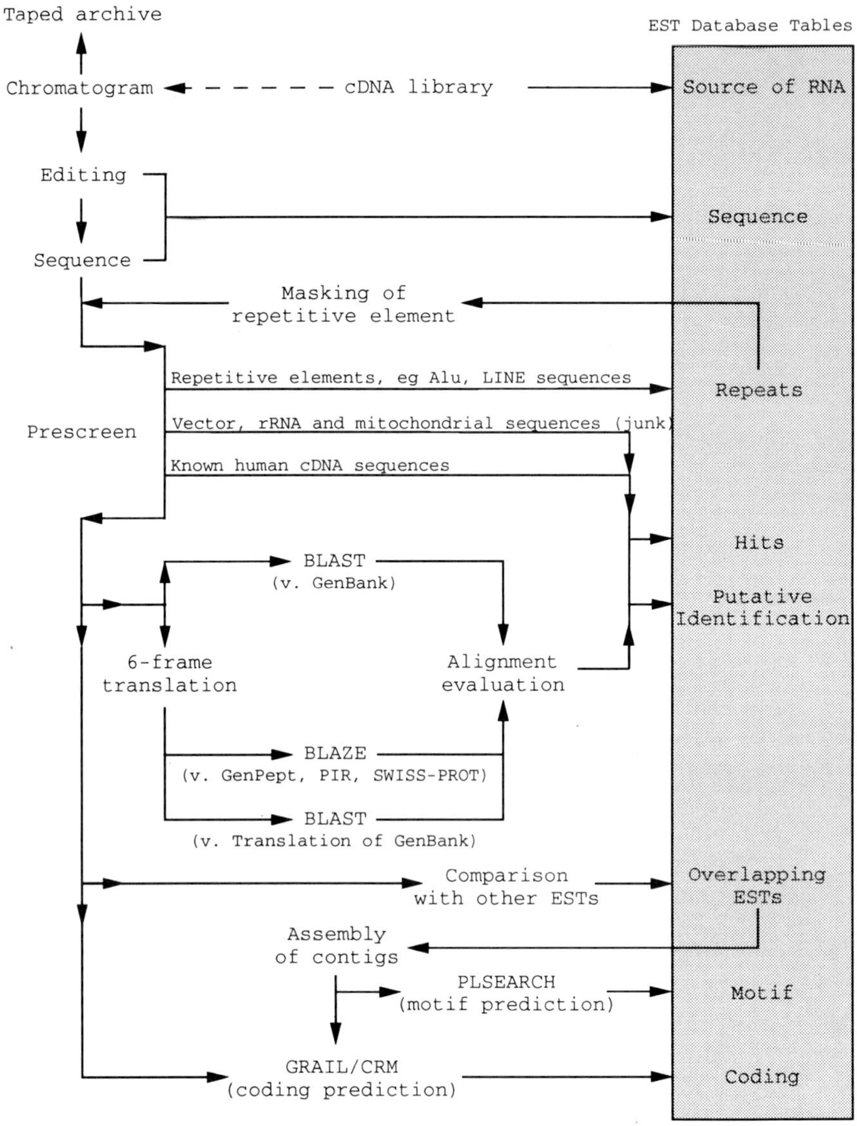

FIG. 2 Analysis and storage of EST sequence data. For each clone, the automated sequencers generate a chromatogram and sequence data that is first edited to remove vector sequences and regions of ambiguity. The edited EST sequence is then prescreened by comparing it with data bases of repetitive elements, other "junk" sequences, and known human cDNA sequences. All hits are recorded automatically in tables of the EST database. Sequences that are not retained by prescreening are searched against public databases of DNA and protein sequences. Each alignment

is designed to run on the MasPar parallel processor. Faster, although less thorough, alternatives to the BLAZE method are provided by FASTA (17, 18) or FASTDB (19). It should be noted that unannotated nucleotide sequences in GenBank are often absent from the protein sequence data bases. It is therefore often fruitful to use the translated EST sequences to search a six-frame translation of GenBank by BLAST. The alignments that are generated by the various search algorithms are evaluated manually. Those that are judged to be significant are recorded in the database, together with a putative identification of the relevant EST. Examples of typical BLAZE-generated alignments are illustrated in Fig. 3A. A survey of the putative identifications obtained from many libraries has demonstrated that the combination of abundant gene transcripts (actin, tubulin, etc.) normally represents a small fraction (<5%) of the total mRNA. The bulk of the mRNA appears to be comprised of species that are, individually, expressed at low or moderate levels. For this reason, the random selection of cDNA clones permits the identification of many gene products, such as receptors, that are normally of low abundance in the cell (Fig. 3B) (20).

The prescreened ESTs are also compared with each other by BLAST alignment. Groups of overlapping ESTs can then be built into larger contigs using the Inherit assembly system (Applied Biosystems). For ESTs that have no obvious homologs, this additional information can help provide a putative identification by revealing multiple regions of weak homology within a large domain. In this respect, software tools such as PLSEARCH can be used to search for the characteristic patterns of known protein families. There is limited evidence the PLSEARCH, by using consensus patterns, can uncover homologs that are not detected by comparison to individual family members (21). Another analysis involves the Coding Recognition Module of GRAIL (22), which can be used to predict protein-coding sequence within ESTs. These data provide a first step toward identifying structural features of unidentified sequences and are of value for comparing the coding potential of different libraries (20). GRAIL can be accessed at the Oak Ridge National Laboratory via an e-mail server.

The combination of tools that are now available for sequence comparison and analysis has greatly increased the rate at which new members of gene

---

is evaluated for significance by experienced curators, and any putative identifications are entered manually into the EST data base. The ESTs that identify novel human cDNA sequences (i.e., those that pass prescreening) are also compared to each other in order to build larger contigs. These larger contigs can be assessed for protein-coding potential and specialized motifs using several predictive computer programs.

A

EST03666 v. RAT GLYCINE RECEPTOR BETA CHAIN (P20781)
Percent Identity = 81.2   Percent Similarity = 85.9 Gaps = 0

```
30        40        50        60        70        80        90
MKFLLTTAFLILISLWVEEAYSKEKSSKKGKGKKKQYLCPSQQSAEDLXRVPANSTSNIXNRLL
||| |  :| ||:||  |:| ||||||||||||||||||||||||||| ||| |||||| ||||
MKFSLAVSFFILMSLLFEDACSKEKSSKKGKGKKKQYLCPSQQSAEDLARVPPNSTSNILNRLL
X       10        20        30        40        50        60   X
```

EST06580 v. HUMAN LOW-DENSITY LIPOPROTEIN RECEPTOR (P01130)
Percent Identity = 38.3   Percent Similarity = 46.8 Gaps = 1

```
     X  10        20        30        40
RRCRHCQPGNFRCRDEKCVYETWVCDGQPDCADGSXEWD-CCYVLPR
     | | :|:|    |: : | ||  ||| ||| ||   |  |
CPVLTCGPASFQCNSSTCIPQLWACDNDPDCEDGSDEWPQRCRGLYV
X    150       160       170       180      X
```

B

Alpha-2-adrenergic receptor
Cannabinoid receptor
c-erbA-alpha-2
Corticotropin-releasing factor binding protein
Endothelin receptor
Enhancer of split
Epidermal growth factor receptor
ERK1-kinase
Fibroblast growth factor receptor
Glutamate receptor, metabotropic
HEK tyrosine kinase
Inositol-1, 4, 5-triphosphate receptor
Insulin-like growth factor receptor
JTK14 tyrosine kinase
KDEL receptor
LDL receptor-related protein
Lamin B receptor
Leukemia inhibitory factor receptor
Leukemia virus cell receptor
Macrophage scavenger receptor
Neuromedin B receptor
N-formylpeptide receptor
NMDA receptor subunit
PDGF receptor
Receptor tyrosine kinase (ERBB4)
Retinoid X receptor
Ryanodine receptor
Somatostatin receptor
trkB
VLDL receptor
ADP/ATP translocase
Anion exchanger homologue AE3
Bib protein
Ca^{2+}-transporting ATPase
Ca^{2+}-transporting ATPase 2
Ca^{2+}-transporting ATPase 3
Ca^{2+} channel, N-type, alpha 1
Ca^{2+} ATPase, cardiac
Glucose transporter-like protein
Glutamate-aspartate transporter
Glvr-1
H^+ ATPase
H^+ -transporting ATP synthase
Memebrane transport superfamily
Mitochondrial phosphate carrier
Neutral amino acid transporter
Na^+/Ca^{2+}-exchange protein
Na^+/K^+ ATPase, alpha subunit
Na^+/K^+ ATPase, beta subunit
Potassium channel
Proton pump protein
Sodium channel II
Taurine transporter

FIG. 3 (A) Alignments obtained from BLAZE comparison of translated ESTs with a database of protein sequences. The upper line of each alignment is the translated EST sequence. The first example identifies EST03666 as encoding a human glycine receptor β-subunit. The divergence of the sequences at the N terminus suggests that this protein, like related receptor subunits, can use alternative signal sequences. The second example illustrates a weaker alignment involving EST06580. Although the overall homology is relatively low, the conserved spacing of cysteine residues is characteristic of low-density lipoprotein receptors, and EST06580 may therefore encode a new homolog of this protein. (B) A selection of receptors and transporters that were found to exhibit significant sequence homology with ESTs after random sampling of 6000 clones from human brain (2, 20).

families can be identified from anonymous sequences. It is no longer necessary to rely solely on the cross-hybridization of nucleotide sequences for such discoveries. Future advances, particularly in the ability to predict tertiary protein structure from nucleotide sequence, will undoubtedly help to identify many of the more distant family relationships.

References

1. M. D. Adams, J. M. Kelley, J. Gocayne, M. Dubnick, M. H. Polymeropoulos, H. Xiao, C. R. Merril, A. Wu, B. Olde, R. F. Moreno, A. R. Kerlavage, W. R. McCombie, and J. C. Venter, *Science* **252,** 1651 (1991).
2. M. D. Adams, M. Dubnick, A. R. Kerlavage, R. Moreno, J. M. Kelley, T. R. Utterbeck, J. W. Nagle, C. Fields, and J. C. Venter, *Nature* (*London*) **355,** 632 (1992).
3. P. Chomczyniski and N. Sacchi, *Anal. Biochem.* **162,** 156 (1987).
4. J. R. Nevins, *in* "Methods in Enzymology" (S. L. Berger and A. M. Kimmel, eds.), Vol. 152, p. 234. Academic Press, San Diego, CA, 1987.
5. S. L. Berger, *in* "Methods in Enzymology" (S. L. Berger and A. M. Kimmel, eds.), Vol. 152, p. 227 Academic Press, San Diego, CA, 1987.
6. B. M. Mechler, *in* "Methods in Enzymology" (S. L. Berger and A. M. Kimmel, eds.), Vol. 152, p. 241. Academic Press, San Diego, CA, 1987.
7. H. D. Lerach, J. M. Diamond, J. M. Wozney, and H. Boedtker, *Biochemistry* **16,** 4732 (1977).
8. J. Sambrook, E. F. Fritsch, and T. Maniatis, "Molecular Cloning: A Laboratory Manual," 2nd ed. Cold Spring Harbor, Lab., Cold Spring Harbor, NY, 1989.
9. E. Hara, T. Kato, S. Nakado, S. Sekiya, and K. Oda, *Nucleic Acids Res.* **19,** 7097 (1991).
10. E. Hornes and L. Korsnes, *Genet. Anal. Technol. Appl.* **7,** 145 (1990).
11. M. W. McDonell, M. N. Simon, and F. W. Studier, *J. Mol. Biol.* **110,** 119 (1977).
12. M. R. Ponce and J. L. Micol, *Nucleic Acids Res.* **20,** 62 (1992).
13. F. Sanger, S. Nicklen, and A. Coulson, *Proc. Natl. Acad. Sci. U.S.A.* **74,** 5463 (1977).
14. L. M. Smith, J. Z. Sanders, R. J. Kaizer, P. Hughs, C. Dodd, C. R. Connell, C. Heiner, S. B. H. Kent, and L. E. Hood, *Nature* (*London*) **321,** 674 (1986).
15. V. Murray, *Nucleic Acids Res.* **17,** 8889 (1989).
16. S. F. Altshul, W. Gish, W. Miller, E. W. Myers, and D. J. Lipman, *J. Mol. Biol.* **215,** 403 (1990).
17. W. R. Pearson and D. J. Lipman, *Proc. Natl. Acad. Sci. U.S.A.* **85,** 2444 (1988).
18. W. R. Pearson, *Genomics* **11,** 635 (1991).
19. D. L. Brutlag, J. D'Autricourt, S. Maulik, and J. Relph, *Comp. Appl. Biosci.* **6,** 237 (1990).
20. M. D. Adams, A. R. Kerlavage, C. Fields, and J. C. Venter, *Nat. Genet.* **4,** 256 (1993).
21. R. F. Smith and T. F. Smith, *Proc. Natl. Acad. Sci. U.S.A.* **87,** 118 (1990).
22. E. Uberbacher and R. Mural, *Proc. Natl. Acad. Sci. U.S.A.* **88,** 11261 (1991).

Section II

Expression and Characterization

[8] Ligand Binding and Second-Messenger Assays for Cloned G_q/G_{11}-Coupled Neuropeptide Receptors: The GnRH Receptor

R. P. Millar, J. Davidson, C. Flanagan, and I. Wakefield

Introduction

The identification and characterization of cloned G protein-coupled neuropeptide receptors require methodology for monitoring ligand binding and coupling to intracellular signaling pathways in transfected cells. The availability of both monitoring systems is a prerequisite for mutagenesis studies directed at the molecular dissection of ligand-binding sites, molecular switches mediating agonist activation of the receptor, and domains involved in G protein binding and activation for signal transduction.

Neuropeptidergic neurons and their targets in the central and peripheral nervous systems represent a major regulatory component which is even more prevalent than classical biogenic amine neurons and their targets. Neuropeptides also encompass a far more diverse array of molecular signaling structures, whose importance in the regulation of neural and endocrine pathways is reflected in the increasing quest for neuropeptide analogs by the pharmaceutical industry. Neuropeptides include the hypothalamic regulators of pituitary function, gonadotropin-releasing hormone (GnRH), growth hormone-releasing factor, corticotropin-releasing factor, thyrotropin-releasing hormone (TRH), and somatostatin; the neurohypophyseal hormones arginine, vasopressin, and oxytocin; the opioids which include enkephalins and endorphins; the tachykinin family of peptides, which includes substance P, substance K, neuromedin K, kassinin, eledoisin, and neurokinin; the pancreatic polypeptide family [NPY, PYYD, and the glucagon-related peptides (e.g., VIP, PHI, PHM)]; neurotensin; melanocyte-stimulating hormone; the bombesin-CCK family of peptides; angiotensin; calcitonin gene-related peptide; and several others.

All of these appear to regulate target cells through G protein-coupled receptors (GPCRs). These receptors are characterized by a single polypeptide chain which traverses the cell membrane seven times and activates the heterotrimeric GTP-binding proteins (G proteins) through their intracellular loop domains. The α-subunit of the G protein then dissociates to regulate

Methods in Neurosciences, Volume 25

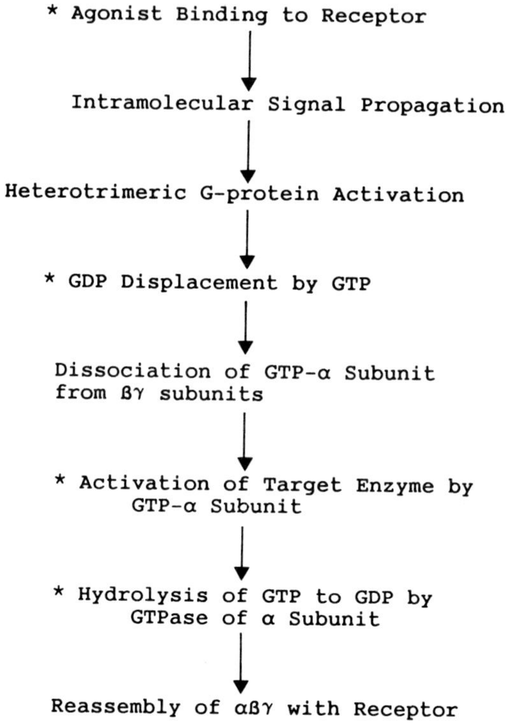

FIG. 1 Cascade of agonist-induced events in G protein-coupled receptors. Asterisk indicates current sites at which monitoring assays are conducted.

enzyme targets such as adenylate cyclase (through G_s and G_i) and phospholipase C (G_q/G_{11}) and ion channels (G_i and G_o).

The cascade of events is summarized in Fig. 1. The most proximal events are the activation of G proteins which may be monitored by GTP displacement of GDP, GTPase activity, and the activation of target enzymes (e.g., phospholipase C and adenylate cyclase). Intermediary events are the generation of diacyl glycerol and phosphoinositols (e.g., IP_3 which mobilizes intracellular Ca^{2+}), the production of cyclic AMP, and the activation of cell membrane ion channels (e.g., voltage-sensitive and voltage-insensitive Ca^{2+} channels). These lead to the activation of protein kinases (e.g., PKC, PKA, and calmodulin-activated protein kinase) and phosphorylation of target proteins. Distal events include the exocytosis of cellular signaling compounds (e.g., neurotransmitters and hormones) and regulation of gene expression.

Since a comprehensive description of the methodology for ligand-binding receptor assays and postreceptor signal-transducing assays is not feasible for the multitude of neuropeptides, this chapter confines itself to a detailed description of these assays for the GnRH receptor which activates phospholipase C. The GnRH receptor provides a good example of the principles and assay methodologies which are widely applied with minor modifications for other neuropeptide GPCRs.

Ideally, the researcher would wish to monitor most of the intermediate steps shown in Fig. 1 as mutations of the receptor may have discrete and complex effects on these. However, in practice, screening the effects of mutations on ligand binding and a distal event such as inositol phosphate production is adequate. The principles of these assays are generally applicable to other G_q/G_{11}-coupled neuropeptide receptors.

Expression of Cloned Receptor

As the methods for expression of receptors in mammalian cells is covered in detail in Chapter 9, a methodology for the expression of cloned GnRH receptor is covered only briefly here. First, GnRH receptor cDNA clones are subcloned into a suitable expression vector. We have obtained good expression of receptor in COS-1 cells, with the pcDNA/AMP vector containing the SV40 early promoter. Transfection may be accomplished by a variety of means including the DEAE–dextran method, electroporation, retrovirus infection, calcium phosphate, and liposome encapsulation. The most effective agent, concentrations required, time of exposure, DNA dose, etc., vary between cell types. An example of the effects of different DNA doses on GnRH receptor expression is shown in Fig. 2. For transfection of COS-1 cells we have found the inexpensive DEAE–dextran procedure to be very effective and result in high receptor expression. The procedure is summarized in Table I.

Receptor-Binding Assays

The essential components of a neuropeptide receptor assay are a high specific activity and high-affinity radiolabeled ligand, a relatively rich source of receptor (membrane-associated or solubilized), and a method for separating bound and free ligand. The most critical of these in the development of a GnRH receptor-binding assay was the introduction of a high-affinity less-degradable GnRH agonist for radioiodination. Concerted attempts over several years to develop assays with the labeled native peptide could not effectively demonstrate the high-affinity receptor and were plagued with low-affinity and high nonspecific binding (1). Analogs of GnRH which are commonly used for

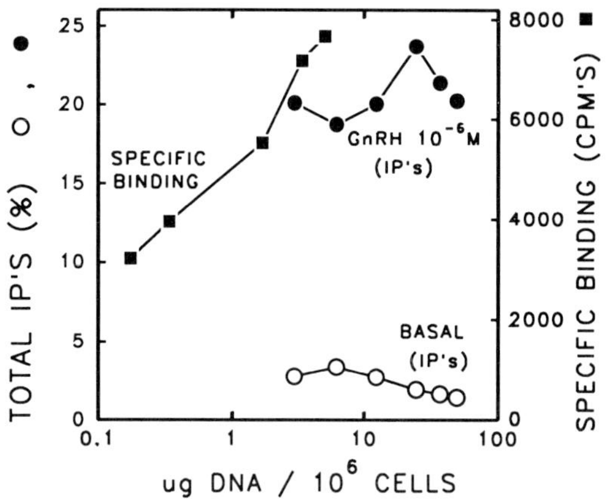

FIG. 2 Effect of DNA dose on GnRH receptor expression in COS-1 cells. Receptor binding studies utilized cells transfected in 10-cm plates while inositol phosphate assays used cells transfected in 12 well plates as described under Method in Table I. Minimal DNA requirements for maximal expression were determined and found to be about 5 $\mu g/10^6$ cells. The inositol phosphate assay exhibited maximal response to GnRH at this concentration.

radioiodination are [D-Ala^6,*N*-Me-Leu^7, Pro^9-NHEt]GnRH, [D-Trp^6]GnRH, and [D-*tert*-butyl Ser^6,Pro^9-NHEt]GnRH.

Radiolabeling of Neuropeptides

Radioiodination of GnRH Analogs

Radioiodination of ligands is the preferred method of radiolabeling as it is relatively inexpensive, easy to perform, and provides a high specific activity ligand. Disadvantages include oxidative damage to the peptide and the large atomic size of iodine which may substantially affect the conformation and properties of the ligand and render it nonbiologically active.

The procedure involves oxidative iodination whereby the iodine atom is usually incorporated into the meta position of tyrosine. Although less reactive, histidine is also a target for oxidative iodination. The classical method described by Greenwood *et al.* (2) employed large quantities of chloramine-T (100 μg) as oxidizing agent which may cause damage to the peptide and the generation of di- and triiodinated products. This led to the development of a variety of more gentle oxidation procedures employing lactoperoxidase with H_2O_2 or with glucose oxidase and glucose, electrochem-

TABLE I Tranfection of COS-1 Cells Using DEAE–Dextran

Materials
COS-1 cells are cultured in DMEM-10 (DMEM with 10% fetal calf serum and 0.1% glucose)
HBS buffer
  137 m*M* NaCl
  5 m*M* KCl
  0.7 m*M* NaH_2PO_4
  20 m*M* HEPES
  pH to 7.4 with NaOH
HEPES–DMEM: 10 m*M* HEPES, pH 7.4, in DMEM (without bicarbonate)
HBS/DEAE–dextran stock
  3 mg/ml in HBS buffer
  Filter-sterilize and store at 4°C
HBS/DEAE–dextran/DNA/DMEM
  Dilute HBS/DEAE-dextran stock 10 times in DMEM (serum-free, with bicarbonate). Add DNA at 15μg/4 ml for 10-cm dishes or 5 μg/ml for 12-well plates (10-cm dishes are used for cells for receptor-binding studies; 12-well plates for inositol phosphate assays)
Chloroquine DMEM
  200 μ*M* chloroquine in DMEM containing 2% fetal calf serum
Method
1. Plate 3 × 10^6 cells/10-cm plate (or 2 × 10^5 cells/well for a 12-well plate) in culture medium overnight. For IP assay, 12-well plates are precoated with poly-D-lysine (Sigma).
2. Wash twice with HEPES-DMEM.
3. Add 4 ml (for 10-cm plates) or 0.5 ml/well (for 12-well plates) of HBS/DEAE-dextran/DNA/DMEM. Leave at 37°C for 4–5 hr in a 5% CO_2 incubator.
4. Aspirate off medium and wash once with DMEM.
5. Add chloroquine–DMEM to cover cells, and incubate for 1 hr at 37°C in CO_2.
6. Wash gently once with HEPES–DMEM, as cells are fragile and easily detached.
7. Add DMEM-10 and leave overnight.
8. Change medium the next day.
9. Do experiment the following day.

ical procedures, and the iodogen methodology. Our laboratory has utilized these alternatives but found considerable variation in ^{125}I incorporation and we reverted to the chloramine-T method. In addition we introduced a modified method in which stoichiometric amounts of chloramine T are used to minimize damage and produce a high specific activity monoiodo GnRH analog which approaches the theoretical specific activity of 1600 μCI/μg. We have also largely abandoned the conventional use of the reducing agent (sodium metabisulfite) to terminate the reaction (2) as this agent may also cause damage to the peptide and disrupt sensitive disulfide bonds present in some neuropeptides. Instead the principle employed is one of brief exposure to chloramine-T, quenching by dilution, and rapid separation on high-

TABLE II Radioiodination and Purification of GnRH Analogs

Materials
0.5 *M* Phosphate buffer (PB), pH 7.4
30 mg chloramine-T/10 ml PB (3 mg for stoichiometric iodination)
6 mg $Na_2S_2O_5$/5 ml PB
10% BSA (0.1 g fraction V/ ml distilled water)
Iodination procedure
10 μl 5μg GnRH analog in PB
15 μl PB
10 μl ^{125}I-Na (1.0 mCi)
10 μl chloramine-T
Add the above in order and mix for 10 sec. 50 μl $Na_2S_2O_5$ then added to stop reaction in non-stoichiometric iodination; reaction mixture diluted with PB and placed immediately on HPLC column for stoichiometric method.
HPLC Purification
Solvent A : 0.01 *M* ammonium acetate (NHAc), pH 4.6
Solvent B : 60% CH_3CN/40% 0.01 M NHAc pH 4.6 (use 80% CH_3CN for GnRH antagonist)
Run a gradient from 0 to 100% B over 30 min followed by 100% B for 10 min at a flow rate of 1.5 ml/min.
Collect 1-min fractions in tubes containing 10 μl 10% bovine serum albumin or gelatin to prevent adsorption.
Dilute label in PB so that aliquots are approximately 2×10^6 cpm in 200 μl. Aliquot into BSA coated vials. Snap-freeze at −70°C.
Store at −20° or −70°C
Notes
(a) For purification of different peptides, solvent B is selected to be from 40–80% depending on the hydrophobicity of the peptide
(b) Not all peptides may be satisfactorily iodinated using the gentle (stoichiometric) chloramine-T method.
QAE Sephadex Purification
Presoak QAE Sephadex A25 and Sephadex G25 (medium) (Pharmacia) in H_2O. Cut the top off a disposable 10-ml plastic pipette. Plug the tip with a little glass wool. Pour and pack about 10 ml QAE Sephadex A25 and then 2 ml Sephadex G25 on top. Equilibrate with 0.1 *M* NH_4HCO_3, pH 9.2.
Place iodination mixture on column, elute with 0.1 *M* NH_4HCO_3 and collect 1-ml fractions into glass tubes containing 10 μl 10% BSA and 400 μl 0.1 *M* acetic acid and count diluted aliquots. Test the two fractions with highest counts for binding (Bo and NSB) using rat pituitary membranes.
Make 25-μl aliquots of the fractions with best properties and store frozen at −70°C

pressure liquid chromatography (HPLC) (see below). The procedure is detailed in Table II.

The alternative methods of radioiodination are therefore not detailed here but may be gleaned from descriptions by Clayton (1). The methods for specific activity determination are also described there.

After purification on HPLC and collection and neutralization of fractions (see below), the peak of radioactivity is pooled and aliquots containing sufficient ^{125}I-peptide for an assay are snap-frozen and stored at −70°C. The ^{125}I-labeled peptides exhibit good binding for at least 3 months although it is common practice in our laboratory to use the ligand for only one month. Clayton (1) has reported that ^{125}I-peptides are satisfactorily stored at pH 4.5 at 4°C for at least 4 weeks.

On a cautionary note it should be recognized that hydrophobic/aromatic side chains of peptide residues can result in adsorptive losses especially in polystyrene containers, and polypropylene or glass containers are preferred. Adsorption and losses may be exacerbated if peptides are not snap-frozen so that a concentrated solution is presented to the wall of the container during slow freezing. Similarly, severe losses and dilutional errors can result from adsorption onto pipet tips.

Design of a Specific GnRH Analog for Radioiodination

Another aspect for consideration in radioiodinating a neuropeptide is the site at which the ^{125}I is incorporated. The molecular size of the adduct is similar to that of many amino acids and its incorporation into tyrosine or histidine will alter the size and properties of these residues and may affect ligand binding considerably. This is further affected if diiodotyrosine is generated. Tyrosine in position five of GnRH is thought to play a role in stabilizing the conformation of the peptide and in its interaction with the receptor (3). Substitution of a D-amino acid for Gly^6 enhances the affinity of the peptide by constraining it in a β-II-type bend conformation which allows the NH_2 and COOH termini of the peptide to interact with the receptor (3, 4). The side chain of the D-amino acid is thought to be oriented away from the receptor and fairly bulky residues are tolerated in this position (3, 4). We considered, therefore, that an ideal position to radioiodinate the peptide would be in position 6. We synthesized an analog in which Gly^6 was substituted with D-Tyr and Tyr^5 was substituted by His (as in chicken GnRH II). This radioiodinated peptide exhibited enhanced binding (about two fold) when compared with conventional ligands (Fig. 3).

Comments on Radiolabeling Neuropeptides in General

While a number of neuropeptides have tyrosine or histidine residues in the native peptide and may be directly radioiodinated, its suitability for radioreceptor assay will depend on whether the incorporation of the large iodine atom perturbs the structure of the molecule such that it is no longer biologically active (see above). Iodinated TRH (pGlu-His-Pro-NH_2) does not bind its receptor well and it is necessary to employ lower specific activity tritrated analogs, [^{3}H]TRH or an [^{3}H](3-Me-His-TRH) (5). On the other hand other neuropeptides such as arginine vasopressin and oxytocin are notori-

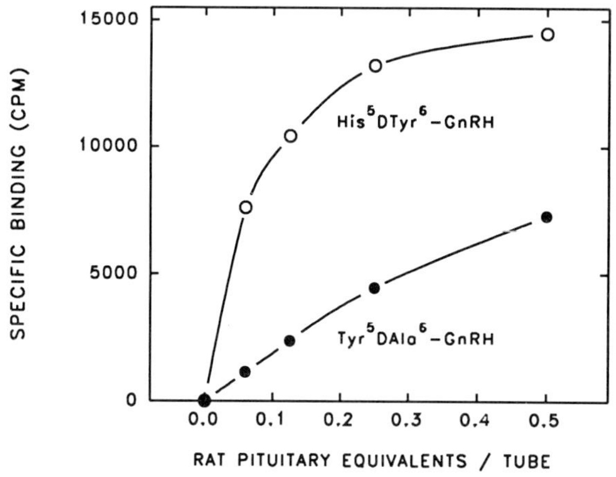

FIG. 3 Comparative binding of ^{125}I-D-Ala^6, *N*MeLeu^7, Pro^{9}NEt GnRH (●) and ^{125}I-His^5-D-Tyr^6 GnRH (○) to rat pituitary membranes. For both ligands 50,000 cpm was added and identical incubation conditions employed.

ously difficult to radioiodinate due to the susceptibility of the disulfide bridge to oxidation. Consequently gentle conditions for radioiodination and good separation techniques have been devised.

Other neuropeptides such as somatostatin and substance P lack a tyrosine or histidine residue and it is necessary to synthesize analogs in which these residues are added or substituted at sites which are not critical for binding to the receptor.

Yet other neuropeptides contain residues such as methionine and tryptophan which become oxidized during iodination and disturb receptor interaction. This problem has been addressed by synthesizing analogs with substitutions for these sensitive residues (e.g., norleucine for the C-terminal methionine in substance P). Thus we used a synthetic analog Tyr^{8}Nle^{11} substance P which introduces a tyrosine for iodination and replaces the terminal methionine which is susceptible to oxidation.

Finally both the lack of a residue for radioiodination and/or the presence of oxidation-sensitive residues may be overcome using the Bolton–Hunter reagent (6) in which a prelabeled ^{125}I-adduct is conjugated to neuropeptides by the α amino or ε amino (e.g., in lysine) moieties.

Purification of Radiolabeled Peptide

The objectives of purification are the removal of unincorporated ^{125}I (usually the sole accomplishment of early methodologies using gel filtration), the separation of ^{125}I-peptide from unlabeled peptide (to obtain maximum

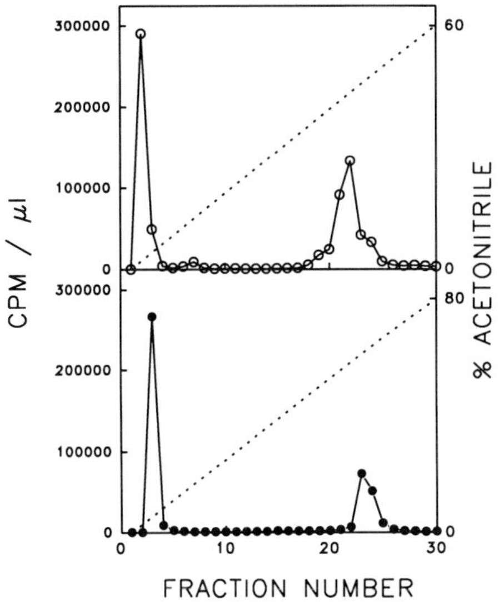

FIG. 4 Purification of ^{125}I-labeled GnRH agonist (above) and GnRH antagonist (below) on HPLC. The free ^{125}I is not retained on the column while the ^{125}I-labeled peptide is eluted later. Broken line shows gradient of B in A (see text). Note that a higher concentration of acetonitrile is required to elute the ^{125}I-GnRH antagonist.

specific activity), and the removal of damaged and diiodo peptide. The earliest methodology utilized gel filtration (e.g., Sephadex G-25 for GnRH analog). In some instances adsorptive properties of G-25 were inadvertently but effectively employed (1). This was followed by both cation (CM Cellulose) and anion (QAE-Sephadex A25-120) exchange purification which are most effective. The preliminary removal of unreacted ^{125}I is often executed by treating the reaction mixture with Dowex 1-X10 resin. Over the years we have utilized all of these procedures but now routinely utilize HPLC purification because of the ability to rapidly separate reactants and products and the potential to separate ^{125}I-peptide from unreacted peptide and obtain maximal specific activity. As this is the method of choice it is described in detail here (Table II) and the reader referred to the comprehensive descriptions of the other methods by Clayton (1).

After radioiodination the reaction mixture is injected directly into a dedicated HPLC system using a reverse-phase column (ODS18) and gradient of increasing acetonitrile (Table II, Fig. 4).

Receptor-Binding Assay Using Radioiodinated GnRH Agonist

The GnRH receptor-binding assay for transfected COS-1 cells is based on our modifications (7, 8) of the original pituitary membrane assay (9). In addition to expression in COS-1 cells we have also described the expression of the receptor in oocytes (see Chapter 4). COS-1 cells are transfected with receptor constructs as described in Table I and also in Chapter 9. Transfected COS-1 cells are detached from culture dishes in HEPES binding buffer, homogenized with a Dounce homogenizer, and centrifuged at 10,000*g* for 40 min at 4°C to pellet plasma membranes along with other particulate cell components (Table III). This crude membrane pellet is then resuspended in binding buffer and incubated (about 10^6 cell equivalents/tube) with approximately 50,000 cpm ^{125}I-[D-Ala^{6}, N-Me-Leu^{7},Pro^{9}-NHEt]GnRH and varying concentrations of unlabeled test peptides for 60 to 90 min on ice to achieve equilibrium. The incubation is terminated by the addition of 3 ml phosphate-buffered saline washing buffer and immediate filtration through glass-fiber filters (GF/C, Whatman) presoaked in washing buffer containing 1% BSA. The filters are then washed twice in washing buffer and the retained radioactivity is counted. Nonspecific binding is estimated in the presence of 10^{-7} *M* unlabeled [D-Ala^{6},N-Me-Leu^{7},Pro^{9}-NHEt]GnRH. Total specific binding ranges from 6000 to 20,000 cpm/tube for the wild-type receptor and nonspecific binding from 1500 to 3000 cpm/tube. Assays are usually performed in triplicate.

To compensate for lower total binding exhibited by mutant receptors of lower affinity, higher concentrations of membranes are required and a larger number of transfected cells are used for homogenization (10). In addition, to avoid dissociation of the labeled ligand from lower-affinity receptors, the dilution step at the end of the assay is eliminated and the filters are washed four times under vacuum with washing buffer to remove nonspecifically bound ^{125}I-[D-Ala^{6},*N*-Me-Leu^{7},Pro^{9}-NHEt]GnRH (10). The on-rate for lower affinity mutant receptors may be slower and incubation times may have to be increased accordingly to achieve equilibrium.

Examples of receptor-binding assays on the mouse wild-type and a mutant GnRH receptor are shown in Fig. 5. In order to characterize the pharmacology of the receptor, the ability of a series of GnRH analogs to displace the ^{125}I-labeled ligand is compared (Fig. 6). These pharmacological characteristics may then be compared in mutant receptors when trying to identify interactions between specific residues of the ligand and receptor.

TABLE III GnRH Receptor-Binding Assay Using Transfected COS-1 Cell Membranes

Materials

HEPES binding buffer stock (A)

10 m*M* HEPES

1 m*M* EDTA

pH to 7.4 using NaOH

Store for up to 2 weeks at 4°C.

HEPES binding buffer (B)

0.1% BSA (fatty acid-free) in binding buffer stock (A)

Do not store longer than 24 hr.

Washing buffer and phosphate-buffered saline (C)

40 m*M* $NaH_2PO_4 \cdot 2H_2O$

150 m*M* NaCl

0.1% BSA

pH 7.4, using NaOH

Buffer for soaking filters

1% BSA instead of 0.1% BSA in buffer C

Radioiodinated peptide

50,000 cpm ^{125}I-GnRH analog in 50μl buffer B

Filters

Whatman GF/C filters are presoaked in washing buffer C for at least 1 hr.

Method

(Except where mentioned, all procedures are done at 4°C)

1. Wash dishes once with buffer B.
2. Incubate with buffer B (2 ml/dish) with shaking at room temperature until cells start to detach.
3. Transfer the cells in buffer B to a Dounce homogenizer and homogenize by 15 strokes with plunger A and 15 strokes with plunger B.
4. Centrifuge homogenate at 10,000*g* for 40 min.
5. During the centrifugation set up the incubation tubes according to the scheme below. Thaw the label just before completion of centrifugation, as it may deteriorate once thawed.

Tube	Binding buffer	Peptide dose	^{125}I-peptide (μl)
TC	—	—	50
Bo	50 μl	—	50
NSB	—	50 μl 10^{-6} *M* GnRH A	50
Assay tubes	—	50 μl peptide standards	50

(Doses of peptides are made up in binding buffer at 10 times the required final concentration).

6. Resuspend the membrane pellet in buffer B (2×10^6 cell equivalents/ml). The resuspension may be performed in the Dounce homogenizer, using 10 strokes of plunger B or, for small volumes, by drawing through a hypodermic needle (22 gauge) and reexpelling 10–20 times.
7. Add 400 μl of cell membrane suspension to Bo, NSB, and assay tubes.
8. Incubate for 60 to 90 min depending on time required to reach equilibrium (5 hr for antagonists).
9. Add 3 ml ice-cold buffer C to each tube.
10. Filter immediately through presoaked Whatman GF/C glass fiber filters (under vacuum).
11. Wash twice with 3 ml buffer C
12. Place filters in numbered tubes and count radioactivity.

Precautions

Salts decrease binding—avoid them or ensure that there are uniform concentrations thoughout, i.e., *do not* make up peptide doses in saline or tissue culture buffers.

Reconstitution of frozen aliquots of ^{125}I-peptide should be done shortly before assay.

Some peptides adsorb strongly to plastics especially in low concentration (e.g., ^{125}I peptide). This can also result in "carry over" when diluting standards, especially GnRH antagonists.

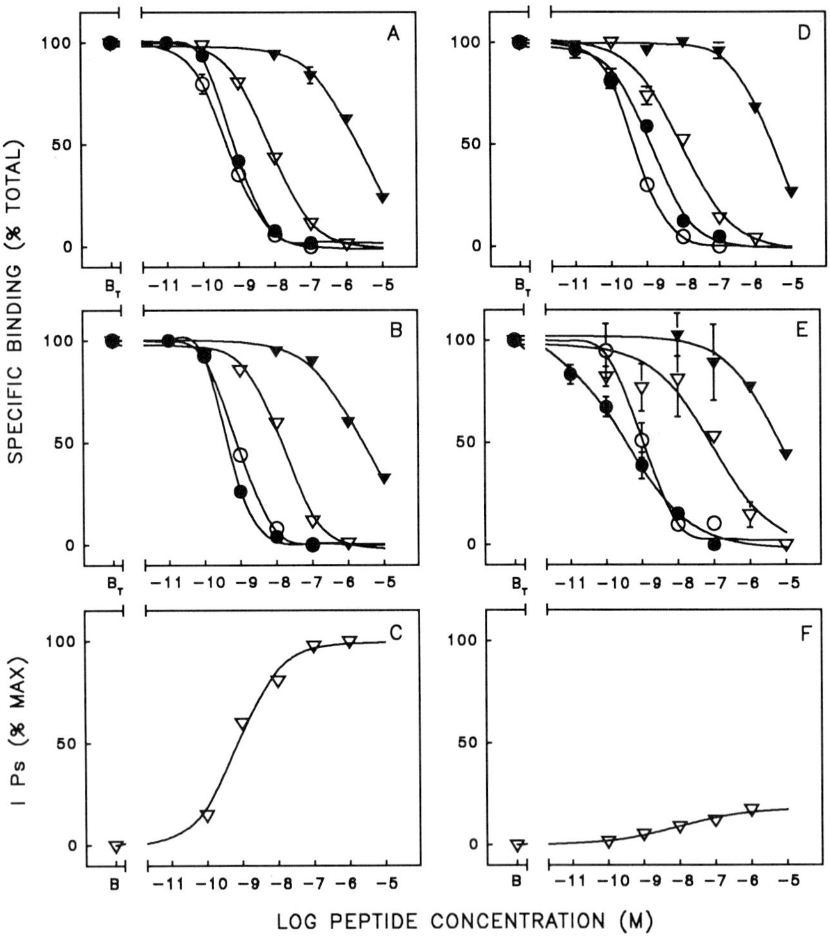

FIG. 5 Receptor binding and ligand stimulation of inositol phosphate production in COS-1 cells transfected with wild-type (left panel) and the $Asp^{87}Asn^{318}$ reciprocal mutant (right panel) GnRH receptor constructs. (Top) Competition binding of GnRH peptides with ^{125}I-GnRH agonist. (Middle) Competition binding of GnRH peptides with ^{125}I-GnRH antagonist. (Bottom) Stimulation of total inositol phosphate production by GnRH. Each symbol represents one GnRH peptide: GnRH agonist (○), GnRH antagonist (●), GnRH (▽) and Gln^8 GnRH (▼). Reproduced with permission from Zhou *et al.* (10). Copyright 1994 Williams and Waverly.

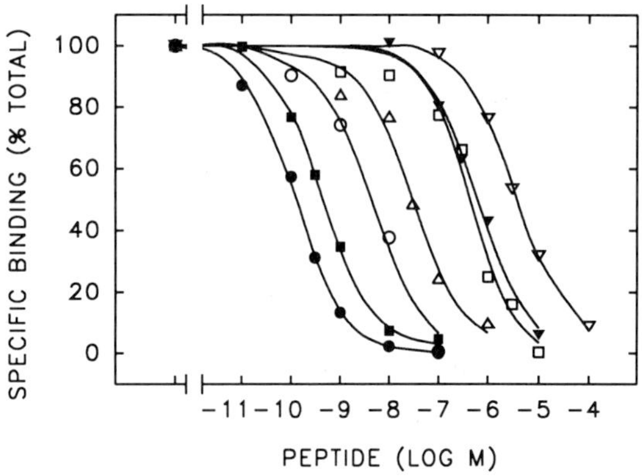

FIG. 6 Pharmacology of GnRH analogs. Displacement of ^{125}I-GnRH agonist bound to membranes of COS-1 cells transfected with the recombinant sheep GnRHR. GnRH (○), GnRH agonist (●), Gln^8 GnRH (▽), $His^5Trp^7Tyr^8$ GnRH (▼), [D-Trp^6,Trp^7,Tyr^8]GnRH (□), antagonist 5 (■), and antagonist 13 (△). Reproduced with permission from Illing *et al.* (8).

Receptor-Binding Assay Using Radioiodinated GnRH Antagonist

Since receptor activation by agonist leads to dissociation of the G protein and assumption of a lower-affinity conformation of the receptor, radiolabeled antagonists have generally been employed as ligands in many of the biogenic amine receptor assays. This problem is overcome in part by doing assays at 4°C when using labeled GnRH agonist. When assays are performed at 37°C binding of radiolabeled agonist declines almost 10-fold. This decline is completely reversed when membranes are recooled to 4°C. In addition certain mutations may have differential effects on agonist and antagonist binding.

We have, therefore developed a receptor-binding assay utilizing radioiodinated GnRH antagonist. Since antagonists generally incorporate substitutions with D-amino acids containing large hydrophobic/aromatic side chains in as many as 6 of the 10 amino acids in GnRH, these analogs tend to have very marked hydrophobic properties which result in extremely high nonspecific binding (e.g., 20,000 cpm). We have therefore screened a number of selected antagonists and found that one of these ([Ac-D-4-Cl-$Phe^{1,2}$-D-Trp^3-D-Lys^6-D-Ala^{10}-NH_2]GnRH) (10) gave an acceptable nonspecific binding of 5000–7000 cpm and a maximal specific binding of 3000–9000 cpm.

The procedure for radioiodination and purification is as described for the GnRH agonist but naturally a higher concentration of acetonitrile is required to elute the peptide from the reverse-phase HPLC column (Table II, Fig. 4).

Some modification of the receptor binding assay is also required in view of the higher nonspecific binding and slower "on" and "off" dissociation rate constants. The procedure involves a preliminary 500*g* centrifugation to remove the nuclear pellet and decrease nonspecific binding and a longer incubation time of 5 hr due to the slower on-rate of the antagonist.

A comparative example of receptor binding assays using ^{125}I-GnRH agonist and ^{125}I-GnRH antagonist is shown in Fig. 5. The greater error of points for the mutant receptor when using ^{125}I-GnRH antagonist results from the relatively high nonspecific binding.

Inositol Phosphate Assays

As indicated in the Introduction, GDP dissociation, GTPase, and cell-free phospholipase C assays may be useful adjuncts in mutagenesis studies on receptor structure and function. We have explored the use of these assays but find them much more difficult to perform and less robust and sensitive than the quantification of inositol phosphate production by intact cells. A major problem of assays of GDP displacement by GTP and the GTPase assays is the high background activity from other G proteins in the cells which is not associated with G_q/G_{11} activation by the ligand. The most standard inositol phosphate assay utilizes the prelabeling of cells with [^{3}H]myoinositol, the addition of ligand in the presence of LiCl (to inhibit the degradation of inositol phosphates), and the termination of the reaction by addition of perchloric acid/phytic acid, followed by separation of inositol phosphates on Dowex columns. Elution conditions may be employed to elute total inositol phosphate or differential elution of IP_1, IP_2, and IP_3 (see Table IV). When studying the latter it is necessary to utilize brief incubation periods of 1–2 min. The method described in Table IV is a synthesis of a number of previously published methods (11–16). This method gives an excellent GnRH stimulation of inositol phosphate production (about 10- to 30-fold over basal) (Fig. 7). [^{3}H]IP production is usually expressed as a percentage of total [^{3}H]inositol incorporation into perchloric acid-insoluble cellular material (see Fig. 8).

The value of using inositol phosphate assays in conjunction with receptor-binding assays in assessing effects of mutation of amino acids in receptors is exemplified in Fig. 5. Here a reciprocal mutation of putative interacting residues N87 and D318 in the mouse GnRH receptor restores binding to the receptor but the mutant receptor is now poorly coupled to G protein such that inositol phosphate production is diminished (10).

TABLE IV Inositol Phosphate Assay

Day 1

Plate out and transfect COS-1 cells in 12-well plates as described in Table I

Day 3

(Evening) wash on shaker for 1 min with Medium 199 containing 2% fetal calf serum to remove inositol present in tissue culture media.

Add 0.5 ml Medium 199 containing 2% fetal calf serum to each well and 2% μCi/ml myo-[2-^{3}H]inositol (Amersham) and leave overnight in 5% CO_2 incubator.

Day 4

Pour off labeling medium and wash twice with buffer I (140 m*M* NaCl, 4 m*M* KCl, 20 m*M* HEPES, 8.6 m*M* glucose, 1 m*M* $CaCl_2$, 1 m*M* $MgCl_2$, 0.1% BSA, pH 7.4). Add ligand or control medium plus 10 m*M* LiCl and incubate for 20–60 min (1–2 min if individual inositol phosphates are to be determined). Aspirate medium and stop incubation with 0.5 ml of PCA [0.5 *M* perchloric acid, 5 m*M* EDTA, 1 m*M* diethylenetriamine pentaacetic acid (DTPA)]. Add 50 μl of 1.8 mg/ml phytic acid to each well.

(Chelators are to prevent formation of Mg-IPx complexes).

Place at 4°C for 5 min to allow extraction of inositol phosphates.

Neutralize with 0.5 *M* KOH (about 0.65 ml—determine exact amoung by a test neutralization with phenol red). Leave at 4°C for 20 min, *then vortex well and stand for a further* 10 *min* to ensure complete precipitation of $KClO_4$. Apply a fixed aliquot of supernatant to columns.

After neutralization samples may be stored at −20°C. Total incorporation of ^{3}H into phospholipids is determined by washing cell residue 3× with cold 0.5 *M* PCA. Drain well and dissolve residue in 0.5 ml of 0.1 *M* NaOH, then neutralize with 50 μl of 6.6% glacial acetic acid (to avoid chemiluminescence) and determine radioactivity.

Dowex chromatography

Preparation of columns

Dowex 1 × 8–200 (in chloride form) is converted to the formate form by suspending in 3 *M* ammonium formate for 1 hr. Decant fines twice and replace with 3 *M* ammonium formate.

Pour columns (Bio-Rad Econo-columns) with 0.5 ml of resin (i.e. 1 ml of a 50% slurry). Wash with 3 ml of 3 *M* ammonium formate/0.1 *M* formic acid 10 ml distilled water 5 ml of 5 m*M* myoinositol/0.1 *M* formic acid

Apply samples. Elute ^{3}H-inositol with 10 ml distilled water 5 ml 0.1 *M* formic acid/5 m*M* myoinositol

To separate out different forms of inositol phosphate add

1. 3 × 3 ml of solution A (5 m*M* Sodium tetraborate, 60 m*M* sodium formate)
2. 2 × 3 ml of solution B (0.1 *M* formic acid, 0.2 *M* ammonium formate (collect IP_3)
3. 2 × 3 ml of solution C (0.1 *M* formic acid, 0.6 *M* ammonium formate) (collect IP_2)
4. 2 × 3 ml of solution D (0.1 *M* formic acid, 1 *M* ammonium formate) (collect IP_1)

To elute total inositol phosphates in a single step add 3 ml solution D.

Reconstitute columns by adding 3 ml of 3 *M* ammonium formate/0.1 *M* formic acid and store.

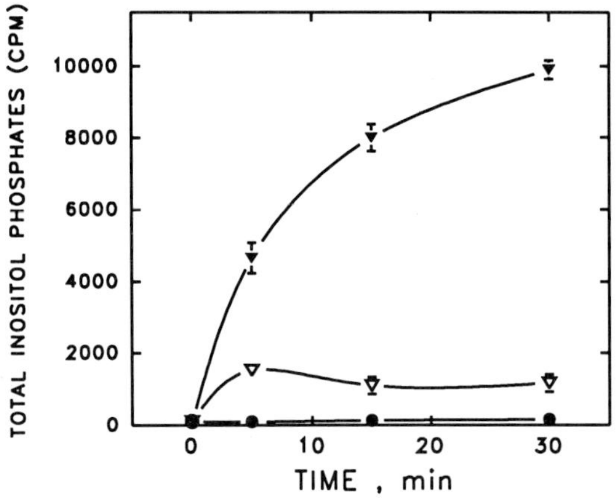

Fig. 7 Time-course of GnRH-stimulated inositol phosphate production. Cells prelabeled with [^{3}H]myoinositol were incubated with buffer only (○, ●) or with 10^{-6} *M* GnRH (▽, ▼) for 0, 5, 15, and 30 min. Solid symbols are incubations in the presence of 10 m*M* LiCl. Basal values (absence of GnRH) were unaffected by LiCl treatment so that symbols are superimposed.

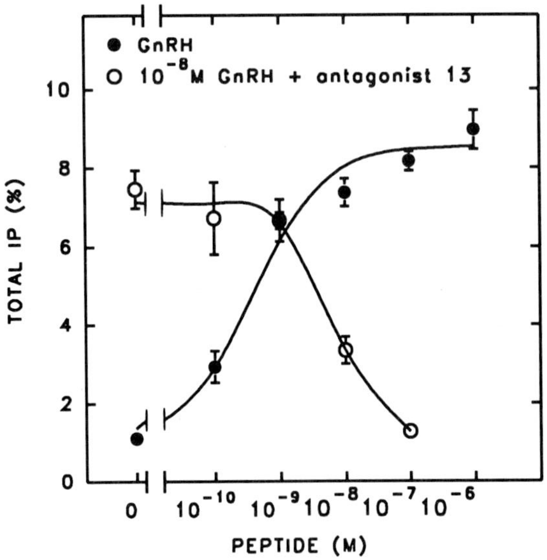

Fig. 8 Effects of GnRH and GnRH antagonist on inositol phosphate production in COS-1 cells transfected with pSV2A-human GnRHR. Reproduced with permission from Chi *et al.* (7).

In experiments using antagonists, it is necessary to preincubate the cells with antagonist for 10 min prior to addition of GnRH due to the relatively slow "on-rate." The duration and conditions of prelabeling cells and the time of exposure to ligands will depend on the degree of receptor expression and G protein coupling in the COS-1 cells and the specific activity attained in the inositol pools. Care must be taken to avoid depletion of the labeled pools in highly active cells.

Finally, it should be noted that a valuable tool for screening and characterizing G_{11}/G_q-coupled receptors is to monitor IP_3 stimulation of Ca^{2+} mobilization in transfected cells using Ca^{2+} indicators such as Fura 2 (17, 18).

Data Analysis

Peptide concentrations required to stimulate half-maximal IP production (ED_{50}s) and to half-maximally inhibit binding of ^{125}I-[D-Ala^6,*N*-Me-Leu^7,Pro^9-NHEt]GnRH (IC_{50}'s) are estimated by four-parameter nonlinear curve fitting using Sigmaplot (Jandel Scientific, Corte Madera, CA). K_d's are calculated from IC_{50}'s using the Munson and Rodbard (19) correction.

K_d values for antagonists may be derived from inhibiton of agonist-stimulated [^{3}H]IP production, assuming competitive interaction at a single site (20).

Conclusions

Neuropeptides are a large and important family of signaling molecules which are receiving increasing attention as targets for pharmaceutical intervention. The majority of neuropeptides signal through G protein-coupled receptors. Most of these are coupled to phospholipase C and adenylate cyclase, while a minority couple through ion channel regulation. In this section detailed procedures for the establishment of receptor-binding assays for GnRH agonists and antagonists are described as a thoroughly researched example for the development of other neuropeptide assays. This methodology is relatively straightforward and has proved to be sensitive and robust. A suitable assay for monitoring coupling of receptor activation to phospholipase C is also described for cells expressing the GnRH receptor as an example of the prinicples for G protein coupling assays and as a specific description of the procedure for quantifying Gq or G_{11} coupling by receptors. It is anticipated that simple, specific, and sensitive procedures for monitoring very proximal events (GDP dissociation, subunit disassembly, GTPase activity, etc.) will be established in the near future as important adjuncts in the investigation of the functioning of these receptors.

References

1. R. N. Clayton, *in* "Methods in Enzymology" (P. Conn, ed.), Vol. 103, p. 32. Academic Press, New York, 1983.
2. F. C. Greenwood, W. M. Hunter, and J. S. Glover, *Biochem. J.* **89,** 114 (1963).
3. R. P. Millar, J. A. King, J. S. Davidson, and R. C. deL. Milton, *S. Afr. Med. J.* **72,** 748 (1987).
4. M. J. Karten and J. E. Rivier, *Endocr. Rev.* **7,** 44 (1986).
5. R. L. Taylor and D. R. Burt, *Neuroendocrinology* **32,** 310 (1981).
6. A. E. Bolton and W. M. Hunter, *Biochem. J.* **133,** 529 (1973).
7. C. Chi, W. Zhou, A. Prikhozhan, C. Flanagan, J. S. Davidson, M. Golembo, N. Illing, R. P. Millar, and S. C. Sealfon, *Mol. Cell. Endocrinol.* **91,** R1 (1993).
8. N. Illing, G. F. M. Jacobs, I. I. Becker, C. A. Flanagan, J. S. Davidson, A. Eales, W. Zhou, S. C. Sealfon, and R. P. Millar, *Biochem. Biophys. Res. Commun.* **196,** 745 (1993).
9. R. P. Millar, C. A. Flanagan, R. C. deL. Milton, and J. A. King, *J. Biol. Chem.* **264,** 21007 (1989).
10. W. Zhou, C. Flanagan, J. A. Ballasteros, K. Konvicka, J. S. Davidson, H. Weinstein, R. P. Millar, and S. C. Sealfon, *Mol. Pharmacol.* **45,** 165 (1994).
11. T. F. J. Martin, *in* "Methods in Enzymology" (P. Conn, ed.), Vol. 124, p. 424. Academic Press, New York, 1986.
12. J. Horwitz and R. L. Perlman , *in* "Methods in Enzymology" (P. Conn and A. Means, eds.), Vol. 141, p. 169. Academic Press, Orlando FL, 1987.
13. W. R. Huckle and P. M. Conn, *in* "Methods in Enzymology" (P. Conn and A. Means, eds.), Vol. 141, p. 149. Academic Press, Orlando, FL. 1987.
14. A. Imai and M. C. Gershengorn, *in* "Methods in Enzymology" (P. Conn and A. Means, eds.), Vol. 141, p. 100. Academic Press, Orlando, Fl, 1987.
15. K. A. Wriggett and R. F. Irvine, *Biochem. J.* **245,** 655 (1987).
16. J. S. Davidson, I. K. Wakefield, U. Sohnius, P. A. van der Merwe, and R. P. Millar, *Endocrinology (Baltimore)* **126,** 80 (1990).
17. M. K. Perrin, L. M. Bilezikjian, C. Hoeger, C. J. Donaldson, J. Rivier, Y. Haas, and W. W. Vale, *Biochem. Biophys. Res. Commun.* **191,** 1139 (1993).
18. J. Reinhart, L. M. Mertz, and K. J. Catt, *J. Biol. Chem.* **267,** 21281 (1992).
19. P. J. Munson and D. Rodbard, *J. Recept. Res.* **8,** 533 (1988).
20. F. M. Leslie, *Pharmacol. Rev.* **39,** 197 (1987).

[9] Receptor Expression in Mammalian Cells

Rachael L. Neve and Kim A. Neve

Introduction

The cloning explosion of cDNAs encoding receptors has stimulated the rapid development of methodology for expressing these cDNAs in mammalian cells. The rationales for expressing receptor cDNAs in mammalian cells are many. Such genetically engineered lines provide a system for studying the pharmacology of specific subtypes of receptors—and even specific spliced forms of a given subtype—in clonal, homogeneous populations of cells expressing a given receptor cDNA. The cell lines chosen for receptor cDNA expression normally do not express the endogenous gene for the transfected receptor cDNA, allowing the investigator to analyze receptor point mutations in a milieu uncluttered by expression of the wild-type gene. Such a cellular environment also favors the analysis of deletion mutants of receptors to determine the function of specific molecular domains, and of chimeras in which protein segments of different receptors are exchanged to pinpoint the role of these domains in ligand binding and signal transduction. Finally, use of transient, high-expression vectors such as the vaccinia virus or baculovirus allows the production of large amounts of receptor for purification and biochemical characterization, as well as for the production of antibodies.

We have used this type of genetic intervention approach to study the pharmacology of dopamine receptors, to identify the structural determinants within these receptors of ligand binding and modulation of signal transduction pathways, and to seek out the differences in ligand binding and intracellular signaling among dopamine receptor subtypes and among the products of differentially spliced mRNAs for given receptor subtypes. We describe below the methodological details of genetically engineering mammalian cells to express dopamine receptors.

Practical Considerations

The expression of receptors in genetically engineered mammalian cells entails the following pragmatic considerations:

i. How will the gene in question be cloned?
ii. What type of expression vector should be used?
iii. How will the recombinant vector be transfected into cells?

Methods in Neurosciences, Volume 25

iv. Should the analyses be done on transiently transfected populations of cells or on clonal stable transfectants?
v. How are stable transfectants selected and analyzed?
vi. How are mutant receptor cDNAs constructed?

In the following sections, we describe methodologies that address these practical considerations for the expression of dopamine receptors in mammalian cells.

Cloning Dopamine Receptor Genes

The gene or cDNA of interest can be cloned *de novo;* alternatively, if the nucleotide sequence is published, the cDNA can be retrieved from a cDNA library using synthetic oligonucleotide probes based on the published sequence or it can be cloned directly from an RNA preparation by reverse transcription (RT) coupled with the polymerase chain reaction (PCR). We describe below our synthesis of a cDNA encoding the rat dopamine D_3 receptor using rat olfactory tubercle RNA as template for such RT-PCR (1).

Design of Primers for the Reverse Transcription Polymerase Chain Reaction

The D3 receptor cDNA is synthesized in two overlapping segments using RT-PCR. First-strand cDNA for the 3′ segment is synthesized using a primer (P4, Fig. 1) complementary to a 25-nucleotide segment of the rat D3 receptor sequence (2) that begins with the stop codon and extends into the 3′ untranslated region (3′ UTR). The design of the primer includes a *Hin*dIII restriction site on its 5′ end. The 3′ half of the cDNA is amplified by PCR using primer P4 together with a 25-base sense primer (P3) beginning with nucleotide 786 of the sequence published by Giros *et al.* (2). First-strand cDNA for the 5′ segment of the D3 receptor cDNA is synthesized using a primer (P2) comprising the reverse complement of P3. It is useful to design P2 and P3 so that they span a natural restriction site in the cDNA; this site can be utilized to splice the two halves of the cDNA together after amplification and purification of the PCR products. The 5′ half of the D3 receptor cDNA is amplified by PCR using primer 2 and primer 1, which represents the 25 bases immediately before the translational start codon, with an *Eco*RI site designed into the 5′ end.

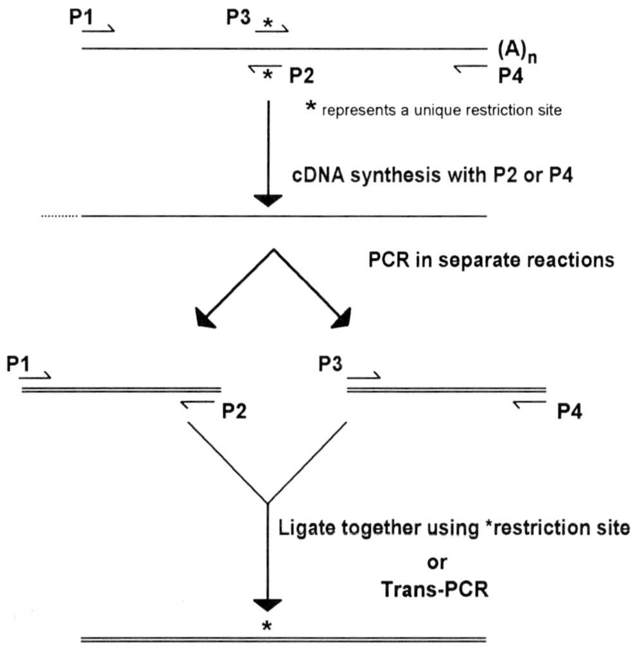

FIG. 1 Scheme for using RT-PCR to clone a dopamine D3 receptor cDNA. Overlapping PCR fragments representing two halves of the D3 receptor cDNA were amplified by PCR. The separate PCR fragments can be joined by conventional ligation or by *trans*-PCR, in which the two overlapping fragments are combined with the two outer primers in a PCR reaction. To minimize the generation of errors during PCR, it is advisable to join the fragments by ligation, or to use error-free heat-stable polymerases for the PCR reactions. This protocol can also be used to generate point mutations in a cDNA, in which case the mutations should be incorporated into P2 and P3 and trans-PCR used to join the two fragments (see Construction of Mutant Dopamine Receptor cDNAs). It may be necessary to determine the sequence of the entire PCR-amplified portion of the cDNA, to verify that random mutations have not occurred.

cDNA Synthesis

RNA is isolated from frozen rat olfactory tubercle using the guanidine isothiocyanate method as described by Neve *et al.* (3). For synthesis of cDNA complementary to the rat dopamine D3 receptor, 2 μg of total RNA is resuspended in 12.1 μl of RNase-free water, to which 3 μl of 100 m*M* methylmercury hydroxide is added to relieve RNA secondary structure. This is allowed to stand at room temperature for 7 min, after which 1.55 μl of 0.7 *M* 2-mercaptoethanol is added, and the mixture is incubated at room temperature for 5 min. The following components are then added to give a final reaction volume of 25 μl: 25 units of RNAsin (Promega Biotec), deoxynucleotide triphosphates (Promega Biotec) to a 500 μM final concentration, 1 μg of the 3′ primer (either P4 or P2), and 25 units of avian myeloblastosis virus reverse transcriptase (Life Sciences). The reaction is incubated at 42°C for 2 hr, at which point the reverse transcriptase is inactivated by incubation at 65°C for 10 min. cDNA is stored at −20°C until it is used as a template for PCR.

Polymerase Chain Reaction

The PCR is performed by adding 1 μl of cDNA to 49 μl of a reaction mixture containing the following: 10 mM Tris–HCl, pH 8.3 (at 25°C), 50 mM KCl, 1.5 mM $MgCl_2$, 0.001% gelatin, 10% DMSO, 200 μM deoxynucleotide triphosphates, 1.25 units of native Taq polymerase (Perkin–Elmer), and 0.5 μg each of the 3′ end and 5′ end primers. The DNA is amplified in a Perkin–Elmer thermal cycler under the following conditions: 94°C for 2 min, 60°C for 2 min, 72°C for 3 min for 40 cycles. Note that, although we use 10% DMSO in these particular PCR reactions, it is not a routine constituent of our PCRs. Inclusion of DMSO enhances some PCR amplifications, but can be deleterious to others. The requirement for DMSO is determined empirically for each set of primers.

Cloning of the Polymerase Chain Reaction Product

The volume of each PCR reaction is brought to 200 μl with water, and NaCl is added to 100 mM. The resultant aqueous mixture is extracted once with an equal volume of water-saturated and Tris-buffered phenol, once with an equal volume of 1 : 1 phenol : SEVAG (chloroform : isoamyl alcohol 24 : 1), and once with an equal volume of SEVAG. Next. 2.5 volumes (500 μl) of cold 95% ethanol is added, and the mixture is vortexed and spun at 14K rpm in a microfuge for 30 min. The supernatant is removed with a fine-tipped pipet, and the pellet is briefly dried under vacuum and resuspended in 20 μl of water.

Each purified PCR product is then digested with the appropriate enzymes. The 3′ fragment is digested with *Bam*HI and *Hin*dIII. The 5′ fragment is cut with *Eco*RI and then partially digested with *Bam*HI. The two fragments are electrophoresed on a 0.8% low-melting-point agarose gel in Tris–acetate–EDTA buffer, and the appropriate bands are cut out of the gel. Gelase (Epicentre Technologies) is used to digest the agarose, after which the DNA is ethanol-precipitated in the presence of ammonium acetate as recommended by the manufacturer. The two fragments are ligated and subcloned into the appropriate expression vector (pRSV, in the case of the D3 receptor cDNA).

Choice of Expression Vector

The expression vector is chosen on the basis of the cell type to be used, the type of expression desired (transient or stable), and the experiments to be performed (e.g., purification of the recombinant protein or binding assays). All other factors being equal, a vector with convenient restriction enzyme sites for the cDNA to be cloned is preferable to one without such sites, for which the cDNA cohesive ends would need to be altered with adaptors or by filling in and ligating on synthetic restriction site linkers.

For maximally efficient transient expression assays of cloned receptors, an SV40-based vector can be transfected into COS cells. COS cells, which have an integrated copy of the early region of SV40 and which express SV40 T antigen, support the replication of vectors that carry the SV40 origin of replication. The resultant multiple copies of recombinant vector in the cell allow for high levels of transient expression of the recombinant gene, but are ultimately deleterious to the survival of the transfected cells.

Stable expression in other types of cells can be achieved with a wide variety of vectors. Some neuronal cells, such as PC12 cells, do not support high levels of expression of most of the commonly used mammalian vector promoters. To identify optimal promoters for the expression of neural genes in PC12 cells, we systematically evaluated seven different promoters for their efficacy in directing the production of chloramphenicol acetyltransferase (CAT) in transiently transfected PC12 cells (4). The simian cytomegalovirus (SCMV) and Rous sarcoma virus (RSV) promoters were more efficient than any of the other promoters tested by almost an order of magnitude. We have found that these two promoters also direct robust expression in many other cell lines, although the clear advantage of CMV and RSV promoters that is observed in PC12 cells is not as prominent in nonneuronal cell lines.

Our analyses of wild-type and mutant dopamine D2 receptors were carried out using COS-7 cells transfected transiently with recombinants made in the vector pRSV (5) modified by the addition of a polylinker containing several unique restriction sites. These same receptors were expressed stably in C6 glioma cells (6, 7). It was important, for maintenance of a stable and consistent phenotype, to clone each transfectant by limiting dilution.

Expression systems that direct the transient synthesis of large amounts of recombinant protein for purification or for other biochemical studies are also available. Two of the most widely used are the vaccinia virus (8) and baculovirus (9).

In deciding whether to use stably or transiently expressed receptors, the following considerations may be helpful. For many purposes, such as the expression of receptors for purification, for the development of antibodies, or for radioligand-binding studies, the primary consideration is to express recombinant receptors at as high a density as possible. Most systems for transient expression of receptors attain higher levels of expression than can be attained by stable expression of receptors. Transient expression may yield results more rapidly than stable expression, since it eliminates the month required to generate, isolate, and characterize transfected cell lines. On the other hand, when the ratio of receptors to other cellular components is important, such as for studies of signal transduction or regulation of receptor responsiveness, the stable expression of recombinant receptors provides two advantages. First, when a clonal cell line is used, the ratio of receptors to cellular components can be maintained at a consistent level throughout a set of experiments. In contrast, transient expression typically involves the use of tissue in which an unknown percentage of the cells express recombinant receptors at variable and unknown densities. Furthermore, transfection efficiency can vary from one preparation of DNA to another. The second advantage, related to the first, is that the lower level of expression typically observed for stably expressed receptors may mimic endogenous expression more closely than does the high, variable expression of transient systems. Cells that express endogenous G protein-coupled receptors, such as dopamine or adrenergic receptors, rarely express one subtype of receptor at a density greater than 1000 fmol of receptor/mg of membrane protein; in fact, receptor densities are usually considerably lower than this. To mimic endogenous expression, it seems sensible to maintain the density of receptors at less than 1 pmol/mg of protein. In transient systems, the density of receptors on a given cell is often much higher than this, or is unknown.

Transfection of the Recombinant Vector into Cells

Calcium Phosphate Precipitation

A number of methods for introducing recombinant DNA into eukaryotic cells are available. We used calcium phosphate precipitation (10) to introduce dopamine receptor recombinants into COS-7 and C6 glioma cells. COS-7 cells, used for transient expression of dopamine D2 receptor cDNAs, are maintained at 37°C in Dulbecco's modified Eagle's medium supplemented with 8% iron-supplemented calf bovine serum and 2% fetal bovine serum.

Cells are seeded in 20 ml of medium in a 150-mm-diameter plate 24 hr prior to transfection. Transfection is carried out by calcium phosphate precipitation. Plasmid DNA (30 μg) is mixed with 1.0 ml of 0.24 *M* $CaCl_2$. An equal volume of 2× BES-buffered saline [50 m*M* *N,N*-bis(hydroxyethyl)-2-aminoethanesulfonic acid, 280 m*M* NaCl, 1.5 m*M* Na_2HPO_4, pH 6.95] is added, and the mixture is incubated at room temperature for 15 min before being added by drops to the medium in the plate. The plate is incubated 20–24 hr in an atmosphere of 3% CO_2/97% air, washed with phosphate-buffered saline, and then refed and incubated in an atmosphere of 10% CO_2 for 48 hr before being harvested for analysis of RNA or for receptor-binding assays.

Stable expression of mutant and wild-type dopamine D2 receptors in C6 glioma cells is obtained by calcium phosphate precipitation exactly as described for the transient transfections, except that the transfected plasmid DNA (15 μg) is comprised of $pRSVD_2$ and RSVneo (7:2). On the second day after addition of DNA, the cells are split into 10 plates and incubated overnight before selection with G418 (600 μg/ml, Gibco–BRL). Clonal G418-resistant cells are isolated after approximately 2 weeks of G418 treatment by trypsinization within a 10-μl cloning ring, transferred to 96-well plates, and expanded into duplicate 60-mm plates. Cell lines positive for D2 receptors, determined by binding of [^{125}I]epidepride, are expanded into 10-cm plates for further characterization.

Some stable transfections utilize selection by resistance to puromycin, in which case D2 receptor cDNA (15 μg) is cotransfected with pBABE Puro (2 μg) (11). Puromycin-resistant clones are isolated by selection at a concentration of 2 μg/ml, a concentration that kills control C6 cells within 1 week.

Electroporation

We have also used electroporation to deliver recombinant DNAs into mammalian cells. For this procedure, cells should be approximately 50–80% confluent and should have been fed or split the previous day. For each electroporation, 20 μg of recombinant plasmid (and 2 μg of plasmid with selectable marker if cotransfections are being performed) and 380 μg of sonicated [to <1000 base pairs (bp)] salmon sperm DNA are precipitated with ethanol by bringing the volume of DNA to 200 μl with water, adding NaCl to 100 m*M*, adding 2 volumes of 95% ethanol, and centrifuging the mixture at highest speed (14K rpm) in a microfuge for 15–30 min. The ethanol is aspirated off and the pellet allowed to air-dry briefly.

The pellet is resuspended in 400 μl of water, and an equal volume of 2× HeBS (40 m*M* HEPES, pH 7.05, 274 m*M* NaCl, 10 m*M* KCl, 1.4 m*M*

Na_2HPO_4, 12 m*M* dextrose) is added. Cells are trypsinized and pelleted (suspension cells can be pelleted directly), after which the cell pellet is resuspended to homogeneity in the 0.8-ml 1× HeBS plus DNA. All manipulations are done at room temperature. The cell–DNA suspension is transferred to a Bio-Rad electrode cuvette, and the DNA electroporated into the cells using a Bio-Rad Gene Pulser with capacitance extender to 960 μFD. Optimal voltage should be determined for each cell type: for COS-7 cells we use 0.28 kV, and for PC12 cells we use 0.25 kV.

After the electroporation the cells are allowed to sit in the cuvette for 10 min, after which 10 ml of the appropriate medium is added to the cells and they are pelleted. The cell pellet is resuspended in medium and seeded onto plates in nonselective medium. Cells should be plated to give a confluent monolayer in 2–3 days; this is approximately 1/2 the original area, given that only 20–30% of the cells will survive the electroporation. Cells can be harvested for transient analysis after 48–72 hr, or split into selective medium.

Selection and Analysis of Stably Transfected Cells

Only a percentage of clonal transfectants selected by virtue of their resistance to the appropriate antibiotic will display robust expression of the transgene. These expressing transfectants can be detected at the 96-well stage of expansion of the clones, using RT-PCR. Alternatively, the antibiotic-selected clones can be expanded to 60-mm plates for RNA blot analysis, to evaluate expression of the recombinant RNA, and/or binding assays, to verify expression of the protein product of the recombinant gene.

Reverse Transcription Polymerase Chain Reaction Detection of Transfectants Expressing the Recombinant Gene

The advantage of ascertaining which clonal transfectants express reasonable amounts of the recombinant RNA while the clones are still in 96-well plates is that further expansion can be restricted to those clones displaying robust expression of the transgene. When we opt for this approach, we first split the clones in the 96-well plates, so that a duplicate set of 96-well plates is made. Each well of the plate targeted for RT-PCR analysis should contain approximately 20–200 cells. The cells in each well are lysed in 35 μl of a solution containing 0.1% Triton X-100, 1 m*M* $MgCl_2$, 10 m*M* dithiothreitol (DTT), 1000 U/ml RNAsin (Promega Biotec), and 30 U/ml of RNase-free DNase (Promega Biotec). To pellet nuclei, the lysate is layered over 30 μl of a sucrose solution (0.3 *M* sucrose, 10 m*M* DTT, 1 m*M* $MgCl_2$) and microfuged for 7 min at 14K rpm. Before cDNA synthesis, 12.1 μl of the supernatant is transferred to a fresh tube, incubated at 37°C (to allow the DNase to act), and then heat-inactivated at 68°C for 10 min. At this point 3 μl of 0.1 *M* MeHgOH is added and the remainder of the procedure carried out as detailed above for cloning receptor genes by RT-PCR. Polymerase chain reaction products are electrophoresed on 0.8% agarose gels, blotted onto 1.2-μm Biotrans membrane (ICN Biochemicals), and hybridized with a ^{32}P-labeled oligonucleotide representing a sequence internal to the expected PCR product.

RNA Blot and Radioligand-Binding Analysis of Selected Transfer

RNA Blot Hybridization

Following the initial screening of transfectants using RT-PCR, those shown to have high recombinant gene expression are expanded for RNA blot analyses, to confirm the results of the RT-PCR, and for radioligand-binding analysis, to confirm expression of the recombinant protein.

Methods of RNA isolation, electrophoresis in formaldehyde/agarose gels, and transfer to 0.2-μm Biotrans membrane have been described in detail (3). A 1.3 *Xho*I-*Bgl*II fragment of the D2(415) receptor cDNA or a 0.68-kb *Hin*dIII fragment of the D3 receptor cDNA is ^{32}P-labeled by random hexamer-primed synthesis and used as a probe to detect D2 or D3 receptor RNA, respectively, in transfected cells. Hybridizations are carried out overnight in 5× standard saline citrate (SSC)/50% formamide at 42°C, followed by washes taken to 0.1× SSC at 55°C. The washed membranes are exposed to Kodak X-OMAT film for 24 hr at −70°C.

Radioligand-Binding Assays

Dopamine D2 receptor transfectants are lysed by replacing the growth medium with ice-cold hypotonic buffer (1 m*M* Na^+-HEPES, pH 7.4, 2 m*M* EDTA). After swelling for 10–15 min, the cells are scraped from the plate and centrifuged at 24,000 *g* for 20 min. The resulting crude membrane fraction is resuspended with a Brinkmann Polytron homogenizer at setting 6 for 10 sec in 5 m*M* K^+-HEPES (pH 7.4) and stored at −70°C for receptor-binding experiments. Aliquots of the membrane preparation are added to assay tubes containing the following (final concentrations): 50 m*M* Tris–HCl, pH 7.4, with 0.9% NaCl (Tris-buffered saline) except where indicated, 0.025% ascorbic acid, 0.001% bovine serum albumin, [^{125}I]epidepride (2000 Ci/mmol) or [^{3}H]spiperone, and appropriate drugs. Saturation and competition experiments are carried out in a volume of 0.5 and 0.25 ml, respectively. Incubations are initiated by the addition of tissue, carried out at 30°C for 60 min, and stopped by the addition of 10 ml of ice-cold wash buffer (10 m*M* Tris, pH 7.4, and 0.9% NaCl) to each assay. The samples are filtered through glass-fiber filters (Schleicher & Schuell, No. 30) and washed with an additional 10 ml of wash buffer. The radioactivity retained on the filter is counted using a Beckman LS 1701 scintillation counter or a gamma counter (LKB Clinigamma 1272).

Dopamine D3 receptor transfectants are lysed as described in the previous paragraph and aliquots of the membrane preparation (3–40 μg of protein) are added to duplicate assay tubes containing the following (final concentrations): assay buffer (Tris-buffered saline), 0.001% bovine serum albumin, [^{125}I]epidepride (2000 Ci/mmol), and appropriate drugs. Spiperone (2 or 5 μM) is used to define nonspecific binding for competition or saturation assays, respectively. Incubations are carried out at 37°C for 90 min.

Construction of Mutant Dopamine Receptor cDNAs

Trans-polymerase chain reaction which is used to join DNA fragments that contain a region of overlap (see Fig. 1, above, and accompanying text for the use of RT-PCR to clone the dopamine D3 receptor cDNA in two halves), is used to perform site-directed mutagenesis by employing primers that contain the desired base substitutions or deletions. To create the alanine substitution for Asp-80 in the dopamine D2 receptor cDNA, for example, we made the oligonucleotide 5′-CTTGCTGTGGCTG$\overline{\text{C}}$TCTTCTGGGTG-3′ and its reverse complement, which changes GAT (Asp) to GCT (Ala). To create the glutamate substitution for Asp-80, we made the oligonucleotide 5′-CTTGCTGTGGCTGA$\overline{\text{A}}$CTTCTGGTG-3′ and its reverse complement, altering the GAT to GAA (Glu) (6). A similar strategy can be used to create deletions rather than point mutations (Fig. 2).

Each mutation is carried out in two steps. In the first step, the fragments to be joined are amplified in separate reactions, in which the sense mutated primer (P3; Fig. 1) is paired with a downstream unmutated antisense oligonucleotide primer (P4), and the antisense mutated primer (P2) is paired with an upstream wild-type sense primer. Each 100-μl reaction utilizes 1 ng of D2(415) cDNA as template, 500 ng of each primer, 200 μM dNTPs, and Promega Taq polymerase and buffer. Twenty cycles of PCR are carried out using the following conditions; 94°C, 1 min; 60°C, 2 min; 72°C, 3 min.

The products of these PCR reactions are electrophoresed on a 0.8% low-melting-point gel in Tris–acetate–EDTA buffer. They are excised from the gel and melted for 10 min at 68°C, after which the second step, the trans-

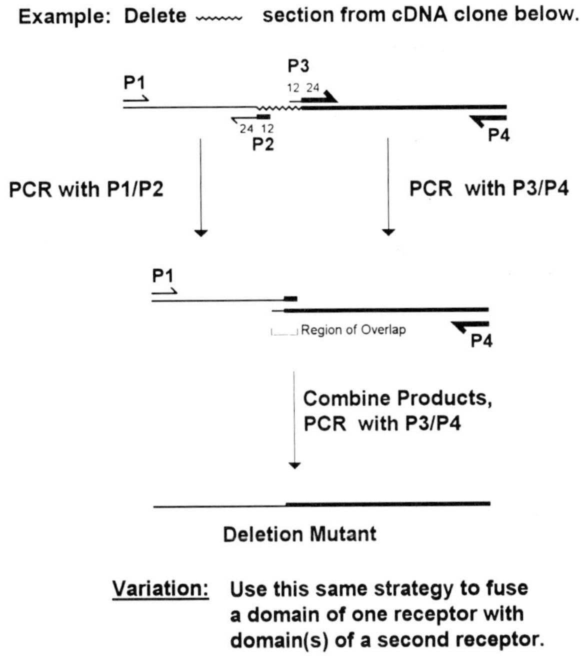

FIG. 2 Scheme for using *trans*-PCR to generate a deletion mutation in a cDNA. This same strategy can be used to fuse a cDNA segment encoding a domain of one receptor with a cDNA segment encoding domain(s) of a second receptor.

PCR reaction, is performed using 2 μl of each of the two gel-purified bands and 500 ng each of P1 and P4 in a 10 μl reaction, repeating the conditions described above. The PCR product is extracted with phenol and chloroform (see above section on RT-PCR for details), precipitated using ethanol, and digested to completion with *Xho*I (within the 5′ end of the PCR fragment) and *Bgl*III (within the 3′ end of the fragment), yielding a 1.3-kb fragment including virtually all of the coding region of D2(415). This fragment is gel-purified and ligated to pRSV-D2(415), which has been cleaved with the same enzymes and gel-purified away from its wild-type *Xho*I-*Bgl*II insert. Confirmation of the mutations is obtained by sequence analysis of the clones.

Summary

The methods outlined above for the cloning of dopamine receptor cDNAs and their expression in mammalian cells allow the investigator to study dopamine receptor pharmacology with a precision and detail not previously possible. These strategies can be used to create stable and clonal genetically engineered cell lines that express a given receptor subtype, a specific spliced form of a receptor subtype, or mutant and chimeric receptors. Such transfected cells can be evaluated to determine how receptors bind ligands in a receptor-selective manner, to learn how receptors transduce the energy of the binding of agonists into a functional response, to ascertain the mechanisms by which specificity or selectivity of coupling of receptors to various G proteins and signaling pathways is attained, and to reveal the molecular basis of regulation of receptor sensitivity.

Acknowledgments

We are grateful to Dr. Frederick Boyce for his considerable helpful advice relating to vectors and to *trans*-PCR methodology. The work is supported by NIH Grants NS28406 (RLN) and MH45372 (KAN), and by the V. A. Merit Review Program (KAN).

References

1. B. A. Cox, M. P. Rosser, M. R. Kozlowski, R. A. Henningsen, S. L. Smiley, K. M. Duwe, R. L. Neve, and K. A. Neve, submitted for publication.
2. B. Giros, M. P. Martres, C. Pilon, P. Sokoloff, and J.-C. Schwartz, *Biochem. Biophys. Res. Commun.* **176,** 1584 (1991).
3. R. L. Neve, P. Harris, K. S. Kosik, D. M. Kurnit, and T. A. Donlon, *Mol. Brain Res.* **1,** 271 (1986).
4. J. A. Donis, M. V.-Michelman, and R. L. Neve, *BioTechniques* **15,** 786 (1993).
5. C. Gorman, R. Padmanabhan, and B. H. Howard, *Science* **221,** 551 (1983).
6. K. A. Neve, B. A. Cox, R. A. Henningsen, A. Spanoyannis, and R. L. Neve, *Mol. Pharmacol.* **39,** 733 (1991).
7. B. A. Cox, R. A. Henningsen, A. Spanoyannis, R. L. Neve, and K. A. Neve, *J. Neurochem.* **59,** 627 (1992).
8. T. R. Fuerst, E. G. Niles, F. W. Studier, and B. Moss, *Proc. Natl. Acad. Sci. U.S.A.* **83,** 8122 (1986).
9. V. A. Luckow and M. D. Summers, *Bio/Technology* **6,** 47 (1988).
10. C. A. Chen and H. Okayama, *BioTechniques* **6,** 632 (1988).
11. J. P. Morgenstern and H. Land, *Nucleic Acids Res.* **18,** 3587 (1990).

[10] Baculovirus Expression of Receptors and Channels

Michael Cascio

Overview

For many receptor and channel proteins, detailed examination of structure–function at a molecular level has been precluded by their low natural abundance. Large-scale production of a pool of active protein which may then be isolated and purified is a prerequisite for subsequent detailed biochemical and biophysical characterization. With the advent of molecular biological technology, the identification and cloning of genes encoding receptors and channels have become fairly routine, and a general expression system would greatly aid in characterizing the translation products. The expression of active receptors and channels in heterologous systems is often complicated in that these proteins must be correctly targeted to the membrane and are often glycosylated or co- or posttranslationally modified, and may require assembly into oligomeric complexes in order to be functional. The baculovirus expression system, which utilizes insect baculoviruses as gene expression vectors in insect cell cultures or larvae, is currently the best candidate for such a general system (1–3). This eukaryotic expression system, in contrast to bacterial systems, provides the greatest likelihood, to date, of obtaining relatively large quantities of target eukaryotic proteins needing extensive co- and posttranslational modification in a biologically active form. Many of the procedures and methods described in this review may be found in a more extended form in the excellent bulletin by Summers and Smith (4) and the more recently published laboratory manual (5) for baculovirus-directed overexpression.

Baculoviruses are a diverse group of viruses with a narrow host range and no known nonarthropod hosts (6) and as such are ideal from the perspective of environmental and biological safety. Additionally, recombinant viruses typically lack the protective coat protein and are therefore not efficient in infecting target organisms naturally (i.e., orally) and do not persist in the environment. Baculovirus-mediated expression of lethal gene products, such as insect-specific neurotoxins, has also proven useful as pesticidal agents (see Refs. 7 and 8, and references therein). This expression system has also been utilized to produce immunogenic agents which may then be used as immunological agents (9–11).

Methods in Neurosciences, Volume 25

Among baculoviruses, the best characterized is the *Autographa californica* multiply embedded nuclear polyhedrosis virus (AcMNPV), and most of the studies referred to in this review utilize this virus. AcMNPV-infected cells may be identified morphologically by the presence of multiple occlusion bodies containing virions encased in a protein matrix within the nucleus. Another well-characterized baculovirus is the *Bombyx mori* nuclear polyhedrosis virus (BmNPV), which is often used to direct protein synthesis in silkworm larvae.

Baculovirus capsids are rod-shaped, and the encapsulated condensed DNA is closed, circular, and double-stranded. Another beneficial aspect of the baculoviral system is that the expandable viral capsid can accommodate very large inserts into its DNA (12). These viruses have a biphasic replication cycle which generates two biochemically and morphologically distinct infectious forms: extracellular budded virus (ECV) and occlusion bodies (OB). Virions typically enter target cells by adsorptive endocytosis, usurp the cellular machinery, and direct synthesis of viral proteins for the production and packaging of ECVs and OB. Extracellular budded viruses are budded and released on lysis and act in initiating secondary infections between cells in a permissive host organism. Baculoviruses also direct production of large nuclear OB. These large aggregates are composed of virions embedded in a crystalline polyhedrosis coat matrix. This matrix affords protection from the environment after death of the host organism and decomposition. These intact aggregates remain as contaminants in the host food source. On ingestion, the polyhedral coat protein is dissociated within the alkaline insect midgut, releasing virions which then act to initiate primary infection.

Infection by AcMNPV results in the complete virtual shutdown of host protein synthesis by 24 hr postinfection. Infection has three general phases: early, late, and very late. During early stages of infection, up to 6 hr postinfection, the host cell is reprogrammed for viral replication. In late phase, occurring between 6 and 20–24 hr postinfection, budded virus particles are produced. Finally, during very late-phase infection occurring from 20 hr postinfection until cell death, expression is almost exclusively viral specific and the large occluded virus particles are produced. The polyhedrin occlusion coat protein (29 kDa) is nonessential for viral propagation in culture and may account for up to 50% of total cellular protein production ($\sim$1 g/l of cells at a density of 10^6 cells/ml). This protein is not a membrane envelope, but rather an external surface coat, a calyx, which is a crystalline (giving rise to its refracticity) carbohydrate-rich protein matrix. To date, most studies have taken advantage of the highly abundant, very late-stage production of the polyhedrin protein by substituting a gene of interest under control of the polyhedrin (*polh*) promoter and using purified recombinant virus to direct cellular expression of target cells to overproduce the heterologous gene product.

The utility of the baculovirus system for any foreign gene must be determined empirically. However, the generic applicability of this system may be evidenced by its many academic and commercial successes in producing large amounts of biologically active products (for a fairly comprehensive list of genes, see Appendix 3 of Ref. 5). Evolutionarily, the translational machinery of eukaryotic cells appears to have been fairly well conserved with respect to synthesis, modification, and targeting of gene products. Insect cell lines have invariably been found to successfully direct translated heterologous proteins to appropriate membranes and organelles. Most of these proteins are glycosylated, phosphorylated, isoprenylated, acylated, amidated, or carboxymethylated in manners analogous to native proteins (for specific examples related to the modifications of receptors and channels expressed in baculovirus systems, see footnotes to Table III and references therein). The targeting and modification of heterologous proteins should also be empirically determined since not all proteins are correctly processed.

In the following sections the methods for isolating and purifying recombinant baculovirus for directing synthesis of selected genes encoding receptors and channels are briefly outlined, along with some strategies for optimizing expression as well as a general troubleshooting guide.

Experimental Design

Purifying Recombinant Baculoviruses: Overview

Due to the large size of the double-stranded circular supercoiled genome of AcMNPV (128 kb), conventional techniques, such as the use of restriction enzyme cleavage sites, for insertion of foreign DNA are inefficient. *Bombyx mori* nuclear polyhedrosis virus, an alternative baculovirus with a narrower host range, is similarly sized (~130 kb) and thus also renders conventional methodology ineffectual. Instead, both systems utilize allelic replacement via homologous recombination between cotransfected viral DNA and an appropriate transfer vector containing the gene of interest to produce recombinant virus. Production and purification may be achieved by a three-step process: (i) engineering the DNA encoding the protein of interest into the appropriate transfer plasmid, (ii) cotransfection of host cell lines with the transfer plasmid construct and the baculovirus into which recombination is desired, and (iii) purifying the recombinant baculovirus. Each of these steps is described in the following sections, and a broad overview is illustrated in Fig. 1. Many of these procedures were first outlined by Summers and Smith (4) and may also be found in a published laboratory manual (5).

Before beginning experiments, an initial choice of baculovirus, transfer vector, and host appropriate to one's needs must be made. The latter two choices are discussed below under Transfer Vector Construction, Infection of Cell Culture and Harvest, respectively. In selecting a baculovirus, most applications have been successful using AcMNPV, and unless specialized needs are anticipated, it is recommended for initial trials. The virus AcMNPV is most useful in infecting insect cell culture since the greatest diversity of transfer plasmids are designed for use with this virus. However, either AcMNPV or BmNPV may be useful for insect larvae infection. Some general features to be considered in selecting either AcMNPV or BmNPV as an infectious agent with respect to cell culture vs larval production are addressed in Table I. Regardless of the virus chosen, the recombinant baculovirus is purified in cell culture as described below.

Pure recombinant baculovirus containing the heterologous gene is then used in infection. Typically, a *Spodoptera frugiperda* (fall armyworm) cell line is used due to its quick doubling time and ease of handling. It has been shown that the transport machinery of baculovirus-infected cells becomes impaired and compromised late in infection (13–15). For secretion products or the production of the extracellular domains of membrane proteins, expression may be higher in infected larvae. Silkworm larvae are a useful host since their handling and rearing in mass culture has been developed over thousands of years. They grow efficiently, cheaply, and extremely rapidly, and the harvesting of hemolymph (containing stored secreted proteins), which comprises ~30% of the total volume of the larvae, is fairly easy. These larvae may be considered as a natural protein manufacturing and storage facility. Maeda and co-workers (16) overexpressed α-interferon in *B. mori* (silkworm) larvae; this was the first reported high-level expression of foreign gene products in a living organism. Analogously, the cytoplasmic domain of the insulin receptor was found to be more efficiently produced in another larval system, with *Trichoplusia ni* (cabbage looper) as a host organism (17). Secreted products are typically recovered by "milking" of the larval hemolymph, but for cellular products the infected organism must be dissected, complicating purification due to contamination from surrounding organs. Another factor is potential increased product heterogeneity since the target protein is being synthesized by a wide variety of cell types in larvae. For initial design of expression systems, it is recommended that production is first tested in cell culture.

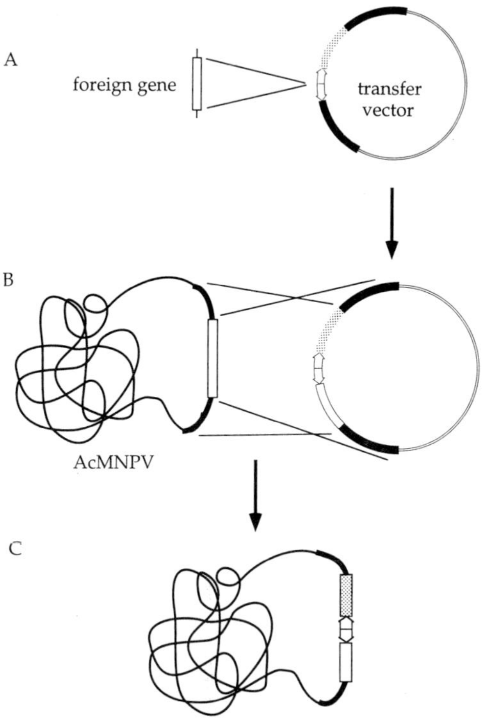

FIG. 1 Schematic overview of the construction of a typical recombinant baculovirus. (A) The foreign gene is inserted into a polylinker site in the transfer vector under control of an appropriate promoter. The orientation of the inserted gene may be checked by restriction mapping or PCR. In this example, an adjacent reporter gene (gray box) under separate promoter control is present in the transfer vector. Regions flanking the site of insertion in the transfer vector which are homologous to sequences in the wild-type virus are represented as black boxes. (B) This transfer plasmid containing the gene encoding the protein of interest is purified and subsequently introduced into host insect cell cultures along with isolated baculoviral DNA (in this example, AcMNPV was used) via cotransfection. (C) On recombination, a recombinant baculovirus containing sequences encoding the target protein, as well as a linked reporter gene, is produced. Selection for recombinant virus is made by loss or aquisition of phenotypic markers in infected insect cells, most typically by screening for loss of refractory nuclear occlusions due to elimination of the *polh* gene. Sequential rounds of serial dilution and plaque purification are used to isolate pure recombinant DNA.

TABLE I Cell Culture vs Larval Production Using AcMNPV or BmNPV

		Source of baculovirus	
Method	General comments	AcMNPV	BmNPV
Cell culture	Easy scale-up, though some specialized equipment may be necessary. Single cell type production results in less heterogeneity. Simpler purification.	Greatest diversity of commercially available baculovirus transfer vectors, host cell lines and promoter control. Quick doubling times (less than 24 hr). Generally successful in overproducing heterologous gene products.	Slower doubling times. No documented larger-scale culturing.
Larvae	Less costly, although must tend to living organisms. Higher efficiency for some proteins. Effectively limited to secreted proteins or extracellular domains (i.e., hemolymph-targeted), otherwise dissection and purification is time consuming. Large organism size.	Variety of transfer vectors available, so wider choice of alternate promoters, hence recombinant virus may retain occ$^+$ phenotype, allowing simple, cost-effective efficient infection via contaminated food sources (although *S. frugiperda* infection is inefficient by this manner). Larger host diversity—may optimize by infecting *S. frugiperda*, *T. ni*, or *H. virescens*.	Larger host—*B. mori* (silkworm) is ~5 g, so increased production. Infection by injection. High hemolymph production, ~0.5 ml/larvae. Used for silk production, so rearing and handling well characterized. Organism cannot survive in wild.

Transfer Vector Construction

All very late genes of baculovirus which have been characterized to date have been found to be unspliced. While some low levels of splicing of heterologous genes have been observed (18, 19), expression levels in these cases were not very strong. It is therefore recommended that the gene of interest not contain any introns, but instead be isolated as cDNA. The following sections focus on the overexpression of mature, unfused protein products. A discussion of fusion products may be found under Optimization and Alternate Methodologies. The 5′ cloning site in the gene should be close to the ATG start codon and the 3′ site should retain any polyadenylation sequences. The cDNA may have to be tailored at its ends for efficient ligation into the desired transfer vector.

Using standard molecular biological techniques (20), the foreign gene is inserted into the engineered transfer vector at a unique restriction site (oftentimes in a polylinker cassette) downstream from the desired viral promoter. Due to the large size of most transfer vector plasmids (typically 5–10 kb without gene insert), ligation is often problematic, so it is essential to isolate very clean and pure insert and plasmid of sufficient quantity to set up multiple ligation reactions at various molar ratios (5 : 1 to 1 : 5). The site of insertion is flanked on both sides with DNA homologous to sequences present in the baculovirus genome where insertion is targeted by recombination. The vector construct and parent baculovirus should be selected such that cells infected with double-recombinant virus exhibit a phenotype which may be distinguished from those infected with either the wild-type virus or the inherently unstable single-recombination virus product. To distinguish single from double recombinants, it is most useful to have markers present on the parent virus which will be lost following a double-recombination event.

A large variety of transfer vectors are now commercially available, with the widest array available from Invitrogen (San Diego, CA) and PharMingen (San Diego, CA). For more information and maps of the transfer plasmids, readers are referred to these vendors and to Chapter 7 of "Baculovirus Expression Vectors" (5). The primary consideration in transfer vector selection is the choice of promoter directing expression of the heterologous gene. The *polh* promoter is most commonly used to direct synthesis of target gene products and is recommended for initial studies. This promoter directs the very late, high overexpression of the occlusion body matrix protein, which is visible as refractory occlusions in cell nuclei and plaques as described above. Another benefit of using the *polh* promoter is that, on double-recombination, the loss of its nonessential polyhedrin coat protein may be easily identified visually under light microscope as cells and plaques which are occ$^-$, i.e., lacking refractory large nuclear occlusion bodies. For ease of selection, some plasmids also provide color selection via chromogenic substrates (e.g., blue/white *lacZ* systems) as an adjunct to occ^- selection (21). In these vectors, a reporter *lacZ* gene encoding β-galactosidase is present adjacent to the polylinker site under separate promoter control.

Another frequently used promoter is *p10,* another very strong, very late promoter. Similar to polyhedrin, the *p10* gene product is also nonessential for viral infection and replication in tissue culture. Unfortunately p10 protein, which is the component comprising large fibrillar structures in infected cells (22), has no readily identifiable phenotype for selection purposes. Most commercial plasmids utilizing this promoter are used in conjunction with an engineered baculovirus carrying a *lacZ* gene at the *p10* locus, so replacement of this gene via recombination with a transfer plasmid containing the heterologous gene may be screened by selecting white plaques (both wild-type infections and single recombinants will appear as blue plaques after development with chromogenic substrate).

Coexpression of more than one gene product is generally achieved by either of two approaches. Multiple genes may be inserted into a single transfer vector, each under control of separate promoters. This type of approach using back-to-back *polh* and/or *p10* promoters has been successful in yielding high expression levels of multiple gene products [see Jarvis *et al.* (15), and references therein]. More recently, this type of approach was used in successfully overexpressing a heteromeric membrane protein; using a transfer vector allowing insertions behind both the *polh* and *p10* promoters, the α- and β-subunits of the H,K-ATPase were overexpressed in insect cells to produce active protein (23). Alternatively, multiple transfer vectors may be constructed, each of which has a single gene insert. Each recombinant baculovirus is then purified independently. The insect cell lines are then coinfected with multiple recombinant viruses, each directing overexpression of a single subunit of the heterolingomeric protein. This approach was utilized in directing coexpression of α- and β-subunits (both subunits under *polh* control) to produce active forms of the interleukin-2 receptor (24) and Na,K-ATPase (25).

The following sections are written with the assumption that the *polh* promoter has been selected to direct heterologous synthesis in *S. frugiperda* cell cultures utilizing a transfer vector designed to recombine with an AcMNPV derivative. For consideration of other promoters, cell lines, and alternate protocols, see Optimization and Alternate Methodologies.

Allelic Replacement

The most common method for incorporating foreign DNA into insect cells is by cotransfection of transfer vector plasmid and wild-type viral DNA via calcium phosphate precipitation (4). Other methods include DEAE–dextran or lipid-mediated transfection (5). Alternatively transfection may be more efficiently mediated (100 to 1000-fold) by electroporation (26). It has been noted that the recombination efficiency may be enhanced through UV irradiation (27). Regardless of the method used, all DNA used in transfections must be very clean since *S. frugiperda* cell lines are very sensitive to contamination. The procedures recommended for eliminating toxic contamination may be found under Troubleshooting. Goswami and Glazer (28) have suggested that the necessity of isolating or purchasing pure viral DNA might be avoided by simply infecting the host cell line with wild-type virus before transfecting with the transfer vector.

Briefly, monolayers of insect cells are cotransfected with both the transfer plasmid containing the desired heterologous cDNA and an appropriate baculovirus. The cells are monitored for signs of infection (i.e., typically occ^- for systems utilizing the *polh* promoter). After 4–6 days postinfection, budded virus (ECV) present in the media is harvested and used to infect monolayers of insect cells for subsequent rounds of virus purification as described in the next section.

Early recombinant virus production relied on *in vivo* recombination of cotransfected transfer plasmids with wild-type virus (via regions of wild-type sequence flanking the foreign DNA insert in the plasmid vector). The frequency of double-recombination events was sufficiently low in our hands and that of others, so that isolation of pure recombinant virus proved difficult and time consuming, requiring multiple rounds of serial dilution and plaque purification. More recently, linearized forms of AcMNPV (29) which require a recombination event for viability have become commercially available. Generally, a unique site is engineered into the viral genome within the area between the homologous sequences where recombination is desired, and the baculovirus is linearized rendering it noninfectious. Cotransfection with a complementary plasmid construct may rescue the lethal deletion (with respect to virus) via recombination, reconstituting viable viral DNA encoding for the heterologous gene product. Two commercially available transfer vectors utilizing this type of approach are the MaxBac (Invitrogen, San Diego, CA) and the BaculoGold (PharMingen, San Diego, CA) systems. Both systems provide linearized AcMNPV and have been reported to significantly reduce the time required for virus purification. In our hands, the use of linearized virus reduced the rounds of plaque purification necessary for viral purification by a factor of ~ 4–5 (M. Cascio, R. L. Grodzicki, and R. O. Fox, unpublished observation).

Another strategy was to generate the recombinant virus by homologous recombination in the yeast *Saccharomyces cerevisiae* for more rapid and easier selection using a baculovirus derivative which could be stably maintained in yeast (30). Using this method, the reported time required for purification of recombinant virus was also significantly shortened.

Purification of Recombinant Baculovirus

Successful generation of a recombinant baculovirus may be identified by one of four general methods:

i. visual identification of occlusion phenotype (if using a *polh*-based system);
ii. morphological selection by loss of or acquisition of linked reporter genes (e.g., *lacZ*-based systems);
iii. identification of specific sequences in isolated baculoviral DNA by PCR amplification or DNA hybridization;
iv. screening for heterologous gene product (e.g., antibody binding or activity assay).

The first two methods are more general, less labor intensive, and recommended for standard use in purifying virus. While the utility of the latter two methods is independent of vector design, they are best utilized for verifying the presence of insert in a given population or the production of a specific gene product, respectively. Importantly, DNA hybridization and product screening, as well as some reporter gene strategies, cannot distinguish between single- and double-recombination events and are of limited use in purification.

The ability to visually screen plates is an exceptionally important skill essential for easy discrimination and successful purification. The polyhedra may be observed as refractory nuclear occlusions having a radius of 1–2.5 μm which may be observed on a dissecting microscope by using an intense illuminating source set at an acute angle to the sample plate. They are easily observable both in infected cells and as residual protein in plaques.

After successful cotransfection, sequential rounds of plaque purification are carried out in order to purify the recombinant virus (4). Since different recombination events from the same transfection are not necessarily equivalent, expression levels may differ between clones. Therefore more than a single recombinant baculovirus should be isolated and carried through purification, and the expression levels compared in order to select the best candidate. Serial dilutions of inoculum are added to monolayers of cells in culture. After incubation (typically 1 hr at 27°C), the cells are overlaid with agarose. If the transfer vector also contains a *lacZ* reporter gene, the chromogenic substrate 5-bromo-4-chloro-3-indolyl-β-D-galactopyranoside (X-Gal), may also be included. After 4–6 days agarose plugs are taken, using a sterile pipet tip, above plaques which are occ$^-$ (also appear blue if *lacZ* is present) and added to small volumes of media. The agarose overlay (which may also include a neutral red dye for ease of visualization) localizes budded virus to the plaque vicinity, and recombinant virus will be present in the plug. Care should be taken in selecting occ^- plaques which are well separated from occ^+ plaques (i.e., taken from most dilute infection producing occ^- plaques). The resultant medium is then used in subsequent serial dilutions. Rounds of serial dilution are continued until the viral inoculum is pure (i.e., all infected cells are occ^-). It is imperative that recombinant viral inoculum be completely pure, otherwise wild-type virus will quickly predominate. The purification of recombinant virus is usually the most time-consuming step in baculovirus preparations.

Once virus is purified, viral stocks may be generated by infecting a suspension culture of insect cells. The culture should be concentrated by gentle centrifugation at 1000*g* for 10 min, incubated with virus for 1 hr at 37°C, and gently resuspended with media to a density of $1.5–2.0 \times 10^6$ cells/ml. After 3–5 days, the cells are pelleted, and the supernatent containing the viral inoculum is collected for later titering. Supernatant from all subsequent infections may be saved for titering and eventual use in infections. Viral stocks should not be frozen, but may be stored at 4°C for prolonged periods of time (no loss in titer for at least 1 year). Virus should not undergo more than five passages.

An alternative general strategy developed for the rapid identification and purification of recombinant baculovirus is the use of fluorescence-activated cell sorting (FACS). Fluorescent β-galactosidase substrate is used to select cells infected with recombinant baculovirus encoding a target gene with an associated reporter *lacZ* gene. FACS is used to quickly purify recombinant baculoviruses encoding for either the LDL receptor or the cystic fibrosis gene product (CFTR) (31). Another strategy developed to more quickly purify virus is the use of magnetic beads coated with monoclonal antibody to immunoselect cells expressing a given surface epitope. This method is used to purify recombinant baculovirus encoding the transferrin receptor (32).

Viral Titering

Virus may be titered by either end-point dilution or plaque assay (4). Of these two methods, end-point dilution is recommended since it is less expensive and easier, especially if there is an associated chromogenic reporter gene. Briefly, end-point dilution involves serial dilution of the viral stock, infection of monolayers in multiwell plates, and determination of the dilution at which 50% of the wells would show evidence of an infectious event ($TCID_{50}$). This is the end-point dilution, and the reciprocal of this value is the titer in infectious doses per unit of inoculum. To determine the plaque-forming units (pfu) per ml, simply divide by the volume of inoculum (in ml) added to the monolayer in the trials and multiply by 0.69.

Alternately, to titer the virus by plaque assay, serial dilutions of the inoculum are used to infect monolayers, and then overlaid with agarose (as in plaque assays conducted for viral purification). By counting the number of plaques in replicate plates, one can calculate the titer (pfu/ml).

Infection of Cell Culture

Lepidopteran cells are commonly grown at 27°C in media that supply basic nutrients and are buffered with sodium phosphate buffer, precluding the need for atmospheric CO_2 required by CO_2-bicarbonate buffer systems. Typically, cells are grown in suspension or monolayer in Graces (sometimes supplemented with lactalbumin hydrolysate and yeastolate), IPL-41, or TC-100; it is recommended that in initial studies all three be completed with 10% fetal bovine serum. All media and supplements may be purchased from Gibco-BRL. Antibiotic and antimycotic agents may also be added. Alternatively, many workers utilize less-expensive low-serum or serum-free media. However, cells grown in serum-free environments are typically very fragile and should not be routinely centrifuged during passage and harvesting. For additional details in culturing of cells, refer to Cell Viability under Troubleshooting.

Many different cell lines are currently available, each with unique properties which may affect the efficiency of expression. Some lines may only be cultured as adherent monolayers, and given the inherent limitations in culture cell volume, these are not recommended except if one is attempting to overexpress a secreted protein. Sf9 cells are recommended for initial studies. These cells may be grown in suspension or monolayers, are easy to manipulate, and have quick (16–22 hr in suspension) doubling times. Alternate cell lines may produce higher yields or less heterogenous products. For example, Hink and co-workers found differential expression of three different proteins in 23 different cell lines (33). Some of the available cell lines are outlined in Table II. The levels of expression obtained in any given cell line cannot be predicted a priori. Since these levels must be empirically determined, comparative studies using multiple permissive cell lines for eventual selection of the most efficient host should be considered.

Typically, cells are infected at a relatively high multiplicity of infection (MOI between 5 and 10). Given a standard Poisson distribution, a MOI of 5 ensures that >99% of the cells in a given population receive at least one infectious unit. To increase the efficiency of infection, cells should be concentrated prior to infection by gentle centrifugation (1000 *g* for 10 min),

TABLE II Selected Available Cell Lines

Organism	Cell line	Source	Virus	Notes
Spodoptera frugiperda	Sf9	Invitrogen and PharMingen	AcMNPV	Most commonly used cell line. Monolayer or suspension.
	Sf21	Invitrogen and PharMingen	AcMNPV	Larger than Sf9, so potentially higher protein production. Monolayer or suspension.
Trichoplusia ni	High five (BT1-TN-5B1-4)	Invitrogen	AcMNPV	Monolayer, limiting cell density. Potentially higher yields of secreted carriers.
Bombyx mori	Bm5 or BmN4	[a]	BmNPV	Reported higher yields than alternate sources. Monolayer or suspension.
Mamestra brassicae		[b]	AcMNPV	Monolayer. Adjunct to larval production.

[a] S. Maeda, *Annu. Rev. Entomol.* **34,** 351 (1989).
[b] L. A. King, S. G. Mann, A. M. Lawrie, and S. H. Mulshaw, *Virus Res.* **19,** 93 (1991).

resuspended in the volume of viral inoculum calculated to give the desired MOI, and incubated for 1 hr at 27°C without agitation. After incubation, the cells are resuspended in fresh media in spinner flasks to give a density between 1.0 and 5.0 × 10^6 cells/ml.

Harvest

Cells are typically harvested 2–3 days postinfection by gentle centrifugation at 1000 *g* for 10 min. For proteins expressed in membranes or intracellularly, cell pellets may then be stored at −70°C until use. When expressing secreted proteins or extracellular domains of membrane proteins, cells may be grown in serum-free media to simplify purification, and the serum and cell pellets assayed separately to determine protein content.

A list of receptors, channels, and miscellaneous membrane proteins which have been successfully (i.e., functionally) overexpressed in insect cells are presented in Table III. The proteins have been broadly divided into three categories: (i) soluble receptors (typically transcription enhancers), (ii) extramembranous domains of receptors and channels, and (iii) holoreceptors, channels, and membrane proteins. Membrane proteins (transporters or signal tranducers) which are neither receptors nor channels have been included in the table since protocols used for these proteins may be applicable and useful to investigators. Unless otherwise noted, the proteins were expressed in AcMNPV-infected *S. frugiperda* cell lines under control of the *polh* promoter.

TABLE III Receptors and Membrane Proteins Overexpressed Using Baculovirus Systems

Protein	Origin	Notes	Ref.[a]
Transcriptional enhancers			
Androgen receptor	Human		1, 2
Estrogen receptor	Human		3
Glucocortocoid receptor	Human	Phosphorylated	4
	Rat		5
Mineralcorticoid receptor	Human		6, 7
Progesterone receptor	Chick		8
	Human		9
β_1 thyroid hormone receptor	Human		10
Vitamin D receptor	Human		11
	Rat		12
Membrane protein receptors—Soluble domains			
CD4 (L3T4) (peptide recognition)	Human	Extracellular, glycosylated	13, 14
CR2 (CD21) Epstein–Barr virus/C3d receptor	Human	Extracellular	15
Epidermal growth factor receptor	Human	Extracellular, glycosylated	16
		Cytoplasmic phosphorylated	17
Growth hormone receptor	Human	Extracellular	18
Insulin receptor	Human	Extracellular, $(\alpha\beta)_2$, glycosylated	19, 20
		Cytoplasmic, β only, phosphorylated	21, 22
		Cytoplasmic, expressed in larvae	23
Interferon γ receptor	Mouse	Extracellular, limited or no glycosylation	24
Interferon γ receptor	Human	Extracellular	61
Interleukin-2 receptor	Human	Extracellular	62
Nerve growth factor receptor	Human	Extracellular, limited or no glycosylation	25
Thyrotropin (TSH) receptor	Human	Extracellular	63
Transferrin receptor	Human	Extracellular	26
Membrane proteins			
Muscarinic acetylcholine receptor, M1 and M2	Human	Glycosylated	27
Muscarinic acetylcholine receptor, M3	Rat	Glycosylated	28
Nicotinic acetylcholine receptor, α-subunit	Bovine	Glycosylated	29

Nicotinic acetylcholine receptor, α-subunit	Chicken	Glycosylated	30
β-Adrenergic β1-AR receptor	Human	Glycosylated	31
β-Adrenergic β1-AR receptor	Turkey	Glycosylated	27
β-Adrenergic β2-AR receptor	Hamster	Glycosylated	32
β-Adrenergic β2-AR receptor	Human	Glycosylated	33
CD23 (low affinity IgE receptor)	Human		34
CD4	Human	Glycosylated	35
Cystic fibrosis gene product (CFTR)	Human	Little or no glycosylation	36
D4 dopamine receptor	Human		37
Epidermal growth factor receptor	Human	Glycosylated	38
Fibroblast growth factor receptor	Human	Glycosylated	39
Fibroblast growth factor receptor	*Xenopus*	Glycosylated	40
GABA receptor, homomeric *Rdl*	*Drosophila*	Extracellular	64
Gap junctions, β_1 liver (connexin 32)	Human		41
GC-A (guanylyl cyclase activity) atrial natriuretic peptide receptor	Rat		42
Glycine receptor, homomeric α receptor	Human	Glycosylated	43
H,K-ATPase, coexpression of α- and β-subunits	Rat	Utilization of both *polh* and p10 promoters in single virus, glycosylated, phosphorylated	44
Insulin holoreceptor, α- and β-subunits	Human	α and β from single-chain precursor under *polh* control, glycosylated, phosphorylated	19
Interferon γ receptor	Human		65
Interleukin-2 receptor, co-expression of α- and β-subunits	Human	Coinfection with baculoviruses each having a single subunit under *polh* control	45
Multidrug transporter (multidrug resistance MDR1 gene product)	Human	Not glycosylated, phosphorylated	46
N-formyl peptide receptor (NFPR)	Human	No observed coupling with exogenous G proteins	66
Neurokinin (NK-2) receptor	Human		47
NMDA receptor	Rat	Heterogeneous glycosylation, His tagged, Ni^2-chelate chromatography purification	67
Opsin	Bovine	Glycosylated	48
Platelet-derived growth factor receptor	Human	Phosphorylated	49
Poliovirus receptor	Human	Glycosylated	50

(continued)

TABLE III (*continued*)

Protein	Origin	Notes	Ref.[a]
Prolactin receptor	Rabbit		51
Na/Ca,K exchanger	Bovine		52
Na^+/glucose cotransporter	Rabbit	Little or no glycosylation	53
Na^+/H^+ antiporter	Human	Not glycosylated, phosphorylated, limited activity	54
Na,K-ATPase, α_1- and β_1-subunits	Rat	Coinfection with baculoviruses each having a single	55
$\alpha 2$, $\alpha 3$		subunit under *polh* control	56
Serotonin ($5-HT_{1A}$) receptor	Human	Couples to endogenous G_0-like protein	68
Shaker H4 K^+ channel	*Drosophila*		57
Synapsin IIa	Rat	Phosphorylated	58
Transferrin receptor	Human	Glycosylated, acylated	59
Tumor necrosis factor receptor	Human		60

[a] 1. Y.-B. Xie, Y.-P. Sui, L.-X. Shan, J. J. Palvimo, D. M. Phillips, and O. A. Jane, *J. Biol. Chem.* **267,** 4939 (1992); 2. C. Chang, C. Wang, H. F. DeLuca, T. K. Ross, and C. C.-Y. Shih, *Proc. Natl. Acad. Sci. U.S.A.* **89,** 5946 (1992); 3. M. Brown and P. A. Sharp, *J. Biol. Chem.* **265,** 11,238 (1990); 4. G. Srinivasan and E. B. Thompson, *Mol. Endocrinol.* **4,** 209 (1990); 5. E. S. Alnemri, A. B. Maksymowych, N. M. Robertson, and G. Litwack, *J. Biol. Chem.* **266,** 3925 (1991); 6. E. S. Alnemri, A. B. Maksymowych, N. M. Robertson, and G. Litwack, *J. Biol. Chem.* **266,** 18,072 (1991); 7. N. Binart, M. Lombes, M.-E. Rafestin-Oblin, and E.-E. Baulieu, *Proc. Natl. Acad. Sci. U.S.A.* **88,** 10,681 (1991); 8. Elliston, J. F., J. M. Beekman, S. Y. Tsai, B. W. O'Malley, and M.-J. Tsai, *J. Biol. Chem.* **267,** 5193 (1992); 9. L. Shemshedini, J. Ji, C. Brou, P. Chambon, and H. Gronemeyer, *J. Biol. Chem.* **267,** 1834 (1992); 10. J. zu Putlitz, S. Datta, L. D. Madison, E. Macchia, and J. L. Jameson, *Biochem. Biophys. Res. Commun.* **175,** 285 (1991); 11. P. N. MacDonald, C. A. Haussler, C. M. Terpening, M. A. Galligan, M. C. Reeder, G. K. Whitfield, and M. R. Haussler, *J. Biol. Chem.* **266,** 18,808 (1991); 12. T. K. Ross, J. M. Prahl, and H. F. DeLuca, *Proc. Natl. Acad. Sci. U.S.A.* **88,** 6555 (1991); 13. R. E. Hussey, N. E. Richardson, M. Kowalski, N. R. Brown, H.-C. Chang, R. F. Siliciano, T. Dorfman, B. Walker, J. Sodroski, and E. L. Reinherz *Nature (London)* **331,** 78 (1988); 14. Y. Morikawa, H. A. Overton, J. P. Moore, A. J. Wilkinson, R. L. Brady, S. J. Lewis, and I. M. Jones, *AIDS Res. Hum. Retroviruses* **6,** 765 (1990); 15. G. R. Nemerov, J. J. Mullen, P. W. Dickson, and N. R. Cooper, *J. Virol.* **64,** 1348 (1990); 16. C. Greenfield, I. Hiles, M. D. Waterfield, M. Federwisch, A. Wollmer, T. L. Blundell, and N. McDonald, *EMBO J.* **8,** 4115 (1989); 17. P. B. Wedegaertner and G. N. Gill, *J. Biol. Chem.* **264,** 11,346 (1989); 18. Y. Ota, A. Asakura, Y. Matsura, H. Kondo, A. Hitoshio, A. Iwane, T. Tanaka, M. Kikuchi, and M. Ikehara, *Gene* **106,** 159 (1991); 19. J. I. Paul, J. Tavare, R. M. Denton, and D. F. Steiner, *J. Biol. Chem.* **265,** 13,074 (1990); 20. J. Sissom and L. Ellis, *Biochem. J.* **261,** 119 (1989); 21. L. Ellis, A. Levitan, M. H. Cobb, and P. Ramos, *J. Virol.* **62,** 1634 (1988); 22. R. Herrera, D. Lebwohl, A. G. de Herreros, R. G. Kallen, and O. M. Rosen, *J. Biol. Chem.* **263,** 5560 (1988); 23. M. Villalba, S. R. Wente, D. S. Russell, J. Ahn, C. F. Reichelderfer, and O. M. Rosen, *Proc. Natl. Acad. Sci. U.S.A.* **86,** 7848 (1989); 24. M. Fountoulakis, E.-J. Schlaeger, R. Gentz, J.-F. Juranville, M. Manneberg, L. Ozmen, and G. Garotta, *Eur. J. Biochem.* **198,** 441 (1991); 25. P. Vissavajjhala and A. H. Ross, *J. Biol. Chem.* **265,** 4746 (1990); 26. S. White, R. Taetle, P. A. Seligman, M. Rutherford, and I. S. Trowbridge, *Cancer Res.* **50,** 6295 (1990); 27. E. M. Parker, K. Kameyama, T. Higashijima, and E. M. Ross, *J. Biol. Chem.* **266,** 519

(1991); 28. S. Vasudevan, H. Reilander, G. Maul, and H. Michel, *FEBS Lett.* **283,** 52 (1991); 29. M. Iizuka and K. Fukuda, *J. Biochem.* (*Tokyo*) **114,** 140 (1993); 30. A. E. Atkinson, F. G. P. Earley, D. J. Beadle, and L. A. King, *Eur. J. Biochem.* **192,** 451 (1990); 31. V. Ravet, N. Blin, J.-L. Guillaume, F. Petitjean, L. Cabanie, and A. D. Strosberg, *J. Rec. Res.* **13,** 541 (1993); 32. S. T. George, M. A. Arabian, A. E. Ruoho, J. Kiely, and C. C. Malbon, *Biochem. Biophys. Res. Commun.* **163,** 1265 (1989); 33. S. T. George, M. A. Arbabian, A. E. Ruoho, J. Kiely, and C. C. Malbon, *Biochem. Biophys. Res. Commun.* **163,** 1265 (1989); 34. K. U. Jansen, J. Shields, J. Gordon, J. Cairns, P. Graber, and J.-Y. Bonnefoy, *J. Rec. Res.* **11,** 507 (1991); 35. N. R. Webb, C. Madoulet, P.-F. Tosi, D. R. Broussard, L. Sneed, C. Nicolau, and M. D. Summers, *Proc. Natl. Acad. Sci. U.S.A.* **86,** 7731 (1989); 36. N. Kartner, J. W. Hanrahan, T. J. Jensen, A. L. Naismith, S. Sun, C. A. Ackerly, E. F. Reyes, L.-C. Tsui, J. M. Rommens, C. E. Bear, and J. R. Riordan, *Cell* (*Cambridge, Mass.*) **64,** 681 (1991); 37. A. Mills, B. Allet, A. Bernard, C. Chabert, E. Brandt, C. Cavegn, A. Chollet, and E. Kawashima, *FEBS Lett.* **320,** 130 (1993); 38. C. Greenfield, G. Patel, S. Clark, N. Jones, and M. D. Waterfield, *EMBO J.* **7,** 139 (1988); 39. J. Xu, M. Nakahara, J. W. Crabb, E. Shi, Y. Matuo, M. Fraser, M. Kan, J. Hou, and W. L McKeehan, *J. Biol. Chem.* **267,** 17,792 (1992); 40. S. A. N. Brown and R. Friesel, *Biochem. Biophys. Res. Commun.* **193,** 1116 (1993); 41. K. A. Stauffer, N. M. Kumar, N. B. Gilula, and N. Unwin, *J. Cell Biol.* **115,** 141 (1991); 42. M. Chinkers, S. Singh, and D. L. Garbers, *J. Biol. Chem.* **266,** 4088 (1991); 43. M. Cascio, N. E. Schoppa, R. L. Grodzicki, F. J. Sigworth, and R. O. Fox, *J. Biol. Chem.* **268,** 22,135 (1993); 44. C. H. W. Klaassen, T. J. F. Van Uem, M. P. De Moel, G. L. J. De Caluwe, H. G. P. Swarts, and J. J. De Pont, *FEBS Lett.* **329,** 277 (1993); 45. C. Lindqvist, C. Engberg, P. Ojala, M. Karp, K. Akerman, and C. Oker-Blom, *Scand. J. Immunol.* **38,** 267 (1993); 46. U. A. Germann, M. C. Willingham, I. Pastan, and M. M. Gottesman, *Biochemistry* **29,** 2295 (1990); 47. D. Aharony, J. Little, S. Powell, B. Hopkins, K. R. Bundell, W. L. McPheat, R. D. Gordon, G. Hassall, R. Hockney, R. Griffin, and A. Graham, *Mol. Pharmacol.* **44,** 356 (1993); 48. J. J. M. Jansen, W. R. Mulder, G. L. J. De Caluwe, J. M. Vlak, and W. J. De Grip, *Biochim. Biophys. Acta* **1089,** 68 (1991); 49. D. K. Morrison, D. R. Kaplan, J. A. Escobedo, U. R. Rapp, T. M. Roberts, and L. T. Williams, *Cell* (*Cambridge, Mass.*) **58,** 649 (1989); 50. G. Kaplan, M. S. Freistadt, and V. R. Racaniello, *J. Virol.* **64,** 4697 (1990); 51. C. Cahoreau, B. Petridou, M. Cerutti, and G. Devauchelle, *Biochimie* **74,** 1053 (1992); 52. H. Reilander, A. Achilles, U. Friedel, G. Maul, F. Lottspeich, and N. J. Cook, *EMBO J.* **11,** 1689 (1992); 53. C. D. Smith, B. A. Hirayama, and E. M. Wright, *Biochim. Biophys. Acta* **1104,** 151 (1992); 54. P. Fafournoux, J. Ghysdael, C. Sardet, and J. Pouyssegur, *Biochemistry* **30,** 9510 (1991); 55. A. W. DeTomaso, Z. J. Xie, G. Liu, and R. W. Mercer, *J. Biol. Chem.* **268,** 1470 (1993); 56. G. Blanco, Z. J. Xie, and R. W. Mercer, *Proc. Natl. Acad. Sci. U.S.A.* **90,** 1824 (1993); 57. K. Klaiber, N. Williams, T. M. Roberts, D. M. Papazian, L. Y. Jan, and C. Miller, *Neuron* **5,** 221 (1990); 58. Y. L. Siow, T. J. Chilcote, F. Benfenati, P. Greengard, and G. Thiel, *Biochemistry* **31,** 4268 (1992); 59. D. L. Domingo and I. S. Trowbridge, *J. Biol. Chem.* **263,** 13,386 (1988); 60. H. Loetscher, Y.-C. E. Pan, H.-W. Lahm, R. Gentz, M. Brockhaus, H. Tabuchi, and W. Lesslauer, *Cell* (*Cambridge, Mass.*) **61,** 351 (1990); 61. D. Moosmayer, E. Gehrlach, H. Bossinger, P. Scheurlich, and K. Pfizenmaier, *Lymphokine Cytokine Res.* **13,** 147 (1994); 62. T. R. Sana, Z. Wu, and T. L. Ciardelli, *Protein Sci.* **3** (Suppl. 1), 68 (1994); 63. G. S. Seetharamaniah, A. Kurosky, A. K. Desai, J. S. Dallas, and B. S. Prabhakar, *Endocrinology* **134,** 549 (1994); 64. H.-J. Lee, T. Rocheleau, H.-G. Zhang, M. B. Jackson, and R. H. ffrench-Constant, *FEBS Lett.* **335,** 315 (1993); 65. O. Quehenberger, E. Prossnitz, C. G. Cochrane, and R. D. Ye, *J. Biol. Chem.* **267,** 19,757 (1992); 66. S, Sydow, A. K. E. Köpke and J. Spiess, *Protein Sci.* **3** (Suppl. 1), 151 (1994); 67. J. G. Mulheron, S. J. Casanas, J. M. Arthur, M. N. Garnovskaya, T. W. Gettys, and J. R. Raymond, *J. Biol. Chem.* **269,** 12,954 (1994).

When reported, maximal expression of receptors and membrane proteins was typically observed 2–3 days postinfection. Either SDS–PAGE or Western immunoblots of preparations should be used to monitor expression and select temporal conditions at which expression is maximal. The yields, for membrane proteins, were typically in the range of 0.5–5 mg/liter of cell culture. While this is much lower than some of the yields reported for secreted proteins, it is orders of magnitude higher than yields obtained from alternate sources. It should be noted that while the levels of protein synthesis may be quite high, the eventual yield of functional purified protein may be significantly lower, and care should be taken in evaluating expressional efficiency. This discrepancy may arise due to the accumulation of subpopulations of protein which are inactive due to misfolding, erroneous cellular targeting, or incorrect modification as a result of overwhelming the cellular machinery by the rate and magnitude of overexpression coincident with the breakdown of the native host cellular machinery with infection.

Membrane protein receptors and channels are typically glycosylated, with the site(s) of glycosylation most often being an asparagine (N-linked). It has been shown that insect cells are capable of glycosylating heterologous proteins at either N-linked or serine- and threonine-linked glycosylation (O-linked) sites, e.g., recombinant human chorianogonadotropin contains both N- and O-linked carbohydrates (34). However, complex glycosylation (the addition of galactose or sialic acid units) of the correctly synthesized core sugars is seldom observed (for review, see Chapter 15 of Ref. 5). As noted in Table III, many of the membrane proteins expressed in the baculovirus system are glycosylated. For many of these proteins the heterogeneity and complexity of glycosylation are undetermined; in most studies the glycosylation of the foreign proteins was examined indirectly by the sensitivity of protein electrophoretic mobility on SDS gels to specific endoglycosidases (which identify the class of glycosidic linkage by their unique cleavage specificities) or to the addition of tunicamycin, a glycosylation inhibitor. The efficiency of glycosylation has also been observed to generally decrease as infection proceeds, probably due to the overwhelming of cellular machinery and/or shutdown of the biosynthesis of the host proteins (which includes proteins necessary for processing). The extent of glycosylation must be determined empirically, and if desired alternate promoters or earlier harvesting times should be chosen accordingly.

Purification

As discussed above, the overproduction of *polh*-directed heterologous gene synthesis often overwhelms the targeting machinery of the cell and viral infection shuts down host biosynthesis, so it is not uncommon to observe proteins sequestered in inappropriate membranes or organelles. For example, overexpressed recombinant α acetylcholine receptor (35), the multidrug resistance protein (36), β_1 gap junction hexamers (37), and the α glycine receptor (38) have been observed in electron micrographs and immunofluorescence studies to be present not only in the plasma membrane of the host insect cells, but also in other cellular membranes. Investigators should also be aware that the conditions for effective solubilization of the membrane protein may differ from published protocols (38, 39). These differences may be a function of the different environment of the insect cell membrane compared to natural tissue sources and/or due to subtle differences in the recombinant protein. This necessitates the development of detergent screening assays. A discussion of the solubilization of membrane proteins is beyond the scope of this review, but interested readers are directed to several excellent reviews (40, 41). Analogously, any reported protocols may have to be modified somewhat in purifying the recombinant protein. Most purification protocols for membrane proteins produced in cell culture involve an initial fractionation of the host cells via lysis and centrifugation to select for the appropriate protein-enriched compartment (e.g., plasma membrane preparations if targeted gene product is sequestered therein). Examples of specific protocols may be found in the references to Table III.

An alternate general strategy to accommodate quick purification is to fuse a series of histidine residues (usually six His) at the amino or carboxy terminus to the heterologous protein and then purify the protein chromatographically using immobilized metal chelate affinity resins (42). This provides a highly efficient single-step purification of the protein. The histidines may be incorporated into the cDNA encoding the protein by using standard molecular biological techniques, or transfer vectors may be purchased which incorporate polyhistidine tags at the amino terminus. Two such vectors are pBlueBacHis (Invitrogen, San Diego, CA) and pAcSG His (PharMingen, San Diego, CA). These vectors also provide proteolytic sites which allow cleavage of the polyhistidine peptide after purification by incubation with enterokinase or thrombin, respectively.

Optimization and Alternate Methodologies

Optimization

As noted previously, empirical observation is necessary to determine the level of expression for a protein of interest. Many of the variables which may be modified have been discussed in previous sections—host cell lines, transfer vector selection, media composition, etc. These modifications, while not dramatically increasing expression levels, can act to boost expression levels. If expression is very low, it may be more time efficient to seek out alternate expression systems. On the other hand, if expression levels are adequate, it may be desirable to systematically alter conditions to optimize expression. In all cases, the level of functional expression should be compared by assaying for activity or ligand binding.

It is known that the integrity of posttranslational modification may decline at very late postinfection times, leading to increased sample heterogeneity. To limit heterogeneity, one may consider alternative promoters with different temporal characteristics [for review of other available promoters, see O'Reilly *et al.* (5)]. In addition, expression and/or secretion of heterologous gene products may be enhanced by inclusion of insect-derived signal peptides or prosequences, but this enhancement is not always seen and must be empirically determined (15).

If expression is low, the sequence should be inspected for TAAG or CTTA sequences since it has been noted that the presence of TAGG sequences on either the sense or antisense strand of the cDNA insert may limit expression (see Chapter 9 of Ref. 5). Both late and very late promoters have TAAG sequences which initiate mRNA transcription, so TAAG sequences on the sense strand may lead to truncated C-terminal translation products. Alternatively, this sequence on the antisense strand (CTTA on coding strand) may result in the presence of interfering antisense transcripts. If present, these sequences should be mutated via alternate codon usage in order to more efficiently express the heterologous gene.

An alternative to producing mature unfused heterologous proteins is to utilize the *polh* promoter of AcMNPV along with varying amounts of 5′ and 3′ viral DNA (1). Many transfer vectors allowing construction of the chimerical proteins are available. The incorporation of possible transcriptional or translational signals, codon usage, or RNA stability encoded in these regions may help to enhance expression.

Electrophysiology

Whole-cell, inside-out, and outside-out patch clamping with attached microelectrodes are very sensitive analytical tools to measure the activity of receptor channels in the plasma membrane of cells (43). Insect cells are relatively easy to patch, and uninfected or wild-type infected cells are electrophysiologically quiet (relatively little endogenous channel activity). Whole-cell currents, therefore, afford a fairly easy method for the detection of functional receptors in the plasma membrane and perfusion systems allow control over external environment, allowing sensitive flux assays to be performed. These studies also allow sensitive assay of protein activity in response to agonist, antagonist, voltage, etc. In addition, single-channel currents may be observed in inside-out or outside-out patches.

For example, electrophysiological studies have been conducted on insect cells expressing the *Shaker* channel (44), the cystic fibrosis gene product (CFTR) (45), and homomeric α glycine receptors (38). In all three cases, large fluctuations of total cell currents were observed in any population of infected cells. These fluctuations are probably due to differential expression levels among individual cells, and this variation of expression was also noted in comparing immunofluorescent staining of infected cells expressing the transferrin receptor (32).

Scale Up

There are three basic strategies employed in scaling up protein production: (i) production in larvae; (ii) setting up multiple cell cultures in spinner flasks (which have limited volume capacity due to O_2 requirements of cells in suspension); or (iii) large-volume cell culture. The benefits and negative aspects of larval production are listed in Table I. For a brief outline of procedures for the rearing, handling, and infection of larvae, see Chapter 18 of "Baculovirus Expression Vectors" (5). The latter two methods are less likely to introduce any additional heterogeneity into the sample, but are more expensive due to serum costs. In comparing these two, the setting up of multiple small cultures is fairly labor intensive, while the latter method requires the purchasing of specialized equipment. The largest obstacle to scale up in suspension culture is providing sufficient oxygenation for the high O_2 requirements of insect cells. Typically, large cell cultures are oxygenated via sparging, but the high turbulence and bubbling at the media surface have been shown to damage insect cells (46). The addition of Pluronic F-68 (see Troubleshooting) offers some, but not sufficient protection. Two specialized methods used successfully in scaling up production involve using stirred bioreactors or airlift fermenters (see Chapters 16 and 17 of Ref. 5, respectively).

Crystallization

Not surprisingly, among the best characterized receptors and channels are those which are naturally abundant. Utilization of the baculovirus system now provides the opportunity to overexpress cloned cDNAs in order to further characterize any eukaryotic protein of interest. In order to provide three-dimensional information at atomic resolution, it is often desirable to crystallize the protein of interest and determine its structure from its X-ray diffraction pattern. While the crystallization of membrane proteins is further complicated by the necessary presence of solubilization and stabilization agents (detergents, organic solvents, phospholipid, or cholesterol), methodologies involving the crystallization of these proteins are currently being developed and expanded (47–49). In addition, soluble domains of many membrane receptors have been overexpressed in the baculovirus system (see Table III for references), and these provide a more tractable crystallographic problem for researchers. A good first sparse matrix approach has been provided by Jancarik and Kim (50), and a kit utilizing this approach is commercially available (Hampton Research, Riverside, CA).

Once crystals are obtained, the next major hurdle is resolving the phase problem. This may be resolved by either of two ways. Conventional chemical techniques can be used to isomorphously substitute heavy metal atoms into the protein, and subsequent crystallization of these multiple forms of the protein may be analyzed to provide phase determination (51). Alternatively, investigators may phase directly from a single crystal by exploiting the appreciable anomalous scattering effects of a few heavier atoms whose bound atomic orbitals resonate with the energy of a tunable source radiation (52). A useful heavy atom for the latter type of analysis is selenium. Selenomethionine has been successfully isomorphously substituted for methionine in recombinant protein (i.e., the recombinant protein is chemically identical to native protein) in prokaryotic systems (53, 54) and provides for phasing and resolution of the protein structure (55). It has been shown that selenomethionine may also be nonlethally incorporated into protein expressed in the baculovirus system (56). As a general phasing vehicle, selenomethionyl substitution into baculovirus-expressed proteins may provide a powerful tool for determining the high-resolution structure of their protein by X-ray crystallographic methods.

Troubleshooting

Passage Number

Since cells are maintained in culture, there is no selective pressure to maintain viability of producing large numbers of occluded virus, as opposed to budded virus, so with prolonged serial passage a viral stock will deteriorate (with respect to yield), often becoming deficient in its ability to overproduce multiple polyhedral nuclear inclusions (and thus the heterologous gene product). The number of times a virus is passed (i.e., generations) should not exceed five, and the yields of expressed protein should be monitored to alert investigators to any potential problems.

Cell Viability

It is essential that cells are healthy (>98%) and doubling every 18–22 hr for good overexpression. The viability and density of cells may be determined by staining with trypan blue (0.04%) and monitoring with a hemocytometer. Healthy cells will appear unstained. The temperature should be strictly maintained at 27 ± 0.5°C. For routine work, the cell density in suspension culture should be maintained between 2.5×10^5 and 3.5×10^6 cells/ml. Cells should be subcultured at least three times per week. In addition, the cells should be periodically (once every few weeks) gently pelleted (1000*g* for 10 min), transferred to a fresh sterile spinner flask, and resuspended in fresh media to prevent accumulation of potentially contaminating by-products. It is essential that careful records are kept with respect to cell density, viability, and age (passage numbers) of the cell line (older cell lines sometimes express gene products inefficiently). If doubling times become too long, the cells should be transferred to new flasks and the media completely replaced (see above). Another potential cause may be insufficient aeration or shearing due to turbulence, so the propeller speed should be systematically adjusted. Typically spinner flasks must be operated at ~70–80 rpm for smaller flasks, and slightly higher (80–90 rpm) for larger flasks to ensure adequate aeration.

Transfection

If there are problems with transfection, impurities introduced with the plasmid and viral DNA are probably the causative agent. All foreign DNA must be pure, otherwise transfected cells will not be viable. Contaminants normally found in crude plasmid preparations, even after phenol extraction or ethanol precipitation, are toxic to *S. frugiperda* cell lines. Plasmid preparations should be purified by CsCl gradient centrifugation (4). Large viral DNA is also very sensitive to nicking and shearing. Protocols for the purification of budded virus may also be found in the bulletin of Summers and Smith (4).

Antibiotics and Fungicides

Though not essential, it is recommended that in larger cultures 50 μg/ml of gentamycin and 25 μg/ml of amphotericin B (Fungizone) be added.

Shear Stress

Lepidopteran cells are very sensitive to shear stress, especially after infection due to cell swelling (46). Pluronic F-68 (BASF, Inc.), a copolymer of polyoxypropylene and polyoxyethylene, is a useful protective agent (57). It is an essential additive at larger volumes, such as during scale-up, since the oxygen requirements of the cells (decreased surface-to-volume ratio) necessitates mechanical intervention and greater external stress.

Summary

Baculovirus expression systems have provided investigators with a powerful tool for the overexpression of eukaryotic proteins. It is an especially useful experimental approach in studies examining receptors and channels. These proteins are generally extensively modified, and standard bacterial expression systems typically fail to produce functional products. Extensive biochemical and biophysical characterization of receptors and channels has historically been hampered by their low natural abundance. Optimization of a baculovirus-mediated expression system provides a general system to produce sufficient functional quantities of a given protein for subsequent study.

References

1. V. A. Luckow and M. D. Summers, *Bio/Technology* **6,** 47 (1988).
2. L. K. Miller, *Annu. Rev. Microbiol.* **42,** 177 (1988).
3. S. Maeda, *Annu. Rev. Entomol.* **34,** 351 (1989).
4. M. D. Summers and G. E. Smith, *Tex., Agric. Exp. Stn.* [*Bull.*] **B-1555,** (1987).
5. D. R. O'Reilly, L. K. Miller, and V. A. Luckow, "Baculovirus Expression Vectors: A Laboratory Manual." Freeman, New York, 1992.
6. A. Groner, *in* "The Biology of Baculoviruses" (R. R. Granados and B. A. Federici, eds.), Vol. 1, p. 177. CRC Press, Boca Raton, FL, 1986.
7. M. D. Tomalski and L. K. Miller, *Nature (London)* **352,** 82 (1991).
8. L. M. D. Stewart, M. Hirst, M. L. Ferber, A. T. Merryweather, P. J. Cayley, and R. D. Possee, *Nature (London)* **352,** 85 (1991).
9. M. D. Moore, M. J. Cannon, A. Sewall, M. Finlayson, M. Okimoto, and G. R. Nemerov, *J. Virol.* **65,** 3559 (1991).
10. G. Kaplan, M. S. Freistadt, and V. R. Racaniello, *J. Virol.* **64,** 4697 (1990).
11. S. White, R. Taetle, P. A. Seligman, M. Rutherford, and I. S. Trowbridge, *Cancer Res.* **50,** 6295 (1990).
12. M. J. Fraser, *J. Ultrastruct. Mol. Struct. Res.* **95,** 189 (1986).

13. J. Kupsch, K. M. Saizawa, and K. Eichmann, *Immunobiology* **186,** 254 (1992).
14. K. U. Jansen, J. Shields, J. Gordon, J. Cairns, P. Graber, and J.-Y. Bonnefoy, *J. Recept. Res.* **11,** 507 (1991).
15. D. L. Jarvis, M. D. Summers, A. Garcia, Jr., and D. A. Bohlmeyer, *J. Biol. Chem.* **268,** 16754 (1993).
16. S. Maeda, T. Kawai, M. Obinata, H. Fujiwara, T. Horiuchi, Y. Saeki, Y. Sato, and M. Furusawa, *Nature (London)* **315,** 592 (1985).
17. M. Villalba, S. R. Wente, D. S. Russell, J. Ahn, C. F. Reichelderfer, and O. M. Rosen, *Proc. Natl. Acad. Sci. U.S.A.* **86,** 7848 (1989).
18. K. T. Jeang, K. M. Holmgren, and G. Khoury, *J. Virol.* **61,** 1761 (1987).
19. K. Iatrou, R. G. Meidinger, and M. R. Goldsmith, *Proc. Natl. Acad. Sci. U.S.A.* **86,** 9129 (1989).
20. J. Sambrook, E. F. Fritsch, and T. Maniatis, "Molecular Cloning: A Laboratory Manual," 2nd ed. Cold Spring Harbor Lab. Press, Cold Spring Harbor, NY, 1989.
21. J. Vialard, M. Lalumiere, T. Vernet, D. Briedis, G. Alkhatib, D. Henning, D. Levin, D. Richardson, and C. Richardson, *J. Virol.* **64,** 37 (1990).
22. F. Van Der Wilk, J. W. M. Van Lent, and J. M. Vlak, *J. Gen. Virol.* **68,** 2615 (1987).
23. C. H. W. Klaassen, T. J. F. Van Uem, M. P. De Moel, G. L. J. De Caluwe, H. G. P. Swarts, and J. J. De Pont, *FEBS Lett.* **329,** 277 (1993).
24. C. Lindqvist, C. Engberg, P. Ojala, M. Karp, K. Akerman, and C. Oker-Blom, *Scand. J. Immunol.* **38,** 267 (1993).
25. A. W. DeTomaso, Z. J. Xie, G. Liu, and R. W. Mercer, *J. Biol. Chem.* **268,** 1470 (1993).
26. S. G. Mann and L. A. King, *J. Gen. Virol.* **70,** 3501 (1989).
27. T. Peakman, M. Page, and D. Gewert, *Nucleic Acids Res.* **13,** 5403 (1989).
28. B. B. Goswami and R. I. Glazer, *BioTechniques* **10,** 626 (1991).
29. P. A. Kitts, M. D. Ayres, and R. D. Possee, *Nucleic Acids Res.* **18,** 5667 (1990).
30. G. Patel, K. Nasmyth, and N. Jones, *Nucleic Acids Res.* **29,** 97 (1992).
31. S. Peng, M. A. Sommerfelt, G. Berta, A. K. Berry, K. L. Kirk, E. Hunter, and E. J. Sorscher, *BioTechniques* **14,** 274 (1993).
32. D. L. Domingo and I. S. Trowbridge, *J. Biol. Chem.* **263,** 13386 (1988).
33. W. F. Hink, D. R. Thomsen, D. J. Davidson, A. L. Meyer, and F. J. Castellino, *Biotechnol. Prog.* **7,** 9 (1991).
34. W. Chen, Q.-X. Shen, and O. P. Bahl, *J. Biol. Chem.* **266,** 4081 (1991).
35. A. E. Atkinson, F. G. P. Earley, D. J. Beadle, and L. A. King, *Eur. J. Biochem.* **192,** 451 (1990).
36. U. A. Germann, M. C. Willingham, I. Pastan, and M. M. Gottesman, *Biochemistry* **29,** 2295 (1990).
37. K. A. Stauffer, N. M. Kumar, N. B. Gilula, and N. Unwin, *J. Cell Biol.* **115,** 141 (1991).
38. M. Cascio, N. E. Schoppa, R. L. Grodzicki, F. J. Sigworth, and R. O. Fox, *J. Biol. Chem.* **268,** 22135 (1993).
39. M. Iizuka and K. Fukuda, *J. Biochem. (Tokyo)* **114,** 140 (1993).

40. L. M. Hjelmeland and A. Chrambach, *in* "Methods in Enzymology" (W. Jakoby, ed.), Vol. 104, p. 305. Academic Press, New York, 1984.
41. L. M. Hjelmeland, *in* "Methods in Enzymology" (M. Deutscher, ed.), Vol. 182, p. 253. Academic Press, San Diego, CA, 1990.
42. J. Porath, *Protein Express. Purif.* **3,** 263 (1992).
43. O. P. Hamill, E. Marty, E. Heher, B. Sakmann, and F. J. Sigworth, *Pfluegers Arch.* **391,** 85 (1981).
44. K. Klaiber, N. Williams, T. M. Roberts, D. M. Papazian, L. Y. Jan, and C. Miller, *Neuron* **5,** 221 (1990).
45. N. Kartner, J. W. Hanrahan, T. J. Jensen, A. L. Naismith, S. Sun, C. A. Ackerly, E. F. Reyes, L.-C. Tsui, J. M. Rommens, C. E. Bear, and J. R. Riordan, *Cell (Cambridge, Mass.)* **64,** 681 (1991).
46. J. Tramper, J. Williams, D. Joustra, and J. Vlak, *Enzyme Microb. Technol.* **8,** 33 (1986).
47. W. Kuhlbrandt, *Q. Rev. Biophys.* **21,** 429 (1988).
48. H. Michel, *in* "Crystallization of Membrane Proteins" (H. Michel, ed.), p. 73. CRC Press, Boca Raton, FL, 1991.
49. F. R. Reiss-Husson, *in* "Crystallization of Nucleic Acids and Proteins" (A. Ducruix and R. Giege, eds.), p. 175. IRL Press, New York, 1992.
50. J. Jancarik and S. Kim, *J. Appl. Crystallogr.* **24,** 409 (1991).
51. D. W. Green, V. M. Ingram, and M. F. Perutz, *Proc. R. Soc. London, Ser. A* **255,** 287 (1954).
52. W. A. Hendrickson, J. L. Smith, and S. Sheriff, *in* "Methods in Enzymology" (H. Wyckoff *et al.,* eds.), Vol. 115, p. 41. Academic Press, New York, 1985.
53. W. Yang, W. A. Hendrickson, E. T. Kalman, and R. J. Crouch, *J. Biol. Chem.* **265,** 13553 (1990).
54. W. A. Hendrickson, J. R. Horton, and D. M. LeMaster, *EMBO J.* **9,** 1665 (1990).
55. W. Yang, W. A. Hendrickson, R. J. Crouch, and Y. Satow, *Science* **249,** 1398 (1990).
56. W. Chen and O. P. Bahl, *J. Biol. Chem.* **266,** 9355 (1991).
57. D. W. Murhammer and C. F. Goochee, *Bio/Technology* **6,** 1411 (1988).

[11] Real-Time Measurements of Receptor Activity: Applications of Microphysiometric Techniques to Receptor Biology

John A. Salon and John C. Owicki

Introduction

In this chapter we report the use of a silicon-based biosensor which can detect the activation of cellular receptors by monitoring changes in metabolism which occur secondary to receptor activation. The method does not depend on a specific transduction pathway and can therefore be used to monitor a wide variety of receptor types in assorted cellular environments.

Neurotransmitters and neuromodulators, as well as hormones, cytokines, and most drugs, initiate their biological action by interacting with specific receptor macromolecules. The receptor conveys this signal to the cell through conformational changes that may involve a variety of ancillary proteins and second messengers. The ultimate effect of the signal transduction can range from modulation of neuron firing rates to regulation of cell growth or differentiation.

The transduction process typically displays both amplification cascades and regulation by negative feedback. Components of the regulatory systems may include ion channels and transporters, enzymes that modulate levels of intracellular second messengers, and phosphorylation of catalytic proteins. The physiological phenotype of a given receptor/cell system is influenced by the cellular ensemble of signaling pathways capable of coupling to the receptor and would seem to reflect the biological job that the cell must perform. The role of a given receptor in cellular signaling is thus determined not only by the molecular properties of the receptor itself but also by the specific features of the cell in which it operates. In this fashion the exquisitely specific lock-and-key interaction of the ligand and receptor is translated into a variable but limited repertoire of cellular effects. It is the specificity of the ligand–receptor interaction and the biochemical impact of this interaction that form the basis for the pharmacological characterization of a receptor.

This characterization can encompass two fundamentally different types of assays: ligand binding and functional response. Binding assays are usually performed by measuring the ability of a test substance to displace a radiolabeled ligand that is known to bind to the receptor. Functional assays are

Methods in Neurosciences, Volume 25

a varied lot and generally are constructed around the signal-transduction mechanism or physiological response in the system under study. Examples include measurements of membrane potential and conductance, of second-messenger systems (e.g., cAMP, phosphatidylinositol (PI) turnover, or $[Ca^{2+}]_i$), of protein phosphorylation, and of more integrative responses such as muscle contraction or cellular proliferation. In both ligand binding and functional approaches a pharmacological profile is generated to rank agonist potency and efficacy and antagonist affinity. These rank-order profiles then form the basis for assignment of receptor class and subtype.

Both traditional ligand binding and functional approaches have limitations. Binding assays fail to reveal whether a substance is an agonist or antagonist. Furthermore, they cannot be performed unless a displaceable radiolabeled ligand of suitably high affinity is available. In addition, the intact coupling of a receptor to the cell's ancillary signaling mechanisms can impact binding studies, since these molecular components can contribute significantly to the affinity states of the receptor's ligand binding sites.

Functional assays are generally more complicated and difficult to perform than binding assays. The selection of an appropriate functional assay requires prior knowledge of the signal-transduction pathway or physiological response implicated with the receptor in question. Unfortunately, the specific requirements of a given receptor's coupling and the compatibility of a host cell's transduction mechanisms are not always known. It therefore may be necessary to examine a considerable variety of possible responses to any given test ligand. As a matter of practice, one must focus on each system, one at a time, assaying for the anticipated effect. This approach can be compromised by the real possibility that a cloned and heterologously expressed receptor does not couple effectively to the expected effector system in the transfected host cell. Thus, if an exhaustive effort is not mounted to examine all the possible effects of receptor activation, it is possible to miss a ligand-induced signal. A comprehensive screen of all possible functional responses to all possible drugs is seldom possible.

A thesis of this article is that it is frequently appropriate to soften one's focus and trade off some mechanistically specific information for broader information about the status of receptor activation. This can be accomplished with a microphysiometer, an instrument that uses a semiconductor-based chemical sensor to detect metabolic changes in living cells induced by receptor activation essentially regardless of the particular signal-transduction mechanism.

In the remainder of this chapter we first give some background on the sensor and on the cell biology of extracellular acidification. Then we describe the operation of a microphysiometer in some detail. We conclude with some selected examples of pharmacological applications of the technology and with a broader discussion of its utility.

Background and Theory of Operation

The extensive interconnections among different biochemical processes of a cell assure that the effects of any given physiological change ripple outward from the pathways directly involved with that change, altering, to a greater or lesser extent, many other cellular processes (Fig. 1). Therefore, it might be possible to detect a response by monitoring cellular events not normally thought to be associated with the primary stimulus. The integrative role of

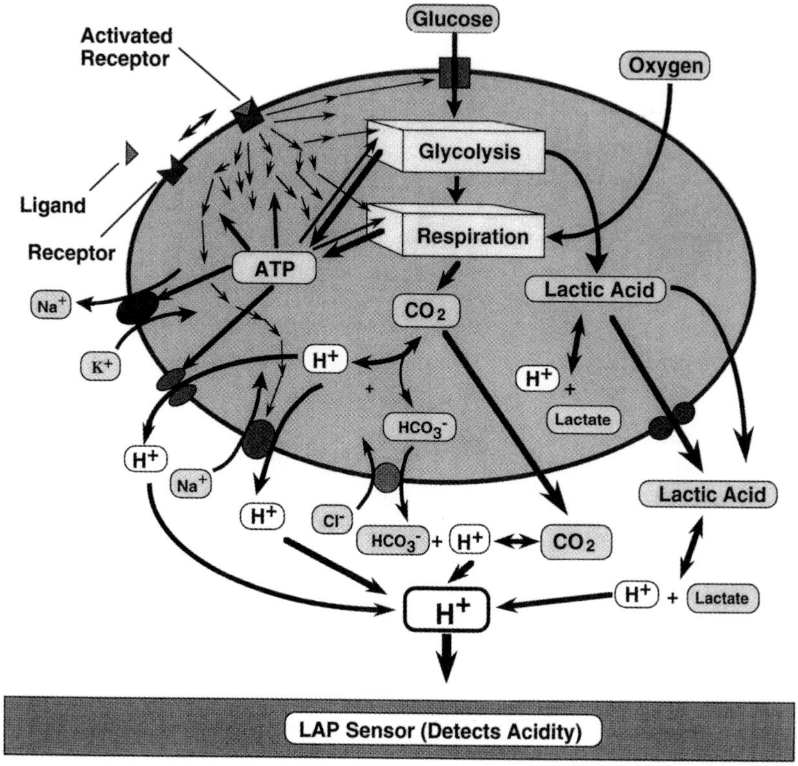

FIG. 1 The cell biology of extracellular acidification. Schematic of the principal features of the cellular proton economy. [From McConnell *et al.* (17), copyright 1992, AAAS].

catabolism makes it an excellent candidate for this indirect detection; few cellular processes are uninvolved in the demand for (or supply of) metabolic free energy.

It is easy to posit such a scheme for detecting receptor activation in principle, but can it be accomplished in practice? The answer depends on the kinetics and amplitudes of such effects compared to the ability of experimental methods to resolve them. Several familiar techniques are available: oxygen or glucose consumption, lactate or CO_2 excretion, and microcalorimetry all can be used to measure catabolic rate. Microphysiometry measures the rate of extracellular acidification, which reflects the rate of excretion of acidic metabolic products such as lactate and CO_2, and it has proved to have technical advantages over the other methods. It is now well established that ligand–receptor interactions can produce changes in cellular catabolic rates, transient or sustained, that can be measured quantitatively by the silicon microphysiometer.

The Light-Addressable Potentiometric Sensor (LAPS)

The heart of the microphysiometer is the flow chamber (Fig. 2), wherein a small population of cells is exposed to a stream of medium that flows through a fluid channel in diffusive contact with a silicon-based electrochemical pH sensor, the LAPS (1). In this section, we provide some fundamental background, briefly describing how the LAPS works and discussing some of the cell biology of extracellular acidification. We defer a more complete description of the microphysiometer until the following section of the article.

The LAPS employs a silicon chip covered on one face with a thin layer of Si_3N_4 insulator, which is put in contact with the aqueous sample. The surface reversibly binds protons, rendering its surface potential pH-dependent in a Nernstian fashion (61 mV per pH unit at 37°C). The external circuitry of the sensor includes a controlling electrode in solution and a contact to the back of the chip. Thus, the surface potential acts as a bias potential applied across the chip and creates a pH-dependent electric field within the silicon near the junction to the surface insulator. The LAPS employs the photoelectric effect to detect the electric field in the chip: when light is absorbed by the silicon, the mobile electrons and holes that are formed create a capacitative current in the external circuit to the extent that they move under the influence of the field. Thus, photocurrent depends on electric field, electric field depends on surface potential, and surface potential depends on pH.

More details of LAPS principles and electronics can be found elsewhere (1, 2). Here we make just three additional points. First, the pH is reported

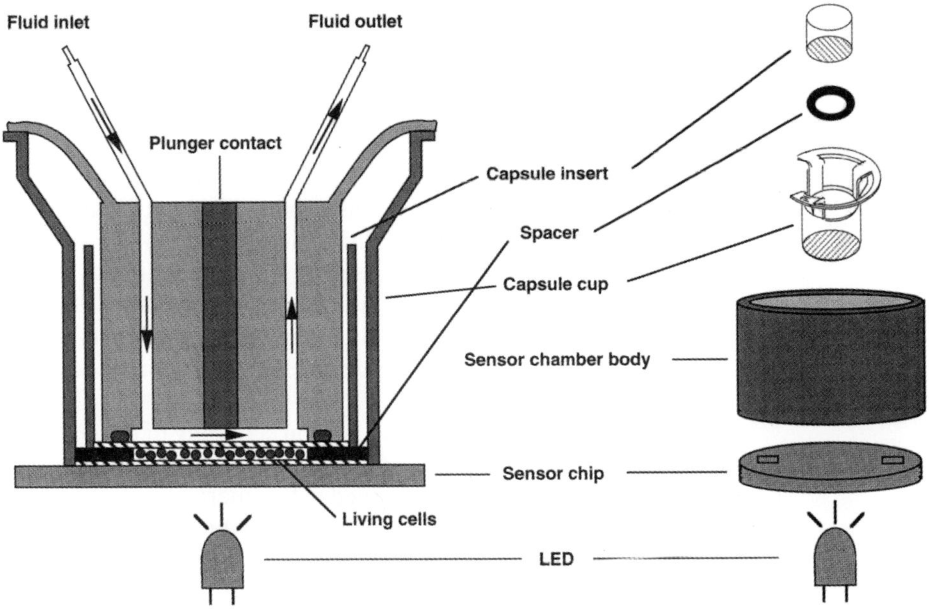

FIG. 2 Cell capsule and flow chamber. Cells are retained in a disk-shaped region 50 μm high and 6 mm in diameter between two track etched microporous polycarbonate membranes. The sensing surface of the LAPS contacts the lower membrane, and culture medium flows across the surface of the top membrane. An LED illuminates the bottom of the LAPS to activate the sensing region.

only from illuminated regions of the chip, which imparts spatial selectivity to the measurements without the need for additional electronics. Either side of the chip may be illuminated, and a circle 2 mm in diameter is typical. Second, the impedance of the signal in the LAPS is low (kΩ), which facilitates measurements in small volumes. Although the DC impedance across the insulator is high (GΩ), the light is flashed at kHz frequency so that the signal is carried by an AC photocurrent, to which the chip presents lower impedance. Third, the LAPS is mechanically well suited for low-volume measurements. This is partly due to the fact that potentiometric measurements depend on concentration rather than number of molecules. It is also partly due to the geometric characteristics of the LAPS: in its basic form its surface is a featureless plane; it is easy to create small volumes by closely apposing a parallel plane. Furthermore, the semiconductor processing steps that are necessary for the LAPS electronics are compatible with those that are required for micromachining, so that it is possible to etch fluidic structural components into the surface of the sensor.

Mechanisms of Extracellular Acidification

Consideration of the balanced equations for the catabolism of the major carbon sources (sugars, lactate, pyruvate, glutamine, and fatty acids) near neutral pH shows that all produce protons as long as the ATP that has been produced has been hydrolyzed. However, different pathways and carbon sources vary in the number of protons produced per ATP synthesized and hydrolyzed: glycolysis of glucose to lactic acid produces 1 H^+/ATP, respiration of glucose to CO_2 ~0.17, and respiration of glutamine to CO_2 and NH_3 only ~0.11. Some insight into the sources of the protons detected in the microphysiometer can be obtained by manipulating nutrients in the culture medium, adding specific metabolic poisons, or making auxiliary biochemical measurements.

Although such an accounting of the proton economy of the cell is biochemically well founded, it is inadequate for understanding the changes in extracellular acidification rates that are observed in the microphysiometer on receptor activation. It does not specify the processes for which ATP demand or supply may be modified. Furthermore, it ignores the role of transient changes of acidification rate that may have nothing to do with catabolic rate. As can be seen in Fig. 1, the protons that are detected in the microphysiometer are those that are transported across the cell membrane. The bulk of the cell volume enclosed by the plasma membrane has a substantial pH buffer capacity, which permits the extracellular acidification rate to be decoupled transiently from the catabolic rate.

For example, if a cell has a cytoplasmic volume of 2 pl and buffer capacity of 20 m*M* per pH unit, then the transport of 2.4×10^9 H^+ across the membrane will alter the intracellular pH (pH_i) by 0.1 unit. This is a relatively small change in pH_i, but the number of protons involved is considerable: the equivalent of 24 sec of basal acidification for a cell that excretes 1×10^8 H^+/sec. Since changes in pH_i greater than a few tenths of a unit are rare, generally the processes that regulate proton transport and pH_i can directly cause significant changes in extracellular acidification only as transients of at most a few minutes. Sustained changes in extracellular acidification rate must be linked to catabolic rate. To complicate the issue somewhat, changes in catabolic rate can produce transients; likewise, a transient due to a change in pH_i may be followed by a sustained change in acidification rate due to effects of the new pH_i on catabolic rate.

What is the experimental evidence regarding the mechanisms of the observed changes of acidification rate on receptor activation, which is typically manifested as an increase of >10% over a few minutes? There are probably many different mechanistic contributions in the various receptor/host cell combinations. A few points are, however, clear.

First, there is strong evidence that glycolysis, rather than respiration, is frequently involved in the response. For example, the response of TF-1 human erythroleukemia cells to granulocyte-macrophage colony-stimulating factor (GM-CSF) is large when glucose is the carbon source in the culture medium but nearly absent when the cells are fed pyruvate (3). Similar observations, based on microphysiometry and lactate measurements, have been made for m_1 muscarinic receptors expressed in Chinese hamster ovary (CHO) cells (4). Whether this reflects a direct effect on glycolytic activity or is secondary to an increased energy demand is not yet known.

One can speculate that phosphorylation of key glycolytic control enzymes could increase glycolytic activity, and many factors could increase the demand for ATP. Although there are few experimental data on the fluxes through second-messenger pathways (as opposed to net concentration changes), the direct cost of generating the second-messenger cascades probably does not cause detectable changes in extracellular acidification rate. For example, the estimated rate of ATP hydrolysis necessary for the production of cAMP in response to β_2-adrenergic receptor stimulation falls four orders of magnitude short of explaining the energy demand required to produce the metabolic effects seen with the microphysiometer (5, 6). If the metabolic enhancements produced by β-adrenergic receptors are absolutely dependent on cAMP production, then tertiary biochemical events such as those mediated by protein kinases must be involved.

Second, a large part of the energy consumed by resting cells maintains the asymmetry of transmembrane ion concentrations (7, 8) and it is likely that some of the response observed in the microphysiometer reflects energy demand required to reestablish electrochemical gradients that have been partially dissipated either directly by the receptor or indirectly by a second-messenger system. For example, patch-clamp studies of the m_1 receptor transfected in A9 L cells have demonstrated K^+ currents of 140 pA in response to acetylcholine, equivalent to 8×10^8 K^+/sec/cell (9). If Na^+, K^+-ATPase were used to pump K^+ back into the cell then 4×10^8 ATP/sec/cell would be used to maintain homeostasis. This value is large enough to account for the >30% increase in acidification rate observed after muscarinic activation of a closely related system (6). Moreover, the onset of K^+ currents begins 3 sec after the application of acetylcholine, consistent with the rapid onset of changes observed with the microphysiometer.

Finally, perhaps the best established mechanism of receptor-mediated changes in extracellular acidification involves the sodium-proton exchanger (NHE-1). The involvement of this protein can be determined by inhibiting it during receptor activation, using amiloride derivatives or by removing extracellular sodium. There are instances of G protein-coupled receptors for which receptor-mediated increases in extracellular acidification rates are essentially completely dependent on NHE activity (10, 11). In microphysiometry of TF-1 cells, a minor component of the response (an early brief pulse) appears to be due to similar exchanger activity (12). These experiments do not distinguish whether the intrinsic activity of the NHE-1 is regulated, perhaps by phosphorylation, or whether instead the NHE-1 is merely a passive conduit for an increased internal supply of protons: by analogy to a water faucet, whether the flow is faster because the faucet has been opened farther or because the water pressure is greater.

At present our understanding of the receptor pharmacology of the phenomena is greater than our understanding of the intracellular mechanisms that couple receptor activation to proton flux. However, the details of the underlying mechanisms are not needed to employ this technique to assess the potency and efficacy of agonists or the specificity and affinity of antagonists. In this respect the microphysiometric measurement is as valid as any traditional measurement of receptor activity and can be viewed as a broad-based functional assay.

Experimental Technique

Microphysiometric methods are well suited to the examination of receptor pharmacology. The common objective of determining ligand specificity and pharmacological constants derived from agonist and antagonist dose–response curves is straightforward.

Setting Up a Microphysiometer Lab

Several versions of the microphysiometer have been constructed, differing in the construction of the flow chamber and in other ways. The Cytosensor microphysiometer (Fig. 3), which we have described here, is the only device

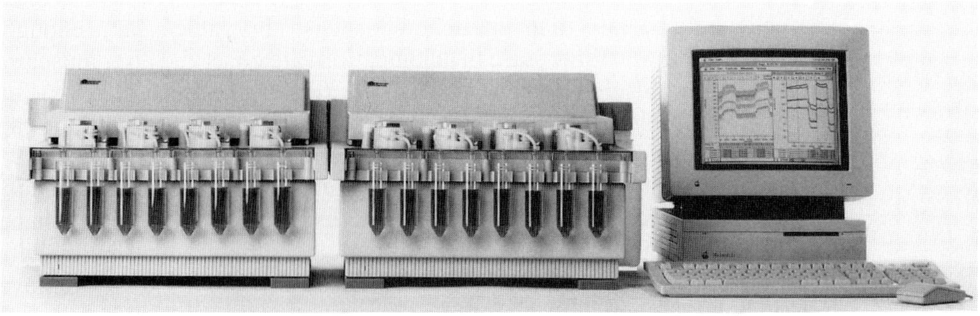

FIG. 3 The Cytosensor microphysiometer.

that is commercially available. Manufactured and sold by Molecular Devices Corp., the full eight-channel model, with attendant Macintosh computer and Hewlett–Packard printer, requires about 6 feet of bench space. Some components of the fluidics pathway are disposable, most importantly the capsule in which cells are immobilized. In most cases, standard cell culture media or balanced salts can be used, though without added buffers (see below). A basic cell culture facility is required to provide the experimental cells, though for some applications the use of cryopreserved cells may permit this to be distant from the microphysiometer (13). Otherwise, no special equipment is required beyond that found in a typical biochemistry or molecular biology laboratory.

Preparing Cells for Microphysiometry

A variety of cell types have proven amenable to microphysiometric recording, including mammalian fibroblasts, hematopoietic cells, neurons, and muscle cells, as well as insect and microbial cells (2). These comprise cell lines that are both adherent and nonadherent, as well as primary cultures. Cells expressing both endogenous and heterologous cloned receptors (stably or transiently expressed) are suitable for microphysiometric analysis.

The nature of the receptor/cell system and specific requirements for its growth will dictate how it must be accommodated by the microphysiometer. Cells are typically enclosed in a cell capsule specially designed for the microphysiometer flow chamber. As shown in Fig. 2, the cells are sandwiched between two disposable track-etched microporous polycarbonate membranes attached to plastic carriers; a 50-μm-thick annular spacer sets the distance between the membranes. These sterile disposable membranes are attached to sterile culture cups arranged in a multiwell plate, which facilitates handling of the cells and assembly of the flow chambers.

The cells are maintained and prepared for microphysiometric recording via standard culture methods. These routine procedures extend to standard protocols for transient and stable transfection of receptor cDNAs. For transient transfections, the crucial issue is the efficiency of expression. For example, if after transfection 20% of the cells express the receptor and activating the receptor increases the acidification rate of these cells by 10%, the effect in the overall population is diluted to only 2%, which is probably not reliably detectable. Methods are available to enrich the populations, using sorting based on a coexpressed cell-surface antigen or with a replicating episomal vector that allows a brief selection period (14).

Adherent cells can be grown directly on the lower membrane, which is 12 mm in diameter and fits into a 12-well cell culture plate. Cells are seeded at a density that is projected to produce a 60–90% subconfluent layer by the time of measurement, typically 5×10^4 to 5×10^5 cells per flow chamber, depending on the size and metabolic activity of the cells. Depending on the cell type and experimental conditions, the cells are seeded as early as several days or as late as a few hours before the experiment.

Nonadherent cells must be immobilized against the flow of medium in the microphysiometer, for example, by embedding in a fibrin or agarose gel within the cell capsule just before use. It has been shown that a nonadherent cell line (TF-1) can be cryopreserved in capsule cups, then thawed and used in the microphysiometer within minutes, preserving receptor-mediated acidification responses (13). Adherent lines, such as CHO, can be treated similarly (15).

Operation of the Microphysiometer

The cell capsule, loaded with cells, is transferred to the flow chamber on the microphysiometer. As shown in Fig. 2, the floor of the flow chamber is formed by the LAPS; the fluid inlet and outlet as well as the controlling electrode are embedded in the ceiling of the chamber, 135 μm above the floor. Culture medium flows across the top of the upper membrane of the cell capsule. The sensor is in contact with the bottom of the lower membrane, 10 μm away from the cells. The height of the flow chamber is crucial to the time resolution of the instrument: the time for vertical diffusional equilibration of protons in the chamber is only a few seconds. The volume of the chamber is about 4 μl, but data are collected only from a central circular region about 2 mm in diameter that is illuminated from below and corresponds to about 300 nl.

The flow of culture medium through the chamber is provided by a peristaltic pump and is typically 10 to 100 μl per minute. A valve controls sample application by switching between two fluidic streams or, alternatively, via an injection loop. The fluid path contains two additional components. First, upstream of the flow chamber there is a debubbler/degasser assembly that eliminates air bubbles from the fluid input and decreases the concentrations of dissolved gases so that bubbles are not nucleated in the chamber when room-temperature medium is raised to the temperature at which the cells are held, typically 37°C. Second, there is a reference electrode downstream of the flow chamber to improve the stability of the LAPS signal.

The current Cytosensor version of the microphysiometer is equipped with eight parallel flow chambers that can function independently, except that their fluid pumps must operate in tandem. The associated dedicated personal computer controls the pumps and valves, as well as the acquisition, real-time display, and analysis of data. The fluidic system of the instrument is treated with a sterilizing/depyrogenating solution prior to each use, and antibiotics are commonly included in the flow medium.

Although mammalian cells are usually cultured in a bicarbonate buffer, perhaps with added HEPES, neither buffer is included in the flow medium that is pumped through the microphysiometer. They are excluded primarily to reduce the buffer capacity of the medium, enhancing the pH change caused by the cellular excretion of protons. Otherwise, standard culture media can generally be used and typically have a buffer capacity of 0.5 to 1.0 m*M* per pH unit, which arises primarily from phosphate and the α-amino groups of amino acids. Serum or a carrier protein such as serum albumin can be included, or the cells simply can be bathed in balanced salts plus a carbon source. Once the cells have been loaded into the microphysiometer flow chambers, the basal metabolic rates generally stabilize within 10–60 min. Challenge with drugs can then be carried out in the desired flow chambers.

The rate at which cells excrete metabolic acids is determined by measuring the rate of acidification of the culture medium bathing the cells during periodic interruptions of the flow of medium through the chamber, expressed as the slope of a least-squares fit (Fig. 4). A nearly confluent monolayer usually produces acidification rates within threefold of 100 μV/sec. This corresponds to $\sim 4 \times 10^8$ H^+/sec for a typical cell (16). The length of the flow interruption is usually adjusted to make the total excursion of pH fall between 0.02 and 0.10 units. At 61 mV per pH unit and a typical buffer capacity of 1 m*M* per pH unit, an acidification rate of 100 μV/sec corresponds to a flow-off period of 12–60 sec. Flow must subsequently remain on long enough to reestablish the starting pH, typically at least 10–30 sec.

Thus, the measurements made with the microphysiometer are nondestructive: the pH excursions are no more than the arterial/venous variation in blood, and flow is not left off long enough to deplete oxygen completely. This allows the experimenter to carry out an extended or complex series of drug studies on single populations of cells. The culture conditions within the flow chamber are good, and cells have been maintained in the apparatus in viable condition for up to 6 days (17).

Performance of the Microphysiometer

The noise in successive acidification-rate measurements is usually 1–3%, enabling one to resolve rapid responses as low as ~5% of basal levels. Over longer periods the acidification rates drift up or down to an extent dependent on the nature of the cell (e.g., its growth rate and acclimation to the culture conditions within the instrument). This drift can limit the resolution of low-

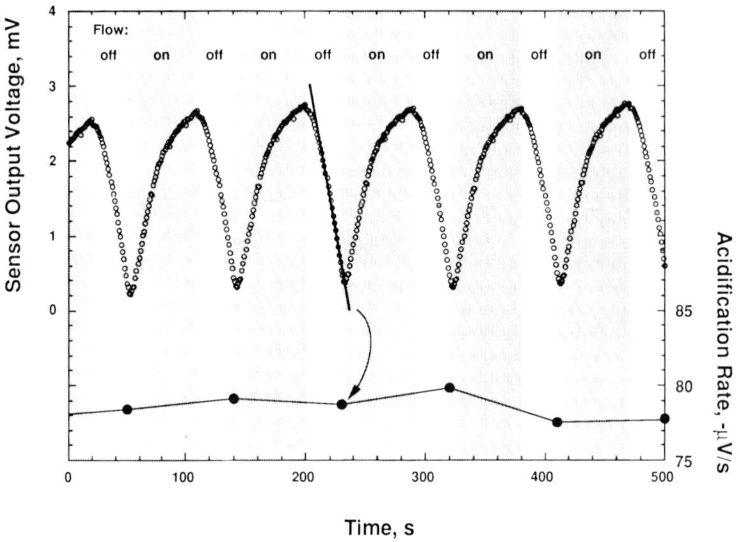

FIG. 4 Data transformation: determining rates of acidification. Top, plot of sensor output voltage versus time (61 mV ≈ 1 pH unit). During each flow-off period shown in the sensor, data are fit to a straight line with the use of a least-squares procedure. Bottom, the slope of this line is reported as the acidification rate and plotted versus time to correlate with challenge of the cells.

amplitude slow responses. Typically, however, responses for most major classes of receptors (excepting steroids) are apparent within a few minutes of the stimulation with agonist.

In most applications, the analysis of data is based on the percentage change in acidification rate evoked by a treatment within a chamber, not the absolute change in rate. Acidification rates are the product of two factors: the average acidification rate per cell and the number of cells in the small 2.5-mm^2 sensing area in the center of the chamber (18). In a region of this size, it can be difficult to produce replicate cultures with cell numbers that are uniform (say, to within +10%). Since one is usually interested in events on a per-cell basis, most of the chamber-to-chamber variability of results can be eliminated by within-chamber normalizing of data to basal acidification rates. Any residual chamber-to-chamber variability is caused principally by other variations in cell biology. For example, some systems show density-dependent expression of receptors.

Hence, we find that precision is often optimized when concentration–response curves are generated sequentially within a chamber rather than by comparing the responses of different chambers to separate concentrations of drug. The validity of such sequential or cumulative concentration–response curves depends on factors such as the importance of receptor desensitization during the experiment.

The time resolution of the instrument depends on the time interval for a pump off/on cycle and on the agility of the fluidics system in changing concentrations within the flow chamber. Under favorable conditions (rapidly acidifying cells and efficient fluid transport through the cell layer), the time resolution of the present instrument is about 30–60 sec.

In terms of throughput for screening applications, the Cytosensor microphysiometer occupies the middle ground between radioligand binding and electrophysiological methods. For microphysiometry, throughput is largely determined by whether exposure to a given compound is likely to alter the responsiveness of the cells to subsequent challenges. If so, cells must be reloaded into the chamber after each test. If not, a single chamber of cells rapidly can be subjected to repeated challenges with a series of materials to be tested. In the favorable case, an eight-channel instrument can probably be used to perform 50–100 tests per day.

Pharmacological Applications: Specific Examples

Since the first paper was published on the technique in 1989, the utility of microphysiometry for monitoring a broad range of signal transduction systems has been documented, including protein kinase C and tyrosine kinase activity (12, 17), activation/inhibition of adenylyl cyclase (10, 19), modulation of [Ca^{2+}] fluxes (20), regulation of phosphatidyl inositol metabolism (6), Na^+/H^+ exchange (11, 21), Na^+/K^+-ATPase activity (22), and less well-characterized transduction systems such as that for ciliary neurotrophic factor (23). These reports encompass approximately 25 receptors of different types and classes and do not include many more cases that are unpublished (24). A few examples of particular interest in neuroscience are discussed below.

Dopaminergic Receptors

Microphysiometric responses for the dopamine family of receptors, including D1, D2, D3, and D4 subtypes, have been reported in stably or transiently transfected cells, and differences in the kinetics of metabolic responses have been reported for several subtypes. In Ltk^- cells heterologously expressing human D1 and D2 receptors (which classically couple through G_s and G_i

pathways, respectively), exposure to 1 μM dopamine caused a rapid increase in acidification rates by up to 75%. This increase could be specifically blocked by either previous or concomitant exposure to subtype-specific antagonists (25). Comparison of D1 and D2 receptors stably expressed in fibroblastic Ltk^- hosts revealed responses that differed in terms of magnitude and kinetics of onset and duration (Fig. 5). The metabolic response to D2 activation was more acute than to D1, which was somewhat slower to rise and longer in duration. The magnitude of the D2 response was greater than that of D1, even though the stable D2 transfectants expressed 20-fold less receptor than did the D1 transfectants, suggesting that the number of cellular receptors alone does not account for the magnitude of the response (26).

As with all functional assays, perhaps the most fundamental limitation on the ability to observe receptor activation in the microphysiometer is whether receptor activation is significantly coupled to the physiology of the cell. We have observed several cases in which a receptor failed to cause a detectable change in acidification rate when expressed in one host cell but did so when expressed in another. For example, while the metabolic activation by D2

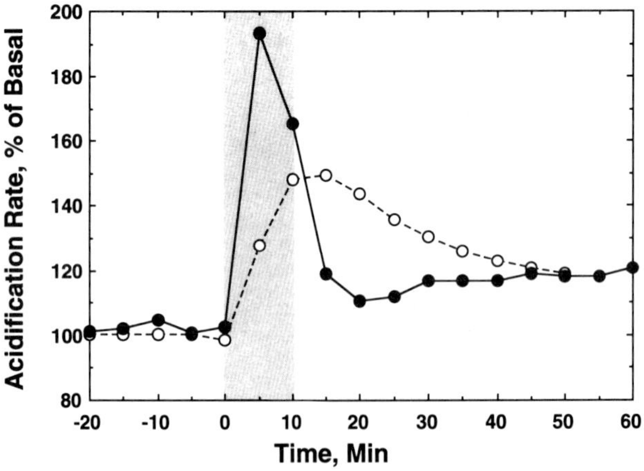

FIG. 5 Detecting the activation of D1 and D2 receptors in Ltk^- cells with microphysiometry. Metabolic response of Ltk^- cells stably transfected with human D1 receptor (open circles) or human D2 receptor (closed circles). Cells were challenged with 100 nM dopamine. Inhibition of the individual responses was complete in the presence of the D1 specific antagonist SCH23390 or the D2 specific antagonist spiperone (data not shown). The D1 response reached a maximum level of 150% of basal rates approximately 20 min after the agonist had reached the cells. The D2 response, on the other hand, reached a maximum level of 175% within the first 5 min of exposure (typical data are shown for $n = 3$).

receptor in fibroblastic Ltk^- cells is robust, a direct microphysiometric response of D2 stimulation was not detectable in the GH4 line of somatotrophic cells. Although the D2 receptor inhibits adenylyl cyclase in both cell lines, it was additionally able to activate PI pathways only in Ltk^-. This suggests that the basis of the metabolic activity may reside in the PI pathway and that this pathway may not significantly couple to D2 in GH4 hosts (25).

Examination of C6 glioma cells also expressing the cloned D2 receptor revealed that, at least in part, the acidification response to dopamine agonists could be attributable to Na^+/H^+ exchange (10). Indeed this idea is supported by experiments in which increases in extracellular acidification rate induced by the D2-specific agonist quinpirole were inhibited by Na^+/H^+ exchange blockers such as amiloride at concentrations that did not alter D2 receptor-mediated inhibition of adenylyl cyclase (11).

The D3 receptor, due to its low expression level *in vivo,* has been studied only by heterologous expression *in vitro,* and even there progress has been slow because of difficulties in demonstrating functional coupling of the receptor. It is noteworthy that Huff and colleagues (27) were able to detect agonist-specific increases in acidification rates for CHO cells expressing the D3 receptor (EC_{50} for dopamine 1.2 n*M;* ~20% increase in acidification rate). Dopamine was shown to be mitogenic and to inhibit forskolin-induced accumulation of cAMP in the same cells. In another report, Rosser *et al.* (28) reported that quinpirole evoked an amiloride-inhibitable increase in the acidification rate of C6 glioma cells expressing the D3 receptor.

Finally, the D4 receptor has been shown to mediate increases in acidification rates when stably expressed in CHO-K1 and HEK293 cells (29). Here and in the D3-CHO construct cited above, the response seen in the microphysiometer was inhibitable by amiloride analogs and by pertussis toxin.

Cholecystokinin Receptors

Denyer and colleagues have used microphysiometry to obtain concentration–response curves for CCK-related peptides in two systems: pancreatic acinar cells expressing an endogenous CCK receptor and HeLa cells expressing a cloned CCK receptor. As is shown in Fig. 6, the rank order of potency suggested that the two receptors were, respectively, of the CCK_A and CCK_B subtypes (30).

In the CCK_B/HeLa system, the nonpeptide CCK_B antagonists PD134308, L-365,260, and devazepide shifted the concentration–response curve for sulfated CCK-8 with pK_b values of (respectively) 8.64, 8.28, and 6.78, where $pK_b = \log\{(\text{dose ratio} - 1)/[\text{antagonist}]\}$. Measurements of intracellular $[Ca^{2+}]$ were carried out in parallel with microphysiometry in this system, giving

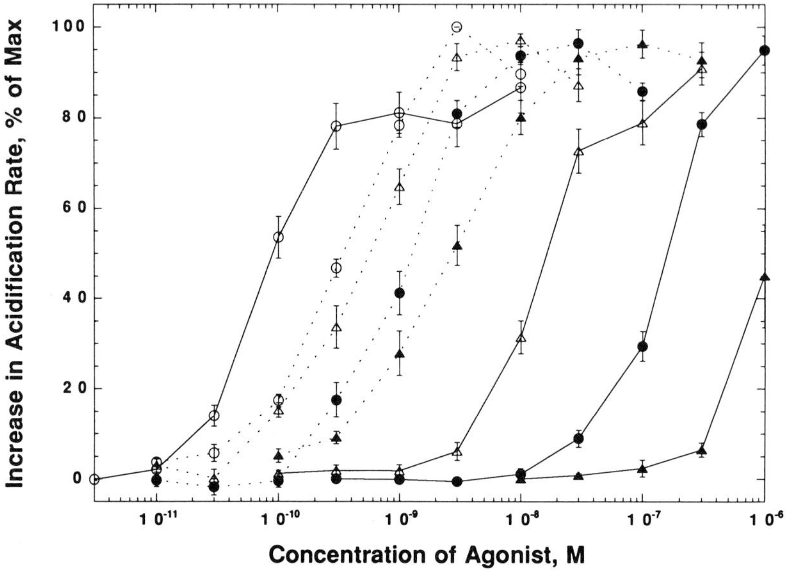

FIG. 6 Detecting the activation of cholecystokinin receptors by microphysiometry. Concentration–response curves evoked by CCK-related peptides are shown for the CCK_A receptor in isolated mouse pancreatic acinar cells (solid lines) and for the human CCK_B receptor expressed in HeLa cells (dotted lines). The peptides are sulfated CCK-8 (open circles), sulfated gastrin-17 (open triangles), pentagastrin (solid circles), and CCK-4 (solid triangles). Data are presented as percentage of the maximum stimulation for each agonist, and error bars represent SEM for $n = 8$ to 24. The data were kindly provided by J. Denyer from the studies reported in Bouvier *et al.* (20) and Rosser *et al.* (28).

slightly higher EC_{50} values but the same rank order of potency for the agonists. However, the $[Ca^{2+}]_i$ assay did not produce parallel concentration–response curves with the antagonists, making it difficult to calculate pK_b values by this method.

Muscarinic Receptors

Four muscarinic systems have been analyzed by microphysiometry in published reports: the m_1 receptor transfected into CHO and fibroblastic B82 cells (6, 17), a putative m_3 receptor in TE671 rhabdomyosarcoma cells (22), and muscarinic receptors in primary cultures of pancreatic acinar cells (21). Agonist-induced increases in acidification rates were evoked more rapidly than the time resolution of the instrument.

To obtain concentration–response curves, each cell population was exposed sequentially to increasing concentrations of agonist. Since the receptor desensitized rapidly in this system, it was crucial to program the fluidics system of the microphysiometer to limit the time of exposure for each agonist concentration to ~1 min and to provide a recovery time (20–30 min) between applications. This worked well because of the rapidity (~1 min) and magnitude (30–100% increase) in the response of the cells. Concentration–response curves show traditional saturation behavior and EC_{50} values consistent with those for second-messenger assays in similar systems. The curves are shifted by the competitive antagonist atropine. In the m_1/CHO system the EC_{50} for microphysiometry was 2 μM, while that for accumulation of IP_3 in the same cells was 3 μM (31). The acinar cells showed an EC_{50} of 200 nM for acetylcholine, which was shifted by atropine with a pA_2 of 9.2.

In the study on muscarinic activation of TE671 cells (22), microphysiometry was used to show that part of the increase in extracellular acidification rate was due to an increase in energy consumption caused by the receptor-mediated activation of the Na^+/K^+-ATPase. Auxiliary measurements of intracellular [Na^+] suggested that the increased activity of the pump was secondary to an increase in the substrate concentration (intracellular [Na^+]) rather than a direct activation of the pump by the signal-transduction pathway of the receptor.

β_2-Adrenergic Receptor

The β_2-adrenergic receptor activates adenylyl cyclase. Increases in acidification rates of 20–40% were seen in CHO cells transfected with human β_2-adrenergic receptor exposed to the β-adrenergic agonist isoproterenol (6). However, the time course of the adrenergic changes was markedly slower than that observed with the m_1 muscarinic receptor in the same host cell, peaking at about 12 min after exposure to the agonist. The response was reversible: removing isoproterenol from the microphysiometer chamber or adding the antagonist propranolol returned the acidification rate to prestimulation values ($t_{1/2}$ ~6 min). Sequential concentration–response curves with isoproterenol gave an EC_{50} ~3 nM that was shifted by propranolol, in agreement with more traditional functional measurements. Isoproterenol also stimulated acidification rate when the β_2-adrenergic receptor was expressed in B82 cells (6).

The Kainate Glutamate Receptor

One study has been published using microphysiometry to study the activation of a ligand-gated ion channel, the kainate-sensitive glutamate receptor (18). Kainate increases the rate of extracellular acidification in primary cocultures of hippocampal neurons and glia from fetal rats, with an EC_{50} of 100 μM, and this stimulation is suppressed by the kainate antagonist kynurenate. Overactivation of glutamate receptors is involved in excitotoxicity, for example, in reperfusion injury after stroke. As shown in Fig. 7, exposure of the hippocampal cultures to toxic levels of kainic acid for 10 min produced an initial elevation in acidification rate and then a gradual decline in metabolism over the next 8 hr that preceded a measurable loss of cell viability (lactate dehydrogenase content), thus delineating a time window for the onset of kainate-induced cell damage.

Growth Factors and Neurotrophic Factors

Among the first responses observed by microphysiometry was the increase in extracellular acidification rate caused by applying epidermal growth factor (EGF) to resting human keratinocytes (6, 16). Approximately 15 min after exposure to EGF this stimulation peaked at 15–90% (depending on the physiological state of the cells); it was specific for EGF, as shown by inhibition with anti-EGF antibody. Subsequently, results have been published for two neurotrophic factors: nerve growth factor (NGF) with PC12 cells and SH-SY5Y cells and ciliary neurotrophic factor with SH-SY5Y cells (17, 32, 33). The EC_{50} for NGF stimulation of PC12 cells was 152 pM (34), indistinguishable from the 148 pM value obtained for neurite outgrowth (35).

Discussion

As mentioned earlier, there are significant limitations to current receptor assay methodologies. Binding assays give no information about function and require a suitable radiolabeled ligand. Functional assays may be slow and tedious, and it is sometimes not clear which of the many signal-transduction paths should be assayed for a particular receptor–cell combination. Furthermore, it is likely that as yet undiscovered effector systems and modes of receptor activation exist, as evidenced by the reported novel ceramide and cADP-ribose systems (36–38).

For these and other reasons it would seem prudent to adopt a methodology

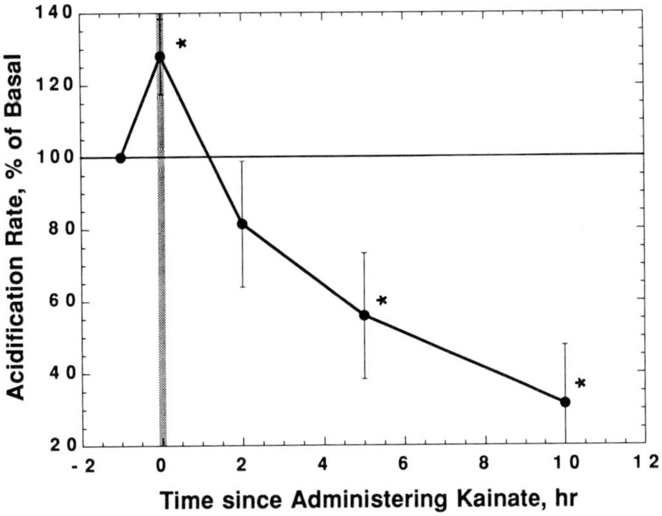

FIG. 7 Effect of exposing a mixed hippocampal culture to an excitotoxic concentration of kainic acid (2.5 mM) for 10 min at $t=0$ (30). An initial stimulation is followed by a progressive decline in metabolic activity. Error bars represent SEM for $n=3$ independent experiments. Asterisks indicate points statistically different from the basal preexposure acidification rate.

that encompasses as many molecular pathways as possible when asking the question, "Does this particular receptor recognize this particular compound?" By exploiting the complicated metabolic web of the cell it may be possible to maximize the chance of detecting a productive ligand-receptor encounter. The scope of applications of these microphysiometric methods continues to be explored, but several uses would seem obvious.

Replacement of Functional Assays That Are Slow, Tedious, or Ineffective

Some assays, such as those based on mitogenesis, require days to detect a result. Others, such as those based on electrophysiology, have notoriously low throughput. For many applications, the combination of real-time response and moderate throughput that the Cytosensor microphysiometer offers may be an attractive alternative. Furthermore, there are cases in which ligands have been identified by binding assays, but no practical mechanistically specific functional assay exists (39). At the date of this writing this is more or less true for the D3 dopamine receptor, and this was an important theme at the British Dopamine Group Workshop held in Harlow in November, 1993. In such a case, microphysiometry may represent the only practical functional assay until the signal-transduction pathway is better understood.

These factors suggest that microphysiometry would be a useful tool for screening new therapeutic drugs. While current microphysiometer technology cannot match the massive throughput of some primary screens ($\sim 10^5$ assays/year), it can come within a factor of 10 of that in favorable circumstances. It may be possible to screen even more compounds by employing divide-and-conquer strategies, using mixtures of natural products or pooled compounds from random/combinatorial or synthetic drug libraries.

Screen for the Ligands of Orphan Receptors

A potential major application of the microphysiometer is the analysis of cloned orphan receptors. While some clues to the identities of the receptors may be provided by sequence homology and tissue distribution, characterization by radioligand binding or functional activity may still be very difficult (39). This problem has at least two roots. First, it is not practical to radiolabel each possible ligand that must be screened. Second, it may be hard to select which of many possible mechanism-based functional assays to use when the mode of signal transduction is in doubt, assuming that an effective assay indeed exists. Expression systems based on electrophysiological methods, such as expressing receptor mRNA in oocytes, can be compromised by the real possibility that the mystery receptor does not couple to an ion channel. The remaining prospect of assaying for any of a dozen second-messenger effects is discouraging, especially when there is no guarantee that the cloned receptor will couple to the expected effectors in the alien environment of the transfection host.

Because of its ability to detect physiological changes in cells relatively independently of their mechanisms, the microphysiometer may circumvent these problems. As we have shown, a detectable microphysiometric response can be sufficient for basic pharmacological characterization and may furthermore provide clues needed for subsequent elucidation of the receptor signal-transduction mechanisms. With respect to heterologous expression, to eliminate those responses caused by endogenous receptors, parallel trials can be run using either nontransfected host cells or host cells transfected with the receptor cDNA inserted into the expression vector in an antisense orientation.

Isolation of Endogenous Ligands

It may be that the endogenous ligands for many receptors, particularly orphans, are compounds that have not yet been identified, irrespective of their function. By employing these receptors in a microphysiometer, one could screen biological homogenates for metabolic effects specific to these receptors. Once active biological extracts were found, traditional fractionation and purification schemes could be employed to enrich and ultimately isolate and characterize the novel bioactive substance.

Excitotoxicity and Ischemia

Microphysiometry can be used to monitor overall cellular metabolic activity as well as to detect receptor activation. This is a natural combination for studying the effects of excitotoxicity and ischemia in cells of the nervous system. As we described above, the method has already shown utility in this regard (18).

Receptor Biology

Signal transduction appears much more complicated than it did a decade or more ago. Instances of one receptor coupling to multiple second-messenger pathways in a single cell or different pathways in different cells are well known. Microphysiometry can help identify known pathways that are activated and may also uncover activation of novel pathways. Particularly useful in this regard is the use of inhibitors of specific pathways, such as pertussis and cholera toxins, kinase inhibitors, and amiloride derivatives to inhibit Na^+/H^+ exchange. In this regard, the activity of an isoform of protein kinase C in a signal-transduction pathway has been identified with antisense oligonucleotides in conjunction with microphysiometric measurements (40). It should likewise be possible to determine the specific coupling pathway for a given G protein-coupled receptor by challenging specific G protein subunits with antisense knock-out reagents.

Other applications of the instrument to receptor biology also follow from its operating characteristics: real-time, noninvasive measurements on living cells in a flow system that provides fine control over the chemical environment of the cells. This system could be used to study questions concerning receptor reserve, biosynthesis and regulation in real-time.

Limitations

Although a powerful tool for many applications, the microphysiometer is no panacea. Receptor activation sometimes fails to cause detectable changes in extracellular acidification rate, the cause of which are only partly understood. For example, no response will be seen if the receptor is not functionally coupled, perhaps due to the absence of some signal-transduction component; the remedy is to supply the component or try a different host cell that may contain it. Alternatively, the cell may already be fully activated metabolically, in which case overnight serum starvation may render the cell more quiescent and able to respond. In some G_i-coupled systems (such as the D2/GH4 system described earlier), it is possible to detect receptor activation only indirectly, as an inhibition of forskolin-stimulated increase in extracellular acidification rate (41). Finally, due to the current time resolution of the cytosensor, if the response is over too quickly (<1 min) and leaves only weak physiological sequelae, it may be missed.

Conclusion

The biological limits of microphysiometry remain to be defined. Initial applications have been primarily pharmacological, and microphysiometry has proven utility as a screen for receptor activation. It is particularly useful where alternative screens are slow, tedious, or ineffective, and it may be especially useful for characterizing orphan receptors and identifying heretofore unknown endogenous ligands. The technique is a viable means of obtaining good quantitative as well as qualitative data. Not only does this method provide a real-time indication of receptor activation, but the agonist EC_{50} and antagonist pA_2 estimates are comparable to those determined by other methods.

While its ability to measure an integrative cellular phenomena is one of its principal strengths, there is evidence that the instrument can also be used to dissect the signal-transduction pathways of receptors. Its nonselective mode of action may in fact provide it with the ability to uncover new pathways. In these respects the method has usefulness in broader questions regarding the impact of receptor-ligand-effector systems on the energy demands of the cell. Thus, microphysiometry may have wide applications in receptor biology.

The technological limits of the method likewise remain to be defined. Researchers at Molecular Devices Corp. are currently exploring experimental versions of microphysiometers that make use of micromachining to increase the parallelism and decrease the sample volumes (42). One aim of this project is to improve the throughput of the system to make it more suitable for large-scale drug screening.

Taken in total, this emerging technique promises to provide a useful new tool with important biomedical applications ranging from the design and testing of new therapeutic compounds to the physiological impact of receptor activation.

References

1. D. G. Hafeman, J. W. Parce, and H. M. McConnell, *Science* **240,** 1182 (1988).
2. J. C. Owicki, L. J. Bousse, D. G. Hafeman, G. L. Kirk, J. D. Olson, H. G. Wada, and J. W. Parce, *Annu. Rev. Biophys. Biomol. Struct.* **23,** 87 (1994).
3. H. G. Wada, R. Indelicato, L. Meyer, T. Kitamura, A. Miyajima, G. Kirk, V. C. Muir, and J. W. Parce, *J. Cell. Physiol.* **154,** 129 (1993).
4. G. Baxter, D. L. Miller, and C. Wang, personal communication (1993).
5. C. M. Fraser, C. D. Wang, D. A. Robinson, J. D. Giocayne, and J. C. Venter, *Mol. Pharmacol.* **36,** 840 (1989).
6. J. C. Owicki, J. W. Parce, K. M. Kercso, G. B. Sigal, V. C. Muir, J. C. Venter, C. M. Fraser, and H. M. McConnell, *Proc. Natl. Acad. Sci. U.S.A.* **87,** 4007 (1990).
7. S. I. Harris, R. S. Balaban, L. Barrett, and L. J. Mandel, *J. Biol. Chem.* **2654,** 10319 (1981).
8. B. Alberts, D. Bray, J. Lewis, M. Raff, K. Roberts, and J. D. Watson, *in* "Molecular Biology of the Cell," 2nd ed., p. 304. Garland, New York, 1989.
9. S. V. P. James, M. R. Barker, J. L. Budal, and M. R. Brann, *Proc. Natl. Acad. Sci. U.S.A.* **85,** 4056 (1988).
10. K. A. Neve, M. R. Kozlowski, and M. P. Rosser, *J. Biol. Chem.* **267,** 25748 (1992).
11. K. A. Neve, M. P. Rosser, and D. L. Barber, Chapter 12.
12. H. G. Wada, R. Indelicato, L. Meyer, T. Kitamura, A. Miyajima, G. Kirk, V. C. Muir, and J. W. Parce, *J. Cell. Physiol.* **154,** 129 (1993).
13. R. Kuo, G. T. Baxter, L. Alajoki, D. L. Miller, J. M. Libby, and J. C. Owicki, *Cryobiology* **30,** 386 (1993).
14. H. G. Wada, S. D. H. Chan, D. Denney, D. M. Antoniucci, K. S. Kok, J. Howison, J. W. Parce, H. M. McConnell, and J. C. Owicki, *Mol. Biol. Cell* **4,** Suppl., 89a (1993).
15. S. Pitchford, personal communication (1994).
16. J. W. Parce, J. C. Owicki, K. M. Kercso, G. B. Sigal, H. G. Wada, V. C. Muir, L. J. Bousse, K. L. Ross, B. I. Sikic, and H. M. McConnell, *Science* **246,** 243 (1989).
17. H. M. McConnell, J. C. Owicki, J. W. Parce, D. L. Miller, G. T. Baxter, H. G. Wada, and S. Pitchford, *Science* **257,** 1906 (1992).
18. K. M. Raley-Susman, K. R. Miller, J. C. Owicki, and R. M. Sapolsky, *J. Neurosci.* **12,** 773 (1992).

19. H. M. McConnell, P. Rice, G. H. Wada, J. C. Owicki, and J. W. Parce, *Curr. Opin. Struct. Biol.* **1,** 647 (1991).
20. C. Bouvier, J. A. Salon, R. A. Johnson, J. R. Bunzow, H. H. M. Van Tol, and O. Civelli, *Soc. Neurosci. Abstr.* **18,** 276 (1992).
21. J. C. Denyer, C. Thorne, C. Bountra, and C. C. Jordan, *J. Physiol.* (*London*) **459,** 390 (1993).
22. D. L. Miller, J. C. Olson, J. W. Parce, and J. C. Owicki, *Biophys. J.* **64,** 813 (1993).
23. R. M. Johnson, P. A. McNeeley, K. De Moor, G. R. Stewart, S. Pitchford, and B. S. Glaser, *Brain Res.,* **646,** 327 (1994).
24. J. C. Owicki, personal communication (1993).
25. C. Bouvier, J. A. Salon, R. A. Johnson, and O. Civelli, *J. Recept. Res.* **13,** 559 (1993).
26. J. A. Salon, personal communication (1993).
27. C. L. Chio, M. E. Lajiness, and R. M. Huff, *Mol. Pharmacol.* **45,** 51 (1994).
28. M. P. Rosser, M. R. Kozlowski, R. I. Neve, and K. A. Neve, *Soc. Neurosci. Abstr.* **17,** 86 (1992).
29. C. L. Chio, G. S. Gill, R. M. Huff, and I. Abraham, *Neurosci. Abstr.* **19,** 735 (1993).
30. J. Denyer, J. Gray, M. Wong, M. Stolz, and S. Tate, *Eur. J. Pharmacol.* **268,** 69 (1994).
31. M. A. Buck and C. F. Fraser, *Biochem. Biophys. Res. Commun.* **173,** 666 (1990).
32. S. Pitchford and B. S. Glaeser, *Soc. Neurosci. Abstr.* **19,** 1309 (1993).
33. R. M. Johnson, P. A. McNeeley, B. S. Glaeser, and S. Pitchford, *Soc. Neurosci. Abstr.* **18,** 617 (1992).
34. S. Pitchford, personal communication (1993).
35. R. O. Messing, M. Henteleff, and J. J. Park, *Brain Res.* **565,** 301 (1991).
36. S. Mathias and R. Kolesnick,, *Adv. Lipid Res.* **25,** 65 (1993).
37. Y. Hannun, L. M. Obeid, and R. A. Wolff, *Adv. Lipid Res.* **585,** 43 (1993).
38. S. Takasawa, K. Nata, H. Yonekura, and H. Okamoto, *Science* **259,** 370 (1993).
39. A. Mills and M. J. Duggan, *Trends Pharmacol. Sci.* **14,** 394 (1993).
40. G. T. Baxter, D. L. Miller, R. C. Kuo, H. G. Wada, and J. C. Owicki, *Biochemistry* **31,** 10950 (1992).
41. C. Bouvier, personal communication (1994).
42. L. J. Bousse, R. J. McReynolds, T. Dawes, P. Lam, W. R. Bemiss, and J. W. Parce, *Int. Conf. Solid State Sens. Actuat., Dig. Tech. Pap., Transducers '93, 7th,* PACIFICO, Yokohama, Japan (1993).

[12] Regulation of Na^+-H^+ Exchange by G Protein-Coupled Receptors

Kim A. Neve, Mary P. Rosser, and Diane L. Barber

Introduction

The Na^+-H^+ exchanger is an electroneutral antiporter that participates in intracellular pH (pH_i) homeostasis, cell volume regulation, and the transepithelial transport of acid-base equivalents. Several populations of Na^+-H^+ exchangers have been identified that differ in distribution, sensitivity to amiloride and amiloride analogs, affinity for Na^+ and H^+, and regulatory properties (1–3). NHE1 is a ubiquitous plasma membrane exchanger that exhibits high sensitivity to inhibition by amiloride and its 5-amino-substituted analogs such as ethylisopropylamiloride, and is completely insensitive to cyclic AMP. The primary function of NHE1 is to regulate pH_i (2, 4). Other more specialized exchangers include NHE2 and NHE3, which are apically restricted exchangers in renal, intestinal, and gall bladder epithelia that function in the transepithelial transport of solutes and fluids.

Through the regulation of pH_i, NHE1 plays either an essential or a permissive role in many cellular processes, including growth, differentiation, and secretion (2, 5). A hormone or neurotransmitter receptor may indirectly modulate these processes by regulating activity of NHE1. Receptor-mediated regulation of Na^+-H^+ exchange has been described for a number of polypeptide growth factors, vasoactive agents, and hormones (1). The receptors mediating these effects include both growth factor receptors with intrinsic tyrosine kinase activity and G protein-coupled receptors. Activation by many of these receptors is associated with phosphorylation of the antiporter on serine residues (6). Although the driving force for Na^+-H^+ exchange is the electrochemical Na^+ gradient generated by the Na^+,K^+-ATPase, cytosolic H^+ allosterically activates the exchanger by interacting with an internal modifier site (7). Receptor-mediated modulation of NHE1 is thought to occur by altering the affinity of this regulatory site for H^+, thus changing the sensitivity of the exchanger to pH_i (8).

Many methods have been used to assess receptor-modulated Na^+-H^+ exchange. These include measuring $^{22}Na^+$ uptake, determining the intracellular/extracellular distribution of a weak acid such as benzoic acid, quantifying regulation of pH_i using H^+-sensitive fluorescent dyes, and monitoring extracellular pH (pH_o). In this chapter we describe and compare two methods

Methods in Neurosciences, Volume 25

used to assess NHE1 activity in populations of cells, the measurement of pH_i using a spectrofluorometer and the measurement of pH_o using a microphysiometer. We also summarize work carried out in our laboratories in which these methods were used to evaluate modulation of NHE1 activity by several G protein-coupled receptors, including β_2-adrenergic, somatostatin, and D2 and D3 dopamine receptors.

Experimental Methods

Measurement of Intracellular pH: Spectrofluorometer

A variety of primary cell cultures and clonal cell lines (9–12) are used to study the regulation of Na^+-H^+ exchange by endogenous and recombinant hormone receptors. We work primarily with adherent cells and restrict our studies to the use of nonpolarized cells, which allows us to focus on NHE1 activity when measuring pH_i. Polarized epithelia can also express apically localized NHE subtypes, which may contribute to receptor regulation of pH_i. Cells are reseeded onto either glass or plastic coverslips 48–72 hr prior to determinations. In most cell types, serum in the culture medium tonically activates NHE1. Stimulatory effects are more pronounced on the quiescent exchanger, and therefore receptor-stimulated exchange activity is studied in growth-arrested cells that have been maintained in the absence of serum for 18–24 hr. In contrast, receptor-mediated inhibition of NHE1 activity is studied in cells maintained continuously in serum.

We use a spectrofluorometer to measure changes in pH_i as an index of Na^+-H^+ exchange activity. With a modification of the methods described below, spectrofluorometric measurements of pH_i are also used to determine activities of the Cl^-/HCO_3^- and Na^+-dependent Cl^-/HCO_3^- anion exchangers (13, 14). To use pH_i as an indirect measure of Na^+-H^+ exchange, determinations are made in a HEPES-buffered, nominally HCO_3-free medium (9) to attenuate effects of the anion exchangers. Cells are washed and then incubated for 15–20 min at 37°C with 1–5 μM of the methyl ester precursor of a fluorescent, H^+-sensitive dye such as 2′,7′-biscarboxyethyl-5(6)carboxyfluorescein (BCECF/AM; Molecular Probes). The methyl ester precursor of the dye is membrane-permeant. Intracellular esterases cleave the precursor, trapping the membrane-impermeant dye within the cell. After cells are loaded with dye, they are washed and placed in a cuvette maintained in a thermostatically controlled (37°C) cuvette holder within a spectrofluorometer. We continuously perfuse the cells at a rate of 2 ml/min, providing the inflow with a peristaltic pump and using gravity for the outflow. Fluorescence intensities of cells loaded with BCECF are measured by exciting the dye alternately at 440 and at 500 nm, with a constant emission at 530 nm. The ratio of fluorescence intensities is then calibrated by the high K^+/nigericin technique (15) to calculate pH_i. Cells are perfused with a solution containing 105 m*M* KCl, 15 m*M* HEPES, 1 m*M* $MgCl_2$, and 10 μM nigericin, adjusted to pH 6.6–7.4 with either KOH or HCl. When intracellular and extracellular [K^+] are equal, the K^+-H^+ exchanger nigericin produces an equilibration between intracellular pH and extracellular pH. This equilibration is made using extracellular solutions of determined pH values, hence the ratio of fluorescence intensities can be calibrated to these values. Several methods can be used to study receptor-regulated changes in pH_i. With each of these methods, we establish whether the changes in pH_i are blocked in the presence of amiloride analogs, such as ethylisopropylamiloride (EIPA; 50 μM), or in the absence of extracellular Na^+ (solutions containing molar equivalents of either *N*-methyl-D-glucamine or choline). These determinations are critical to ascertain whether the changes in pH_i are due to Na^+-H^+ exchange activity. Receptor-mediated effects can be determined by monitoring changes in steady-state pH_i (Fig. 1A) or by evaluating the rate of pH_i recovery (dpH_i/dt) from an acid load (Fig. 1B). Although changes in steady-state pH_i are an easier and more rapid way to measure agonist-induced effects, the rate of pH_i recovery from an acid load, although time intensive, is a more accurate index of the extrusion of acid and Na^+-H^+ exchange. In this latter technique, as described by Boron and De Weer (16), cells are prepulsed with a solution containing 10–30 m*M* NH_4Cl to induce an acid load. With this prepulse, NH_3 rapidly enters cells and combines with cytosolic H^+, inducing a rapid increase in pH_i (Fig. 1B, a and b). The slower entry of NH_4^+ causes a gradual decline in pH_i (Fig. 1B, b and c) from the intracellular formation of NH_3 and H^+. Removing extracellular NH_4Cl traps H^+ within the cell as NH_3 exits rapidly, and pH_i decreases below its resting level (Fig. 1B, c and d). H^+-extrusion mechanisms contribute to the recovery from this acid load (Fig. 1B, d and e), and under the appropriate conditions the rate of H^+ extrusion is an estimate of Na^+-H^+ exchange. Rates of recovery are determined by evaluating the derivative of the slope of the pH_i tracing at a selected pH_i. The exchanger is regulated by H^+ (7), and its activity will therefore change over the pH_i range of the recovery; it is more active at lower pH_i and slows as pH_i approaches neutrality. Hence, when evaluating dpH_i/dt to compare control and experimentally induced recovery rates, it is critical to compare dpH_i/dt at a specified pH_i and not as a function of time (Fig. 2). While dpH_i/dt provides an estimate of Na^+-H^+ exchange activity, the actual H^+ flux (m*M*/min) is calculated by multiplying dpH_i/dt by the intrinsic buffering capacity of the cell (Fig. 3) (13, 14).

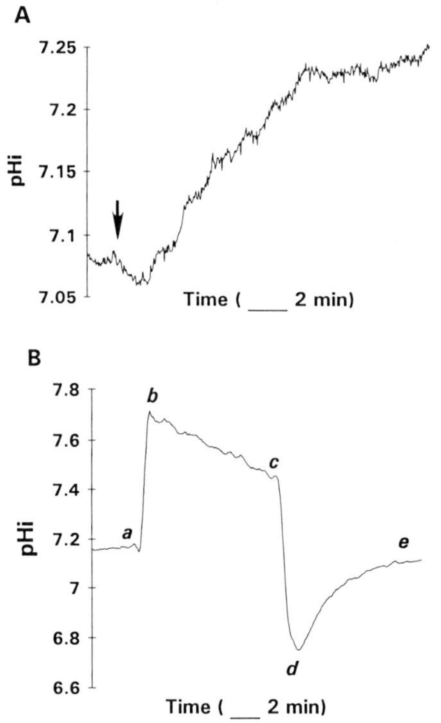

FIG. 1 Approaches for studying receptor-mediated effects on pH_i and acid extrusion. (A) Changes in steady-state pH_i indicate the time course of hormone-induced alkalinization. Fibroblast L cells stably expressing recombinant β_2-adrenergic receptors were perfused with 1 μM isoproterenol at the indicated arrow. (B) pH_i recovery from an acid load induced by 20 mM NH_4Cl is an index of acid extrusion. The indicated tracing represents a control (no receptor stimulation) recovery in fibroblast L cells.

The optimal conditions for studying receptor-regulated pH_i using the spectrofluorometer depend on the cell types being studied. High concentrations of fluorescent dye can contribute to an unstable resting pH_i. We find that 1 μM of BCECF is optimal in most clonal cell lines, whereas primary cells generally require 3–5 μM. Cells are used within 60–90 min after dye loading. Optimal conditions for acid loading with an NH_4Cl prepulse are also cell type-dependent. Fibroblast L cells are quite resistant to acid loading and require a 10- to 15-min pulse with 20 mM NH_4Cl to drive pH_i to 6.65. In contrast, human embryonic kidney (HEK) 293 cells are more sensitive to acid loading, and a 5-min pulse with 20 mM NH_4Cl will yield a similar pH_i.

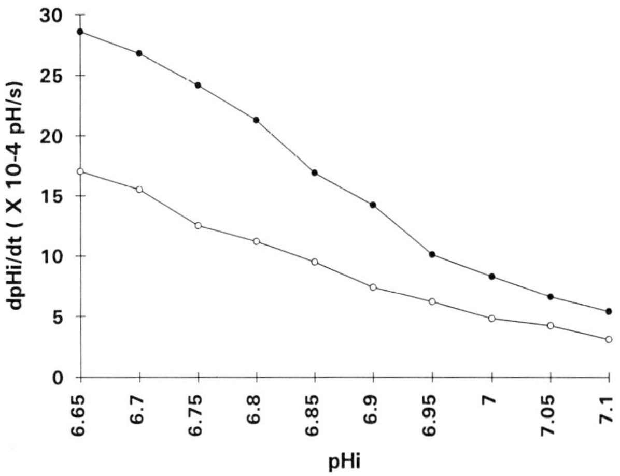

FIG. 2 Effects of pH_i on rates of recovery from an acid load. Fibroblast L cells stably expressing recombinant β_2-adrenergic receptors were acid-loaded by applying and removing 20 m*M* NH_4Cl. The rates of recovery (dpH_i/dt) from this acid load were determined at the indicated pH_i in untreated, control cells (○) and in cells treated with 1 μM isoproterenol (●). Prior to pH_i determinations, cells were maintained in the absence of serum for 18 hr.

Additionally, resting pH_i, steady-state NHE1 activity, and the rate of H^+ extrusion from an acid load vary in different cell types.

Measurement of Extracellular pH: Microphysiometer

The studies described in this chapter that used the microphysiometer were obtained using cells that expressed recombinant D2 or D3 dopamine receptors (17). Cells (either C_6 glioma cells or Ltk^- fibroblasts) are seeded into 12-mm polycarbonate capsule cups (Molecular Devices) at a density of 7.5×10^5 cells/well in Dulbecco's modified Eagle's medium (DME, JRH Scientific) supplemented with 10% fetal bovine serum (FBS). The optimal cell density needs to be determined for each cell type, but cells are typically seeded at densities ranging from 3×10^5 to 1×10^6 cells/well. Cells in capsule cups are incubated for 3 hr in a humidified environment of 95% air/5% CO_2 at 37°C to allow the cells to attach to the membrane of the cell capsule. The medium is then removed and replaced with serum-free DME, and the capsule cups are returned to the incubator overnight (16–18 hr). Incubating the cells in serum-free medium prior to examining them in the microphysiometer

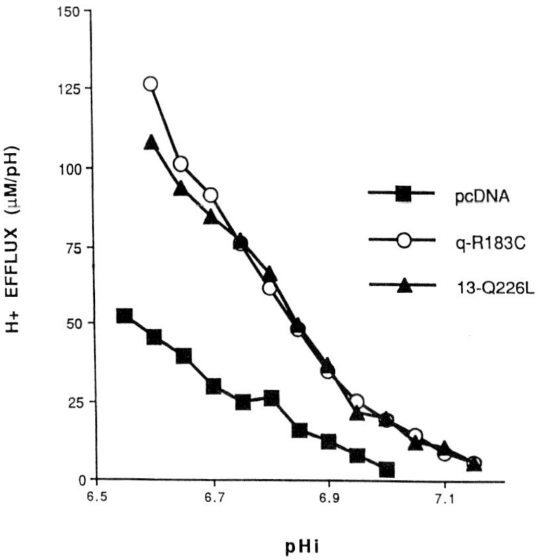

FIG. 3 pH_i dependence of H^+ efflux. Recovery rates from HEK293 cells were multiplied by the intrinsic buffering capacity to determine H^+ efflux at the indicated pH_i. Conditions included cells transiently expressing plasmid vector alone (pcDNA) or expressing mutationally activated constructs of G protein α-subunits α_q-R183C or α_{13}-Q226L. See Results and Discussion section for further experimental description (23).

increased the magnitude of the response of dopamine D2 or D3 receptors in the present studies, but is not necessary for all cell types.

Capsule cups are removed from the incubator and assembled in a laminar air flow hood. When assembling the capsule cup, the spacer and the insert are aseptically placed into the cell capsule. The insert is immersed by adding 0.5 ml of low-buffered medium (see below) into the insert. The cell capsule assembly is then placed in the previously sterilized flow chamber and mounted on the microphysiometer. A plunger containing the inlet and outlet tubing is placed into the flow chamber. It is critical not to introduce bubbles at any step while assembling the capsule cup or when loading the chambers, as bubbles anywhere in the system will result in "noisy" data, error messages, or an inability to calibrate the chamber containing the bubbles. Bubbles can be removed by tapping the reference electrode or tubing, removing the plunger and tapping the chamber, or by removing the insert and aspirating the bubbles. When removing and replacing the inserts, care must be taken not to damage the cell monolayer.

To measure extracellular pH, cells are perfused with culture medium lacking bicarbonate and with a low buffering capacity (1 m*M* sodium phosphate), supplemented with 1 mg/ml bovine serum albumin, 100 units/ml penicillin, and 100 μg/ml streptomycin, hereafter referred to as modified medium. Depending on the experimental conditions or the length of time cells are incubated in the chambers, antibiotics may not be necessary. Modified medium is prewarmed to 37°C, transferred to 15-ml conical centrifuge tubes, and attached to the microphysiometer. The medium is pumped through the tubing and into the plunger before attaching the flow chambers, with care taken not to introduce bubbles. Running the pumps while loading the cell capsules maintains a fluid-to-fluid contact and reduces the formation of bubbles. In the present studies, the flow rate was 50 μl/min, although the rate should be optimized for each type of cell used. The microphysiometer light-addressable potentiometric sensor (LAPS) measures changes in medium acidity in the extracellular environment during periodic interruption of flow through the flow chamber. Acidification is predominantly due to the metabolic production of lactic and carbonic acid and is measured in real time. For the data presented in this chapter, multiple determinations of the rate of acidification were obtained in 4-min cycles of operation. In each cycle, cells were perfused with modified medium for 210 sec. The flow was interrupted for 30 sec, during which the rate of acidification of the modified medium was measured and recorded. The flow was then resumed and the next cycle begun. For purposes of quantification, the rate of acidification produced by medium from a particular reservoir was determined 2500 sec after opening the valve for that reservoir.

The microphysiometer used for most of the studies described in this chapter was a prototype that is not commercially available. A number of improvements on the prototype have been made in the current Cytosensor microphysiometer (Molecular Devices). Some of these improvements are described briefly here, and the Cytosensor is described thoroughly elsewhere (18). The Cytosensor allows the user to expose cells to a drug for as little as 5 sec, at a cell density as low as 2×10^4 cells/capsule. Protocols for multiple exposure to various drugs can be created in the Cytosoft software program that runs on a Macintosh Centris 610. Data are expressed as the rate of acidification in real time and can be normalized against predrug rates. Rate data from all eight chambers can be viewed at one time and plotted in color.

General Considerations

Spectrofluorometers are manufactured and sold by many vendors, with prices ranging from under $20,000 to considerably higher. A spectrofluorometer permits the study of any intracellular response for which a suitably responsive fluorescent dye is available. To study regulation of pH_i, the use of a H^+-sensitive fluorescent dye such as BCECF appears to be a rather unequivocal measure, particularly when combined with the acid-loading technique of activating the exchanger. On the other hand, constructing a dose–response curve using the acid-loading technique, for example, is a time-consuming procedure, requiring extensive cell washing between NH_4Cl pulses and post hoc calibration to a standard pH curve to convert the excitation ratios to pH values. It is often not possible to do more than two NH_4Cl pulses on a single cell preparation.

The Cytosensor microphysiometer is available from only one vendor at a list price of $100,000 for the eight-chamber model. Disposable capsule cups for the Cytosensor cost $5/cup, and the purchase of special low-buffered medium is sometimes necessary, although low-buffered solutions of physiological saline may also be used. The Cytosensor measures one response (extracellular pH) and measures it well. Washing cells between experimental manipulations is an inherent part of the flow-through design of the Cytosensor. Similarly, because of the presence of eight chambers, multiple conditions (e.g., multiple doses of an agonist) can be tested in parallel. The multichamber design, real-time data processing, and ease of data manipulation and output all contribute to the ability to generate data rapidly with the Cytosensor microphysiometer. On the other hand, a change in extracellular acidification is an indirect sign of a change in pH_i. A variety of cellular processes contribute to extracellular acidification, so the conclusion that a change in acidification is due to one process such as Na^+-H^+ exchange requires converging evidence from experiments that assess the reliance of the response on extracellular sodium and the sensitivity of the response to amiloride and amiloride analogs. These control experiments are also important when using a spectrofluorometer to measure Na^+-H^+ exchange. Even after demonstrating involvement of NHE1 in a response measured by the microphysiometer, however, it may be difficult to distinguish direct extrusion of protons by NHE1 from acid extrusion that is a consequence of increased metabolic activity resulting from NHE1-induced intracellular alkalinization (18, 19).

Results and Discussion

Spectrofluorometer

Using spectrofluorometric measurements of pH_i to study receptor and G protein regulation of NHE1, we have determined that endogenously expressed and recombinant β-adrenergic (β-AR) and somatostatin receptors in many cell types divergently regulate adenylyl cyclase and NHE1. The β-AR stimulates adenylyl cyclase through coupling to the stimulatory G protein G_S, but independently of G_S it also activates NHE1, increasing pH_i in a HEPES-buffered medium (9–11). Somatostatin receptors inhibit adenylyl cyclase through coupling to one of several pertussis toxin-sensitive G_i proteins, but they inhibit NHE1 and lower pH_i through a mechanism that is insensitive to pertussis toxin (9, 20). As described under Experimental Methods, we use two different approaches to study receptor regulation of NHE1. In growth-arrested cells, β-adrenergic agonists produce a rapid increase in resting pH_i (Fig. 1A) and they also increase the rate of pH_i recovery (dpH_i/dt) from an acid load induced by an NH_4Cl prepulse, compared to control dpH_i/dt in the absence of ligand (Fig. 2). In cells maintained in the presence of serum, somatostatin effects include a decrease in resting pH_i and an attenuation of control pH_i recoveries from an acid load. Additionally, somatostatin inhibits β-adrenergic stimulation of dpH_i/dt in cells expressing both receptor types.

In addition to studying wild-type receptors, we used spectrofluorometric measurements of pH_i in cells transfected with mutant receptors to determine that different domains of the β-adrenergic and somatostatin receptors divergently regulate adenylyl cyclase and NHE1. Constructs of the β_2-AR with deletions in residues 222–229 and 258–270 of the third cytoplasmic domain are defective in coupling to G_S and adenylyl cyclase, but retain isoproterenol-stimulated NHE1 activity that is similar in time course and magnitude to that observed with the wild-type β_2-AR (12). To evaluate whether other residues in the third cytoplasmic region confer coupling to NHE1, we stably expressed a series of mutant constructs containing small deletions (10–12 residues) that completely spanned this domain. All of these mutant receptors retained the ability to stimulate the exchanger, indicating that none of the individual amino acids in this region is directly involved in coupling to NHE1.

To study the structural basis for inhibition of NHE1 by somatostatin receptors we used chimeric receptors, taking advantage of our observation that there are receptor subtype-specific differences in coupling to the exchanger (20). When expressed stably in Ltk^- cells or transiently in HEK293 cells, two somatostatin receptor subtypes, SSTR1 and SSTR2, inhibit adenylyl cyclase through a pertussis toxin-sensitive G protein. In contrast, only SSTR1 mediated inhibition of Na^+-H^+ exchange by somatostatin (Fig. 4B), and this action was insensitive to pertussis toxin. We therefore generated two chimeric receptors by replacing sequential residues of SSTR2 with cognate sequences of SSTR1 to identify molecular determinants unique to SSTR1 that may confer coupling to the exchanger (Fig. 4). SSTCR4 included a SSTR1 segment encompassing determinants within the fifth and sixth hydrophobic domains and the entire third cytoplasmic loop, whereas SSTCR5

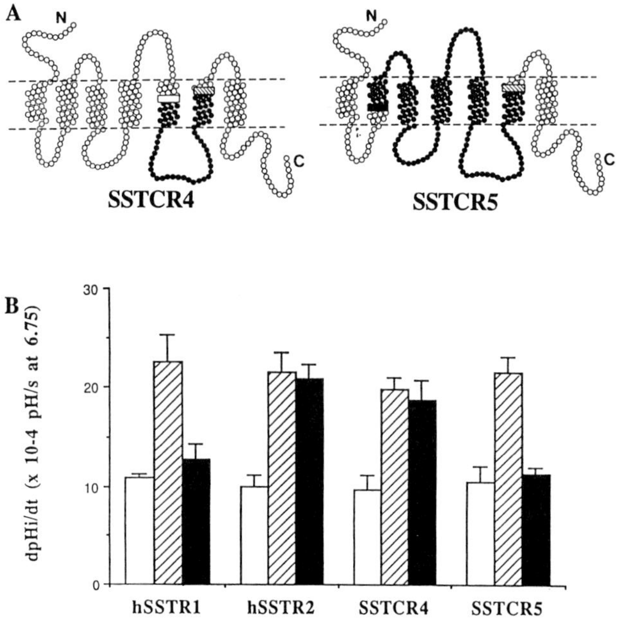

FIG. 4 Effect of wild-type and chimeric somatostatin receptors in regulating pH_i recoveries from an acid load. (A) Schematic diagram of chimeric SSTR2/SSTR1 receptors. SSTCR4 was constructed by subcloning a 188-bp PCR product of human SSTR1 into the *Kpn*I and *Nco*I sites of human SSTR2. SSTCR5 was constructed by ligating a 574-bp PCR product of SSTR1 into the *Sac*I and *Nco*I sites of SSTR2. (B) Somatostatin inhibition of isoproterenol-stimulated pH_i recoveries from an acid load. In HEK293 cells transiently expressing the β_2-adrenergic receptor and SSTR1, SSTR2, SSTCR4, or SSTCR5, dpH_i/dt following a 20 m*M* NH_4Cl prepulse was determined in the absence (open bars) and presence of 1 μM isoproterenol (hatched bars) or 1 μM isoproterenol plus 100 n*M* somatostatin (solid bars). Results are expressed as dpH_i/dt at pH_i 6.75 and represent the means ± SE from three transfections for wild-type somatostatin receptors and from five transfections for chimeric receptors (20).

contained a SSTR1 segment spanning the second through sixth hydrophobic domains, including both second and third cytoplasmic loops (Fig. 4A). Although both chimeric receptors mediated somatostatin inhibition of cAMP accumulation, only SSTCR5 inhibited Na^+-H^+ exchange activity (Fig. 4B), and this action was insensitive to pertussis toxin (20). Our findings with both the β_2-AR and the SSTR1 suggest that coupling to NHE1 may be conferred by determinants outside the third cytoplasmic domain of these receptors.

We also used the spectrofluorometer to study regulation of NHE1 by G proteins, independently of receptor activation. The mutant (del 222-229)β_2-AR described above that is uncoupled from G_S stimulates NHE1 through a guanine nucleotide-dependent mechanism (11), suggesting that a previously unidentified G protein, distinct from G_S, may mediate this effect. To identify G proteins that regulate NHE1, we assessed the regulation of pH_i by mutationally active $G\alpha$ subunits, in which GTPase activity was abolished by single amino acid substitutions. We reasoned that if a specific $G\alpha$ subunit mediates agonist stimulation of an effector pathway, a constitutively activated version of the same $G\alpha$ should stimulate the same pathway in an agonist-independent fashion (21, 22). We expressed a number of wild-type and mutationally active G proteins transiently in HEK293 cells and identified only two α-subunits, α_q and α_{13}, that stimulated NHE1 (23). Expression of either α_q-R183C or α_{13}-Q226L, but not wild-type α_q and α_{13}, increased two indices of NHE1 activity, the steady-state pH_i and the rate of recovery of pH_i following a transient acid load induced by NH_4Cl. These changes were apparently caused by increased Na^+-H^+ exchange, because pH_i recoveries were abolished in the absence of extracellular Na^+ or in the presence of the NHE1 blocker EIPA (50 μM). Because long-term expression of a mutant cDNA might affect pH_i by changing the ability of cells to buffer protons, we also assessed changes in the rate of H^+ efflux, determined by multiplying the rates of pH_i recovery by the intrinsic cellular buffering capacity. Although expression of mutant α_q and α_{13} had no effect on intrinsic buffering capacities compared to untransfected cells or cells transfected with vector DNA alone, it increased the rate of H^+ efflux (Fig. 3). Whether either of these $G\alpha$ subunits mediate β-AR activation of NHE1 remains to be determined.

Microphysiometer

Our use of the microphysiometer to measure receptor-regulated Na^+-H^+ exchange has focused on effects of stimulation of dopamine D2 and D3 receptors. Dopamine D2 receptors, like somatostatin receptors, belong to the subfamily of receptors that interact with inhibitory G proteins. The D2-like dopamine receptors D2, D3, and D4 have more sequence similarity to α_2-adrenergic receptors and the muscarinic receptors than to D1 dopamine receptors (24). One of the characteristics of this subfamily is the capacity of a single class of receptor such as the D2 receptor to modulate a variety of signal transduction pathways, including adenylyl cyclase, K^+ channels, Ca^{2+} channels, and polyphosphoinositide hydrolysis (5). Because of the many similarities between dopamine D2 receptors and α_2-adrenergic receptors, and considerable direct and indirect evidence that some effects of stimulation of the latter receptors result from acceleration of Na^+-H^+ exchange (5, 10), we postulated that D2 receptors would also modulate Na^+-H^+ exchange. This postulate was confirmed when Barber and colleagues demonstrated that D2 receptors and several other G protein-coupled receptors increase or decrease pH_i via acceleration or inhibition of Na^+-H^+ exchange (9, 10).

We reasoned that a change in the rate of proton extrusion by NHE-1 would be reflected in the extracellular pH and, therefore, could be monitored by the microphysiometer, which measures small changes in the pH of low-buffered solutions. This reasoning was supported by modeling studies demonstrating that a change in the rate of proton export by a Na^+/H^+ antiporter resulting in intracellular alkalization of 0.3 pH units would cause detectable extracellular acidification in the microphysiometer (25). Furthermore, prior studies demonstrated that activation of G protein-coupled receptors increases solution acidity, although the molecular basis for the extracellular acidification was not determined (26).

To test the hypothesis that D2 receptors modulate N^+-H^+ exchange, C_6 glioma cells expressing recombinant D2 receptors were tested in the microphysiometer. Initial studies were carried out using the rat short form of D2 receptors, $D2_{415}$. Perfusion of C_6D2_{415} cells with the D2 receptor agonist quinpirole increases the rate of extracellular acidification by the cells (Fig. 5). The effect is concentration-dependent, and a maximal response of approximately 10% over the control rate of acidification was observed at 100 n*M* quinpirole. The dopamine D2 receptor antagonist spiperone has no effect on the rate of acidification but prevents the change in rate induced by 100 n*M* quinpirole, indicating that the response to quinpirole is mediated by D2 receptors (Fig. 5). In addition, C_6 cells transfected with the rat long form of D2 receptors, $D2_{444}$, also respond to 100 n*M* quinpirole by increasing the rate of extracellular acidification, whereas cells not transfected with either D2 receptor cDNA are insensitive to quinpirole (19). Ltk^- cells are murine fibroblasts in which intracellular alkalization in response to stimulation of endogenous PGE_1 receptors or recombinant β-adrenergic receptors has been demonstrated using a spectrofluorometer (10). To compare techniques, Ltk^- cells expressing recombinant D2 receptors (LZR1 cells) (27) were tested in the microphysiometer. As observed for C_6 cells, quinpirole substantially increases the rate of extracellular acidification by LZR1 cells. PGE_1 (10 n*M*) is also very effective at enhancing the extrusion of acid from Ltk^- cells (19).

A number of cellular processes contribute to extracellular acidification, often by the final production and secretion of the principal catabolic products in mammalian cells, lactic acid and CO_2. To determine the signaling pathway involved in D2 receptor-mediated extracellular acidification, the sensitivity of the response to the NHE1 inhibitors amiloride and methylisobutyl-amiloride

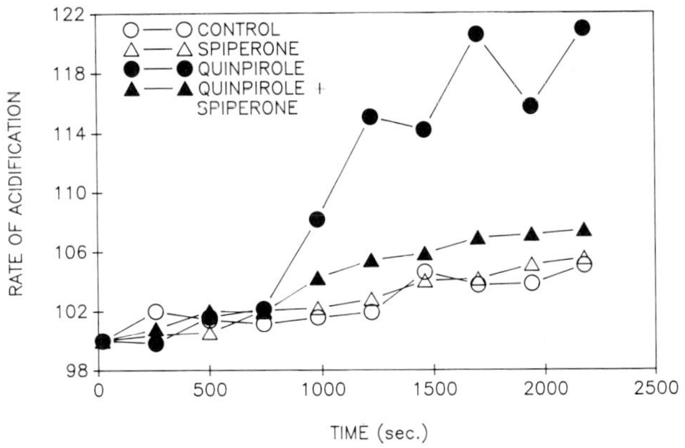

FIG. 5 Effect of quinpirole on rate of extracellular acidification in C_6 cells expressing D2 receptors. Responses are shown for rate of extracellular acidification in the presence of the indicated drugs. Data from a representative experiment using C_6D2_{415} cells are plotted as rate of acidification, expressed as a percentage of predrug baseline, versus time. The first data point indicates the time at which the flow of experimental medium from the reservoir toward the microphysiometer chamber was initiated (19).

(MIA) was determined. Both amiloride (10 μM) and MIA (1 μM) inhibit quinpirole-induced extracellular acidification without altering the basal rate of acidification (Fig. 6). The MIA is more potent than amiloride at inhibiting NHE1, and MIA is also more potent at inhibiting D2 receptor-stimulated extracellular acidification, since 3 μM MIA completely prevents stimulation by 100 nM quinpirole. At considerably higher concentrations, amiloride inhibits radioligand binding to D2 receptors, presumably via a direct effect on the receptors. At the concentrations of amiloride and MIA used in Fig. 6, however, neither amiloride nor MIA have any effect on the binding of [^{125}I]epidepride to D2 receptors, or on inhibition of adenylyl cyclase by the receptors (19).

Inhibition of extracellular acidification by amiloride and MIA argues for involvement of NHE1 in the response. NHE1 is also sensitive to changes in the extracellular concentration of sodium, since the sodium electrochemical gradient is the driving force for Na^+-H^+ exchange. When D2 receptor-mediated extrusion of acid is assessed in a standard salt solution, the response is attenuated compared to the response in growth medium. Removal of sodium from the salt solution abolishes quinpirole-induced extracellular acidification (19).

The lack of response in the absence of extracellular sodium and the sensitivity to amiloride and MIA both indicate that NHE1 has a role in D2 receptor-

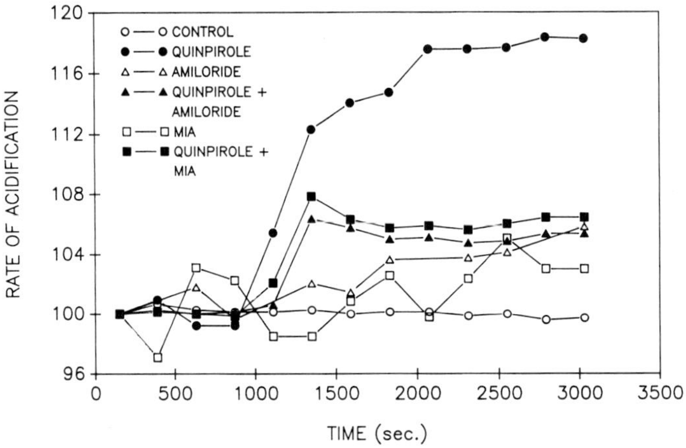

FIG. 6 Effect of amiloride and MIA on quinpirole-induced extracellular acidification. Representative data are shown from one of four (amiloride, 10 μM) or three (MIA, 1 μM) independent experiments in which the effect of the compounds on quinpirole-induced extracellular acidification was determined. Data are expressed as in Fig. 5 (19).

stimulated extracellular acidification. Still, activation of NHE1 could conceivably be a secondary response to some better characterized signaling pathway modulated by D2 receptors, such as inhibition of adenylyl cyclase. Since inhibition of adenylyl cyclase and most other known responses to stimulation of D2 receptors are mediated by pertussis toxin-sensitive G proteins, the sensitivity of D2 receptor-stimulated extracellular acidification to pertussis toxin was assessed (19). Overnight treatment with 100 ng/ml of pertussis toxin completely abolishes inhibition of adenylyl cyclase by D2 receptors, but has no effect on the rate of extracellular acidification by C_6-D2 cells or LZR1 cells (19). These data demonstrate that D2 receptor-stimulated Na^+-H^+ exchange occurs via a signaling mechanism that is distinct from inhibition of adenylyl cyclase and other pertussis toxin-sensitive signal transduction pathways. These data are also consistent with results obtained using spectrofluorometric techniques, demonstrating that modulation of Na^+-H^+ exchange by a number of G protein-coupled receptors is not mediated by a pertussis toxin-sensitive G protein (9, 10).

The use of the microphysiometer to assess Na^+-H^+ exchange has recently been extended to the D3 dopamine receptors, a subtype of D2-like receptors, with results very similar to those observed for D2 receptors. Stimulation of recombinant D3 receptors accelerates extracellular acidification by CHO cells (28) and C_6 cells (29). The response is sensitive to amiloride and MIA and inhibited by D3 receptor antagonists. In C_6 cells, treatment with pertussis toxin has no effect of the D3 receptor-mediated response, whereas treatment with pertussis toxin abolishes D3 receptor-stimulated extracellular acidification in CHO cells.

Summary

Both spectrofluorometric measurements of pH_i and microphysiometric measurements of pH_o can be used with a number of different approaches to study regulation of NHE1 by receptors and G proteins. Both techniques are limited by the requirement for whole cells and an intact cell membrane and are therefore relatively time intensive. Activities of other receptor-regulated effectors such as adenylyl cyclase and phospholipase C can be measured rapidly in a number of samples using membrane and reconstituted preparations. In contrast, a single determination with the spectrofluorometer, including acid loading, pH_i recovery, and nigericin calibration, requires approximately 45 min. Additionally, reagents commonly used for characterizing signal transduction pathways, such as membrane-impermeant guanine nucleotides and function-perturbing antibodies, are difficult to use when studying pH_i in intact cells. Recently available microinjection systems (Eppendorf), designed for mammalian cells, can now be used with ratiometric imaging systems (Universal Imaging; Zeiss) to facilitate the manipulation of transduction pathways. Methods used for ratiometric imaging, including buffer compositions, fluorescent dyes, acid loading, and nigericin calibration of pH_i, are identical to those described for the spectrofluorometer.

Both of these techniques establish that NHE1 activity is regulated by G protein-coupled receptors independently of many other signaling pathways modulated by the receptors. The microphysiometric and fluorometric techniques yield data that are in agreement concerning the effect of PGE_1 on Na^+-H^+ exchange and that are generally in agreement concerning the sensitivity of the response to pertussis toxin, but that differ in one important respect: endogeneous dopamine D2 receptors inhibit NHE1 in primary cultures of rat anterior pituitary cells (10), whereas recombinant D2 receptors stimulate NHE1 in C_6 glioma cells and Ltk^- cells (19). There are several possible explanations for this discrepancy. It could be due to differences between recombinant and endogenous D2 receptors, or differences between the cell lines and primary cultures. Alternatively, the discrepancy could be related to the use of serum-treated cells in the spectrofluorometer to assess inhibition of NHE1 activity and serum-starved cells in the microphysiometer to assess stimulation of NHE1 activity. Although G protein involvement in receptor-stimulated NHE1 activity has been demonstrated, and two constitutively active G proteins stimulate NHE1, the specific G proteins that mediate stimulation and inhibition of NHE1 by G protein-coupled receptors remain to be identified. An additional important area of research concerns the possibility that G proteins directly modulate NHE1 activity, as opposed to modulating other signaling pathways that in turn activate NHE1 by phosphorylation.

Acknowledgments

Supported by the VA Merit Review and Career Scientist Programs, MH45372, DK40259, GM47413, and the American Heart Association.

References

1. D. L. Barber, *Cell Signal.* **3,** 387 (1991).
2. S. Grinstein, D. Rotin, and M. J. Mason, *Biochim. Biophys. Acta* **988,** 73 (1989).
3. J. Orlowski, *J. Biol. Chem.* **268,** 16369 (1993).
4. G. L'Allemain, S. Paris, and J. Pouyssegur, *J. Biol. Chem.* **259,** 5809 (1984).
5. L. E. Limbird, *FASEB J.* **2,** 2686 (1988).
6. C. Sardet, P. Fafournoux, and J. Pouyssegur, *J. Biol. Chem.* **266,** 19166 (1991).
7. P. S. Aronson, J. Nee, and M. A. Suhm, *Nature (London)* **299,** 161 (1982).
8. W. H. Moolenaar, R. Y. Tsien, P. T. van der Saag, and S. W. de Laat, *Nature (London)* **304,** 645 (1983).
9. D. L. Barber, M. E. McGuire, and M. B. Ganz, *J. Biol. Chem.* **264,** 21038 (1989).
10. M. B. Ganz, J. A. Pachter, and D. L. Barber, *J. Biol. Chem.* **265,** 8989 (1990).
11. D. L. Barber and M. B. Ganz, *J. Biol. Chem.* **267,** 20607 (1992).
12. D. L. Barber, M. B. Ganz, P. B. Bongiorno, and C. D. Strader, *Mol. Pharmacol.* **41,** 1056 (1992).
13. G. Boyarsky, M. B. Ganz, R. B. Sterzel, and W. F. Boron, *Am. J. Physiol.* **255,** C844 (1988).
14. W. H. Weintraub and T. E. Machen, *Am. J. Physiol.* **257,** G317 (1989).
15. J. A. Thomas, R. N. Buchsbaum, A. Zimniak, and E. Racker, *Biochemistry* **18,** 2210 (1979).
16. W. F. Boron and P. De Weer, *J. Gen. Physiol.* **67,** 91 (1976).
17. R. L. Neve and K. A. Neve, Chapter 9.
18. J. A. Salon and J. C. Owicki, Chapter 11.
19. K. A. Neve, M. R. Kozlowski, and M. P. Rosser, *J. Biol. Chem.* **267,** 25748 (1992).
20. C. Hou, R. L. Gilbert, and D. L. Barber, *J. Biol. Chem.* **269,** 10, 357 (1994).
21. C. A. Landis, S. B. Masters, S. B. Spada, A. M. Pace, H. R. Bourne, and L. Vallar, *Nature (London)* **340,** 692 (1989).

22. J. M. Lowndes, S. K. Gupta, S. Osawa, and G. L. Johnson, *J. Biol. Chem.* **266,** 14193 (1991).
23. T. Voyno-Yasenetskaya, B. R. Conklin, R. L. Gilbert, R. Hooley, H. R. Bourne, and D. L. Barber, *J. Biol. Chem.* **269,** 4721 (1994).
24. K. G. Mountjoy, L. S. Robbins, M. T. Mortrud, and R. D. Cone, *Science* **257,** 1248 (1992).
25. J. C. Owicki and J. W. Parce, *Biosens. Bioelectron.* **7,** 255 (1992).
26. J. W. Parce, J. C. Owicki, K. M. Kercso, G. B. Sigal, H. G. Wada, V. C. Muir, L. J. Bousse, K. L. Ross, B. I. Sikic, and H. M. McConnell, *Science* **246,** 243 (1989).
27. K. A. Neve, R. A. Henningsen, J. R. Bunzow, and O. Civelli, *Mol. Pharmacol.* **36,** 446 (1989).
28. C. L. Chio, M. E. Lajiness, and R. M. Huff, *Mol. Pharmacol.* **45,** 51 (1994).
29. B. A. Cox, M. P. Rosser, M. R. Kozlowski, R. A. Henningsen, S. L. Smiley, K. M. Duwe, R. L. Neve, and K. A. Neve, submitted for publication.

[13] Receptor-Activated Tyrosine Phosphatases: Activity Assays and Molecular Cloning

Philip J. S. Stork, Anita Misra-Press, and Ming-Gui Pan

Introduction

Receptor-activated tyrosine phosphatases are a diverse group of enzymes involved in signal-transduction pathways that are intimately associated with normal cellular physiology in most cell types and have been shown to also be involved in neuronal differentiation and neuroendocrine cell growth. This ill-defined group can be classified into receptor-linked phosphotyrosine phosphatases (PTPases) and receptor-coupled PTPases. Receptor-linked PTPases are analogous to receptor-linked tyrosine kinases (EGF receptor, NGF receptor/*trk*) and are implicated in neuronal differentiation (1). Receptor-coupled PTPases are downstream effectors of G protein-coupled pathways primarily involved in the inhibition of cell growth (2). The discovery of both groups extends our traditional notion of receptors and G protein-coupled pathways.

Tyrosine Phosphatases

Although a small percentage of total cellular phosphoproteins are phosphorylated on tyrosines (3), tyrosine phosphorylation appears to play an extremely important role in maintaining a balanced cellular environment. Tyrosine phosphorylation is now a well-accepted form of regulation of protein function, most clearly demonstrated by studying the signal-transduction pathways initiated by growth factor receptors that contain tyrosine kinase activity. The study of the biological role of phosphotyrosine phosphatases (PTPases) in maintaining the balance of phosphorylation has begun and already this class of molecules appears to be critical for the normal growth regulation of a cell.

Five years after the isolation of the first PTPase, PTP1B, from human placenta (4), numerous PTPases have been cloned and characterized (5). PTPases are classified based on structural characteristics. According to Fischer *et al.*, the PTPase family can be divided into the transmembrane (receptor-linked) and the intracellular forms (6) (see Figs 1 and 2). The receptor-linked phosphatases can then be further subdivided depending on

Methods in Neurosciences, Volume 25

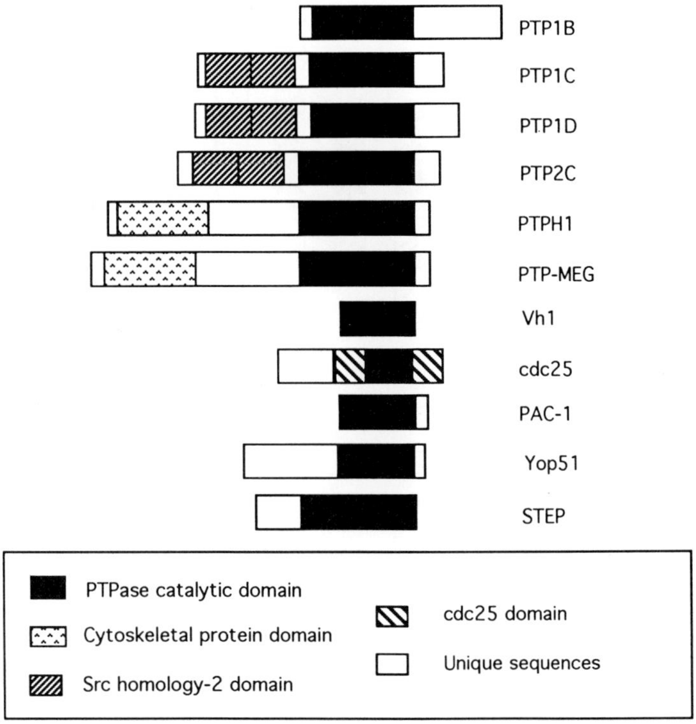

FIG. 1 Structural features of intracellular PTPases.

the organization of the extracellular domain (6). All the PTPases contain an essential cysteine in their catalytic domain that is located within a highly conserved core sequence (I/V)HCXAGXXR(S/T)G (7) which is present in all the PTPases (8).

Characterization of intracellular PTPases led to the isolation of few (PTPIC, PTPID, PTP2C) that contain the SH2 (src-homology domain) domain in their amino terminus (9, 10). (Fig. 1). These phosphatases are involved in the signal-transduction pathway by being coupled to receptors or proteins through protein–protein interactions. For example, PTP1D binds to and dephosphorylates the EGF and PDGF receptors (11). Studies have demonstrated that this phosphatase is activated by binding to these receptors through its SH2 domain (12).

A growing family of intracellular PTPases are the dual specificity PTPases that catalyze the removal of phosphate from both serine/threonine and tyro-

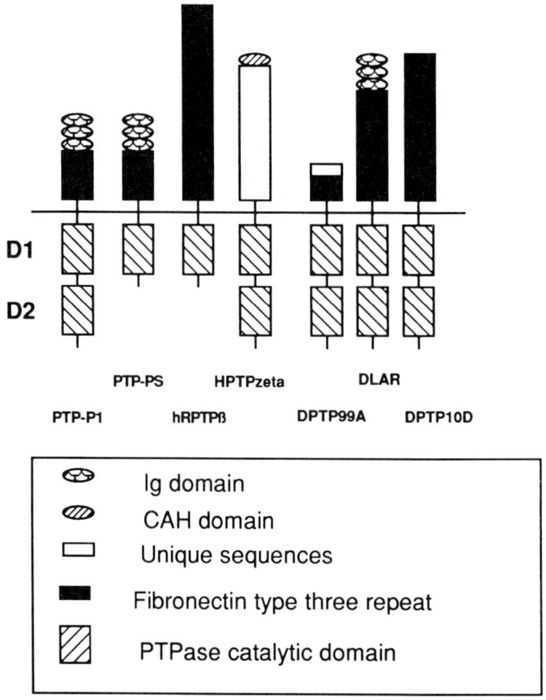

FIG. 2 Comparison between receptor-linked phosphotyrosine phosphatases. The fibronectin type three repeat, immunoglobulin (Ig)-like domain, carbonic anhydrase homology domain, PTPase domain, and unique sequences unrelated to any known proteins are schematically represented. D1 and D2; PTPase domain 1 and 2.

sine residues. The first such PTPase, Vh1, was identified in vaccinia and showed similarity to cdc25, a protein that controls cell entry into mitosis (13). Vh1 homologs from human, mouse (PAC 1), and yeast (Yop51) have also been isolated (14–18) (Fig. 1). Recent data suggest a critical role for this PTPase in signal-transduction pathways involving MAP kinase (19, 20).

Receptor-Linked PTPases

The structure of receptor-linked PTPases appears to be similar to receptor tyrosine kinases such as the EGF receptor, NGF receptor, and FGF receptor. They all contain an extracellular domain, a single transmembrane region, and a cytoplasmic region. The majority of the receptor-linked PTPases have been found to contain a varying number of immunoglobulin-like (Ig-like) domains and fibronectin type III (FN-III) repeats in their extracellular region (21). The Ig-like domains are about 100 amino acids long and contain intrachain disulfide bonds which are thought to function as cell-surface recognition structures and are found in many cell-adhesion molecules such as NCAM(21) (see Fig. 2).

Neuronal Differentiation and PTPases

One of the major issues in the field of neuroscience is understanding the molecular basis of neuronal differentiation. Neuronal differentiation involves cell migration, directional axon growth, synaptogenesis, and selective survival. A variety of studies have suggested that protein phosphorylation is an important mechanism regulating these cellular processes. Discoveries of several protein tyrosine kinases as receptors for NGF and other neurotrophins have highlighted the importance of protein tyrosine kinases in the regulation of neuronal differentiation. However, the role of PTPases that are capable of counteracting the actions of the receptor tyrosine kinases is less understood. Several lines of evidence indicate that PTPases regulate neuronal function and differentiation.

Several PTPases, mostly of the receptor-linked form, have been localized in the nervous system. RPTPβ, was found to be only expressed in the brain, with high levels of expression in the ventricular and subventricular zones of the embryonic mouse brain, suggesting the importance of this PTPase in the development of the central nervous system (22). PTPζ was isolated from a fetal brain cDNA library and is highly expressed in a glioblastoma cell line (22, 23). A PTPase called STEP for "striatum enriched phosphatase" was also found to be highly enriched in the striatum with subcellular localization in dendrites and axonal processes (24).

Several neuronal-specific PTPases have also been isolated from *Drosophila* (see Fig. 2.) Using *in situ* hybridization and immunochemistry, it has been shown that three receptor-linked PTPases, DLAR, DPTP10D, and DPTP99A are expressed in different patterns of cells in the ventral nerve cord, and all three gene products are restricted to axons (25, 26). DLAR and DPTP99A are apparently expressed on most or all axons, while DPTP10D is localized to the anterior commisure and its junction with the longitudinal tracts. All three PTPases possess multiple fibronectin type III repeats, similar to that of NCAM, L1, and fasciclin (21, 27). These extracellular elements may mediate some of the receptor-linked PTPase effects in the nervous system to regulate cell–cell interaction and cell migration.

The identification of a family of receptor-linked PTPases, PTP-P1 and PTP-PS, in PC12 cells (28) that are regulated by NGF (manuscript in preparation) highlights the potential role of receptor-linked PTPases in neural development. One of these PTPases, PTP-P1, is identical to a neural-specific PTPase (PTP-NE3) identified in olfactory neuroepithelium that is transcriptionally regulated during neural development (29).

It is clear that phosphatases will continue to play an ever-growing role in our understanding of neuronal signaling. Their unique structure and limited expression may underlie important roles in neuronal migration and axonal guidance. However, like the family of G protein-coupled receptors, these receptor-linked PTPases remain "orphan" receptors until potential ligands are identified.

Receptor-Coupled PTPases

Another class of PTPases also plays an important role in neuroendocrine signaling: receptor-coupled PTPases. This class of PTPases is activated by ligand binding to G protein-coupled receptors. Many hormones have been shown to stimulate PTPase activity on binding to specific receptors, including somatostatin (2, 30), dopamine (31), and angiotensin II (32). Interleukin-6 (33) and TGF-β1 (34) have been shown to be coupled to PTPase stimulation, as well. Two pathways, involving somatostatin and dopamine, have been studied in our laboratory (2, 31). Both pathways share features characteristic of a defined receptor-coupled pathway. PTPase activation by somatostatin and dopamine are coupled to G proteins and are blocked by pertussis toxin (2, 31). Both dopamine (31) and somatostatin (31a) show receptor subtype specificity.

Dopamine and PTPase Activation

Dopamine activates a variety of physiological responses through second-messenger pathways. All of the known actions of dopamine at D2 receptors are thought to involve the activation of one or more pertussis toxin-sensitive GTP binding proteins (G proteins) (35). Likely candidates are G_{i1}, G_{i2}, and G_{i3}, which act to decrease intracellular cyclic AMP (cAMP) and to increase potassium efflux. Dopamine may also couple to an additional PTX-sensitive G protein to reduce intracellular free calcium (36). All of these changes induced by dopamine act to inhibit the release of preformed secretory vesicles containing specific endocrine and paracrine hormones. This ability to inhibit the release of the growth-promoting hormone prolactin was long thought to

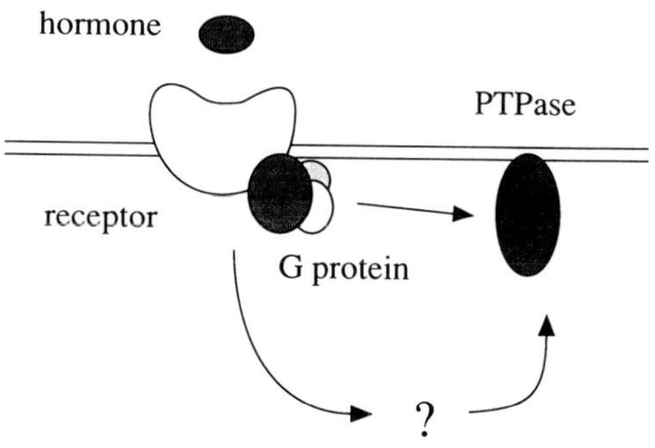

FIG. 3 Schematic representation of G protein coupling to a membrane-associated PTPase. The pathways mediating this activation are not known.

account for the anti-neoplastic action of the dopamine agonist, bromocriptine. The anti-growth properties of bromocriptine were first described in 1975 (37) and have been widely confirmed in humans (38) and in animal studies (39). Bromocriptine therapy is the treatment of choice for prolactin-secreting macroadenomas, with tumor reduction reported in 50–80% of cases (40, 41).

A possible clue into the antiproliferative actions of dopamine and bromocriptine has come from the demonstration that it stimulates PTPase activity in cultured tumor cells and that this stimulation may account for some of its anti-neoplastic actions (31). It is not known whether dopaminergic stimulation of PTPase activity is a consequence of dopamine's effects on cAMP and intracellular Ca^2, or whether it represents a direct activation of PTPases independent of either diffusible second messenger, as proposed for calcium-channel activations (42) (see Fig. 3).

Somatostatin and PTPase Activity

Somatostatin inhibits the growth of many cells including cultured human pancreatic adenocarcinoma cells. This inhibition was shown to be associated with a stimulation of a PTPase activity (43). It was postulated that dephosphorylation of the EGF receptor, a tyrosine kinase, activated by tyrosyl phosphorylation or dephosphorylation of particular substrates might abrogate EGF action in these cells (44). Subsequent studies have shown that this stimulation is coupled to PTPase activation via G proteins and is blocked by pertussis toxin (2). Biochemical studies of phosphatase activities activated by somatostatin in related pancreatic tumor cells suggest that somatostatin stimulates a single PTPase, or a subset of PTPases, directly rather than activating all PTPases through an indirect mechanism (45).

Methods of Assaying PTPase Activity

Introduction

All assays for PTPase activity require the measurement of a phosphorylated substrate. This substrate can be an endogenous protein, like the EGF receptor, whose phosphorylation state can be identified using specific antibodies (see below). However, since the substrate specificity of most PTPases has not been established, more general phosphorylated substrates are required.

Protein and peptide substrates represent the msot physiological substrates employed. The major disadvantage of their use is the need to prelabel the substrate with [^{32}P]phosphate *in vitro* using a (partially) purified tyrosine kinase prior to their use. One nonisotopic substrate, *p*-nitrophenylphosphate (*p*Npp) is widely used. It is an analog of phosphotyrosine that reveals a spectrophotometrically detectable yellow product, nitrophenol, on hydrolysis. It is a substrate for both tyrosine and ser/thr phosphatases, so care must be made to ensure that the ser/thr phosphatase activity is blocked by the use of specific inhibitors (okadaic acid or microcystine-LR) in the assay buffer (2). In addition, it is prudent to determine the vanadate-sensitivity of the PTPase in question before making claims to the type of phosphatase that *p*Npp is measuring. Summarized below are the available methods of assaying for PTPase activity.

p-Nitrophenylphosphate Assay

Membranes containing 5–20 μg protein prepared from untreated cells, or hormone-treated cells where appropriate, are incubated for 5 min at 30°C in 80 μl containing 20 μl of 5× reaction buffer [250 m*M* HEPES (pH 7.2), 50 m*M* DTT, 25 m*M* EDTA] in the presence of microcystine-leucine-arginine (M-LR) (20 n*M*) and $ZnCl_2$ (10 m*M*) to inhibit serine/threonine phosphatases. The M-LR is a specific inhibitor of the major serine-threonine protein phosphatases 1 and 2A (46). These serine/threonine phosphatases are present in all cells and can be stimulated by hormone (47). At micromolar concentrations, $ZnCl_2$ inhibits cytosolic PTPases (48) more effectively than membrane PTPases. The PTPase assay is initiated with the addition of *p*Npp substrate (50 m*M*, final concentration) for 30 min at 30°C and terminated with the addition of 0.9 ml of 0.2 *N* NaOH. The absorbance of the samples at 410 nm provides an index of the production of the yellow nitrophenol following hydrolysis by the phosphatase.

PTPase Assay Using a Radiolabeled Peptide as Substrate

The advantage of radiolabeled peptides is that they provide a reproducible source of phosphotyrosine substrate. A multitude of peptides containing single tyrosine residues (without serine and threonine residues) provide absolute control over phosphatase specificity and therefore are preferred over *p*NPP. In addition, the measurement of released [^{32}P]phosphate is more sensitive than the spectrophotometric assay described above.

Choice of Peptides

The ability to examine the dephosphorylation of multiple peptides by crude or purified PTPases allows one to address specificity of substrates. The use of multiple peptides has been invaluable for us to differentiate the hormonal regulation of PTPase activity by somatostatin, as compared to dopamine, for example, in pituitary cell extracts. The peptides we routinely employ include Raytide (Oncogene sciences), RR-src (Sigma), angiotensin (Sigma), and the peptide derived from cdc2 (28). Each peptide varies in its ability to be phosphorylated by tyrosine kinases (see below) and dephosphorylated by PTPases during the assay. In our hands, Raytide is roughly four times more efficiently phosphorylated by either src kinase or lck kinase, and is roughly two times more easily dephosphorylated during the assay. In addition, we often use myelin basic protein (MBP)(Sigma) as a protein substrate of tyrosine kinases. Other protein substrates include the EGF receptor, lysozyme, histones, and poly-Glu-Tyr. The reader is referred to "Methods in Enzymology," Vol 48 (1991), for a detailed description of the preparation of these protein substrates.

Choice of Kinase

Tyrosine kinases, whether partially purified or recombinant enzymes, are required to phosphorylate the peptides on tyrosyl residues. Therefore, the use of crude extracts of A431 cell membranes containing EGF receptor is not recommended because they often have serine kinase activity as well. If proteins like MBP or histones or the EGF receptor itself are used as a substrate for dephosphorylation, then this contaminating activity may obscure the experimental results. In addition, commercially available kinases [src kinase (UBI) and abl-kinase (Oncogene Sciences)] are unstable and do not provide consistent results. Therefore, it is advantageous to prepare kinases freshly following their expression in mammalian cells.

The approach used by our laboratory utilizes a cell line that overexpresses the *c-src* cDNA in 10T 1/2 fibroblast cells (provided kindly by Sally Parsons, University of Virginia). These cells produce moderate levels of *c-src* that can be immunoprecipitated using standard techniques from crude cell extracts using a src-specific antibody. In addition, the preparation of crude membranes without further purification provides sufficient membrane-associated src kinase to phosphorylate substrate peptides directly, to produce a few million counts of peptide substrate.

Kinase Reaction

Peptide substrate (20 μg) and [γ-^{32}P]ATP (500 μCi, 10 μCi/μl) are preincubated in 60–100 μl of 1× kinase buffer at 30°C for 3 min. Freshly prepared kinase, whether extract or immunoprecipitate (in 1× kinase buffer), is added in two stages: once at the end of the preincubation and again after 30–40 min. The kinase is very unstable at 30°C. Therefore, the double dose ensures that active kinase is present in the reaction for at least 1 hr. Following a 1-hr incubation, the reaction is terminated by the addition of 10 μl of 1 *M* phosphoric acid. The entire reaction volume (roughly 150 μl) is then spotted on P-81 Whatman paper and air-dried. The filters are subsequently washed three times in 300–500 ml of 0.5% phosphoric acid and then air-dried again. The filters are individually placed into a scintillation vial and counted (Cherenkoff counts) to estimate incorporation. We generally recover 50–70% of these counts in the subsequent elution step.

Elution of Peptide

The peptide is eluted from the P-81 filters using ammonium carbonate (0.5 *M*). One milliliter is added to each filter, in the scintillation vial, and gently agitated for 6 hr (or overnight) at room temperature. The liquid is collected and lyophilized to dryness. While this lyophilization is proceeding, an additional 1 ml of ammonium carbonate is applied to the filters, as described above. When the first aliquots are lyophilized the additional 1 ml is applied to the dried white powder that remains in each lyophilization tube. When this additional sample is lyophilized, the combined white pellets are dissolved in water. Then 1 μl is counted in scintillation fluid to determine total counts. A total of 10–20 million cpm should be recovered when labeling Raytide and 2–5 million cpm when labeling src peptide. The percentage of free radioactivity is estimated by measuring the counts that remain in the supernatant following the precipitation of peptide with charcoal. An aliquot (1 μl) of peptide is added to both a control tube containing 700 μl water (total cpm) and a sample tube containing 700 μl stop solution (free cpm) (see below). Following a brief spin in a microcentrifuge, 400 μl of the supernatant in each tube is added to 2.5 ml of scintillation fluid. Detected cpm should correspond to roughly 10–20,000 cpm/μl of labeled peptide (assuming resuspension of total peptide prepared in 0.5–1.0 ml) with free counts per minute constituting no more than 1.5–2.0%. This ratio of unincorporated to incorporated counts per minute must be low since all phosphatase assays are performed under conditions of dephosphorylation ranging from 10 to 20% of total added peptide. Therefore any free phosphate that is much more than 2% of the total may obscure the measurement of enzymatic dephosphorylation (see below).

Preparation of Cell Membranes for PTPase Assay

Choice of Cells

The choice of cells is dictated by the experimental design. To examine receptor-coupled PTPase activity we generally examine cells that have been stably transfected with the appropriate receptor subtypes (31). We also routinely examine endocrine cell lines and other cells that express endogenous receptors. The amount of protein assayed will depend on the cell line used. It is important to passage the cells at least 2 days prior to assay to ensure proper expression of receptors and PTPases. It is important to harvest cells that are less than 60% confluent to eliminate the density-dependent induction of PTPase activity seen in many cells.

Drug Treatment

Drug (or hormone) is added 1 hr prior to harvest and the cells are maintained at 37°C in a 5% CO_2 atmosphere. The timing of drug treatment may vary among cell lines used and must be determined empirically to achieve maximal percentage stimulation or inhibition by the drug of choice.

Preparation of Cell (Membrane) Extracts

Cells are harvested in membrane solubilization buffer (see below) using 1–2 ml per 10-cm plate. Membranes are prepared from unstimulated cells grown to 80 to 90% confluence, washed twice in phosphate buffered saline (PBS), and mechanically scraped in membrane buffer [0.32 *M* sucrose, 10 m*M* Tris (pH 7.5), 5 m*M* EGTA, and 1 m*M* EDTA]. Nuclei are removed after centrifugation at 2000*g* for 10 min at 4°C and the membrane fraction is sedimented following centrifugation at 15,000*g* for 45 min at 4°C. The membranes are resuspended in membrane preparation buffer [20 m*M* HEPES (pH 7.5), 150 m*M* NaCl, 1 m*M* PMSF, Bacitracin, 20 μg/ml] and assayed for protein content using a Bradford assay (Protein Determination Kit, Bio-

Rad). Stock solutions of proteins are prepared in membrane solubilization buffer at a concentration of roughly 1 μg/μl. These samples can be assayed immediately or stored in 10% glycerol at −70°C.

Assay of Extracts

Pharmacological regulation of phosphatase activity is routinely performed using crude membrane preparations. However, total cellular extracts as well as nuclear extracts can be substituted where appropriate. Initial phosphatase assays are performed using various amounts of protein extracts (using unstimulated cell extracts) to determine the optimal conditions of the assay. It is imperative that the assay be carried out under conditions during which the PTPase activity is proportional to released counts. This is usually achieved by adjusting the added proteins to achieve 10–20% dephosphorylation of labeled peptide. At higher levels of dephosphorylation a plateau is reached and the percentage stimulation of PTPase activity does not reflect the percentage of released peptide. The dephosphorylation reaction often approaches a plateau at roughly 50–70% dephosphorylation and dephosphorylations occuring within this region of the curve are not linear (see Fig. 4). Initial samples containing 0.1, 0.5, 1.0, and 5.0 μg of protein are assayed in 1× phosphatase buffer (see below) and roughly 20–50,000 cpm of labeled peptide, in a reaction volume of 100 μl. The reaction is carried out for 10 min at 30°C and terminated with the addition of 700 μl of stop solution (see below). Following a brief spin, 400 μl of the supernatant is removed, leaving the charcoal behind, added to scintillation fluid, and counted. The released counts should be

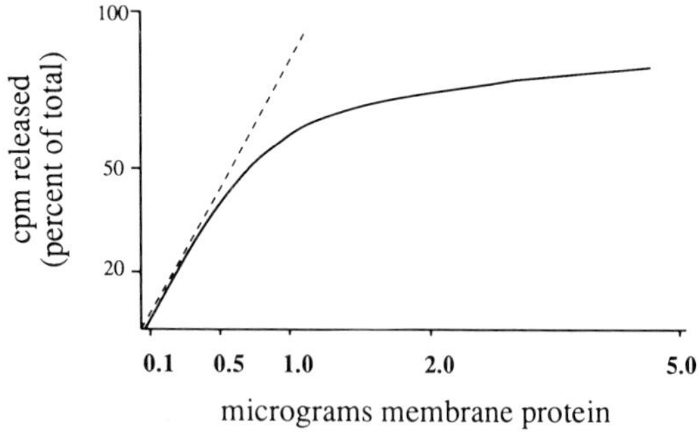

FIG. 4 Measurement of PTPase activity. PTPase activity is shown as the percentage of [^{32}P]$_i$phosphate released into the supernatant following hydrolysis of radiolabeled peptide. The curve approaches linearity (dotted line) between 10 and 20% dephosphorylation of the available substrate, corresponding to less than 0.5 μg of protein in this figure.

proportional to the protein content (see Fig. 4). If not, the plateau phase of the assay has already been reached and less protein should be assayed. Total counts and free counts should be measured at this time, as described above. The concentration of protein that releases 10–20% of the [^{32}P]phosphate should be used in subsequent experiments using hormone-stimulated and unstimulated cells. All assays should be carried out in duplicate and experimental conditions in duplicate or triplicate to ensure significance of the results.

Evaluation of the Data

The resultant counts can be recorded in a number of ways for analysis. The easiest way is to calculate the percentage dephosphorylation of added peptide. This will allow a determination of whether the data points fall within the linear range of the assay (see Fig. 4). Quantitative assay of PTPase activity requires the conversion of counts per minute to nanomoles of released phosphate. This is easily calculated from the specific activity of the [γ-^{32}P]ATP used in the kinase reaction. Units of PTPase specific activity are defined as nanomoles/milligram protein/minute reaction time (nmol/mg/min).

Controls that should be used include assaying untreated cells, untransfected cells, and cells incubated in the presence of vanadate, as well as the inclusion of vanadate in the assay. Treatment of cells with a range of hormone doses is important to establish a dose–response curve of phosphatase stimulation. Pretreatment of cells with pertussis toxin is necessary to define the mechanism of G protein coupling to hormone-stimulated PTPases. For the study of transfected receptors, useful data can be achieved using CHO cells, indicating that these cells contain the appropriate G proteins and PTPases.

Buffers Used

Cell lysis buffer (1X):	1% NP-40, 10 m*M* Tris (pH 7.4), 0.1 *M* NaCl.
Kinase buffer (1X):	25 m*M* HEPES, pH 7.4, 5 m*M* $MnCl_2$, 5 m*M* $MgCl_2$, 100 μM vanadate.
PTPase buffer (1X):	50 m*M* Tris (pH 7.4), 10 m*M* DTT, 1 m*M* EDTA.
STOP solution:	0.9 *M* HCl, 90 m*M* sodium pyrophosphate, 2 m*M* NaH_2PO_4, 4% charcoal.

Monitoring Dephosphorylation of Receptor Kinases

Another useful technique of monitoring PTPase activity is to measure the dephosphorylation of a receptor kinase that is stimulated to autophosphorylate on tyrosine residues. A good example is the EGF receptor that is enriched in the membranes of A-431 cells. The EGF receptors can be isolated from membrane vesicles of this cell line, incubated with EGF to activate the intrinsic tyrosine kinase activity of the receptor, and then used as a substrate for membrane-associated tyrosine phosphatases. The dephosphorylation of the EGF receptor then is used as a measure of PTPase activity.

A-431 membrane vesicles enriched for the EGF receptor are prepared as described (45). Membrane vesicle proteins (2 μg) are incubated in 200 μl of kinase buffer [30 m*M* HEPES (pH 8.0) 15 m*M* $MnCl_2$, Nonidet-P40 (0.75%), and 15 m*M* ATP] in the presence of EGF (100 n*M*) and 60 μCi[γ^{32}P]ATP (5000 Ci/mmol) for 10 min at 37°C. Portions (40 μl) are immediately added to four separate reactions containing membranes to be assayed for PTPase activity (20 μg) alone or membranes in the presence of pharmacological agents. The amount of protein used in each assy tube is determined as described above. Reactions are incubated at 30°C for 30 min in a final volume of 60 μl, which includes 50 m*M* HEPES (pH 7.2), 1 m*M* EDTA, 5 m*M* DTT, 20 n*M* M-LR, 10 m*M* $ZnCl_2$, 2.5 m*M* $MgCl_2$, and 0.8 m*M* ATP, and the reaction is terminated with the addition of STOP solution (see above). Then, 5× Laemmli buffer (12 μl) is added to each tube and the samples are separated by SDS–polyacrylamide gel electrophoresis (7% gel) and the phosphotyrosyl EGF receptor (170 kDa) is detected by autoradiography (2).

Immunoblotting with Phosphotyrosine Antibodies

A more simplified but crude method of detecting PTPase activity is to prepare cell lysates and perform Western blot analysis using a specific phosphotyrosine antibody (UBI). The advantage of this method is that specific bands can be identified that are targets for an increase or decrease in a particular PTPase activity and the whole array of proteins that are phosphorylated with a particular treatment can be visualized.

Molecular Cloning of Phosphotyrosine Phosphatases

Low-Stringency Hybridization Screening

Few examples are found of cloning PTPases using the traditional method of screening a library with a homologous gene as a probe (49). RPTP-k, a new member of the receptor-linked PTPase family, was cloned from a mouse brain cDNA library under low-stringency conditions using a probe that consisted of the intracellular domain of human CD45, a PTPase with two tandem catalytic

PTPase domains (49). However, the polymerase chain reaction (PCR) methodology described below has been utilized more extensively in cloning PTPases.

Biochemical Purification and Subsequent Cloning

Initial cloning procedures all involved protein purification. PTP1B was purified over 10,000-fold from placental extracts before sufficiently pure protein was obtained for sequencing (4). Similar purification, involving anion exchange and affinity chromatography, was employed in the purification of brain (50, 51) and kidney PTPases (52). Phosphotyrosyl protein affinity columns have also been used successfully (53). Methodological considerations are discussed elsewhere (48).

Polymerase Chain Reaction Using Degenerate Oligonucleotides

The use of PCR has been a very popular and powerful method to clone novel PTPases (22, 28, 54). As shown in Fig. 5, alignment of the cloned PTPases shows clearly that there are regions that are conserved throughout the PTPase members including the catalytic core "HCS" sequences. This knowledge led to the design of degenerate oligonucleotides directed toward the conserved PTPase regions for use as primers in reverse-transcription (RT)/PCR reac-

PTP-P1	DYINA NYVDGYRRQNAYIATQGPLPETFGDFWRMVWEQRSATVVMMTRLEE KSRVKCDQYWPNRGTETYGFIQVTLLDT
HPTP-d	DYVNA NYIDGYRKQNAYIATQGSLPETFGDFWRMIWEQRSATVVMMTKLEE RSRVKCDQYWPSRGTETHGLVQVTLLDT
LAR	DYINA NYIDGYRKQNAYIATQGPLPETMGDFWRMVWEQRTATVVMMTRLEE KSRVKCDQYWPARGTETCGLIQVTLLDT
PTP-1B	DYINA SLIKMEEAQRSYILTQGPLPNTCGHFWEMVWEQKSRGVVMLLNRVMEKGSLKCAQYWPQKEEKEMIFEDTNLKLTLISEDI
HPTPa	DYINA SFINGYQEKNKFIAAQGPKEETVNDFWRMIWEQNTATIVMVTNLKERKEC KCAQYWPDQGCWTYGNIRVSVEDVTVLVDY
CD45	NYINA SYIDGFKEPRKYIAAQGPRDETVDDFWRMIWEQKATVIVMVTRCEEGNRN KCAEYWPSMEEGTRAFGDVVVKINQHKRCP
PTP-1C	DYINANYVKNQLGPDENSKTYIASQGCLDATVNDFWQMAWQENTRVIVMTTREVEKGRN KCVPYWPEVGTQRVYGLYSVTNSREHDT
PTP-P1(D2)	DYINA SFIDGYRQQKAYIATQGPLAETTEDFWRALWENNSTIVVMLTKLREMGREEKCHQYWPAERSARYQYFVVDPMAEYN
HPTP-d(D2)	DYINA SFIDGYRQQKAYIATQGPLAETTEDFWRMLWEHNSTIVVMLTKLREMGREEKCHQYWPAERSARYQYFVVDPMAEYN
LAR(D2)	DYINA SFLDGYRQQKAYIATQGPLAESTEDFWRMLWEHNSTIIVMLTKLREMGREEKCHQYWPAERSARYQYFVVDPMAEYN

PTP-P1	MELATFCVRTFSLHKNGSSEKREVRHFQFTAWPDHGVPEYPTPFLAFLRRVKTC NPPDAGPVVVHCSAGVGRTGCFIVID
HPTP-d	VELATYCVRTFSLYKNGSSEKREVRQFQFTAWPDHGVPEHPTPFLAFLRRVKTC NPPDAGPMVVHCSAGVGKHGCFIVID
LAR	VELATYTVRTFSLHKSGSSEKRELRQFQFMAWPDHGVPEYPTPILAFLRRVKAC NPLDAGPMVVHCSAGVGKHGCFIVID
PTP-1B	KSYYTVRQLELENLTTQETREILHFHYTTWPDFGVPESPASFLNFLFKVRESGSLSLPEHGPVVVHCSAGIGRSGTFCLAD
HPTPa	TVRKFCIQQVGDMTNRKPQRLITQFHFTFWPDFGVPFTPIGMLKFLKKVKAC NPQYAGAIVVHCSAGVGRTGTFVVID
CD45	DYIIQKLNIVNKKEKATGREVTHIQFTSWPDHGVPEDPHLLLKLRRRVNAF SNFFSGPIVVHCSAGVGRTGTYIGID
PTP-1C	AEYKLRTLQISPLDNGDLVREIWHYQYLSWPDHGVPSEPPGGVLSFLDQINQRQESLPHAGPIIVHCSAGIGRTGTIIVID
PTP-P1(D2)	MPEYILREFKVTDARDGQSRTVRQF..TDWPEQGAPKSG EGFIDDFIGQVHKTKEQFQDPISVHCSAGVGRTGVFITLSIVL
HPTP-d(D2)	MPQYILREFKVTDARDGQSRTVRQFQFTDWPEQGAPKSG EGFIDDFIGQVHKTKEQFQDPISVHCSAGVGRTGVFITLSIVL
LAR(D2)	MPQYILREFKVTDARDGQSRTIRQFQFTDWPEQGAPKTG EGFIDDFIGQVHKTKEQFQDPITVHCSAGVGRTGVFITLSIVL

FIG. 5 Alignment of several receptor-linked and intracellular PTPase catalytic domains. Homologies are shaded. The essential cysteine residue is marked with an asterisk. All PTPase sequences represent domain 1 unless indicated by "(D2)".

PCR Primers for PTP Amplication

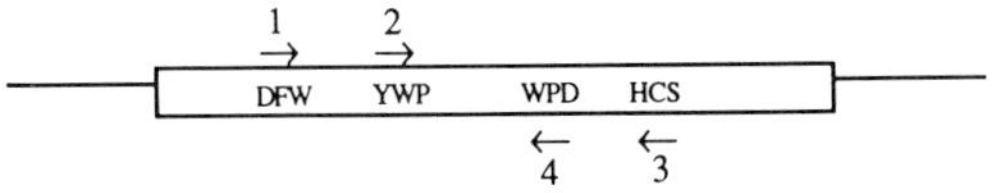

Oligo Name	Sequence (5' to 3')
1 DFW	TGGMIIATGITBTGGSARCA
2 YWP	CCAGTCGACAAITGYGMHCARTAYTGGCC
3 HCS	SCCSACSCCIGCISWICAATG
4 WPD	AAAGAATTCGGSACSCCRWIRTCIGGCCA

FIG. 6 Schematic diagram of nested PCR strategy to clone PTPases. The numbers correspond to the sequences listed below. The degenerate oligonucleotides used in PCR cloning are shown below. Degenerate positions are denoted by the following letters: I, inosine; M, A+T+C; B, G+T+C; R, A+G; Y, C+T; H, A+T+C; S, G+C; W, A+T.

tions. Figure 6 shows the design of degenerate oligonucleotides for use in cloning PTPases. The 5′ primers corresponding to the "ET(F/V/T)GDFW" and "KC(A/D/H)QYWP" sequences (shaded areas of homology shown in Fig. 5) and 3′ primers corresponding to the "WPDHGVP" and "VHCSAG" sequences have been generated. For the RT/PCR reactions, RNA was isolated from cells according to traditional methods (55) and 10 μg of total RNA was subjected to first-strand cDNA synthesis using random hexamers as the primers. A total of 100 ng of random hexamers were heat-annealed to 10 μg of total RNA. Reverse transcription was carried out at 37°C for 1 hr in a total volume of 20 μl. The reaction contained 50 m*M* Tris–HCl (pH 8.3), 75 m*M* KCl, 3 m*M* $MgCl_2$, 10 m*M* DTT, 500 μ*M* each dNTPs, and 50 Units of Superscript RT. The reaction was terminated by ethanol precipitation of the cDNA. A typical yield of first-strand cDNA should be approximately 25–35% of the total amount of input mRNA. Polymerase chain reaction was then carried out using 5 μl of the cDNA, the remainder can be kept frozen at −20°C and reused. Polymerase chain reaction was performed using the traditional procedure and with different primer pairs. We amplified for 30 cycles (each cycle: 94°C denaturation, 1 min; 55°C annealing, 1 min; and 72°C elongation, 1 min). The reactions were analyzed on 1.5–2% agarose gels. Shown in Figure 7 are RT/PCR products obtained from two different cell lines, AtT20 and GH_4C_1 cells. The primer pairs used are schematically shown in Figs. 6 and 7. In each case, the reactions showed the expected size product and in some cases there were additional specific bands that probably correspond to related PTPases. Restriction enzyme linkers added on the ends of each of the primar pairs facilitates cloning of these products into the vector of choice for sequencing to determine their identity. The amplified RT/PCR product obtained is more than likely to be a mixture of

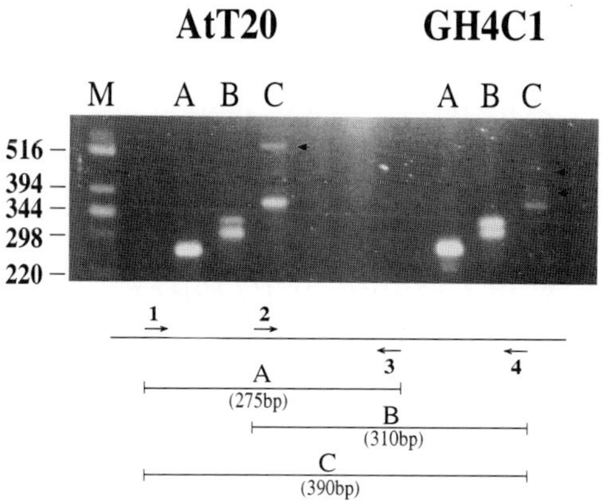

FIG. 7 The PCR products amplified from GH_4C_1 cDNA and AtT20 cDNA. Preliminary comparisons of GH_4C_1 and AtT20 cells suggest that they express distinct sets of PTPases (see arrows). The oligonucleotides used are described in the text. A schematic of a hypothetical cDNA encoding a PTPase catalytic domain showing the oligonucleotides used is shown below.

numerous PTPases that contain the conserved domains used to perform the amplification. Therefore, on sequencing individual clones, a number of different PTPases will be identified.

To clone a transcriptionally regulated PTPase, the amplified RT/PCR product obtained as described above can be digested with restriction enzymes to attempt to differentiate between the different PTPase populations within the group. These fragments are then analyzed by electrophoresis through an acrylamide gel under native or denaturing conditions to ensure detection of small differences in size of the amplified and digested fragments (56). Induction of specific bands that appear to be induced or inhibited by drug treatment can then be followed by eluting the band of interest, cloning, and sequencing.

Advantages and Disadvantages of Polymerase Chain Reaction Cloning

The ease of cloning PTPases has established a large family of related cDNAs. Continued use of PCR techniques will reveal additional members of this family. However, the major disadvantage of this approach is that, by using a particular set of degenerate oligonucleotides, one is preselecting the PTPases that will be identified. Additional families of PTPases that cannot be cloned with the set of PCR oligos described above include the VH1 family of dual-specificity phosphatases (19, 20). Expression cloning strategies for PTPases, as for receptors, are laborious but remove these biases. One method of expression cloning is described below.

Bacterial Expression Cloning of PTPases

The premise for this technique is that an anti-phosphotyrosine antibody will not recognize plaques carrying a clone encoding an active PTPase when a phagemid expression cDNA library is used to infect bacteria that are expressing an active tyrosine kinase (15). This procedure was used successfully to clone a dual-specificity phosphatase from human lung fibroblasts. Briefly, phage containing the cDNA library was used to infect bacteria carrying the *bek* tyrosine kinase and grown for 4 hr at 37°C. The plates were overlaid with nitrocellulose filters which had been soaked in 10 m*M* IPTG and were further incubated for 4 hr at 37°C. The filters were washed in PBS–0.05% Tween-20 and were immunoblotted using an anti-phosphotyrosine antibody. Phage showing decreased signals were isolated and further analyzed for an active PTPase. Plasmids containing active PTPase were rescued from isolated phagemid clones. Using this method Ishibashi *et al.* isolated a novel PTPase showing both serine/threonine and tyrosine phosphatase activity (15).

References

1. L. F. Aparicio, I. Ocrant, J. M. Boylan, and P. A. Gruppuso, *Cell Growth Differ.* **3,** 363 (1992).
2. M. G. Pan, T. Florio, and P. J. S. Stork, *Science* **256,** 37 (1992).
3. T. Hunter and J. A. Cooper, *in* "The Enzymes" (P. D. Boyer and E. G. Krebs, eds.), Vol. 17, Part A, p. 192. Academic Press, New York, 1986.
4. N. K. Tonks, C. R. Diltz, and E. H. Fischer, *J. Biol. Chem.* **263,** 6722 (1988).
5. D. A. Pot and J. E. Dixon, *Biochim. Biophys. Acta* **1136,** 35 (1992).
6. E. H. Fischer, H. Charbonneau, and N. K. Tonks, *Science* **253,** 401 (1991).
7. M. Streuli, N. X. Krueger, T. Tahi, M. Tang, and H. Saito, *EMBO J.* **9,** 2399 (1990).
8. K. L. Guan and J. E. Dixon, *J. Biol. Chem.* **266,** 17026 (1991).
9. S. H. Shen, L. Bastien, B. I. Posner, and P. Chrétien, *Nature (London)* **352,** 736 (1991).
10. G.-S. Feng, C.-C. Hui, and T. Pawson, *Science* **259,** 1607 (1993).
11. R. M. Freeman, Jr., J. Plutzky, and B. G. Neel, *Proc. Natl. Acad. Sci. U.S.A.* **89,** 11239 (1992).

12. R. J. Leichleider, S. Sugimoto, A. M. Bennett, A. S. Kashishian, J. A. Cooper, S. Shoelson, C. Walsh, and B. G. Neel, *J. Biol. Chem.* **268,** 21478 (1993).
13. A. Kumagai and W. G. Dunphy, *Cell (Cambridge, Mass.)* **70,** 139 (1992).
14. K. Guan, R. J. Deschenes, H. Qiu, and J. E. Dixon, *J. Biol. Chem.* **266,** 12964 (1991).
15. T. Ishibashi, D. P. Bottaro, A. Chan, T. Miki, and S. A. Aaronson, *Proc. Natl. Acad. Sci. U.S.A.* **89,** 12175 (1992).
16. P. J. Rohan, P. Davis, C. A. Moskaluk, M. Kearns, H. Krutzch, U. Siebenlist, and K. Kelly, *Science* **259,** 1763 (1993).
17. C. H. Charles, A. S. Abler, and L. C. Lau, *Oncogene* **7,** 187 (1992).
18. S. M. Keyse and E. A. Emslie, *Nature (London)* **359,** 644 (1992).
19. C.-F. Zheng and K.-L. Guan, *J. Biol. Chem.* **268,** 16116 (1993).
20. D. R. Alessi, C. Smythe, and S. M. Keyse, *Oncogene* **8,** 2015 (1993).
21. L. Patthy, *Cell (Cambridge, Mass.)* **61,** 13 (1990).
22. J. B. Levy, P. D. Canoll, O. Silvennoinen, G. Barnea, B. Morse, A. M. Honegger, J.-T. Huang, L. A. Cannizzaro, S.-H. Park, T. Druck, T. Huebner, J. Sap, M. Ehrlich, J. M. Musacchio, and J. Schlessinger, *J. Biol. Chem.* **268,** 10573 (1993).
23. N. X. Kruger and H. Saito, *Proc. Natl. Acad. Sci. U.S.A.* **89,** 7417 (1992).
24. P. J. Lombroso, G. Murdock, and M. Lerner, *Proc. Natl. Acad. Sci. U.S.A.* **88,** 7242 (1991).
25. X. Yang, K. T. Seow, S. M. Bahri, S. H. Oon, and W. Chia, *Cell (Cambridge, Mass.)* **67,** 661 (1991).
26. S. Tian, P. Tsoulfas, and K. Zinn, *Cell (Cambridge, Mass.)* **67,** 675 (1991).
27. B. A. Cunningham, J. J. Hemperly, B. A. Murray, E. A. Prediger, R. Brackenbury, and G. M. Edelman, *Science* **236,** 799 (1987).
28. M.-G. Pan, C. Rim, K. P. Lu, T. Florio, and P. J. S. Stork, *J. Biol. Chem.* **268,** 19284 (1993).
29. K. Walton, K. Martell, S. P. Kwok, J. E. Dixon, and B. L. Largent, *Neuron* **11,** 387 (1993).
30. M. T. Hierowski, C. Liebow, K. du Sapin, and A. V. Schally, *FEBS Lett.* **179,** 252 (1985).
31. T. Florio, M. G. Pan, B. Newman, R. E. Hershberger, O. Civelli, and P. J. S. Stork, *J. Biol. Chem.* **267,** 24169 (1992).
31a. T. Florio, C. Rim, M. G. Pan, and P. J. S. Stork, *Mol. Endocrinol.* **8,** 1289 (1994).
32. S. P. Bottari, I. N. King, S. Reichlin, I. Dahlström, N. Lydon, and M. De Gasparo, *Biochem. Biophys. Res. Commun.* **183,** 206 (1992).
33. D. Zafriri, M. Argaman, E. Canaani, and A. Kimchi, *Proc. Natl. Acad. Sci. U.S.A.* **90,** 477 (1993).
34. P. A. Gruppuso, R. Mikumo, D. L. Brautigan, and L. Braun, *J. Biol. Chem.* **266,** 3444 (1991).
35. L. Vallar and J. Meldolesi, *Trends Pharmacol. Sci.* **10,** 74 (1989).
36. P. M. Lledo, V. Homburger, J. Bockaert, and J. D. Vincent, *Neuron* **8,** 455 (1992).
37. M. O. Thorner, G. M. Besser, A. Jones, J. Dacie, and J. Jones, *Br. Med. J.* **4,** 694 (1975).

38. W. Winkelmann, B. Allolio, D. Heesen, D. Kaulen, E. Keymer, and R. Mies, *Acta Endocrinol.* (*Copenhagen*) **102,** 37 (1983).
39. H. M. Lloyd, J. D. Meares, and J. Jacobi, *Nature* (*London*) **255,** 497 (1975).
40. M. E. Molitch, R. L. Elton, R. E. Blackwell, B. Caldwell, R. J. Chang, R. Jaffe, G. Joplin, R. J. Robbins, J. Tyson, and M. O. Thorner, *J. Clin. Endocrinol. Metab.* **60,** 698 (1985).
41. J. H. Wass, J. Williams, M. Charlesworth, D. P. E. Kingsley, A. M. Halliday, I. Doniach, L. H. Rees, W. I. Mcdonald, and G. M. Besser, *Br. Med. J.* **284,** 1908 (1982).
42. A. M. Brown and L. Birnbaumer, *Annu. Rev. Physiol.* **52,** 197 (1990).
43. M. T. Lee, C. Liebow, A. R. Kamer, and A. V. Schally, *Proc. Natl. Acad. Sci. U.S.A.* **88,** 1656 (1991).
44. C. Liebow, M. T. Lee, A. R. Kamer, and A. V. Schally, *Proc. Natl. Acad. Sci. U.S.A.* **88,** 2244 (1991).
45. B. Colas, C. Cambillau, L. Buscail, M. Zeggari, J.-P. Esteve, F. Thomas, N. Vaysse, and C. Susini, *Eur. J. Biochem.* **207,** 1017 (1992).
46. C. MacKintosh, K. A. Beattie, S. Klumpp, P. Cohen, and G. A. Codd, *FEBS Lett.* **264,** 187 (1990).
47. R. W. White, A. Schonbrunn, and D. L. Armstrong, *Nature* (*London*) **351,** 570 (1991).
48. N. K. Tonks, C. D. Diltz, and E. H. Fischer, *in* "Methods in Enzymology" (T. Hunter and B. Sefton, eds.), Vol. 201, 427. Academic Press, San Diego, CA, 1991.
49. Y.-P. Jiang, H. Wang, P. D'Eustachio, J. M. Musacchio, J. Schlessinger, and J. Sap, *Mol. Cell. Biol.* **13,** 2942 (1993).
50. S. W. Jones, R. L. Erikson, V. M. Ingebritsen, and T. S. Ingebritsen, *J. Biol. Chem.* **264,** 7747 (1989).
51. T. S. Ingebritsen, *J. Biol. Chem.* **264,** 7747 (1988).
52. C. L. Shriner and D. L. Brautigan, *J. Biol. Chem.* **259,** 11383 (1984).
53. L. Mei and R. L. Huganir, *J. Biol. Chem.* **266,** 16063 (1991).
54. B. J. Mosinger, U. Tillmann, H. Westphal, and M. L. Tremblay, *Proc. Natl. Acad. Sci. U.S.A.* **89,** 499 (1992).
55. J. M. Chirgwin, A. E. Przybyla, R. J. MacDonald, W. J. Rutter, *Biochemistry* **18,** 5294 (1979).
56. T. Boehm, *Oncogene* **8,** 1385 (1993).

Section III

Studies of Functional Domains of Receptor and Channels

[14] Site-Directed Mutagenesis

Tung Ming Fong, Mari R. Candelore, and
Catherine D. Strader

Theoretical and Design Considerations

Before the development of recombinant DNA technology, the structure–function relationships of proteins were explored by chemical modification of the protein. This approach required the availability of large quantities of pure protein, as well as considerable detective work to characterize the site(s) of modification. Once the pioneering work in DNA chemistry had led to the routine synthesis of oligonucleotides, and when various DNA-modifying enzymes became readily available, it was realized that the amino acid sequence of a protein could be easily modified by altering the sequence of the DNA encoding the protein. This chapter reviews the design and implementation of site-directed mutagenesis approaches to elucidating the structure–function relationships of proteins.

The first consideration in experimental design is the identification of amino acid residues to be targeted for mutagenesis. In many cases, initial deletion of large domains of a protein or the construction of chimeras of two related proteins allows one to "scan" the entire sequence to identify functionally important (and unimportant) regions. Single-residue substitutions can then be constructed to more precisely characterize the role of individual residues within key regions of the protein. While several successful studies have designed site-directed mutagenesis experiments based on chimeric or deletion mutant receptors (see Specific Examples), it should be pointed out that chimeric proteins and deletion mutants may adopt distorted conformations. This is especially true when the splicing junction is not located in a structurally flexible region. Therefore, caution should be exercised in interpreting the data from these mutants to distinguish direct effects of the mutations on protein function from indirect effects on protein structure. For example, the construction of chimeric neurokinin receptors suggested a potential role for the divergent residues of the third extracellular segment in antagonist binding. However, subsequent single-residue substitutions indicated that none of the divergent residues capable of contributing electrostatic interactions in this region are required for binding, suggesting a possibility of structural effect of the chimeric mutant (1).

Other strategies can be employed to select residues for site-directed muta-

Methods in Neurosciences, Volume 25

genesis. For example, the results of previous chemical modification studies can be a useful guideline. This approach has been useful in the analysis of the nicotinic acetylcholine receptor where a large body of chemical modification data was accumulated before the advent of site-directed mutagenesis (2–4). Another strategy for selecting residues is based on the sequence alignment of several homologous proteins, from which conserved or divergent residues can be identified. A conserved residue may serve the same catalytic function in several related proteins or play a structural role in maintaining a similar conformation throughout the protein family. Finally, residue selection can be based on primary sequence analysis, in which the variations in hydrophobicity are used to predict the solvent-exposed surface and the interior of the protein, as well as such structural features as α helices and β sheets (5).

A second design consideration is the nature of the specific amino acid substitution to be produced. It is critical that the structural changes introduced by mutagenesis be as small as possible to avoid potential conformational effects (6). In the absence of three-dimensional structural data on the mutant protein, the interpretation of mutagenesis data often requires the assumption of a lack of conformational changes as a result of the amino acid substitution, and a great deal of effort can go into investigating potential conformational effects. Therefore, the new amino acid should be as close in structure to the wild-type residue as possible. Substitution of an Ala at the position will give an initial indication of whether the residue is required for the activity of the protein. In fact, alanine-scanning mutagenesis, in which adjacent residues within a region are sequentially replaced with Ala residues, is useful for rapidly defining key residues within a region of interest. Substitution of a residue with Ala will completely eliminate strong noncovalent interactions at that position, but the cavity created by the Ala substitution, especially if a large interior residue is being replaced, sometimes leads to unstable proteins (7). The substitution of a closely related residue will not usually create a cavity, although the information obtained from conservative substitutions may be more difficult to interpret. In any case, accumulation of a body of data from multiple substitutions at a single position is beneficial in controlling for conformational effects as well as providing hints as to what type of noncovalent interactions are in effect at that position (8).

Even when substituting an amino acid with a closely related residue, other factors need to be considered. For example, substituting Glu and Gln cannot be regarded as simply removing a negative charge from the protein. The carboxylate anion of Glu is a proton acceptor, while the amide of Gln can serve as both a proton acceptor and a proton donor (9). Therefore, a Gln for Glu substitution not only removes a potential negative charge, but may also introduce a new interaction by being a hydrogen bond donor. Another example of a multifactorial replacement is the Cys-to-Ser substitution or vice versa. Although both Cys and Ser can participate in hydrogen bonds (when the Cys is known not to form a disulfide bond), the conformation of the sulfhydryl is significantly different from that of the hydroxyl. The resulting 0.4 Å displacement of the hydrogen bond can be sufficient to cause a conformational change in the active site (10). In these cases, comparison of two substitutions for a wild-type Ser residue (a Cys substitution and an Ala substitution) may indicate an anomaly that requires cautious interpretation.

Another issue in the interpretation of data from site-directed mutagenesis experiments is whether the observed effect on mutation is due to an alteration of direct ligand-receptor interactions, to a local conformational change within that region of the protein, or to a more global, long-range conformational change (6). A conformational effect of an amino acid substitution is technically impossible to rule out in the absence of a three-dimensional structure of the mutant protein. However, clear experimental design can often make structural artifacts less likely. In constructing mutant β-adrenergic receptors, we relied on analysis of receptor glycosylation patterns to identify mutant receptors in which the overall structure of the protein was altered (11). The processing and correct glycosylation of the receptor was determined to be sensitive to the overall folding of the protein, providing a measure of the structural integrity of the mutant receptor in the absence of structural data. In studies aimed at identifying ligand-binding sites, if a mutation affects the binding affinity of one ligand but not another structurally related ligand, it is likely that the mutation alters a specific interaction with the first ligand, rather than causing a conformational change in the receptor. Thus, employing multiple assays in which at least one assay is not affected by the mutation provides a nonstructural means of arguing against large-scale conformational effects. On the other hand, the existence of local conformational effects can also be exploited to support the role of other previously identified interactions. For example, the His-48–substrate interaction in tyrosyl-tRNA synthetase has been further investigated by substituting a nearby residue (Thr-51) with Pro, thereby affecting the precise position of the histidine residue that interacts with substrate (12). When the catalytic activity of membrane proteins was studied, conformational changes become part of the process to be investigated. For example, substitution of an Ala residue in the third intracellular loop of the α_{1B}-adrenergic receptor leads to a permanently activated receptor (13). In this case, multiple amino acid substitutions at this position had similar effects on the activity of the receptor, suggesting that the hyperactivity of the mutant receptors results from a conformational change in this region, rather than from a new interaction between the substituted residue and the G protein.

Specific Examples

In the following sections, important considerations in experimental design and interpretation are illustrated with two examples of mutagenesis, using the β_2-adrenergic receptor (β-AR) and the neurokinin-1 receptor (NK1R). These receptor proteins belong to the family of G protein-coupled receptors which are located in the plasma membranes of various cells. These receptors, which act by stimulating members of the family of guanine nucleotide binding regulatory proteins (G proteins), share common structural characteristics, although the endogenous ligands for these receptors vary considerably, ranging from small biogenic amines to peptides and larger glycoprotein hormones. Members of the G protein-coupled receptor family are characterized by seven stretches of hydrophobic amino acids, postulated to form transmembrane α helices, connected by alternating extracellular and cytoplasmic loops. Most of the primary sequence homology is located within these transmembrane domains, with the hydrophilic connecting loops being more divergent. In our laboratories, we have used site-directed mutagenesis to probe the binding sites for two of these receptors: the β-AR, which binds catecholamine agonists and β-blocker antagonists, and the NK1R, which binds the undecapeptide substance P, as well as recently discovered small molecule antagonists.

β-Adrenergic Receptors

Deletion mutagenesis showed that none of the hydrophobic clusters of amino acids (the putative transmembrane helices) could be deleted without loss of binding (14). In contrast, most of the connecting loops could be deleted without affecting the ligand-binding properties of the receptor, indicating that these hydrophilic loops are not required for ligand binding to the receptor and suggesting that the ligand-binding pocket is located predominantly within the transmembrane domain of the protein. Deletions in the connecting loops that were large enough to encompass the entire loop led to steric problems, resulting in incorrect processing of the protein. Certain connecting loop deletion mutations, however, did lead to loss of functional activation of adenylyl cyclase by the receptor. For example, deletion of the carboxy terminal region of the third intracellular loop attenuated the ability of the receptor to activate adenylyl cyclase, and deletion of the amino terminal portion of this loop abolished cyclase activation (11). Moreover, the agonist-binding isotherms for these mutants displayed a single affinity site, suggesting altered G protein interactions. Since these mutants also retain their functional activation of Na-H exchange, which is mediated through a different G protein, the deletions appear not to result in gross structural perturbations of the receptor, suggesting that the changes seen in adenylyl cyclase activation are due to alteration of a specific G protein interaction (15). Subsequent amino acid replacements in the third intracellular loop confirmed the role of this region in G protein interaction (16).

Single-residue substitutions of charged and polar residues within the hydrophobic α helices have implicated Asp-113 in transmembrane helix 3, Ser-204 and Ser-207 in transmembrane helix 5, and Phe-290 in transmembrane helix 6 in direct interactions with the biogenic amine ligands (17). Mutation of Asp-113 to Glu leads to a 100-fold loss in the affinity for the agonist isoproterenol as measured by cyclase activation, whereas substitution with an Asn residue causes a 10,000-fold decrease in affinity (18). The direct interaction of this residue with the protonated amine group of the ligand can be inferred by the results of the substitution of this residue with a Ser. The Ser-113 mutant receptor binds and is activated by catechol esters and ketones, although these compounds do not activate the wild-type receptor (19). The demonstration that a new ligand-binding specificity can be acquired in a predictable way in response to a single amino acid substitution strongly suggests that this residue forms part of the catecholamine-binding site in the wild-type receptor. The stringent requirement for an Asp residue at this position suggests that the side chain of Asp-113 provides the counterion for the protonated amine group of the ligand.

A similar combination of mutagenesis of the receptor and structural alteration of the ligand has implicated the hydroxyl side chains of Ser-204 and Ser-207 in the fifth transmembrane domain in hydrogen-bonding interactions with the catechol hydroxyl groups of the agonist (20). An analysis of the additivity of the effects of substitution of each of these Ser residues with an Ala with the effects of removal of either the *meta* or the *para* hydroxyl groups of the ligand would suggest the existence of two specific hydrogen bonds between the ligand and the receptor, one linking the *p*-OH group of the catecholamine to the side chain of Ser-207 in the recepor and another between the *m*-OH group of the ligand to Ser-204 in the receptor. The position of these two Ser residues, one helical turn apart in transmembrane helix 5 of the receptor, suggests that simultaneous binding of the catechol hydroxyl groups to both of these Ser residues would serve to orient the catechol ring of the agonist in the binding pocket of the receptor. The proximity of this region of the receptor to the region at the bottom of helix 5 that was determined to be critical for receptor activation suggests that the formation of these interactions may serve as a trigger for receptor activation.

In searching for the specific residue that interacts with the β-hydroxyl group of the endogenous ligand, many mutant receptors have been analyzed. Thus far, we have been unable to identify a specific interaction between the β-hydroxyl and an amino acid side chain in the receptor. Whether the β-

hydroxyl moiety interacts with the protein backbone is unclear; this hypothesis could, in principle, be tested by incorporating unnatural protein building blocks to alter the backbone (21). Other residues, such as Ser-165, remain likely candidates, but cannot be mutated without affecting the structure of the protein.

Neurokinin Receptors

Neurokinin receptors bind small peptides, including substance P, neurokinin A, and neurokinin B. Since these peptides are substantially larger than β agonists and antagonists, it is possible that the peptide-binding domain includes residues from both the transmembrane domain and the extracellular domain. To test this hypothesis, the extracellular segments of the NK1R were substituted with the corresponding sequences from another receptor subtype. These studies implicated the first, second, and third extracellular segments in peptide binding (22). However, subsequent site-directed mutagenesis identified specific residues within the first and second, but not the third, extracellular segments in interactions with the peptide agonists (1). These studies illustrate both the usefulness and limitations of chimeric receptors in locating key regions for subsequent site-directed mutagenesis experiments.

Another successful example of combining a chimeric receptor approach with site-directed mutagenesis is the identification of residues responsible for the observed species selectivity of nonpeptide antagonists. The rat and human NK1R bind various peptide agonists with essentially the same affinity and rank order of potency. However, the recently described synthetic nonpeptide antagonists exhibit a profound species selectivity, with the rank order of potency CP95,345 > RP67580 for the human NK1R and RP67580 > CP96,345 for the rat NK1R, even though the two receptors are 95% identical. Chimeric mutants based on the rat and human NK1R indicated that one divergent residue in the seventh transmembrane helix in combination with one or more residues in helices 2–6 form the basis of the antagonist species selectivity. Subsequent site-directed mutagenesis demonstrated that substitution of two divergent residues at positions 116 and 290 is both necessary and sufficient to reverse the species selectivity of CP96,345 and RP67580 (23).

Within the transmembrane domains, chimeric receptors between different receptor subtypes have proven to have conformational effects, probably because of the more stringent requirements of amino acid compatibility for interhelical packing. To identify ligand-interacting residues within the transmembrane domain of the NK1R, the hydrophobicity pattern of transmembrane sequence was subject to Fourier transform analysis. Putative interior residues were investigated for their role in ligand binding. This approach has led to the identification of His-197 and His-265 in the NK1R as specific sites of interaction with antagonists (8, 24) and several hydrophilic residues in helix 2 as specific sites of interaction with peptides (25).

At position 197 of the human NK1R, various amino acid substitutions do not alter the binding affinities of either peptide agonists or of the nonpeptide antagonist RP67580, suggesting a lack of conformational changes in these mutant receptors. However, the binding affinity of the quinuclidine antagonist CP96,345 is significantly affected. Although an Ala substitution at this position leads to a 30-fold reduction in binding affinity, Gln is able to substitute for His-197 in its interaction with CP96,345. Thus, the aromatic characteristics of residue 197 are not absolutely required for CP96,345 binding. Phe substitution at this position leads to a slight reduction in affinity, while Ser or Tyr substitution leads to 20-fold reduction in affinity, suggesting a lack of conventional hydrogen-bonding interaction between residue 197 and CP96,345. Therefore, multiple substitutions at position 197 clearly indicate that the amino group of His-197 contributes a significant interaction with CP95,345. A survey of known protein structures and ligand–protein interactions would suggest that an amino–aromatic interaction between His-197 and Cp96,345 is likely. This hypothesis is supported by further experiments using modified analogs of CP96,345, which indicate that a phenyl ring of CP96,345 interacts with His-197 (8).

Data Interpretation and Caveats

Site-directed mutagenesis has been proven to be a powerful technique in understanding how receptors function, how ligands interact with the receptor, and how receptors interact with downstream effectors. There is no doubt that site-directed mutagenesis will continue to be at the center stage of structure–function relationship studies until, and even after, the three-dimensional structures for these receptors are available. Nonetheless, it should be realized that the structural information that can be deduced from mutagenesis studies is low resolution at best and the deduced interactions cannot be proved based on mutagenesis experiments alone. Therefore, mutagenesis data should be interpreted cautiously, especially for those studies that rely on a single mutant and a single functional assay. Multiple mutations at any given position in the primary protein sequence and multiple indicators of receptor function will certainly reduce the risk of overinterpreting the data. In addition, complementary "mutations" in the ligands by synthetic manipulation of the ligands themselves, or complementary mutations of target effector proteins, serve to clarify the interpretation of the data. The ability to design a mutant receptor to accommodate a specific ligand provides a powerful "reverse medicinal chemistry" approach to understanding the molecular interactions that link a ligand to its receptor. The caveat in all of these approaches is that we can determine the functional changes resulting from altering the primary structure of the receptor (i.e., the functional half of structure–function relationship), but the structural half is more speculative. The correlation of primary structure and function can easily be established, while the three-dimensional structure based on primary sequence can only be extrapolated. However, understanding these limitations, site-directed mutagenesis allows us to get an early look at the structure–function relationships of a protein at a stage when three-dimensional structure determination is unapproachable. The full power of site-directed mutagenesis is revealed when both the three-dimensional structure and the biochemical function of a mutant protein can be determined, as has been demonstrated in the studies of catalytic mechanism of nuclease (26), substrate binding to the tRNA synthetase (12), and photoactivation mechanism of bacteriorhodopsin (27).

Overview of Site-Directed Mutagenesis Methodologies

The original method of oligonucleotide-directed mutagenesis involved (i) hybridizing an oligonucleotide with one or more nucleotide mismatches to a single-stranded M13 bacteriophage DNA containing the cDNA encoding the protein of interest, (ii) synthesizing a new strand of DNA primed by the mutating oligonucleotide, (iii) ligating the new strand, and (iv) transforming *Escherichia coli* with the heteroduplex DNA (28, 29). On replication of the heteroduplex DNA in bacteria, half of the double-stranded DNA should theoretically be the wild type and half should be the desired mutant. In fact, the frequency of obtaining a desired mutation is typically around 5%, based on the original protocol developed by Smith, Hutchson, and colleagues (29).

Improvements in oligonucleotide-directed mutagenesis techniques have been made in two major areas. One improvement is the use of a phagemid instead of M13 phage as the mutagenesis vector. A phagemid is a bacterial plasmid containing an M13 origin of replication, such as pBluescript for *E. coli* expression (available from Stratagene) and pCDNA-Amp (available from Invitrogen) or pCDM9 (30) for mammalian cell expression. A helper M13 phage, which has been mutated to reduce the efficiency of self-replication, is then used to produce single-stranded DNA (ssDNA) from the plasmid containing the M13 replication origin. The ease of plasmic manipulation and the fact that the mutant plasmid can be used immediately in most protein expression studies have made phagemids the vectors of choice for site-directed mutagenesis.

The other major area of improvement addresses the frequency of achieving mutations. Two methods have been highly successful in increasing mutation frequency, both relying on destroying the wild-type template DNA strand and preserving the mutated strand. The uracil replacement method first described by Kunkel relies on the use of single-stranded DNA containing uracil instead of thymidine (produced in a special strain of *E. coli*) as a template. The complementary strand synthesized *in vitro*, which contains the mismatched oligonucleotide, will contain the normal thymidine. When the heteroduplex DNA is transformed into a normal strain of *E. coli*, the uracil-containing DNA will be destroyed, thus enriching the mutated DNA (31). A commercial kit based on this method is available from Bio-Rad, and the actual protocol we used satisfactorily in many studies is listed in Protocol A. Another method of removing the wild-type strand was based on incorporation of deoxynucleoside 5′-[α-thio]triphosphate into the newly synthesized DNA strand (32). The newly synthesized phosphorothioate-containing strand is resistant to restriction endonucleases and exonuclease III. Therefore, the wild-type template can be nicked by an appropriate restriction endonuclease and subsequently removed by exonuclease III. A second DNA synthesis then uses the phosphorothioate-containing strand as a template to generate double-stranded DNA for *E. coli* transformation. This technology has also been commercialized by Amersham.

With the recent development of the polymerase chain reaction (PCR), PCR-based oligonucleotide-directed mutagenesis has been used in many studies (33). This technique is especially beneficial for GC-rich sequences because their high degree of secondary structure is usually diminished at high temperature. In the PCR-based mutagenesis, four oligonucleotides are needed to generate one mutant (see Protocol B). Because the mutated region is encoded by the priming oligonucleotide and a PCR product is obtained only when the priming is successful, the mutation frequency in the PCR product can be as high as 100%. However, the error rate of nucleotide incorporation performed at high temperature is intrinsically higher than that catalyzed by other DNA polymerases, and this high error rate was confirmed in our studies. Therefore, it is necessary to confirm all protein-encoding sequences derived from PCR, while in the cases of uracil replacement or phosphorothioate replacement it is not necessary to sequence the entire protein-coding sequence except in the mutated region. The need to sequence all protein-encoding sequences is one major disadvantage of PCR-based mutagenesis techniques.

While oligonucleotide-directed mutagenesis can insert, delete, or substitute amino acids at any position of a protein sequence, the substitution is limited by the 20 natural amino acids. To reach the ultimate goal of placing any kind of structure at any position in a protein, Schultz and colleages have developed an *in vitro* site-specific mutagenesis method to incorporate unnatural amino acids (34). In this method, the codon encoding the amino acid to be changed is mutated to a stop codon by traditional oligonucleotide-directed mutagenesis as described above. In parallel, a suppressor tRNA (e.g., a tRNA with a CUA anticodon recognizing the TAG stop codon) is chemically acylated with an unnatural amino acid. When the DNA is transcribed *in vitro*, followed by *in vitro* translation in the presence of the suppressor tRNA, the unnatural amino acid is incorporated into the growing polypeptide chain. This method has been successfully applied to soluble proteins, with a typical yield of 0.02 mg of active enzyme per ml of reaction mixture. The same procedure should theoretically be applicable to membrane-bound proteins as long as there is a sensitive functional assay for the *in vitro* translation products. The *in vitro* translation and assembly of membrane proteins can be accomplished in the presence of microsomal membranes (35), and sensitive techniques such as patch clamping have allowed analysis of *in vitro* transcribed ion channels (36). Thus, the incorporation of unnatural amino acids is technically feasible for membrane receptors, as well as for soluble proteins.

Protocols for Mutagenesis

Protocol A: The Uracil-Replacement Method of Single-Stranded DNA-Based Mutagenesis

Preparation of Single-Stranded DNA

1. Insert the gene of interest into a phagemid such as pCDM9 (30).
2. Transform 1 ng of the phagemid (prepared by standard plasmid mini-prep methods) into a special *E. coli* strain CJ236, and incubate the plate (containing ampicillin) overnight (31, 37, 38).
3. Pick a single colony, inoculate 3 ml Luria-Bertani media (containing chloramphenicol), and incubate overnight with shaking.
4. Inoculate 50 ml of 2× YT media containing chloramphenicol with 1 ml of the overnight culture and continue shaking until OD_{600} = 0.3 (approximately 2–4 hr), corresponding to 1 × 10^7 cells/ml. Add helper phage R408 (Strategene, San Diego, CA) to achieve a multiplicity of infection (MOI) of 20 (i.e., 20 phage/cell). Incubate overnight with shaking.
5. Centrifuge the overnight culture (17,000*g*) and transfer the phage-containing supernatant to a fresh tube. Repeat the centrifugation once. Precipitate the phage by adding 1/4 volume of 20% PEG-6000/2.5 *M* NaCl.

6. Prepare the ssDNA from the phage pellet by standard phenol chloroform extraction and ethanol precipitation.

7. Determine the ssDNA concentration and quality by running a sample in an agarose gel with standard ssDNA markers. This ssDNA can be used for all mutagenesis experiments because only the oligonucleotide needs to be changed for different mutants.

Mutagenesis

8. Synthesize an oligonucleotide complementary to the ssDNA prepared above, but encode the desired mutation. A 25–30 mer with one to three base mismatches in the middle is usually appropriate. The identity of the mismatched nucleotides will be determined by the mutation intended. The GC content of the oligonucleotide should preferentially be within the range of 40–60%. A string of four or more identical nucleotides should be avoided.

9. Purify the oligonucleotide using a NENSORB cartridge (NEN, Boston, MA) and phosphorylate the 5′-OH with T4 polynucleotide kinase.

10. Anneal the oligonucleotide (6 pmol) with ssDNA (0.3 pmol) in T7 polymerase buffer by placing the reaction tube in 200 ml of 70°C water in a beaker, and let the beaker cool down to room temperature (this should take approximately 30 min). DNA can then be extended from the oligonucleotide with T7 DNA polymerase (0.05 units/μl), followed by ligation with T4 DNA ligase (0.3 units/μl).

11. Transform the heteroduplex DNA into an *E. coli* strain such as SURE cells (Stratagene, San Diego, CA), pick four to five colonies, and prepare miniprep DNA from each of them.

12. Sequence the region from which the oligonucleotide is derived to confirm the success of mutation. Double-stranded DNA sequencing using the Sequenase kit (USBC, Cleveland, OH) is fast and convenient. If all clones are wild type, sequence another 20 clones or screen more by hybridization using the ^{32}P-labeled mismatch oligonucleotide.

Expression

13. Prepare a larger quantity of the mutated DNA (usually a 100- to 200-ml culture is sufficient for most transient expression in COS).

14. Transfect 10 μg of DNA into 1×10^7 COS cells (or other appropriate mammalian cells) by electroporation (39), followed by appropriate assays for the receptor of interest.

Problem Solving

It is important that the single-stranded DNA templates used in site-directed mutagenesis be of high quality. Contaminating fragments of DNA or RNA may act as primers, thereby inhibiting the hybridization of the mutant oligonucleotide. Contaminating polyethylene glycol or salts may also prevent the annealing of the oligonucleotide. The best way to ensure a high-quality ssDNA preparation is to achieve complete separation at each step of the protocol. With a final preparation of ssDNA, only proteins and RNA can be easily removed but not other contaminating DNA. The quality of single-stranded DNA can be gauged by using it as a template for dideoxy sequencing. High background or the appearance of extra bands in the sequencing gel would indicate that new ssDNA needs to be prepared.

One common problem encountered in site-directed mutagenesis is that some positions in the DNA sequence are difficult to mutate. This problem can often be resolved by varying the oligonucleotide concentration in the mutagenesis reaction. Difficulties in achieving mutagenesis can also arise if the mutating oligonucleotide hybridizes to itself. One solution is to introduce another silent mutation simultaneously, thereby changing the sequence of the oligonucleotide without affecting the sequence of the resulting mutant protein.

Difficulty in generating mutants may also arise from a high GC content at those positions of the cDNA or the presence of secondary structure in the single-stranded DNA. A balance of achieving hybridization and reducing secondary structure can be accomplished by experimenting with several annealing temperatures around the range of the calculated melting temperature (40–42). If single-stranded DNA-based mutagenesis cannot produce the desired mutant, PCR-based mutagenesis should be used.

Protocol B: Polymerase Chain Reaction-Based Mutagenesis

Synthesis of Oligonucleotides

1. Four oligonucleotides are needed to generate one mutation. Two are the mismatch oligonucleotides encoding the mutation, one being sense (A) and one being antisense (B). The other two are perfectly matching oligonucleotides, one being sense and upstream of A/B (C) and one being antisense and downstream of A/B (D). The sequence between C and D should contain two unique restriction sites (site *x* and site *y*) so that the mutated PCR fragment can be exchanged with the corresponding wild-type fragment. The oligonucleotide design considerations are the same as those described in step 8 of Protocol A. In addition, one should avoid consecutive G/C at the 3′ end of any oligonucleotide to prevent primer–dimer formation.

Mutagenesis

2. The first PCR uses C and B as primers (1 μM final concentration) and a plasmid containing wild-type cDNA as template (1 μg) in 100 μl of 10 mM Tris, pH 8.3, 50 mM KCl, 1.5 mM $MgCl_2$, 800 μM dNTP, and 2.5 U of Taq

DNA polymerase. Standard PCR conditions recommended by Perkin–Elmer should be followed (e.g., denature at 94°C for 1 min, anneal at 37°C for 1 min extend at 72°C for 1 min, for a total of 20–30 cycles).

3. A second PCR uses A and D as primers and a plasmid containing the wild-type cDNA as template.

4. Gel-purify the PCR products from steps 2 and 3, and recover the PCR fragments using GeneClean (Bio101, La Jolla, CA).

5. The third PCR uses the first and second PCR products from steps 2 and 3 as template and C and D as primers. Extract the final PCR product with phenol/$CHCl_3$, followed by ethanol precipitation and resuspension in water. Cleave the PCR product with enzymes recognizing sites *x* and *y*. Gel-purify the correct fragment, which should contain the mutated sequence.

6. Cleave the plasmid containing the wild-type cDNA with enzymes recognizing sites *x* and *y*. Gel-purify the fragment that does not contain the sequence corresponding to the PCR fragment in step 5.

7. Ligate the two purified fragments from steps 5 and 6. Transform the ligation product into *E. coli*. Pick two colonies and prepare miniprep plasmid DNA.

8. Sequence the entire region between sites *x* and *y* to confirm the presence of the designed mutation and the absence of any random mutation caused by the PCR.

Expression

9. Prepare a larger quantity of the mutated DNA (usually 100–200 ml culture is sufficient for most transient expression in COS).

10. Transfect 10 μg of DNA into 1×10^7 COS cells (or other appropriate mammalian cells) by electroporation, followed by appropriate assays for the receptor of interest.

Problem Solving

Although PCR-based mutagenesis usually will eliminate the problem of secondary structure as mentioned previously, nucleotide analogs such as 7-deaza-dGTP can also be used in PCR reactions to further destabilize the secondary structure. Anecdotal reports state that the last round of PCR should be done with GTP only, in order to increase the efficiency of subcloning, although we have been successful in cloning fragments containing deaza-GTP. Denaturing agents such as formamide and DMSO can also be used if one is trying to amplify a DNA with a high GC content. Formamide should be titrated between 1 and 5% and DMSO between 2.5 and 15%; the purity of the reagents is critical. In order to reduce the secondary structure of the template, "hot start" PCR can be used instead of nucleotide analogs or denaturing agents. This technique involves melting the template DNA in buffer in the presence of primers, followed by a quick cool before adding the nucleotides and polymerase.

Because PCR-based site-directed mutagenesis requires four oligonucleotides for any given mutation, the hybridization of two different oligonucleotides may prevent a successful PCR amplification. Several computer programs exist that offer some assistance in trying to avoid some of the more common problems such as sequence complementarity between primers and secondary structure of the primers. We have used several of these programs with equivalent results.

References

1. R. R. C. Huang, H. Yu, C. D. Strader, and T. M. Fong, *Mol. Pharmacol.* **45,** 690 (1994).
2. L. Pradier, A. S. Yee, and M. G. McNamee, *Biochemistry* **28,** 6562 (1989).
3. R. J. Leonard, C. G. Labarca, P. Charnet, N. Davidson, and H. A. Lester, *Science* **242,** 1578 (1988).
4. J.-P. Changeux, *Trends Pharmacol. Sci.* **11,** 485 (1990).
5. D. C. Rees, L. DeAntonio, and D. Eisenberg, *Science* **245,** 510 (1989).
6. G. K. Ackers and F. R. Smith, *Annu. Rev. Biochem.* **54,** 597 (1985).
7. A. E. Erikson, W. A. Baase, X. J. Zhang, D. W. Heinz, M. Blaber, E. P. Baldwin, and B. W. Matthews, *Science* **255,** 178 (1992).
8. T. M. Fong, M. A. Cascieri, H. Yu, A. Bansal, C. Swain, and C. D. Strader, *Nature (London)* **362,** 350 (1993).
9. W. H. J. Ward, D. Timm, and A. R. Fersht, *Trends Pharmacol. Sci.* **11,** 280 (1990).
10. J. J. He and F. A. Quicho, *Science* **251,** 1479 (1991).
11. R. A. F. Dixon, I. S. Sigal, E. Rands, R. B. Register, M. R. Candelore, A. D. Blake, and C. D. Strader, *Nature (London)* **326,** 73 (1987).
12. D. M. Lowe, A. R. Fersht, and A. J. Wilkinson, *Biochemistry* **24,** 5106 (1985).
13. M. A. Kjelberg, S. Cotecchia, J. Ostrowski, M. G. Caron, and R. J. Lefkowitz, *J. Biol. Chem.* **267,** 1430 (1992).
14. R. A. F. Dixon, I. Sigal, M. R. Candelore, R. B. Register, W. Scattergood, E. Rands, and C. D. Strader, *EMBO J.* **6,** 3269 (1987).
15. D. L. Barbar, M. B. Ganz, P. B. Bongiorno, and C. D. Strader, *Mol. Pharmacol.* **41,** 1056 (1992).
16. A. H. Cheung, R. R. C. Huang, and C. D. Strader, *Mol. Pharmacol.* **41,** 1061 (1992).
17. C. D. Strader, I. Sigal, and R. A. F. Dixon, *FASEB J.* **3,** 1825 (1989).
18. C. D. Strader, I. S. Sigal, M. R. Candelore, E. Rands, W. S. Hill, and R. A. F. Dixon, *J. Biol. Chem.* **263,** 10267 (1988).
19. C. D. Strader, T. Gaffney, E. E. Sugg, M. R. Candelore, R. Keys, A. A. Pachett, and R. A. F. Dixon, *J. Biol. Chem.* **266,** 5 (1991).

20. C. D. Strader, M. R. Candelore, W. S. Hill, I. S. Sigal, and R. A. F. Dixon, *J. Biol. Chem.* **264,** 13572 (1989).
21. J. A. Ellman, D. Mendel, and P. G. Schultz, *Science* **255,** 197 (1992).
22. T. M. Fong, R. R. C. Huang, and C. D. Strader, *J. Biol. Chem.* **267,** 25664 (1992).
23. T. M. Fong, H. Yu, and C. D. Strader, *J. Biol. Chem.* **267,** 25668 (1992).
24. T. M. Fong, H. Yu, Cascieri, D. Underwood, C. J. Swain, and C. D. Strader, *J. Biol. Chem.* **269,** 2728 (1994).
25. R. R. C. Huang, H. Yu, C. D. Strader, and T. M. Fong, *Biochemistry* **33,** 3007 (1994).
26. J. K. Judice, T. R. Gamble, E. C. Murphy, A. M. de Vos, and P. G. Shultz, *Science* **261,** 1578 (1993).
27. A. K. Mitra, L. J. W. Miercke, G. J. Turner, R. F. Shand, M. C. Betlach, and R. M. Stroud, *Biophys. J.* **65,** 1295 (1993).
28. C. A. Hutchison, III, S. Phillips, M. H. Edgell, S. Gillam, P. Jahnke, and M. Smith, *J. Biol. Chem.* **253,** 6551 (1978).
29. M. Smith, *Annu. Rev. Genet.* **19,** 423 (1985).
30. T. M. Fong, S. A. Anderson, H. Yu, R. R. C. Huang, and C. D. Strader, *Mol. Pharmacol.* **41,** 24 (1992).
31. T. A. Kunkel, *Proc. Natl. Acad. Sci. U.S.A.* **82,** 488 (1985).
32. J. R. Sayers, W. Schimdt, and F. Eckstein, *Nucleic Acids Res.* **16,** 791 (1988); D. B. Olsen, J. R. Sayers, and F. Eckstein, *in* "Methods in Enzymology" (R. Wu, ed.), Vol. 217, p. 189. Academic Press, San Diego, CA, 1993.
33. D. H. Jones, K. Sakamoto, R. L. Vorce, and B. H. Howard, *Nature* (*London*) **344,** 793 (1990).
34. S. J. Anthony-Cahill, M. C. Griffith, C. J. Noren, D. J. Suich, and P. G. Schultz, *Trends Biochem. Sci.* **14,** 400 (1989).
35. M. Friedlander and G. Blobel, *Nature* (*London*) **318,** 338 (1985).
36. R. L. Rosenberg and J. E. East, *Nature* (*London*) **360,** 166 (1992).
37. "Muta-Gene Phagemid In Vitro Mutagenesis Instruction Manual." Bio-Rad, Richmond, CA.
38. M. Russel, S. Kidd, and M. R. Kelley, *Gene* **45,** 333 (1986).
39. J. Sambrook, E. F. Fritsch, and T. Maniatis, "Molecular Cloning," 2nd ed. Cold Spring Harbor Lab., Cold Spring Harbor, NY, 1989.
40. J. G. Wetmur and N. Davidson, *J. Mol. Biol.* **1,** 349 (1968).
41. M. K. Eghtedarzadeh and S. Henikoff, *Nucleic Acids Res.* **14,** 5115 (1986).
42. J. Meinkoth and G. Wahl, *Anal. Biochem.* **138,** 267 (1984).

[15] Receptor Chimers

Sankuratri Suryanarayana and Brian K. Kobilka

Introduction

Site-directed mutagenesis has been a powerful tool for identifying functionally important domains of proteins for which cDNA clones are available. When a mutation disrupts a specific function, the mutated domain is likely to be directly involved in carrying out that function. Yet, site-directed mutations may disrupt the function of a distant structural domain through allosteric effects. Another approach to studying structure–function relationships is the construction of chimeric proteins. The cDNAs encoding structurally related proteins with distinguishable functional properties can be used to construct chimeric genes. By correlating the functional properties of the resulting chimeric proteins with their structural components, it is possible to identify functional domains. This approach differs from site-directed mutagenesis in that the goal is to change the specificity of a functional property rather than to disrupt it. The chimeric approach to mutagenesis involves initially exchanging large structural domains and progressively narrowing down the size of the domain until the smallest structural unit capable of modifying the functional specificity is found. This approach has been successfully used to define the functional domains of small molecule binding receptors such as adrenergic, serotonin, dopamine, mascurinic, and glutamate receptors (1–8) as well as peptide (9–16) and polypeptide (17–21) binding G protein-coupled receptors. This article describes in detail various approaches used for the construction and analysis of chimeric α_2/β_2-adrenergic receptors that have been used to define the domains responsible for ligand binding and G protein-coupling specificity and to identify intramolecular relationships that exist in the three-dimensional structure of the protein. These approaches could be adapted to define the functional domains of any homologous proteins.

The α_2- and β_2-adrenergic receptors are ideal for chimeric protein studies as they have distinct pharmacologic properties, and activate different G proteins (1, 22). The α_2 receptor inhibits adenylate cyclase through the inhibitory G protein, G_i; while the β_2 receptor activates adenylate cyclase through the stimulatory G protein, G_s.

Methods in Neurosciences, Volume 25

Construction of α_2/β_2 Chimeric Receptors

An initial step in the construction of chimeric receptor is the alignment of the two wild-type receptor protein sequences and the identification of convenient restriction sites on the DNA sequences. The alignment can be achieved manually when the homology between the two proteins is relatively high (>50% identity). For the alignment of the sequences with lower homology, any of the commercially available computer software can be used. We have found that the optimal site to join two receptor sequences (which we refer to as splicing) is between membrane-spanning domains. The simplest method of constructing chimeric receptors is by ligating DNA fragments from the two receptors at a position having a common restriction site. For example, the human β_2-adrenergic receptor and the human 5HT1A receptor both have a *Pst*I site in the DNA sequence encoding the hydrophilic domain between the first and second hydrophobic domains. It is therefore possible to make a chimeric receptor by simply splicing these sequences together at the *Pst*I site (Fig. 1). However, in practice it is rare to find a convenient restriction site at the same location in both the receptors. For example, no convenient restriction site is present at the same location in human α_2C10 and β_2-adrenergic receptors despite their overall amino acid sequence identity of 34%. Table I lists the methods used for the construction and expression of G protein-coupled receptor chimers. The methods described below have been employed in our laboratory for the generation of α_2/β_2 chimers.

The genes encoding human α_2C10 and β_2-adrenergic receptors are cloned into the multiple cloning site of either pSP65 vector (Amersham Corp.) (23) or pGEM-3Z vector (Promega Biotec.). The construction and the sequencing of the chimeric molecules are performed on these vectors. To ensure uniformity in the expression of the chimeric receptors and the wild-type receptors, the 3′ and 5′ untranslated regions of all genes are derived from the β_2-adrenergic receptor cDNA (24). The chimers are identified by restriction analysis and the authenticity of the sequences of the ligated junctions is verified by sequence analysis. Both strands of plasmid DNA obtained by a miniprep procedure (Maniatis) are sequenced using Sequenase version 2.0 (U.S. Biochemical Corp.).

Adapter-Mediated Ligation of Restriction Fragments

The restriction endonuclease fragments encoding the desired structural domains of the α_2- and β_2-adrenergic receptors are obtained by preparative agarose gel electrophoresis. The fragments are eluted from the agarose gel either by electroelution or by Geneclean II kit following manufacturer's

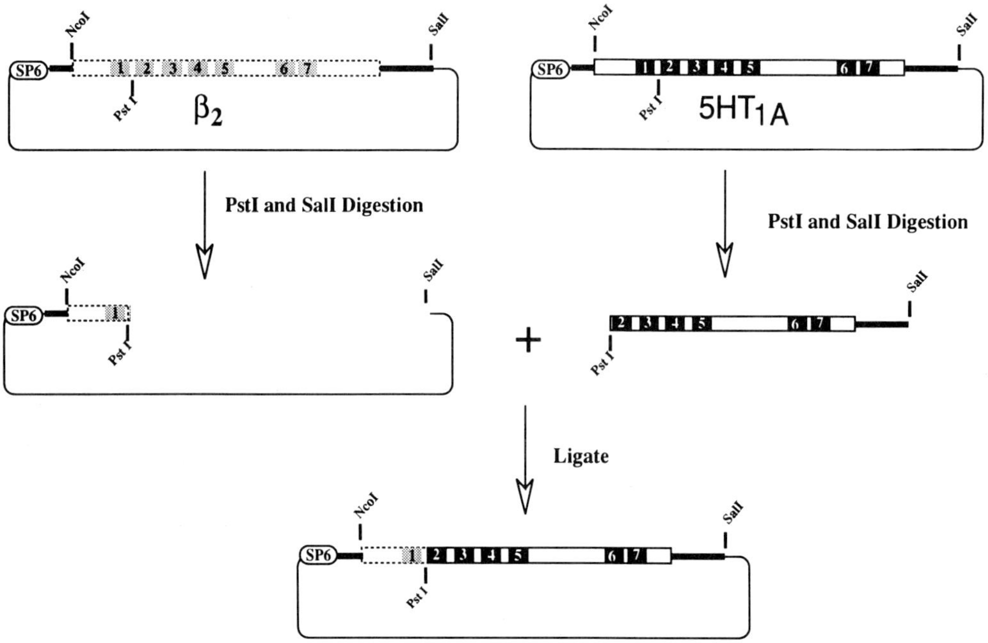

FIG. 1 Construction of a β_2/5HT1A chimeric receptor by ligation at *Pst*I restriction site. The plasmids pSPβ_2 and pSP5HT1A were digested with restriction enzymes, and the purified fragments were ligated as shown. The large rectangle in each plasmid diagram represents the coding sequence of β_2 (broken line) and 5HT1A (solid line) receptors. The areas numbered 1 to 7 indicate the sequences corresponding to the hydrophobic domains of the receptors.

instructions (Bio 101, Inc., La Jolla, CA). Electroelution of the electrophoresed DNA fragments is accomplished as follows. The DNA bands are viewed under a long-wavelength (>300 nm) ultraviolet lamp. A well is cut immediately below the band of interest on the anode side. The level of the liquid in the electrophoresis chamber is reduced such that only half of the thickness of the gel will be in the buffer (the buffer did not cover the top surface of the gel). The well is filled with 75 m*M* ammonium acetate and the gel is returned to the electrophoresis chamber. Electrophoresis is performed until the band of interest has migrated into the well. The liquid is removed from the well and extracted with an equal volume of phenol, three or four times with isobutanol and once with chloroform. One-half volume of 7.5 *M* ammonium acetate and two volumes of ethanol are added to the sample to precipitate the DNA fragment which is washed with 70% ethanol and dried. This electroelution procedure works equally well with any size fragments while the Geneclean method is less suitable for the isolation of fragments <500 base pairs and for very large fragments (>10 kb).

As restriction fragments obtained from α_2 and β_2 receptors have noncomplementary cohesive ends, they are ligated with oligonucleotide adapters sandwiched between them (Fig. 2). These adapters are double-stranded synthetic oligonucleotides with cohesive ends, each end being complementary to one of the two receptor fragments. The adapter also contains sequence coding for the part of the chimeric receptor. The length of the adapters varies and contains a minimum of about 10 bases double-stranded with the cohesive tails on either end. The double-stranded adapters are made by heating the unphosphorylated or 5′ phosphorylated single-stranded synthetic oligonucleotides for 5 min at 65°C in 10 m*M* Tris–Cl. 1 m*M* EDTA, pH 8.0, and cooling down to room temperature over a period of about 20 min. As the oligonucleotides are synthesized without the 5′ phosphate, they can be phosphorylated using T4 polynucleotide kinase. Although the efficiency of ligation is better with phosphorylated adapters compared to the unphosphorylated adapters, there is a potential for the formation of concatamers (multiple adapters ligated end-to-end) with phosphorylated adapters.

Construction of Chimeric Receptors Using Polymerase Chain Reaction

The use of the linker-adapter approach for constructing chimeric receptors limits the splice junction to areas around which convenient restriction sites can be found (unless one is able to make very large adapters). Using the polymerase chain reaction (PCR) technique it is possible to produce chimeric proteins by ligating the DNA fragments at any desired site. We used PCR in several ways to generate α_2/β_2 chimeric molecules. The examples considered below provide two basic techniques which could be applied to produce chimers of any proteins. In most cases, PCR reactions are performed using AmpliTaq polymerase kit from Perkin–Elmer Cetus Corp. by following the standard protocol enclosed with the kit. A standard reaction mixture (100 μl) consists of 10 m*M* Tris–Cl, pH 8.3 (25°C); 50 m*M* KCl; 1.5 m*M* $MgCl_2$; 200 μM each of dATP, dCTP, dGTP, and dTTP; 1 μM each of the primers; 2 ng of the linearized wild-type α_2 or β_2 receptor in pGEM-3Z vector in 10 m*M* Tris–Cl; 0.1 m*M* EDTA, pH 8.0; and 2.5 units of AmpliTaq polymerase. The oligonucleotides with or without base mismatches are synthesized on an Applied Biosystems 394 DNA/RNA synthesizer by using β-cyanoethyl phosphoramidite chemistry. The deprotected oligonucleotides are purified on either OPC cartridges (Applied Biosystems) or Nensorb car-

TABLE I Methods Used for the Construction and Expression of G Protein-Coupled Receptor Chimers

Receptor chimer(s)	Mode of construction	Expression		Reference[d]
		Cell line	Transfection	
α_2/β_2-adrenergic	Oligonucletide adapters	*X. laevis* oocytes	Microinjection of mRNA	1, 2
	Polymerase chain reaction	COS-7	DEAE–dextran	
		Raji	Electroporation	
		HEK 293	Calcium phosphate	
Mouse α_{2A}/human α_{2A}-adrenergic	Oligonucletide adapters	COS-7	DEAE–dextran	3
β_1/β_2-adrenergic	Taylor's procedure[a]	*X. laevis* oocytes	Microinjection of mRNA	4, 5
β_1/β_2-adrenergic	Oliogonucletide adapters	*E. Coli*	Transformation	6
D2 dopamine/m1 muscarinic	Polymerase chain reaction	CHO-K1	Electroporation	7
m2/m5 muscarinic	Polymerase chain reaction	COS-7	Calcium phosphate	8
5HT2/5HT1C	Polymerase chain reaction	COS-7	DEAE–dextran	9
D1/D2 dopamine	Polymerase chain reaction	COS-7	DEAE–dextran	10
mGluR1/mGluR2 glutamate	Kunkel's procedure[b]	*X. laevis* oocytes	Microinjection of mRNA	11, 12
FPR/FPR2 fMLP	Polymerase chain reaction	Mouse L cell fibroblasts	Calcium phosphate	13
SPR/SKR tachykinin	Kunkel's procedure[b]	COS-7	Calcium phosphate	11, 14
		COS-m6	Calcium phosphate	
ET_A/ET_B endothelin	Kunkel's procedure[b]	CHO	DEAE–dextran	11, 15
Formylpeptide/C5a	"Altered sites" procedure[c]	COS-7	Lipofectin	16
Human NK1/rat NK1	Taylor's procedure[a]	COS-7	DEAE–dextran	4, 17
NK1/NK3 neurokinin	Kunkel's procedure[b]	CHO	Calcium phosphate	11, 18
		COS-7	Calcium phosphate	

Human/canine CCK-B/gastrin	Kunkel's procedure[b]	COS-7	DEAE–dextran	11, 19
Neromedin B/GRP-R	Altered sites procedure[c]	Balb 3T3 fibroblasts	Calcium phosphate	20
TSH-LH/CG	Kunkel's procedure[b]	CHO	Calcium phosphate	11, 21, 22
Leutropin/β-adrenergic	Polymerase chain reaction	COS-7	Calcium phosphate	23
		COS-7	Electroporation	
LH/CG-FSH	Native restriction sites	HEK 293	Calcium phosphate	24
HDEL-DDEL/HDEL	"Sticky feet" method	*S. Cerevisiae*	Transformation	25, 26

[a] An *in vitro* mutagenesis kit based on Taylor's procedure is available from Amersham Corp. (Arlington, IL).

[b] An *in vitro* mutagenesis kit based on Kunkel's procedure is available from Bio-Rad (Richmond, CA).

[c] An *in vitro* mutagenesis kit from Promega Biotec (Madison, WI).

[d] Key to references: 1. B. K. Kobilka, T. S. Kobilka, K. Daniel, J. W. Regan, M. G. Caron, and R. J. Lefkowitz, *Science* **240,** 1310 (1988); 2. S. Suryanarayana, M. von Zastrow, and B. K. Kobilka, *J. Biol. Chem.* **267,** 21991 (1992); 3. R. Link, D. Daunt, G. Barsh, A. J. Chruscinski, and B. Kobilka, *Mol. Pharmacol.* **42,** 16 (1992); 4. J. W. Tayler, J. Ott, and F. Eckstein, *Nucleic Acids Res.* **13,** 8765 (1985); 5. T. Frielle, K. W. Daniel, M. G. Caron, and R. J. Lefkowitz, *Proc. Natl. Acad. Sci. U.S.A.* **85,** 9494 (1988); 6. S. Marullo, L. J. Emorine, A. D. Strosberg, and C. Delavier-Klutchko, *EMBO J.* **9,** 1471 (1990); 7. B. P. England, M. S. Ackerman, and R. W. Barrett, *FEBS Lett.* **279,** 87 (1991); 8. J. Wess, D. Gdula, and M. R. Brann, *Mol. Pharmacol.* **41,** 369 (1992); 9. M. S. Choudhary, S. Craigo, and B. L. Roth, *Mol. Pharmacol.* **42,** 627 (1992); 10. R. G. MacKenzie, M. E. Steffey, A. M. Manelli, N. J. Pollock, and D. E. Frail, *FEBS Lett.* **323,** 59 (1993); 11. T. A. Kunkel, J. D. Roberts, and R. A. Zakour, *in* "Methods in Enzymology" (R. Wu and L. Grossman, eds.), Vol. 154, p. 367. Academic Press, Orlando, FL, 1987; 12. K. Takahashi, K. Tsuchida, Y. Tanabe, M. Masu, and S. Nakanishi, *J. Biol. Chem.* **268,** 19341 (1993); 13. O. Quehenberger, E. R. Prossnitz, S. L. Cavanagh, C. G. Cochrane, and R. D. Ye, *J. Biol. Chem.* **268,** 18167 (1993); 14. Y. Yokota, C. Akazawa, H. Ohkubo, and S. Nakanishi, *EMBO J.* **11,** 3585 (1992); 15. M. Adachi, Y.-Y. Yang, A. Trzeciak, Y. Furuichi, and C. Miyamoto, *FEBS Lett.* **311,** 179 (1992); 16. H. D. Perez, R. Holmes, L. R. Vilander, R. R. Adams, W. Manzana, D. Jolley, and W. H. Andrews, *J. Biol. Chem.* **268,** 2292 (1993); 17. B. S. Sachais, R. M. Snider, J. A. Lowe, III, and J. E. Krause, *J. Biol. Chem.* **268,** 2319 (1993); 18. U. Gether, T. E. Johansen, R. M. Snider, J. A. Lowe, III, S. Nakanishi, and T. W. Schwartz, *Nature (London)* **362,** 345 (1993); 19. M. Beinborn, Y.-M. Lee, E. W. McBride, S. M. Quinn, and A. S. Kopin, *Nature (London)* **362,** 348 (1993); 20. Z. Fath, R. V. Benya, H. Shapira, R. T. Jensen, and J. F. Battey, *J. Biol. Chem.* **268,** 14622 (1993); 21. Y. Nagayama, H. L. Wadsworth, G. D. Chazenbalk, D. Russo, P. Seto, and B. Rapoport, *Proc. Natl. Acad. Sci. U.S.A.* **88,** 902 (1991); 22. Y. Nagayama, D. Russo, H. L. Wadsworth, G. D. Chazenbalk, and B. Rapoport, *J. Biol. Chem.* **266,** 14926 (1991); 23. W. R. Moyle, M. P. Bernard, R. V. Myers, O. M. Marko, and C. D. Strader, *J. Biol. Chem.* **266,** 10807 (1991); 24. T. Braun, P. R. Schofield, and R. Sprengel, *EMBO J.* **10,** 1885 (1991); 25. T. Clackson and G. Winter, *Nucleic Acids Res.* **17,** 10163 (1989); 26. J. C. Semenza and H. R. B. Pelham, *J. Mol. Biol.* **224,** 1 (1992).

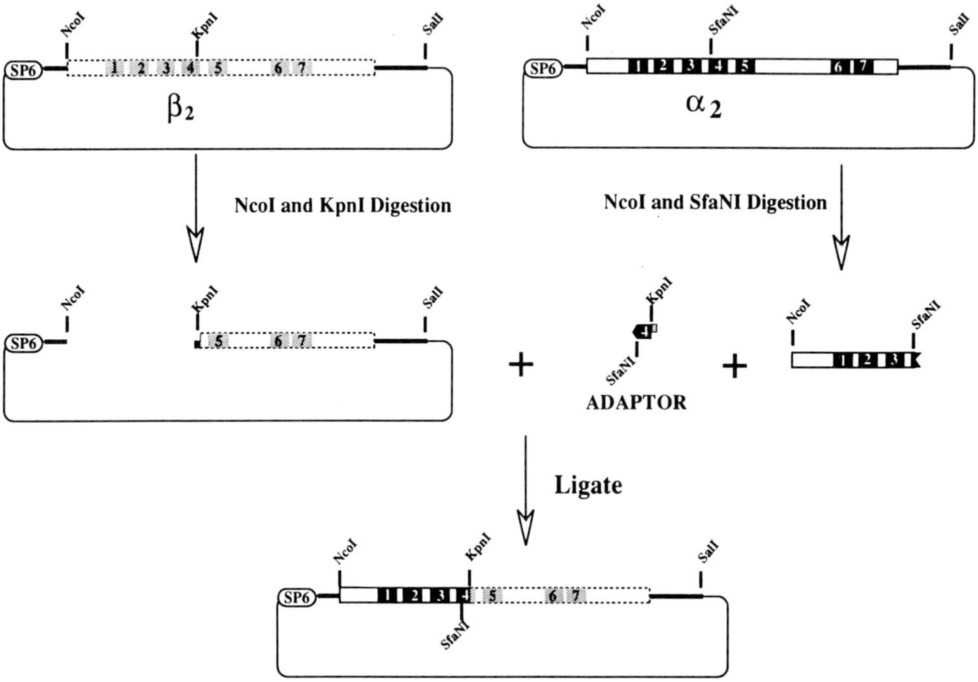

FIG. 2 Construction of α_2/β_2 chimeric receptor by adapter-mediated ligation of restriction fragments. The plasmids pSPα_2 and pSPβ_2 were digested with restriction enzymes, and the purified fragments were ligated along with the linker as shown. The large rectangle in each plasmid diagram represents the coding sequence of α_2 (solid line) and β_2 (broken line) receptors. The areas numbered 1 to 7 indicate the sequences corresponding to the hydrophobic domains of the receptors.

tridges (NEN-DuPont) by following the manufacturer's instructions. The purified oligonucleotides are dried and redissolved in water. Before the thermostable polymerase is added, the reaction mixture is heated at 94°C for 5 min and cooled to the annealing temperature over a period of another 5 min in a Perkin–Elmer Cetus thermocycler. The tubes are layered with about 50 μl of mineral oil after the enzyme is added. The standard PCR cycling conditions are denaturation at 94°C for 1 min, annealing at 42°C for 1.5 min, and polymerization at 72°C for 2 min. The cycle is repeated 25 times and is followed by a polymerization at 72°C for another 10 min.

Most of the oil from the PCR reaction tube is carefully pipetted out and the aqueous sample is collected by pipetting from the bottom of the tube. The remaining oil from the sample is removed by extraction with an equal volume of chloroform. The sample is extracted further with phenol and chloroform and precipitated with isopropanol. The pellet is carefully washed with 70% ethanol, dried, and redissolved in 20 μl of 10 mM Tris–Cl, 1 mM EDTA, pH 8.0. An aliquot of 4 μl is electrophoresed on an agarose gel. If the reaction products contain only the expected fragment, it is directly subjected to restriction digestion. On occasions where the reaction products contain other nonspecific fragments, the desired fragment is isolated by preparative agarose gel electrophoresis as described above.

In some cases, PCR reactions are performed by using either Vent DNA polymerase (New England Biolabs, Inc.) or *Pfu* DNA polymerase (Stratagene). The reaction mixture for Vent DNA polymerase is similar to that of AmpliTaq polymerase with the exception of the reaction buffer which consisted of 20 mM Tris–Cl, pH 8.8, 10 mM KCl, 10 mM $(NH_4)_2SO_4$, 2 mM $MgSO_4$, 0.1% Triton X-100, and 10 μg of bovine serum albumin/100 μl of reaction mixture. The reaction mixture for *Pfu* DNA polymerase is also similar to that of AmpliTaq polymerase. The reaction buffer consists of 20 mM Tris-Cl, pH 8.8, 10 mM KCl, 6 mM $(NH_4)_2SO_4$, 1.5 mM $MgCl_2$, and 0.01% Triton X-100. The denaturation and the polymerization temperatures used for *Pfu* DNA polymerase are 95 and 75°C, respectively.

The manufacturers of Vent DNA polymerase and *Pfu* DNa polymerase claim that, in addition to $5' \rightarrow 3'$ polymerization, both these enzymes also possess $3' \rightarrow 5'$ proofreading exonuclease activity which might result in a reduced rate of unexpected mutations in the PCR product. However, in our hands both these enzymes failed to produce any product in some instances, for example, when oligonucleotide primers differing by more than 10 nucleotides in length were used. In all these instances, the use of Taq polymerase resulted in the desired product. Moreover, we did observe some unexpected mutations when Vent polymerase was used although at a lower frequency compared to Taq polymerase (S. Suryanarayana and B. K. Kobilka, unpublished results). It is essential to confirm the sequence of the PCR-generated portion(s) and the ligated junctions of the final construct. We have noted that the number of unexpected mutations in the PCR product can be reduced to a large extent by (i) preheating and cooling the reaction mixture before adding the thermostable DNA polymerase (see above), (ii) choosing the highest possible annealing temperature in the PCR reactions, and (iii) minimizing the number of PCR cycles to 25. It has also been suggested that the use of reduced concentration deoxynucleotides (<200 μM) might increase the fidelity of Taq polymerase (25).

Generation of a DNA Fragment with a Desired Restriction Site at the 5′ (or 3′) End

If one of the two DNA fragments to be ligated contains an unique restriction site at its 3′ (or 5′) end, the same site can be introduced in to the 5′ (or 3′) end of the second fragment by PCR. This strategy can be exemplified by the construction of chimer S6F^{412}→N, which consists of the amino terminus to the end of hydrophobic domain 1 from the β_2-adrenergic receptor and the second hydrophobic domain to the carboxy terminus from the α_2-adrenergic receptor with a point mutation of F^{412}→N in the seventh hydrophobic domain (Fig. 3). The point mutation is introduced into the wild-type α_2-adrenergic receptor as described previously (22). The chimer S6F^{412}→N is constructed by ligating the *Nco*I–*Sal*I fragment of the vector, *Bss*HII–*Sal*I of the

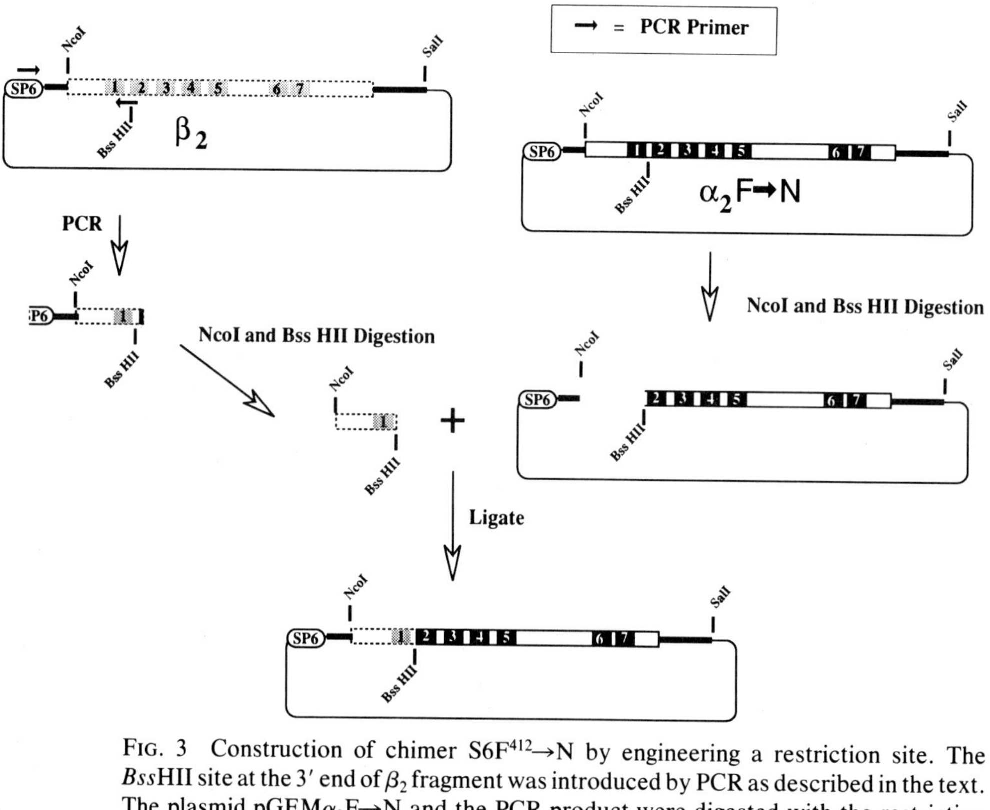

FIG. 3 Construction of chimer S6F^{412}→N by engineering a restriction site. The *Bss*HII site at the 3′ end of β_2 fragment was introduced by PCR as described in the text. The plasmid pGEMα_2F→N and the PCR product were digested with the restriction enzymes and ligated as shown. For explanation of other notation, see legend to Fig. 2.

α_2F[412]→N and *Nco*I–*Bss*HII fragment of a PCR product which codes for the amino terminus to the end of the first hydrophobic domain of β_2-adrenergic receptor. This fragment is generated by PCR using an oligo of 19 nucleotides (5′ GATTTAGGTGACACTATAG 3′) complementary to the antisense strand of the SP6 promoter of pGEMβ_2 and the second oligo of 30 nucleotides (5′ CTTGAGCGCGCGGAACTTGGCAATGGCTGT 3′) of which the first 12 are complementary to the sense-strand of α_2-adrenergic receptor around its *Bss*HII site and the rest are complementary to the sense strand of β_2-adrenergic receptor sequences immediately after the first hydrophobic domain. The PCR is performed using pGEMβ_2 as a template, the products are digested with *Nco*I and *Bss*HII, and the desired fragment is purified as described above. Two important factors which should be considered while designing an oligo with a new restriction site at its 5′ end are (i) it should contain a minimum of about 18 nucleotides complementary to the template, and (ii) there should be at least two to three bases 5′ to the recognition sequence for efficient digestion of the PCR product with the restriction enzyme.

Creating a New Restriction Site at the End of One or Both the Strands by Introducing Silent Mutation(s)

A restriction site can be created by changing one or more nucleotides without changing the amino acid it codes for. As the number of commercially available restriction enzymes is high (over 300), it is not hard to create a restriction site at a desired position without altering the amino acid sequence. This approach is used to construct chimer S7F[412]→N (CRS7F→N) which consists of the amino terminus to the end of the third hydrophobic domain from the β_2-adrenergic receptor and the fourth hydrophobic domain to the carboxy terminus from the α_2-adrenergic receptor with a point mutation F[412]→N in the seventh hydrophobic domain. This chimeric receptor is obtained by designing a site for the restriction enzyme *Dra*I at the 3′ end of the β_2 fragment and at the 5′ end of the α_2 fragment (Fig. 4). An oligo of 19 nucleotides complementary to the antisense strand of the SP6 promoter (see above) and a second oligo of 30 nucleotides (5′CTG GTA tTT aAA AGG TGA AGT AAT GGC AAA) complementary to the sense strand in the second intracellular loop of the β_2-adrenergic receptor are used to produce the β_2 fragment. The silent mutations introduced to create a site for *Dra*I at the 3′ end of the PCR product are shown in small letters. A *Dra*I site at the 5′ end of α_2 fragment is created by using an oligo (CAG GCC tTt aAa TAC AAC CTG AAG CGC ACG) complementary to the antisense strand in the second intracellular loop and a second oligo (ACGGCGTCCGGACCCCGG) complementary to the sense strand in the third intracellular loop of the α_2-adrenergic receptor. The β_2-template-generated PCR fragment is digested with *Dra*I

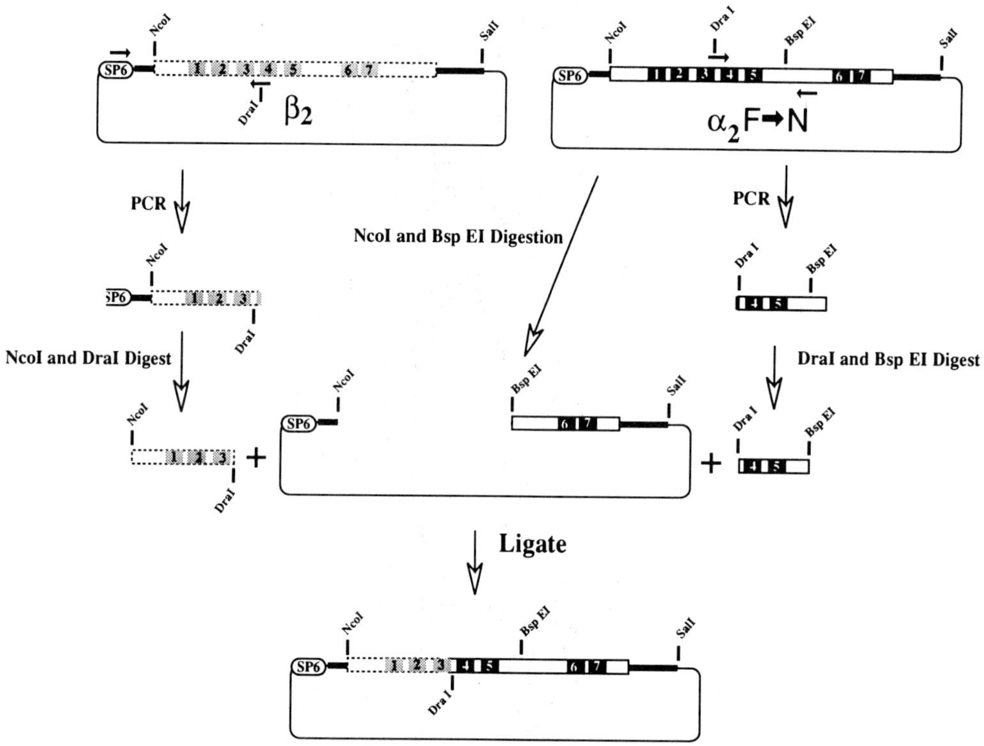

FIG. 4 Construction of chimer S7F^{412}→N by engineering silent mutations. Restriction sites for *Dra*I were created at the 3′ end of β_2 receptor fragment and at the 5′ end of α_2 receptor fragment by PCR as described in the text. The PCR-generated fragments and the plasmid pGEMα_2F→N were digested with the restriction enzymes and ligated as shown. For explanation of other notations, see legend to Fig. 2.

plus *Nco*I and the α_2 template-generated PCR fragment is digested with *Dra*I plus *Bsp*EI. The resulting fragments are ligated as shown in Fig. 4.

Expression of Chimeric Receptors in Mammalian Cells

The functional properties of chimeric molecules can be studied by expressing them in tissue culture cells. The system used for the expression should be simple, rapid, and efficient. For expression of adrenergic receptor chimers, we have used *Xenopus laevis* oocytes, COS-7 cells, Raji cells, and HEK 293 cells.

Expression in Xenopus laevis Oocytes

Chimeric receptor genes ligated into the pSP65 vector (1) are expressed in *Xenopus* oocytes as described for wild-type α_2- and β_2-adrenergic receptors (23, 26). The vector is linearized with *Hin*dIII or *Sal*I, which cuts at the 3′ end of the untranslated sequences, and mRNA is prepared using SP6 RNA polymerase (Boehringer-Mannheim) according to the method of Melton *et al.* (27). The transcription mixture consists of 20 m*M* Tris–Cl, pH 7.5; 50 m*M* NaCl; 10 m*M* $MgCl_2$; 2 m*M* spermidine-HCl; 10 m*M* dithiothreitol; 100 μg/ml bovine serum albumin; 0.6 m*M* each of ATP, CTP, and UTP; 0.05 m*M* GTP; 0.5 m*M* $m^7G(5')pppG$ sodium salt (Pharmacia); 100 μg of linearized template DNA/ml; 1 unit/μl RNasin (Promega); and 900 units of SP6 RNA polymerase/ml. The reaction is carried out at 37°C for 3 hr, then extracted with phenol and chloroform, and ethanol precipitated. The RNA (~10 μg/25 μg template) along with the template DNA is dissolved in diethylpyrocarbonate-treated water for oocyte injection.

The oocytes are obtained from adult female *X. laevis* supplied 3–4 weeks after hCG injection (NASCO, Fort Atkinson, WI). Frogs are sacrificed within 8 weeks of injection and the harvested oocytes are treated with 2 mg/ml collagenase type 1a (Sigma) in sterile Barth's solution (28), for 1–2 hr at room temperature. Oocytes were allowed to sit overnight at room temperature prior to selection of stage V-VI oocytes for injection. The oocytes are injected with 50–100 μl of RNA solution (5–10 ng of RNA) and then allowed to sit for 48 hr in Barth's solution supplemented with pyruvate (0.01 m*M*), bovine γ-globulin (0.05 mg/ml), penicillin (100 μl/ml), and streptomycin (100 μg/ml). The oocytes (500/ml) are homogenized in ice-cold assay buffer (75 m*M* Tris–Cl, pH 7.4, 12.5 m*M* $MgCl_2$, 1 m*M* EDTA) containing 30% (wt/vol) sucrose, using 30 strokes of a Dounce homogenizer. The pigment granules of the homogenate are removed by centrifugation at 3000*g* for 10 min and the supernatant is centrifuged again at 10,000*g* for 10 min. The supernatant thus obtained is saved and the pellet is resuspended in 1 ml of the same buffer and recentrifuged at 10,000*g* for 10 min. This supernatant is combined with the one saved in the previous step and centrifuged at 400,000*g* for 30 min in a Beckman TL-100 ultracentrifuge. The pellet is washed twice with the assay buffer and then resuspended in the same buffer at a concentration of 1000 oocytes per 0.5–1 ml.

Expression in COS-7 Cells

Chimeric receptors are expressed transiently in COS-7 cells using DEAE–dextran-mediated transfection (1, 29). The genes coding for the chimeric receptors are transferred from pSP65 or pGEM vector into the *Nco*I and *Sal*I sites of the modified eukaryotic expression vector pBC12BI (29).

COS-7 cells are maintained in Dulbecco's modified Eagle's medium (DMEM) supplemented with 10% fetal calf serum (FCS) and 25 mg/liter gentamicin (Boehringer-Mannheim). One day before transfection, the medium is aspirated from the flask (150 cm^2) and cells are rinsed with 10 ml of sterile phosphate-buffered saline (PBS) at room temperature. The cells are harvested by adding 5 ml of PBS and 0.5 ml of 10× trypsin-EDTA (Sigma); the flasks are rocked intermittently until all the cells are detached from the surface (requires about 10 min). An equal volume of DMEM is added to the cell suspension and mixed gently by pipetting up and down to break all the clumps of cells. An aliquot is taken and the number of cells per milliliter is determined by counting on a hemocytometer after staining with trypan blue. An aliquot of 10×10^6 cells is pipetted into each of the 150-cm^2 flasks; 30 ml of DMEM is added and left overnight at 37°C.

On the next day, all the medium is aspirated from the flask and rinsed with 10 ml of PBS at room temperature. The transfection mixture [10 ml PBS + 50 μg DNA in 10 mM Tris–Cl, 1 m*M* EDTA, pH 8.0 + 0.25 ml of sterile DEAE–dextran (Sigma, average molecular weight of 500,000) stock (100 mg/5 ml PBS)] is added and incubated at 37°C for 30 min. Then, 9 ml of the transfection mixture is removed from the flask, and 10 ml of chloroquine medium [5.16 mg/100 ml of DMEM, made from a 258 mg/5 ml sterile stock solution of diphosphate salt of chloroquine (Sigma) in PBS] is added and incubated for 2 hr at 37°C. After 2 hr, the chloroquine medium is removed and 10 ml of 10% sterile dimethyl sulfoxide (DMSO) in DMEM is added and incubated at 37°C for 2.5 min. The DMSO medium is removed and replaced with 30 ml of regular DMEM with FCS and gentamicin and incubated at 37°C overnight, and the medium is changed the next day. The cells are harvested 3 days after transfection and the membranes are prepared for binding assays as described below.

The transfected cells are rinsed twice with 5 ml of cold PBS and then 5 ml of ice-cold lysis buffer (10 m*M* Tris–Cl, pH 7.4, 1 m*M* EDTA) is added. The cells are scraped off the flask with a disposable scraper and collected into a 50-ml centrifuge tube. The flask is rinsed with 5 ml more of lysis buffer and the washings are pooled. The cells are homogenized by 4× 5-sec bursts at full speed using a polytron (Virtis Co. Inc., New York). The nuclei are pelleted by centrifugation at 200*g* for 5 min at 4°C. The supernatant is transferred to another tube and centrifuged at 16,000 rpm in a Sorvall SS34 rotor for 30 min. The pelleted membranes are resuspended in an appropriate volume of binding buffer (75 m*M* tris, 12.5 m*M* $MgCl_2$, 1 m*M* EDTA, pH 7.4). For freezing, the membrane pellet is resuspended in a minimum volume of binding buffer, aliquoted into 1-ml screw-capped Eppendorf tubes, and stored at −80°C or in liquid nitrogen.

Expression in Raji Cells

The Raji line of lymphoblast-like cells is established from a Burkitt lymphoma (ATCC CC1 86). We found no evidence of the presence of either α_2- or β_2-adrenergic receptor as judged by radioligand binding or the stimulation of adenylate cyclase by β-adrenergic receptor agonists in untransfected Raji cells. Wild-type and chimeric receptor genes are transferred from the pGEM vector into the *Hin*dIII and *Sal*I sites of the eukaryotic expression vector pWS290 (a gift from Bill Sugden, University of Wisconsin, Madison). This expression vector is 7.8 kilobase pairs in length and is a derivative of pHEBo (30). The wild-type or the chimeric receptor is under the control of a strong promoter regulatory region of human cytomegalovirus (31). The vector contains a *cis*-acting element, *ori P* of Epstein–Barr virus (EBV), which is responsible for the maintenance of the vector as a plasmid when introduced into EBV-transformed lymphoblasts. In addition, the plasmid contains the Amp^r gene and *hph* gene (which confers hygromycin B resistance to transfected eukaryotic cells) under the control of the transcriptional promoter and polyadenylation signals of the thymidine kinase gene of herpes simplex virus type 1.

Raji cells are transfected with the plasmid DNA by electroporation. The cells are maintained in 75-cm^2 flask using RPMI 1640 medium supplemented with 10% FCS and 25 mg/liter gentamicin, until a density of ~500,000 cells/ml is reached. Cells are counted and pelleted by centrifugation at 200*g* for 5 min. Then, the cells are resuspended in RPMI medium with FCS and gentamicin to a density of 4.0×10^7 cells/ml. Plasmid DNA (20 μg) in 10 m*M* Tris–Cl, 1 m*M* EDTA, pH 8.0, and 300 μl of this cell suspension are added to a prechilled electroporation cuvette (0.4 cm electrode, Bio-Rad), mixed well, and left on ice for 10 min. Then the cells are electroporated with a capacitance of 960 μF and a voltage of 0.20 kV using a Bio-Rad gene pulser. This gives a resistance–capacitance time constant of 50–70 msec. It is important to mix the cells immediately before electroporation. The electroporated cells are placed in 50 ml of medium in a 75-cm^2 flask. After 24 hr at 37°C, 100 μl of hygromycin B solution is added. To prepare the hygromycin B stock solution, 1 g of solid hygromycin B (Boehringer or Calbiochem) is dissolved in 5 ml of sterile PBS and 200 μl of glacial acetic acid is added. The pH of the solution should be between 6 and 8. Extreme care is required while handling hygromycin B as it is very toxic to humans. Seven days after electroporation, control cells electrophoresed without DNA begin to die while cells transfected with pWS290 vector are viable. The cells are collected 7 days after transfection by centrifugation at 200*g* and membranes for binding assays are prepared as described for COS cells. The expressed β_2-adrenergic receptor can be detected readily on the third day of transfection and the levels of expression peak around Day 9 (32). Although the expression level drops slightly after the ninth day, it plateaus at 6–8 pmol/mg protein for 3 weeks. Thereafter, the expression continues to decrease; however, β_2 receptor binding can be detected up to 50 days following electroporation. No specific binding of radioligand is observed in Raji cells transfected with the vector alone. This quasi-stable expression system is very useful for the expression of chimeric receptors. The levels of expression obtained in Raji cells are similar to that in COS cells. Normally about 10% of the transfected COS cells express the receptor. In contrast, the Raji cells harboring the transfected plasmid can be selected by their resistance to hygromycin B. Like stably transfected cell lines, transfected Raji cells can be frozen in liquid nitrogen and can be revived for future use. As described before (32), expressed β_2-adrenergic receptor is functionally coupled to adenylate cyclase system in Raji cells; therefore, this cell line can be used to express chimeric receptors in order to delineate the domains involved in signal transduction.

Expression in HEK 293 Cells

Chimeric receptors are expressed transiently in human kidney HEK 293 cells (ATCC CRL 1573) using calcium phosphate-mediated transfection (33). The genes coding for the chimeric receptors are transferred from pSP65 or pGEM vector into the *Nco*I and *Sal*I sites of the modified eukaryotic expression vector pBC12BI (29). HEK 293 cells are maintained in DMEM supplemented with 10% FCS and 25 mg/liter gentamicin (Boehringer-Mannheim). As described for COS-7 cells, the cells are trypsinized and an aliquot of 10×10^6 cells is pipetted into each of the 150-cm^2 flasks; 30 ml of DMEM is added and left overnight at 37°C.

On the next day, 20 μg of plasmid DNA in 0.5 ml of Tris–EDTA (1 m*M* Tris, 0.1 m*M* EDTA, pH 8.0) is mixed with 568 ml of 2 × HEPES-buffered saline (280 m*M* NaCl, 10 m*M* KCl, 1.5 m*M* $Na_2HPO_4 \cdot 2H_2O$, 12 m*M* dextrose, 50 m*M* HEPES, pH adjusted to 7.05 with NaOH solution). An aliquot (70.4 ml) of 2 *M* aqueous solution of $CaCl_2$ is added slowly to sample mixture and incubated for 30 min at room temperature. Meanwhile, all the medium is aspirated from the flask containing the HEK 293 cells and rinsed with 10 ml of PBS at room temperature. After 30 min, the sample is added with 11.2 ml of DMEM medium with 10% fetal calf serum, transferred to the flasks and incubated overnight at 37°C. The medium is changed the next day and the cells are harvested 3 days after transfection as described for COS-7 cells.

Ligand-Binding Assays

Binding experiments are performed in 0.5- or 2-ml volumes of 75 mM Tris–Cl, 12.5 mM $MgCl_2$, and 1 mM EDTA, pH 7.4, for 60 or 90 min at room temperature, using [^{125}I]cyanopindolol (2200 Ci/mmol, DuPont-NEN), [^{3}H]dihydroalprenolol (110 Ci/mmol, Amersham Corp.), or [^{3}H]yohimbine (70.5 Ci/mmol, DuPont-NEN). The assays are performed in 15–25 μg of oocyte membrane protein (equivalent to about three to five oocytes, 50–100 μg of COS or HEK 293 membrane protein, or 5–10 μg of Raji membrane protein). The protein content of the membrane preparations is determined by using the Bio-Rad dye-binding reagent. The bound radioactivity is separated from free by filtration through GF/C filters (Whatman). Saturation isotherms are performed by incubating the membranes with varying concentrations of the radioligand. Competition experiments are carried out by incubating membranes with varying concentrations of competing ligand and a concentration of a radioligand at or below its K_d for the receptor. Nonspecific binding is determined by adding 1 μM (−)alprenolol or yohimbine. Equilibrium dissociation constants are determined from saturation isotherms and competition curves. Saturation isotherm data are analyzed by a nonlinear least-square curve-fitting technique, and competition data are analyzed according to a four-parameter logistic equation to determine EC_{50} values using InPlot software (GraphPAD Software Inc., San Diego, CA).

Determination of Adenylate Cyclase Activity

Isoproterenol- or epinephrine-stimulated adenylate cyclase activity of oocyte membranes, prepared as described under Expression of *Xenopus laevis* Oocytes above, from oocytes injected with wild-type β_2-adrenergic receptor or chimeric receptors, is determined as described previously (34). Each determination is performed on membranes from 25–35 mRNA-injected oocytes using 2.5 μCi of [α-^{32}P]ATP (0.12 mM) per assay. The separation of [^{32}P]cAMP from the substrate and other ^{32}P-containing compounds is achieved by sequential chromatography on columns of Dowex cation exchange resin and aluminum oxide as described in Salomon *et al.* (35). The protocol which is currently being followed in our laboratory for the determination of adenylate cyclase activity in Raji cells is described in detail below.

The total volume of the assay mixture is 50 μl and consists of 10 μl of agonist, NaF, or forskolin stock solution, 20 μl of cyclase cocktail, and 20 μl of membrane preparation. The cyclase cocktail consists of 6.75 mM mono(cyclohexylammonium) phosphoenolpyruvate, 132.5 μM GTP·3Na·$3H_2O$, 250 μM cAMP Na salt, 0.5 IU pyruvate kinase, 2.5 IU myokinase,

120 μM ATP (pH 7.0, Pharmacia), and 12.5 μCi [α-^{32}P]ATP (3000 Ci/mmol, DuPont-NEN) per 100 μl of cocktail. Before cold and radioactive ATP are added, the pH of the cocktail is adjusted to ~7 by adding sodium hydroxide solution. The cocktail without the cold and radioactive ATP can be stored at −80°C in smaller aliquots. Both cold and radioactive ATP have to be added immediately before use.

Raji cells are centrifuged at 200*g* for 5 min at 4°C and the pellet is resuspended in a small volume of ice-cold assay buffer (75 m*M* Tris, 25 m*M* $MgCl_2$, 2 m*M* EDTA, pH 7.4). The cell suspension is transferred to an Eppendorf tube and recentrifuged at 2000 rpm at 4°C for 5 min. The pellet is again resuspend in ice-cold assay buffer to get a cell density of one-half million per 20 μl and the cell suspension is saved on ice. Immediately before being added to the reaction mixture, the cells are homogenized about 15 times in a 2-ml glass homogenizer. After addition of the membrane preparation, the tubes are incubated in a shaking water bath at 37°C for 10 min. The basal activity of the system is determined by substituting water for the ligand solution.

The reaction products are processed as described by Alvarez and Daniels (36). The reaction is stopped by the addition of 20 μl of stop solution [2.2 *N* HCl containing 0.005 μCi [^{3}H]cAMP (DuPont-NEN) per 20 μl]. Twenty microliters of stop solution is also added to a scintillation vial to determine the efficiency of column elution. The reaction tube is centrifuged at 12,000 rpm for 30 sec in a microfuge and 65 μl of the supernatant is transferred onto a disposable column (E & K Scientific Products, Inc., Campbell, CA) filled with 1.3 g of neutral alumina (Sigma). cAMP is eluted from the columns with 4 ml of 0.1 *N* ammonium acetate, pH 7.0, and the eluate is directly collected into a scintillation vial containing 12 ml of Aquasol (DuPont-NEN). The contents in the vial are thoroughly mixed and counted for ^{32}P and ^{3}H in a scintillation counter. More than 80% cAMP elution efficiency can be achieved by this single column procedure. The cyclase activity data are expressed as a fraction of the basal activity, that is, the difference between the agonist stimulated value (X) and the basal value (Y) divided by the basal value $[(X-Y)/Y]$.

No functional coupling of either α_2- or β_2-adrenergic receptor to the adenylate cyclase was observed in COS cells. Unlike β_2-adrenergic receptor, a functional interaction of adenylate cyclase with the α_2-adrenergic receptor expressed in *Xenopus* oocytes has not been observed (1). The adenylate cyclase activity of the Raji cells transfected with β_2-adrenergic receptor can be stimulated by agonists in a dose-dependent manner (32). As expected, the order of potency of agonists in stimulating adenylate cyclase through β_2 adrenergic receptor is isoproterenol>epinephrine>>*p*-aminoclonidine. The functional coupling of α_2-adrenergic receptor with the adenylate cyclase system of Raji cells has not yet been tested.

Immunofluorescence Staining of Transfected COS-7 Cells

The receptor protein in transfected COS cells grown on glass coverslips is localized by indirect immunofluorescence (22, 37–39). Primary antibody is a rabbit antiserum directed to the carboxyl terminal 15 residues of the human β_2 receptor. The α_2 and $\alpha_2F^{412}{\rightarrow}N$ were tagged with an epitope of nine amino acids (sequence YPYDVPDYA) at the amino terminus by ligating an oligonucleotide linker adapter into the *Nco*I site located at the 5′ end of the receptor coding sequence. The epitope-tagged receptors are localized in transfected COS-7 cells using the monoclonal antibody 12CA5 (BABCO, Berkeley, CA) as described before (37, 39). The secondary antibody is a Texas red conjugate of goat anti-rabbit IgG (Jacson ImmunoResearch, West Grove, PA). The transfected cells on coverslips are fixed with 4% formaldehyde (freshly prepared from paraformaldehyde in phosphate-buffered saline). Fixed specimens are permeabilized at room temperature for 20 min with 0.2% Nonidet-P40 in a blocking buffer containing 5% nonfat dry milk in 50 m*M* Tris–Cl, pH 7.6. Primary antibody, diluted 1/600, is applied to the specimens in the same blocking buffer. Secondary antibody (goat anti-rabbit F(ab′)2 fragment of IgG conjugated to Texas red) is diluted 1/500 and applied in the same buffer and incubate at room temperature for 20 min. Immunofluorescence microscopy is performed using standard epifluorescence optics (Zeiss Axiophot) or by a confocal laser scanning microscope (Noran Odyssey) equipped with a Nikon 60X/NA 1.4 objective.

Concluding Remarks

Identification of Functional Domains

The construction and analysis of chimeric receptors can be a very effective approach to defining functional domains of structurally homologous proteins. The application of the chimeric protein approach to human α_2- and β_2-adrenergic receptors resulted in the identification of the seventh hydrophobic domain as the site of receptor-specific binding of ligands, especially antagonists (1). A comparison of the amino acid sequences of the seventh hydrophobic domain of all the known α- and β-adrenergic receptors revealed that an Asn residue is conserved in all β receptors and a Phe residue is conserved in all α receptors. Site-directed mutagenesis of Phe to Asn in α_2-adrenergic receptor by using an oligonucleotide with base mismatches and the polymerase chain reaction technique described above, resulted in a mutant receptor ($\alpha_2F^{412}{\rightarrow}N$) with at least 300 times higher affinity for a class of β receptor antagonists compared to the wild-type α_2 receptor and considerably reduced affinity for α receptor antagonists such as yohimbine, rauwolscine, and atipamezole (22). Analysis of the binding of several β antagonists to $\alpha_2F \rightarrow N$ suggested that Asn^{312} in the β-adrenergic receptor is involved in forming a hydrogen bond with the phenoxy oxygen of β receptor antagonists such as alprenolol, propranolol, and pindolol. This hypothesis was supported by the observation that the β antagonist sotalol, which does not contain this phenoxy oxygen, did not show high affinity binding to $\alpha_2F \rightarrow N$ (22). The change in free energy $[\Delta(\Delta G) = RT \ln [K_d (\alpha_2)/K_d (\alpha_2F \rightarrow N)] = 3.3\text{-}4.7$ kcal/mol] associated with the increased affinity of $\alpha_2F \rightarrow N$ for β-adrenergic receptor antagonists is consistent with formation of a hydrogen bond between the Asn^{312} residue and the ligand. Moreover, when the Asn residue in the seventh hydrophobic domain of the β_2-adrenergic receptor was substituted with structurally different amino acid residues, only those residues capable of forming a hydrogen bond through their side chain (Gln or Thr) supported the antagonist binding but not Ala or Phe (40). Similarly experiments performed on β_2 receptor suggested that Phe^{412} in the homologous position of the seventh hydrophobic domain of α_2 receptor confers specificity for yohimbine binding (38, Fig. 5).

Deriving Structural Information from Chimeric Receptors

It has been observed by us and several others that not all chimers constructed from two different receptors are functional (they do not bind ligands or couple to G proteins) (1, 38, S. Suryanarayana and B. K. Kobilka, unpublished results). This is most likely due to a major structural incompatibility between α_2/β_2 receptor domains resulting in improper folding and retention of the chimer in the endoplasmic reticulum. Retention of the chimer in the endoplasmic reticulum can be identified by immunocytochemical techniques or by determining the sensitivity of protein to endoglycosidase H (22, 38, 41, see below). Until the three-dimensional structure of both of the parent receptors is known it is not possible to predict the compatibility of structural domains from two receptors. An interesting example is that a chimer (CR3) consisting of hydrophobic domains 1 through 5 from α_2 adrenergic receptor and 6 and 7 from β_2 adrenergic receptor when expressed in mammalian cells was able to bind to ligands (1). A complementary chimer (CRS1) consisting hydrophobic domains 1 through 5 from β_2 adrenergic receptor and 6 and 7 from α_2 adrenergic receptor did not show binding to any ligands and was incapable of mediating agonist-stimulated adenylylcyclase. It is occasionally possible to make further structural changes to these nonfunctional, improperly folded chimers to produce functional molecules. This may lead to the

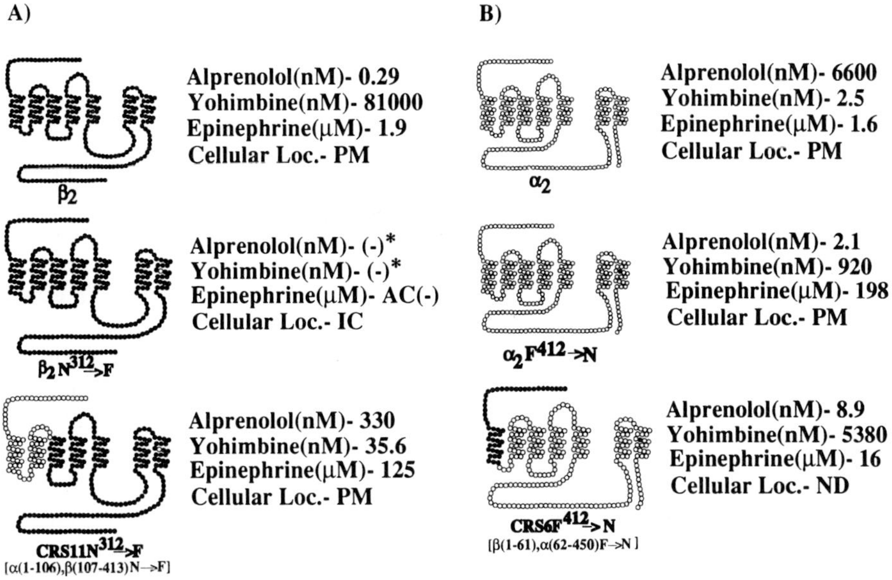

FIG. 5 Pharmacological properties and subcellular localization of human α_2, β_2-adrenergic receptors, the mutants $\alpha_2F^{412}{\rightarrow}N$ and $\beta_2F^{312}{\rightarrow}N$, and the α_2/β_2 chimeric receptors expressed in Raji and COS-7 cells (38). The numbers are equilibrium dissociation constants (K_d or K_i). The receptor protein in transfected COS-7 cells was localized by immunofluorescence microscopy as described for Figure 5. Cellular Loc., subcellular localization of the receptor. *No specific binding of [^{3}H]dihydroalprenolol or [^{3}H]yohimbine was observed up to a concentration of 64 nM. Epinephrine competition studies on wild-type α_2 receptor, $\alpha_2F^{412}{\rightarrow}N$, and CRS6F$^{412}{\rightarrow}$N were performed in the presence of 100 μM Gpp(NH)p to prevent the influence of G proteins on agonist binding affinity. PM and IC, the protein was localized to plasma membrane or an intracellular compartment, respectively; AC(−), no stimulation of adenylylcyclase activity was observed in presence of (−) isoproterenol, (−) epinephrine, or p-aminoclonidine up to a concentration of 1 mM; ND, not determined.

identification of the incompatible domains and thus provide evidence for intramolecular interactions and insights into the three-dimensional arrangement of the hydrophobic domains. This approach is illustrated below.

We recently reported that when Asn^{312} in the seventh hydrophobic domain of the β_2 receptor is replaced by Phe ($\beta_2N^{312}{\rightarrow}F$), the amino acid found in the homologous region of the α_2 receptor, there is a complete loss of function (22). Not only did this mutant receptor fail to bind agonist and antagonist ligands, but immunocytochemical localization in transfected cells revealed that, unlike wild-type α_2 and β_2 receptors, $\beta_2N^{312}{\rightarrow}F$ was not delivered to the plasma membrane and was retained in a reticular intracellular compartment (Figs. 6A and 6B). These observations suggest that a structural incompatibility exists between Phe at position 312 and an adjacent structural domain in $\beta_2N^{312}{\rightarrow}F$, the amino acid sequence of which is different in α_2 and β_2 receptors. If this hypothesis is correct, it should be possible to improve the folding of this β_2 receptor mutant by replacing the structural domain adjacent to the Phe^{312} in $\beta_2N^{312}{\rightarrow}F$ with sequence from the α_2 receptor.

We searched for such an interacting domain by constructing a series of chimeric receptors based on the $\beta_2N^{312}{\rightarrow}F$ mutation and containing different hydrophobic domains derived from the α_2 receptor (38). We found that replacing the first two hydrophobic domains of the mutant receptor with corresponding α_2 receptor sequence ($CRS11N^{312}{\rightarrow}F$) resulted in restoration of plasma membrane localization (Fig. 6C) and recovery of both agonist and antagonist binding (Fig. 5). Studies performed on additional chimers and the converse mutation in α_2-adrenergic receptor described below suggested a specific interaction between the seventh and the first hydrophobic domains.

We have previously reported that changing Phe^{412} in the seventh hydrophobic domain of the human α_2 receptor to Asn ($\alpha_2F^{412}{\rightarrow}N$), which is found in the homologous position of the β_2 receptor, led to a 300- to 3000-fold increase in affinity for a class of β receptor antagonists represented by alprenolol (22, Fig. 5). This mutation also led to 200-fold loss of affinity for epinephrine, an agonist equally active at both α_2 and β_2 receptors. The results presented in Fig. 5 indicate that exchanging only the first hydrophobic domain with the corresponding β_2 receptor sequence can complement the $F^{412}{\rightarrow}N$ mutation in $\alpha_2F^{412}{\rightarrow}N$ resulting in more than 10-fold increase in affinity for epinephrine.

In conclusion, the results of these experiments in which functional abnormalities resulting from a mutation in the seventh hydrophobic domain are complemented by sequence from the first hydrophobic domain are consistent with a structural model in which Asn^{312} of the β_2 receptor and Phe^{412} of the

FIG. 6 Immunocytochemical localization of wild-type β_2 (A), $\beta_2N^{312}{\rightarrow}F$ (B), and $CRS11N^{312}{\rightarrow}F$ (C) receptors expressed in COS-7 cells. The receptor proteins were localized by indirect immunofluorescence using a rabbit antiserum directed to the carboxyl terminus of the β_2 receptor as primary antibody and a Texas red conjugated goat anti-rabbit IgG as secondary antibody as described in the text (38). Specimens were imaged using a Noran Odyssey confocal microscope, with the plane of focus adjusted 2 μm above the surface of the coverslip.

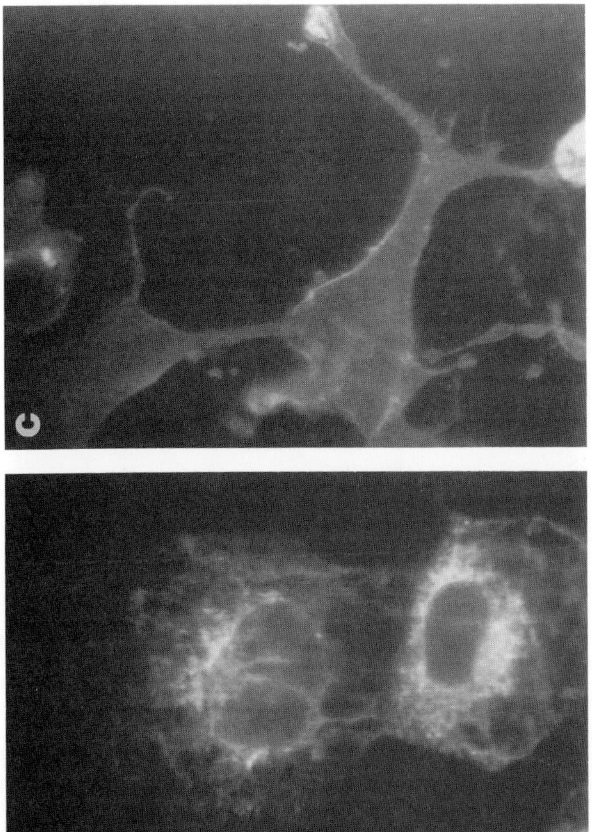
C
B
A

α_2 receptor form important structural interactions with the first hydrophobic domain. The arrangement of transmembrane domains predicted by these experiments is similar to that found in bacteriorhodopsin (42). This approach should facilitate the formulation and testing of more accurate models for this class of membrane proteins.

Acknowledgment

This research was supported by Howard Hughes Medical Institute and National Institutes of Health Grant 5 RO1 NS28471.

References

1. B. K. Kobilka, T. S. Kobilka, K. Daniel, J. W. Regan, M. G. Caron, and R. J. Lefkowitz, *Science* **240,** 1310 (1988).
2. T. Frielle, K. W. Daniel, M. G. Caron, and R. J. Lefkowitz, *Proc. Natl. Acad. Sci. U.S.A.* **85,** 9494 (1988).
3. S. Marullo, L. J. Emorine, A. D. Strosberg, and C. Delavier-Klutchko, *EMBO J.* **9,** 1471 (1990).
4. B. P. England, M. S. Ackerman, and R. W. Barrett, *FEBS Lett.* **279,** 87 (1991).
5. J. Wess, D. Gdula, and M. R. Brann, *Mol. Pharmacol.* **41,** 369 (1992).
6. M. S. Choudhary, S. Craigo, and B. L. Roth, *Mol. Pharmacol.* **42,** 627 (1992).
7. R. G. MacKenzie, M. E. Steffey, A. M. Manelli, N. J. Pollock, and D. E. Frail, *FEBS Lett.* **323,** 59 (1993).
8. K. Takahashi, K. Tsuchida, Y. Tanabe, M. Masu, and S. Nakanishi, *J. Biol. Chem.* **268,** 19341 (1993).
9. Y. Yokota, C. Akazawa, H. Ohkubo, and S. Nakanishi, *EMBO J.* **11,** 3585 (1992).
10. M. Adachi, Y.-Y. Yang, A. Trzeciak, Y. Furuichi, and C. Miyamoto, *FEBS Lett.* **311,** 179 (1992).
11. H. D. Perez, R. Holmes, L. R. Vilander, R. R. Adams, W. Manzana, D. Jolley, and W. H. Andrews, *J. Biol. Chem.* **268,** 2292 (1993).
12. B. S. Sachias, R. M. Snider, J. A. Lowe, III, and J. E. Krause, *J. Biol. Chem.* **268,** 2319 (1993).
13. U. Gether, T. E. Johansen, R. M. Snider, J. A. Lowe, III, S. Nakanishi, and T. W. Schwartz, *Nature (London)* **362,** 345 (1993).
14. M. Beinborn, Y.-M. Lee, E. W. McBride, S. M. Quinn, and A. S. Kopin, *Nature (London)* **362,** 348 (1993).
15. Z. Fath, R. V. Benya, H. Shapira, R. T. Jensen, and J. F. Battey, *J. Biol. Chem.* **268,** 14622 (1993).
16. O. Quehenberger, E. R. Prossnitz, S. L. Cavanagh, C. G. Cochrane, and R. D. Ye, *J. Biol. Chem.* **268,** 18167 (1993).

17. Y. Nagayama, H. L. Wadsworth, G. D. Chazenbalk, D. Russo, P. Seto, and B. Rapoport, *Proc. Natl. Acad. Sci. U.S.A.* **88,** 902 (1991).
18. W. R. Moyle, M. P. Bernard, R. V. Myers, O. M. Marko, and C. D. Strader, *J. Biol. Chem.* **266,** 10807 (1991).
19. Y. Nagayama, D. Russo, H. L. Wadsworth, G. D. Chazenbalk, and B. Rapoport, *J. Biol. Chem.* **266,** 14926 (1991).
20. T. Braun, P. R. Schofield, and R. Sprengel, *EMBO J.* **10,** 1885 (1991).
21. J. C. Semenza and H. R. B. Pelham, *J. Mol. Biol.* **224,** 1 (1992).
22. S. Suryanarayana, D. A. Daunt, M. von Zastrow, and B. K. Kobilka, *J. Biol. Chem.* **266,** 15488 (1991).
23. B. K. Kobilka, C. MacGregor, K. Daniel, T. S. Kobilka, M. G. Caron, and R. J. Lefkowitz, *J. Biol. Chem.* **262,** 15796 (1987).
24. B. K. Kobilka, R. A. F. Dixon, T. Frielle, H. G. Dohlman, M. A. Bolanowski, I. S. Sigal, T. L. Yang-Feng, U. Francke, M. G. Caron, and R. J. Lefkowitz, *Proc. Natl. Acad. Sci. U.S.A.* **84,** 46 (1987).
25. W. Block, *Biochemistry* **30,** 2735 (1991).
26. B. K. Kobilka, H. Matsui, T. S. Kobilka, T. L. Yang-Feng, U. Francke, M. G. Caron, R. J. Lefkowitz, and J. W. Reagan, *Science* **238,** 650 (1987).
27. D. N. Melton, P. A. Krieg, M. R. Rebagliati, T. Maniatis, K. Zinn, and M. R. Green, *Nucleic Acids Res.* **12,** 7035 (1984).
28. J. B. Gundon, *J. Embryol. Exp. Morphol.* **20,** 401 (1968).
29. B. R. Cullin, *in* "Methods in Enzymology" (S. L. Berger and A. R. Kimmel, eds.), Vol. 152, p. 684. Academic Press, Orlando, FL, 1987.
30. B. Sugden, K. Marsh, and J. Yates, *Mol. Cell. Biol.* **5,** 410 (1985).
31. M. F. Stinski and T. J. Roehr, *J. Virol.* **55,** 431 (1985).
32. S. Suryanarayana and B. K. Kobilka, *Methods: Companion Methods Enzymol.* **3,** 193 (1991).
33. F. L. Graham and A. J. van der Eb, *Virology* **52,** 456 (1973).
34. R. J. Lefkowitz, *J. Biol. Chem.* **249,** 6119 (1974).
35. Y.Salomon, C. Londos, and M. Rodbell, *Anal. Biochem.* **58,** 541 (1974).
36. R. Alvarez and D. V. Daniels, *Anal. Biochem.* **187,** 98 (1990).
37. M. von Zastrow and B. K. Kobilka, *J. Biol. Chem.* **267,** 3530 (1992).
38. S. Suryanarayana, M. von Zastrow, and B. K. Kobilka, *J. Biol. Chem.* **267,** 21991 (1992).
39. M. von Zastrow, R. Link, D. Daunt, G. Barsh, and B. K. Kobilka, *J. Biol. Chem.* **268,** 763 (1993).
40. S. Suryanarayana and B. K. Kobilka, *Mol. Pharmacol.* **44,** 111 (1993).
41. A. J. Dorner and R. J. Kaufman, *in* "Methods in Enzymology" (D. Goeddel *et al.,* eds.), vol. 185, p. 577. Academic Press, San Diego, CA, 1990.
42. R. Henderson, J. M. Baldwin, T. A. Ceska, F. Zemlin, E. Beckmann, and K. H. Dowing, *J. Mol. Biol.* **213,** 899 (1990).

[16] Mapping G Protein Coupling Domains by Site-Specific Peptides

Dieter Palm, Gerald Münch, and Daria Malek

Introduction

Transmembrane signaling pathways in general require a signal transduction coupling mechanism to link external signals to cellular responses. More specifically, G protein-coupled receptors (GPCRs) make use of regulatory guanosine nucleotide binding proteins to transmit stimulation of the receptor by the first messenger (hormone, neurotransmitter, agonist) to the effector (enzyme or ion channel) (1–4). In all consecutive steps the signal is transmitted by "coupling" through protein–protein interactions. If two proteins interact it appears reasonable that small peptides derived from the putative polypeptide contact domains might be true substitutes or competitors of the larger entities making them useful tools to study the specific properties of the interaction. "Mapping G protein coupling domains by site-specific peptides" provides the means to establish a functional topography of receptor interaction sites.

Receptors that act via G proteins share substantial structural homology, including rhodopsin which, by its similarity to bacteriorhodopsin, provides the only structural model at the molecular level (5). Rhodopsin consists of a tight assembly of seven hydrophobic, essentially α-helical membrane-spanning segments connected alternatively by internal and external hydrophilic loops. Sequence homology among the receptors is concentrated in membrane-spanning regions. The consensus regions serve to align the family of GPCRs (6). Also parts of the intracellular loops, especially short regions adjacent to the membranes, are conserved, however, to a lesser degree. The highly conserved membrane-spanning regions create the ligand-binding pocket; correspondingly, it can be assumed that the conserved portions of the intracellular loops form at least part of the G protein coupling sites. Since the specificity of signal transmission may be determined by the specificity of G protein coupling, some GPCR regions that are necessary for coupling also determine the specificity of the G protein interaction. This involves conserved receptor and G protein motifs.

G proteins also share a number of conserved structural and functional characteristics (1,3,7–10). Each G protein is a heterotrimer of a guanosine nucleotide-binding α subunit and a tight complex of $\beta\gamma$ subunits. The α, β

Methods in Neurosciences, Volume 25

and γ subunits each constitute families of homologous proteins (9). Sequence and functional similarities of α subunits suggest that they also share definable functional domains, e.g., single or multiple receptor-binding sites (R-domain), nucleotide-binding sites (G-domain), effector-binding sites (E-domain), and $\beta\gamma$-binding sites (BG-domain) (1, 2, 4, 11). Correspondingly, $\beta\gamma$ subunits dispose of receptor (12)- and effector-binding sites as well as α subunit-binding sites (4, 10).

Three types of studies on structural aspects of receptor–G protein coupling have been performed: (i) Deletion and site-directed mutagenesis studies implicate regions that are necessary for coupling. (ii) Chimera experiments delineate the receptor regions that determine the specificity of G protein coupling. Effects caused by the preceding changes could however be indirectly geared by changes in overall structure or ligand interaction. (iii) In contrast, competition of coupling by synthetic peptides can easily be adapted to be performed under physiological conditions and will provide first-hand evidence on peptide segments directly interacting with G proteins.

To relate single steps of the reaction pathway to a function, it will be necessary to review briefly the receptor-dependent G protein activation outlined in Scheme 1 (1):

$$\underset{\mathbf{I}}{\mathrm{G\cdot GDP}} + \mathrm{HR} \xleftrightarrow{\mathrm{Mg^{2+}}} \mathrm{GDP} + \underset{\mathbf{II}}{\mathrm{HR\cdot G}} + \mathrm{GTP} \leftrightarrow$$

$$[\underset{\mathbf{III}}{\mathrm{HR\cdot G\cdot GTP}} \xrightarrow{\mathrm{Mg^{2+}}} \underset{\mathbf{IV}}{\mathrm{HR\cdot G^{*}GTP}}] \leftrightarrow \beta\gamma + [\underset{\mathbf{V}}{\mathrm{HR\cdot \alpha^{*}GTP}}] \leftrightarrow \mathrm{HR} + \underset{\mathbf{VI}}{\alpha^{*}\mathrm{GTP}}$$

Scheme 1

On agonist binding and receptor activation, the receptor interacts with the G protein heterotrimer in its inactive, GDP-bound form (α·GDP,$\beta\gamma$;**I**) to promote a conformational change resulting in GDP dissociation and GTP binding. GTP binding induces again a conformational change so that the α*GTP subunit attains reduced affinity for the receptor and the $\beta\gamma$ subunit. The α and $\beta\gamma$ subunits are free to interact with the respective effectors. The cycle will be closed by the intrinsic GTP hydrolysis of α*GTP and reassociation of α·GDP with $\beta\gamma$.

From the point of view of mapping of mutual contacts between the receptor and G proteins, all steps between binding of α·GDP,$\beta\gamma$ to the receptor and dissociation of α*GTP from the receptor are of major interest. Some care should be taken to differentiate effects of receptor-derived peptides which constitute functional G protein coupling and secondary effects of peptides which indirectly influence receptor–G protein coupling because the parent receptor sites are potentially involved in interactions with regulatory proteins, e.g., receptor kinases.

Besides making specific contacts with G protein subtypes, functional coupling implies that activation of either component be mutually transmitted to both coupling entities. Thus, the result of coupling will be stimulation of guanosine nucleotide exchange in one direction and stabilization of the high-affinity receptor state in the other.

At a time when a very large number of pharmacologically distinct receptors were known without a corresponding number of G protein subtypes or their combinations, we began to study receptor–G protein interaction believing that the specificity of signal transduction could be determined by the specificity in G protein coupling. One of the best-defined systems at this time has been the hormonally (β-adrenergic) stimulated adenylyl cyclase (13). Several prerequisites for a successful application of site-specific peptide competition became available simultaneously: the sequences and a structural model of several β-adrenergic receptors (14, 15) and of the $G_s\alpha$ subunit (16) as well as automated peptide synthesis. This made it possible to synthesize, on a reasonable scale, a significant selection of peptides, about 10–20 amino acids long, from the amino acid sequences of the putative interaction sites anchored to the membrane. Membranes containing the receptor and G_s proteins were incubated with each of these peptides, and adenylyl cyclase activity was measured in response to receptor stimulation. During the initial experiments we observed partly attenuation, partly activation of adenylyl cyclase that turned out to be typical for the intrinsic properties of receptor peptides (17). Extending these studies we could show that three peptides from the second, third, and fourth loop inhibited synergistically activation of the effector while a fourth peptide from the C-terminal region of the third loop apparently stimulated adenylyl cyclase even in the absence of the receptor (18). This was essentially confirmed by other laboratories (19, 20).

At the same time, a similar approach was made with another prototype GPCR by Hamm and Hofmann, who showed that peptides from three internal loops of rhodopsin inhibited synergistically the formation of a stable metarhodopsin II–transducin complex (21).

In view of the steadily increasing number of G protein subtypes and interactions of GPCRs with regulatory proteins the interest in site-specific peptides as tools for studying these interactions is still increasing.

Interest in this approach comes also from the observation that a number of membrane-spanning receptors not related to GPCRs, above all receptor tyrosine kinases, were found to have a coupling potential to G proteins (22, 23). It can be surmised that the structural requirements for coupling to G proteins are shared by all peptides and therefore must account for new receptors and naturally occurring G protein-activating peptides as well (24).

Theoretical Design Considerations

A project for mapping of coupling sites may be realized essentially in three steps consisting of the selection of the alleged or putative site-specific sequences; peptide synthesis and assay of receptors, G proteins, and effectors; or their combinations in the presence of peptides.

Selection of Peptides

Selection of peptides representative for coupling domains is first determined from sterical and structural considerations about which parts of the integral membrane protein might be available to interact with predominantly cytosolic proteins such as G proteins. This presents no problem with all GPCRs because the hydropathy plot clearly delimits membrane segments from extra- and intracellular segments. It has to be observed, however, that the intrinsic properties of a peptide selected from a specific site when released from the restrictions of the higher order structures of the native receptor or any modifications of the synthetic peptide may alter the controlled accessibility of the represented region to a G protein target site with consequences on specificity and function. The portion of intracellular peptide segments from the total polypeptide chain of some GPCRs (96 of 348 residues in the case of bovine rhodopsin) would easily allow one to synthesize all intramolecular segments. These peptides, however, represent only the primary and secondary structure, but not the assembly of composite structures.

The classification and mapping of potential coupling domains always refers to the model of the membrane topography of the prototypic rhodopsin. This structure is composed of the extracellular N-terminal and the intracellular C-terminal chains connected by seven transmembrane segments (TM1–TM7) and three intracellular (i1–i3) and three extracellular loops (e1–e3). While there are only minor differences between receptors in length of i1 (8–14 residues, average 10 residues) and i2 (18–22; average 20), the size of i3 varies between 17 and more than 200 residues (6). Due to its larger size i3 can accommodate at least two coupling domains with the possible exception of the much smaller loops in rhodopsin and odorant receptors. The C-terminal chain distal to TM7 can consist of >100 but not less than 14 residues, which constitute a distinct conserved region (CT1) and a more distal cytosolic region (CT2). Quite notably, in a large number of GPCRs one to two essentially conserved cysteines delimit this region. Even more conspicuously these cysteines might carry a palmitate, which is expected to insert into the membrane forming an additional intracellular loop i4 (25). Peptides covering the position around this cysteine often map for a G protein coupling domain

(18, 21) (see below). According to the sizes of the loops or extensions one would expect that the smaller the loops the less the probability that it will become engaged. Indeed, there has not been any consistent proof for participation of i1, but all other intracellular regions might be involved in receptor coupling or regulation.

The results from several rather complete mapping studies with peptides of five receptors [rhodopsin (12, 21, 26), β-adrenergic receptor (βAR) (18–20), α-adrenergic receptor (α2AR) (27), dopamine D2 receptor (D2R) (28), and the human *N*-formyl peptide receptor (FPR) (29, 30)] show that at least two and up to four (independent) coupling sites out of five potential sequence segments located within i2, i3, i4, and CT become involved or interfere with coupling. Concerning i3, the majority of examples have shown that the N-terminal (i3N) and C-terminal (i3C) segments immediately joining TM5 and TM6 may contain two independent coupling sites. Concerning the C-terminal chain, the CT1 and i4 region mapped for coupling at least in two GPCRs (12, 18, 21). We found that neither palmitoylation nor the freely available -SH residue of the highly conserved Cys is essential for coupling of βAR. Within CT2, immediately distal to the i4 domain, another coupling site was reported for rhodopsin (26) and FPR (29). In all other GPCRs natural or artificial truncation of the C-terminal chain beyond i4 or CT1 appears of no significance for coupling (31). Despite reports on a role of the C-terminal tail in determining G protein specificity (32) coupling sites within the distal part of the C-terminal chain cannot be essential for all receptors since this region is neither conserved (29) nor sufficiently extended in all GPCRs. Furthermore, generally the i3C and CT2 regions are rich in Ser and Thr, which are potential phosphorylation and binding sites for regulatory proteins, which in turn might interfere with coupling.

The second criterion for selection of peptides is conservation of primary or secondary structures in the intracellular domains. As has been pointed out by Birnbaumer (1), "Single G proteins are designed to interact with classes of receptors as opposed to single receptors," with the consequence that the coupling domains of functionally related receptors must match the same site on the corresponding G protein. We found that peptides from loops i2, i3, and i4 of turkey βAR equally inhibited signal transduction of the βAR or human prostaglandin E1 receptor (PGE1R) to G_s even in xenotypic membranes. On the other hand, a peptide representing i2 of a G_i-coupled αAR potently stimulated G_s (33), suggesting that at least in some cases the sequence is not sufficient for determining the coupling specificity, that coupling to one site on the G protein is not sufficient to make the contact selective, or that all G protein subtypes share common recognition sites. The latter might explain why peptides, which act directly on G proteins, might act indiscriminately on otherwise specific signal transduction chains, e.g., muscarinic cholinergic receptor (MChR)/G_o coupling (34) and βAR/G_s coupling (18) were equally inhibited by mastoparan.

These examples in toto point out that potential coupling sites claimed and selected by the above-mentioned criteria might well represent cryptic or dormant sequences kept under control by regulatory elements, but when unburied the corresponding soluble peptides might simulate coupling *in vitro* or compete with it.

From the preceding considerations up to five sites come into question, covering all potential coupling sites of GPCRs that requires an equal number of synthetic peptides to examine their actual contribution. With respect to the sizes of peptides, it appears reasonable to synthesize peptides covering the complete loops i2 and i4, including the immediately adjacent residues, and CT2 consecutive to i4, *in toto,* which is in reach of present standards of peptide synthesis. I3N and i3C should be treated as independent sites to be represented by separate peptides. Inadvertently short peptides often have been also short of function, compare (19) with (18, 20), (21) with (26), and (29) with (30).

Peptide Synthesis

The aim of peptide synthesis in the past was primarily to rebuild natural sequences in which the native material was inaccessible. The first example using site-specific synthetic peptides for recognition of G proteins—rhodopsin coupling with transducin (34)—suffered from the inadequate size of peptides available from preautomated peptide synthesis and cannot be considered representative for the present approach. While competition with peptides from five to eight residues long could be demonstrated for one receptor site, other interactions were missed when compared with experiments using 12–15 mers (12, 21, 26, 34). The larger size compares well with naturally occurring peptides which directly target G proteins like mastoparan or substance P. Furthermore, the close correlation between an amphiphilic secondary structure of model peptides defines one of the future goals in characterizing, synthesizing, and modifying peptides to obtain peptides with a defined secondary structure. Only experiments with defined structures will give insight into the mechanisms of coupling. Solvent peptides structures have been assessed by CD spectra in phospholipid vesicles (35, 36) or with NMR-linked modeling in (deuterated) phospholipid vesicles (37).

The experimental potential of synthetic peptides as compared with their natural counterpart could be significantly increased by adding residues which are part of the transmembrane region (17) or which can promote oriented membrane incorporation like fatty acid side chains [Table 1 in Mousli *et al.*

(24)]. The meaning of this is not entirely clear considering the as yet unsolved mechanism of some directly G protein activating neurokinins. Okamoto (22) went a step further by adding 11 amino acid residues from the membrane-spanning sequence to a G protein coupling motif in the IGF-II/mannose 6-phosphate receptor, 14 amino acids long, by which the potency increased by a factor of 300–1000 times.

Experimental Design Considerations

Assays with site-directed peptides or controls should directly report on coupling effects or should be closely integrated with coupling. The formation of physical complexes and their competition by receptor-derived peptides cannot be considered sufficient as final proof for functional coupling sites. On the basis of the mechanism proposed by Birnbaumer we assume that during G protein activation two steps might be the preferential target of receptor action: (i) Interaction with the inactive G·GDP complex (**I,** Scheme 1) resulting in the formation of a stable HR·G complex (**II**); (ii) The productive binding of GTP that leads to G activation (G*GTP;**IV**) and α*GTP (**V**) dissociation. It is not clear if the sequence of steps from **III** to **V** (Scheme *I*) is solely determined by the binding of GTP or promoted by new or additional HR interactions potentially involving alternate coupling sites (26). Consequently, coupling can be directly assayed from properties of the initial receptor–G protein complex (HR·G, **II**) or from altered structural properties including protein modification or from triggering intrinsic activities of the participating G protein. Finally, coupling can be assayed indirectly by downstream effector activities.

Most "coupling" events affecting G proteins can also be elicited by naturally occurring peptides bypassing the receptor and membrane barriers which require nothing more than promotion of nucleotide exchange of the relevant G protein, probably by binding to a single site (34), which, however, might be composed of three regions of the α subunit (4, 11, 38, 39). In contrast, couplings with conventional GPCRs always involve at least two receptor domains. This emphasizes the fact that one peptide representing the activating site is not sufficient to account for the selectivity, although in many cases it has been shown that single peptides show preferential selectivity. Limitations of this model can be overcome by an alternative mechanism in which two amphipathic helices are required to substitute for the receptor, and both can be mimicked by a single peptide (39). A coupling mechanism using two or more mutal interaction sites does affect the strength of interaction. Since the total affinity is the product of the intrinsic affinities of all sites, competition requires high concentrations of the soluble competitor.

Also, if the activating function and the selectivity function are represented by different peptides, some peptides are able to activate G proteins directly and others are not.

The interaction model further implies that peptides can perform as competitors of the receptor domain or replace the activated receptor. In assays performed in the presence of the receptor, the competitive situation will prevail and generally lead to attenuation of the receptor effect. In contrast, activation above the basal rate should be observed for those peptides which can substitute for the activating domain. The latter experiments do not require a receptor and if studied in membranes should be carried out in the presence of an antagonist or, alternatively, in partially reconstituted phospholipid vesicles or in aqueous systems. A case study will exemplify the differences observed by studying receptor peptides in the presence or absence of a competing receptor.

The dodecamer T284-295 from the highly conserved C-terminal end of loop i3 was the first example of a GPCR domain to display β-adrenergic receptor stimulation in a concentration and G_s-dependent manner independent of the presence of receptor (17). Studied more closely, we found that in the presence of the peptide the basal rates of GTP binding and GTPase activity of G_s were increased and decreased, respectively. The consequent increase in cAMP formation complies with the commonly accepted mechanism of adenylyl cyclase stimulation by $\alpha_s \cdot$GTP subunits. Hormone-independent adenylyl cyclase activation was approximately half the isoproterenol stimulated activity. If, however, in an experiment conducted in native membranes, the hormone-stimulated receptor competed with the same concentration of peptides, activities dropped to the same level as in the absence of the receptor (30). Activation of adenylyl cyclase to a level of 42% of an agonist was also reported for the corresponding peptide from β2AR in S 49 membranes; however, in the latter case, GTPase activity was increased *in vitro* (20). An indiscriminate GTPase stimulation was also reported for the C- and N-terminal peptides of i3 from β2AR when assayed with G proteins in phospholipid vesicles (19) which is at variance with the commonly accepted $\alpha_s \cdot$GTP action on adenylyl cyclase.

Some of the more common assay methods are discussed below:

1. *Changes in receptor parameters:* G protein coupling stabilizes the active conformation of the receptor in the HR·G (**II,** Scheme 1) state (1). Examples are the stabilization of light-activated rhodopsin (21), induction of fluorescence-sensitive conformational changes in the receptor and β,γ subunit (12), or stabilization of the high-affinity binding conformation of GPCRs where the receptor has two affinity states (26). Disruption of physical

coupling by peptides can be detected from velocity sedimentation in sucrose density gradients (29).

2. *Changes in G protein parameters:* A second class of experiments that allow one to follow coupling consists of assaying different steps of the GTPase cycle, with a preference for the assay of GTP-binding or GTPase activity (20, 31). Although direct assays of intrinsic G protein activities appear preferable at first, such experimentation is more prone to artifacts and the evaluation of the data is more difficult for the determination of "unspecific" interactions of peptides with more than one G protein subtype. To perform these assays more selectively, class-specific tools, such an ADP-ribosylation, can be applied. It is obvious that assays linked to nucleotide turnover are more likely handicapped by an unfavorable signal-to-noise ratio and high basal activities compared with assays based on the selectivity and amplification of a specific effector, at least in membranes. Further limitations result from the fact that some of the steps of the GTPase cycle are dependent on Mg^{2+} concentration (35): while receptor or peptide-dependent activation of the guanine nucleotide exchange and steady-state GTPase activity is reciprocally dependent at μM Mg^{2+}, effects of peptides can be blocked at mM Mg^{2+}.

3. *Activation or inhibition of effectors:* Specificity of coupling is also conveyed from the specificity of effector stimulation or inhibition such as changes in adenylyl cyclase (17, 18), phospholipase C, cGMP phosphodiesterase (26), or ion channel activities. The specific contribution of peptides can be determined from the difference they make on agonist and antagonist action. Generally, such studies can be performed with membranes from cells disposing of the complete signaling chain. To study some effects more specifically, single components are deleted or added to the cells by recombinant techniques.

4. *Aqueous system assay:* The assay of peptides can also be conducted in aqueous systems or partially reconstituted phospholipid vesicles containing purified G protein preparations. This approach in principle allows one to study not only the peptide interactions during coupling but also the interactions of the peptide with isolated α (26, 40) or β,γ subunits (12). On the other hand, peptides representing domains purely concerned with subtype selectivity of the GPCR interaction will barely be detected. In view of the generalized direct action of model peptides on G proteins (see below), preferential selectivities of site-directed peptides in aqueous systems or partially reconstituted vesicles in the absence of competing receptor appear ambiguous.

Use of Model Peptides

G protein activation can also be promoted by the direct targeting of G protein signaling pathways through a number of neurokinin peptides and peptide toxins like mastoparan (35, 41, 42). Since their effects are expressed at the inner membrane surface or intracellularly, these peptides might be adequately studied in an aqueous system, but their effects are reported to become potentiated in phospholipid vesicles (35). Mastoparan, substance P, and related peptides can show a graded specificity toward G protein subtypes; permutation of the mastoparan sequence causes changes in specificity (40, 43). This has to be considered when using synthetic peptides structurally related to mastoparan (33).

From the preceding, the assay setup cannot in all cases differentiate if the "peptide" effects truly simulate functional coupling of the receptor in its native environment or if the effects solely characterize the potential of peptides to elicit generalized effects on G proteins or to act as substrate analog inhibitors detached from the specificity provided by the remainder of the receptor. To establish if peptide effects are specific, we suggest that in all cases assays be conducted in the presence of the receptor and the extent of coupling be measured by the difference between agonist stimulation and suppression by an antagonist, which is most easily performed in membranes.

Experimental Procedures Using Peptides

General Principles

The methods described here use short synthetic peptides of an average length of 10–20 amino acid residues. Standard protocols for synthesis, purification, and analysis are found in comprehensive resource and laboratory manuals (44–46). Except for peptide acids representing the C-terminus of a polypeptide we prefer peptide amides with a C-terminal amide residue, resulting typically from the Fmoc-synthesis strategy using Rink-amide resins. Since most peptides represent intracellular sites or regions exposed to cytosolic proteins, they are easily soluble in aqueous buffers.

Customarily, peptides are provided as lyophilized samples, but despite certified homogeneity or purity they still might contain up to 50% of foreign material, mainly salts. A shortcut cleanup that removes salts and contaminants from the cleavage procedure can be performed with Sep-Pak Plus Reverse Phase "360 mg" cartridges (Millipore Corporation, US):

Solvate the bonded phase with 10 ml methanol. Equilibrate with an equal volume of 0.1% aqueous TFA. Load the (acidified) sample to the cartridge. Wash the cartridge with 20 ml 0.1% aqueous TFA. Optionally, elute weakly held components with a less polar solvent. Collect the peptide by elution with 5 ml of acetonitrile/water (80/20) containing 0.1% TFA. Lyophilize the sample.

Stock solutions of the peptides are made up in water or the assay buffer at approximately 5 m*M*, requiring 5–10 mg/ml. If no precise data on the peptide content are available, peptides recovered from a reverse-phase cleanup are approximated by ≥80%.

The following competition assays require the presence of a GPCR and G protein usually contained in membranes. To assay for the action of peptides, membranes or G protein preparations are preincubated with a range of peptide concentrations prior to stimulation of the receptor by its agonist. Although, conceptionally, extended preequilibration times are not required, membranes or phospholipid vesicles together with peptides are incubated for 30–60 min on ice (4°C) or for 15 min at 30°C. We found that peptide-dependent effects were saturated to ≥80% after 15 min at 4°C.

Synergistic effects from the simultaneous addition of two or more peptides representing different receptor sites were observed in two cases (18, 21). Sequential addition of peptide combinations has not been reported to date.

Control experiments are very important in assessing the specificity of peptide action and should include the following: (1) Peptides representing presumed coupling sites and inactive sites from the cytosolic surface or extracellular site [see Theoretical Design Considerations and Nishimoto *et al.* (23, 31, 43)]. (2) Reversal of addition of peptides and receptor agonists in a competitive setup must be considered of limited value due to the reversibility of some steps (see Scheme I) and the GTPase cycle. However, in the course of assays based on downstream effector activation by nonhydrolyzable GTP analogs, reversal of addition of peptides should abolish competition. (3) Controls with G protein-directed peptides such as mastoparan should be efficient in competing with the activating site of GPCPs (18, 35) or other receptors for the "common" receptor binding site of G proteins (11). They might even disrupt physical complexes of GPCRs with G_i or G_s. Similarly, we propose the use of a complete set of site-specific peptides from GPCRs, where class-specific G protein coupling has been already established, to disrupt physical GPCR complexes.

Some basic preparations and assays are outlined below, with preference for those used in the author's lab.

Basic Procedures and Assays

Avian Erythrocyte Membranes

Turkey erythrocyte membranes are useful models for studying G protein-linked regulation of adenylyl cyclase (18) or phospholipase C (48, 49) and are easily obtained in large quantities.

Four-liter portions of fresh turkey blood are collected and stored until used by addition of 1000 ml 75 m*M* glucose–70 m*M* citrate buffer, pH 5.0, containing 50,000 U heparin, 40,000 U penicillin, and 400 mg streptomycin. Five liters of stabilized blood are added to an equal volume of 10 m*M* Tris, pH 7.4, 150 m*M* NaCl, and 1 m*M* EDTA and centrifuged 10 min at 1000*g*. After careful removal of the buffy coat by aspiration, the cells are washed repeatedly with the same buffer. The packed erythrocytes are suspended in 15 volumes of hypotonic lysis buffer (15 m*M* phosphate, pH 7.4 1 m*M* dithiothreitol). After 5 min the hemolysate is made isotonic by addition of an equal volume of 0.3 *M* NaCl and spun for 20 min at 7900*g*. The supernatant is removed and lysis repeated on resuspension in lysis buffer until the pellet is a faint pink. The nucleated ghosts are suspended in 4 liters NPE buffer (150 m*M* NaCl, 25 m*M* NaH_2PO_4, 1 m*M* EDTA, pH 7.4) and centrifuged as above. The pellet is suspended in five times the volume of the original packed cells and homogenized in 125-ml portions in a Buehler blender at maximum speed for 10 sec in an ice bath. The nuclear material is removed by centrifugation at 3000*g* for 15 min. The supernatant containing the membrane fraction is centrifuged at 23,000*g* for 20 min. The pellet is removed carefully from the residual nuclear material, resuspended, and centrifuged in NPM buffer (150 m*M* NaCl, 20 m*M* NaH_2PO_4, 2 m*M* $MgCl_2$, 1 m*M* EDTA, pH 7.4) several times. The final pellet is suspended in NPM buffer and 10% glycerol at 10 mg/ml protein and shock-frozen in liquid nitrogen. Frozen membranes can be stored for several months at −70°C.

293 Cell Membranes

Adenylyl cyclase in human embryonal kidney 293 cells can be stimulated by its own PGE1R. 293 cells are well-suited for transfection with G_i-coupled GPCRs (50).

293 cells are grown in Dulbecco's modified Eagle's F-12 medium (Gibco), supplemented with 10% fetal calf serum, 2 m*M* L-glutamine, 1000 U penicillin, and 1 mg streptomycin per liter. Membranes are prepared from monolayer cultures. The plates are washed with PBS, pH 7.4. After incubation for 5–10 min in EDTA buffer (145 m*M* NaCl, 1.5 m*M* EDTA, 10 m*M* Tris–HCl, pH 7.4), cells are removed from the plates and pelleted at 1000*g* for 5 min, washed twice with PBS, and homogenized with a Polytron homogenizer at setting 6 for 20 sec in ice-cold PBS. The nuclei and debris are removed by centrifugation at 1000*g* for 5 min at 4°C. The membranes are pelleted from the supernatant by centrifugation in a SW-50 rotor (Beckman) at 100,000*g* for 20 min. Aliquots are frozen in liquid nitrogen and stored at −70°C for no longer than 2–3 days. Thereafter a significant decrease in performance is observed.

G Protein Preparations

G_s from turkey erythrocytes is purified according to Hanski *et al.* (51). Purified $G_{i,o}$ protein fractions are obtained according to the original protocol of Sternweis and Robishaw (52) or Codina *et al.* (53). Recombinant $G_{i\alpha}$ and $G_{o\alpha}$ proteins can be obtained according to Linder and Gilman (54).

Phospholipid Vesicles

A total of 15 μl of a phosphatidylethanolamine/phosphatidylserine/cholesterol hemisuccinate (12 : 8 : 5) lipid mixture (15 mg/ml) added to 215 μl buffer (20 m*M* HEPES, 20 m*M* NaCl, 0.1 m*M* EDTA, 0.1 m*M* dithiothreitol, 0.2% Lubrol PX) which contains 3–6 pmol purified G protein. The reconstitution mixture is kept at room temperature for 5 min. Detergent is removed by gel filtration on a 1 × 0.4-cm Sephadex G–25 fine column and the vesicle fraction is collected from the void volume. The amount of G protein incorporated is determined by quantitation of GTPγS bound from 1 μM [^{35}S]GTPγS at 10–50 m*M* Mg^{2+} (see below).

Detailed procedures for reconstitution of receptors and G proteins in phospholipid vesicles are given in Cerione and Ross (55).

Ligand Binding

β-Adrenergic receptor-containing membranes (20–100 μg membrane protein) are incubated with 5–25 n*M* [^{3}H]dihydroalprenolol (50 cpm/fmol) in 10 m*M* Tris, 90 m*M* NaCl, pH 7.4, in a reaction volume of 300 μl at 30°C. After 30 min, aliquots are filtered through Whatman GF/C filter and washed five times with 4 ml cold buffer. The filters are counted for radioactivity. Unspecific binding is determined in the presence of 10 μM DL-propranolol. Displacement experiments are performed with a fixed concentration of [^{3}H]dihydroalprenolol (10 n*M*) and increasing concentrations of L-isoproterenol (0.1 n*M*–10 μM). For effects of peptides on ligand binding, membranes are incubated with synthetic peptides for 60 min at 4°C prior to addition of the ligands.

Dopamine D2 receptor-containing membranes (50 μg per assay) are incubated for 15 min at 37°C in 50 m*M* Tris (pH 7.7) with 0.5 n*M* [^{3}H]spiperone with or without 10 μM haloperidol to calculate the amount of specific binding. The total concentration of receptor-specific binding sites is 3 ± 0.5 pmol/mg membrane protein. Nontransfected cells do not show any specific [^{3}H]spiperone binding.

GTPγS Binding

The binding of GTPγS to the Gα subunit is [Mg^{2+}] dependent. At low [Mg^{2+}] (<1 m*M*) it measures the stimulation by receptor coupling, peptides, or other autacoids. At high [Mg^{2+}] (25 m*M*) it measures maximal GTPγS binding (56).

Membranes or G proteins are incubated with or without peptides for 1 hr at 4°C in the assay buffer followed by hormonal activation. Assays in parallel with the appropriate antagonist serve as a control or to distinguish direct interaction of peptides with G proteins.

For the assay of GTPγS binding in membranes (52) typically 10 μg membrane protein (28) is added to 40 μl of 20 m*M* HEPES, pH 7.4–8.0, 20–100 m*M* NaCl, 0.1 m*M* DTT, 0.1 m*M* EDTA, 0.1–0.5 m*M* Mg^{2+}, and the agonist or antagonist if appropriate. Following addition of [^{35}S]GTPγS (20 cpm/fmol) at a final concentration of 0.3–1 μM, the incubation proceeds for 30 min at 30°C until stopped by dilution with 1 ml of an ice-cold buffer containing 20 m*M* Tris, pH 7.4–8.0, NaCl, and $MgCl_2$ adjusted to the incubation buffer. The samples are filtered directly through BA 85 nitrocellulose filters and washed five times with 4 ml of the same buffer. Unspecific binding is accounted for by controls made in presence of 1 m*M* GTP and soaking the filters before use in washing buffer containing 1 m*M* GTP. The filters are dissolved in 2 ml ethyleneglycolmonomethylether and counted.

Alternatively, assays for determination of the GTPγS binding rate are performed under similar conditions, but at preset times (0–30 min, 30°C) 50-μl aliquots containing ~300 fmol G protein are removed and treated as above.

The total amount of GTPγS binding, equal to the total amount of G protein, is determined in the presence of 25 m*M* Mg^{2+} and 1 μ*M* GTPγS (56).

For *in vitro* assays of G proteins in aqueous solution or G proteins contained in vesicles, conditions similar to those applied to G protein activation by mastoparan appear appropriate (35): The assays are carried out at a concentration of 10 n*M* G protein in 50-μl aliquots in 50 m*M* HEPES, pH 8.0, 1 m*M* EDTA, and 1 m*M* DTT, which contains 1.3 m*M* Mg^{2+} and 0.3 μ*M* [^{35}S]GTPγS (20 cpm/fmol) incubated for 10–60 min at 30°C. Samples are filtered through BA 85 nitrocellulose filters and treated as described above. Total GTPγS binding is determined at 10 m*M* $MgCl_2$ ($G_{i,o}$) or 50 m*M* $MgCl_2$ (G_s).

GTPase Assay

GTPase activity of G proteins is determined from the release of [^{32}P]phosphate from [γ-^{32}P]GTP (57, 58). Depending on the setup, the method indirectly measures the guanine nucleotide exchange reaction or GTP binding.

The assay contains 50 m*M* imidazole-HCl, pH 7.4, 0.1–6 m*M* $MgCl_2$, 0.1–1 m*M* EDTA (for concentrations of Mg^{2+} and EDTA see GTPγS Binding), 1 m*M* DTT, 1 m*M* App(NH)p, 0.1 m*M* ATP, 5 m*M* creatine phosphate, 0.4 mg/ml creatine kinase, 0.05–0.25 μ*M* [γ-^{32}P]GTP (5–20 cpm/fmol) (or 0.1–2 μ*M* [γ-^{32}P]GTP instead of the regenerating system), and 20–100 μg membrane protein in a final volume of 200 μl. The assay is run for 5–10 min at 30°C. The reaction is stopped by taking 80-μl aliquots in duplicate and adding them to 1 ml of a Norit A suspension (5%, w/v) in 50 m*M* phosphate buffer, pH 7.2, at 4°C. The mixture is centrifuged for 8 min at 2000 rpm, 4°C, and 700 μl of the supernatant is mixed with 500 μl of Norit A suspension. After centrifugation, ^{32}P is counted (Cerenkov) from 800 μl of the supernatant.

Synthetic peptides are first incubated with membranes for 60 min at 4°C prior to stimulation with the agonist. Hormone-independent and basal GTPase activities are determined by incubation with the peptides in the presence of the respective antagonists (10 μ*M*), also considering other stimulatory and inhibitory components of the membranes under study.

GTPase in turkey erythrocyte membranes is determined from 40 μg membrane protein at pH 6.7 in the presence of 10 μ*M* L-isoproterenol. GTPase from 293 cell membranes is determined from 100 μg protein at pH 7.4 in the presence of 100 μ*M* dopamine (28). GTPase activity in human platelet membranes is determined in the presence of 1 m*M* $MgCl_2$, 1 μ*M* GTP from 20 μg membrane protein/100 μl (26). GTPase from transducin (15 pmol) is determined in lipid vesicles containing 0.4 to 4 pmol rhodopsin and peptides in presence of 2 m*M* $MgCl_2$, 1 μ*M* GTPγS (26).

For *in vitro* GTPase assays of G proteins in aqueous solution or G proteins contained in vesicles, conditions similar to those applied to G protein activation by mastoparan appear appropriate (35). The reaction mixture contains 50 m*M* HEPES, pH 8, 1 m*M* EDTA, 1 m*M* DTT, 1.1 m*M* $MgSO_4$, and 0.3 μ*M* [γ-^{32}P]GTP (10–30 cpm/fmol). Alternatively the assay is carried out in vesicles containing the same buffer. The samples are incubated with the peptide for 5 min at 20°C.

Adenylyl Cyclase Assay

Following receptor- or peptide-dependent activation of the stimulatory (G_s) or inhibitory (G_i) pathway of hormonally activated adenylyl cyclase, the formation of [^{32}P]cAMP from [α-^{32}P]ATP is determined.

Activation of G_s

When (pre)activation of G_s is determined from β-adrenergic stimulation of adenylyl cyclase, the reaction is initiated by addition of 10 μ*M* isoproterenol (G_s-specific activation is calculated from the differences obtained in the presence of 10μM proterenol) and 0.5 μM GTPγS (or 100 μ*M* GTP) to 150 μg turkey erythrocyte membranes (final concentration 1 mg membrane protein/ml) contained in 20 m*M* HEPES, pH 7.8, 20 m*M* NaCl, 0,1 m*M* DTT, 0.1 m*M* EDTA, and 0.5 m*M* $MgCl_2$. Hormonal preactivation is terminated after 5 min at 37°C by addition of 50 μ*M* propranolol and the assay kept on ice.

Activation of G_i

Inhibition of prestimulated adenylyl cyclase is determined in 293 cells containing an PGE1R. Adenylyl cyclase stimulation is initiated by addition of 1 μM PGE_1 and 50 μM GTP to 50 μg 293 membranes with or without 100 μM dopamine contained in 20 mM HEPES, pH 7.8, 0.1 mM DTT, 0.1 mM EDTA, 0.1 mM $MgCl_2$ in 115 μl. Controls contain 10 μM haloperidol or 10 μM sulpiride and propranolol.

Synthetic peptides are first incubated with membranes for 60 min at 4°C.

Adenylyl Cyclase Activity

These assays are performed in a final volume of 150 μl containing 20 mM HEPES, pH 7.8, 20 mM NaCl, 5 mM $MgCl_2$, 0.1 mM EDTA, 0.1 mM dithiothreitol, 100 μM GTP, 5 mM theophylline, 10 mM creatine phosphate, 0.2 mg/ml creatine kinase, and 0.5 mM ATP, containing 1–2 $\times 10^6$ cpm [α-^{32}P]ATP. The reaction is initiated by addition of preactivated membranes (50–100 μg protein/assay) and incubated for 20 min at 30°C.

[^{32}P]cAMP is isolated and determined according to Johnson and Salomon (59). Reaction of adenylyl cyclase is stopped by the addition of 100 μl 2% sodium dodecyl sulfate, 10 mM Tris, pH 7.4, 4 mM ATP, and 1.5 mM cAMP. The mixture is made up to 1 ml by further addition of 10,000 cpm [^{3}H]cAMP in 750 μl H_2O and transferred onto 1 ml Dowex 50W-X8 in Econo columns (Bio-Rad). After being rinsed with 2 ml H_2O, the columns are washed thoroughly with 7 ml H_2O. The total wash is transferred to 1 ml alumina (Al_2O_3) columns and eluted with 0.1 M imidazole, pH 7.4. The first 1.2 ml are discarded and 3 ml collected into scintillation vials.

For assay of adenylyl cyclase activity in 293 cell membranes, a buffer containing 2.5 mM $MgCl_2$, without NaCl, is used (28).

Discussion

The positive results from the peptide mapping approach prove that a complete account of the topography for coupling interactions can be established, especially if the experiments are performed, as suggested, by competition with the membrane-bound receptor (18, 21, 27).

A somewhat different view was obtained from assays were effects of site-directed peptides were deliberately studied in aqueous systems or in partially reconstituted phospholipid vesicles containing G protein preparations. The latter approach is limited to cover peptides that initiate or modulate the GTPase cycle of available G proteins, equivalent to coupling with the activating domain of GPCRs (19, 20, 33, 47). Peptides which do not alter basal G protein activities (20, 36) are not covered by this approach, which therefore appears less complete than peptide competition or alternative methods. However, this approach turned out to be extremely useful to demonstrate the potentiality of unique peptide regions from nonGPCRs like tyrosine kinase receptors or related proteins to make functional contacts with G proteins (22, 23).

While at least two sites distributed through the second, third, and fourth intracellular loops of GPCRs are required for functional and specific coupling, interaction with one G protein docking site by some natural or synthetic G protein-activating peptides can already cause or modify G protein-dependent activation, sometimes even with preferential selectivity. The obligatory second coupling site of GPCRs, therefore, must contribute essentially to establish or control specificity, but shared contributions of two or more coupling sites are conceivable. While two coupling sites might be sufficient to initiate the GTPase cycle, additional coupling sites might be required, to promote or control effector-oriented activation of G protein subunits (26). There is some reason to believe that some GPCRs might contain cryptic or dormant sequences which partially fulfill the above requirements and might also elicit coupling-associated effects if the peptides were assayed *in vitro*.

Related functional and structural properties of peptides representing activating domains of GPCRs or other receptors reflect a common G protein target site located on the α subunit (4, 11). An essentially amphiphilic character of this class of peptides should allow for oriented membrane integration and presentation of a G protein recognition surface generally believed to be composed of positive charges (24). The G protein $\beta\gamma$-complex is also the target of GPRC coupling (12).

Concluding Remarks

Readily available site-directed receptor peptides and the ease of performance of assays for functional coupling in membranes or *in vitro* render the peptide mapping approach a valuable addition to this field. From the present experience it can be inferred that the methodology will not be limited by the availability of a small number, up to four, of synthetic peptides, provided the complete sequences are available, which is also a prerequisite for other methods like mutagenesis. Besides results comparable to mutagenesis, this approach provides additional flexibility for the application of peptides during and consecutive to the coupling step.

Future Directions

It has been pointed out that GPCR coupling served as a model for other signaling pathways mediated by G proteins, like neurokinins, which might be mimicked not only by venom peptides (38) but also by nonGPCR receptors, which acquire a potential to crosstalk with G proteins. Essentials of common structural principles for interaction with "conserved" sites on G proteins have been proposed. These models have successfully helped in the search for potential interaction sites in receptor tyrosine kinases-like growth hormone-like receptor (22) and related structures (23).

The future of peptide synthesis will consist in preparing ligands for direct targeting of G proteins, based on the structural principles of receptor sites and model peptides. These compounds, after further modification, will be equally useful for identification of coupling sites on a molecular level by affinity labeling and as site-directed drugs. It is expected that a number of limitations from membrane barriers can be overcome by extending the membrane adjacent peptides into the membrane (17), or artificially joining with a segment from the membrane (22). A compilation by Landry *et al.* (24) shows that the hydrophobic extension need not be peptidic. The unique properties of these peptides to target G proteins and to modulate their performance appears promising for drug design in the future.

References

1. L. Birnbaumer, J. Abramowitz, and A. M. Brown, *Biochim. Biophys. Acta* **1031,** 163 (1990).
2. E. M. Ross, *Neuron* **3,** 141 (1989).
3. G. L. Johnson and N. Dhanasekaran, *Endocr. Rev.* **10,** 317 (1989).
4. B. R. Conklin and H. R. Bourne, *Cell (Cambridge, Mass.)* **73,** 631 (1993).
5. R. Henderson, J. M. Baldwin, T. A. Ceska, F. Zemlin, E. Beckmann, and K. H. Downing, *J. Mol. Biol.* **213,** 899 (1990).
6. W. C. Probst, L. A. Snyder, D. I. Schuster, J. Brosius, and S. C. Sealfon, *DNA Cell Biol.* **11,** 1 (1992).
7. A. G. Gilman, *Annu. Rev. Biochem.* **56,** 615 (1987).
8. Y. Kaziro, H. Itoh, T. Kozasa, M. Nakafuku, and T. Satoh, *Annu. Rev. Biochem.* **60,** 349 (1991).
9. M. I. Simon, M. P. Strathmann, and N. Gautam, *Science* **252,** 802 (1991).
10. D. E. Clapham and E. J. Neer, *Nature (London)* **365,** 403 (1993).
11. J. P. Noel, H. E. Hamm, and P. B. Sigler, *Nature (London)* **366,** 654 (1993). D. G. Lambright, J. P. Noel, H. E. Hamm, and P. B. Sigler, *Nature* **369,** 621 (1994).
12. W. J. Phillips and R. A. Cerione, *J. Biol. Chem.* **267,** 17032 (1992).

13. E. J. M. Helmreich and T. Pfeuffer, *Trends Pharmacol. Sci.* **6,** 438 (1985).
14. R. A. Dixon, B. K. Kobilka, D. J. Strader, J. L. Benovic, H. G. Dohlman, T. Frielle, M. A. Bolanowski, C. D. Bennett, E. Rands, R. E. Diehl, R. A. Mumford, E. E. Slater, I. S. Sigal, M. G. Caron, R. J. Lefkowitz, and C. D. Strader, *Nature (London)* **321,** 75 (1986).
15. Y. Yarden, H. Rodriguez, S. K.-F. Wong, D. R. Brandt, D. C. May, J. Burnier, R. N. Harkins, E. Y. Chen, J. Ramachandran, A. Ullrich, and E. M. Ross, *Proc. Natl. Acad. Sci. U.S.A.* **83,** 6795 (1986).
16. T. Nukada, T. Tanabe, H. Takahashi, M. Noda, T. Hirose, S. Inayama, and S. Numa, *FEBS Lett.* **195,** 220 (1986).
17. D. Palm, G. Münch, C. Dees, and M. Hekman, *FEBS Lett.* **254,** 89 (1989).
18. G. Münch, C. Dees, M. Hekman, and D. Palm, *Eur. J. Biochem.* **198,** 357 (1991).
19. A. H. Cheung, R. R. Huang, M. P. Graziano, and C. D. Strader, *FEBS Lett.* **279,** 277 (1991).
20. T. Okamoto, Y. Murayama, Y. Hayashi, M. Inagaki, E. Ogata, and I. Nishimoto, *Cell (Cambridge, Mass.)* **67,** 723 (1991).
21. B. König, A. Arendt, J. H. McDowell, M. Kahlert, P. A. Hargrave, and K. P. Hofmann, *Proc. Natl. Acad. Sci. U.S.A.* **86,** 6878 (1989).
22. T. Okamoto, T. Katada, Y. Murayama, M. Ui, E. Ogata, and I. Nishimoto, *Cell (Cambridge, Mass.)* **62,** 709 (1990).
23. I. Nishimoto, T. Okamoto, Y. Matsuura, S. Takahashi, T. Okamoto, Y. Murayama, and E. Ogata, *Nature (London)* **362,** 75 (1993).
24. M. Mousli, J. L. Bueb, C. Bronner, B. Rouot, and Y. Landry, *Trends Pharmacol. Sci.* **11,** 358 (1990).
25. B. F. O'Dowd, M. Hnatowich, M. G. Caron, R. J. Lefkowitz, and M. Bouvier, *J. Biol. Chem.* **264,** 7564 (1989).
26. W. J. Phillips and R. A. Cerione, *Biochem. J.* **299,** 351 (1994).
27. H. M. Dalman and R. R. Neubig, *J. Biol. Chem.* **266,** 11025 (1991).
28. D. Malek, G. Münch, and D. Palm, *FEBS Lett.* **325,** 215 (1993).
29. R. K. Bommakanti, K. N. Klotz, E. A. Dratz, and A. J. Jesaitis, *J. Leukoc. Biol.* **54,** 572 (1993).
30. R. E. Schreiber, E. R. Prossnitz, R. D. Ye, C. G. Cochrane, and G. M. Bokoch, *J. Biol. Chem.* **269,** 326 (1994).
31. B. K. Kobilka, C. MacGregor, K. Daniel, T. S. Kobilka, M. G. Caron, and R. J. Lefkowitz, *J. Biol. Chem.* **262,** 15796 (1987).
32. T. Namba, Y. Sugimoto, M. Negishi, A. Irie, F. Ushikubi, A. Kakizuka, S. Ito, A. Ichikawa, and S. Narumiya, *Nature* **365,** 166 (1993).
33. T. Okamoto and I. Nishimoto, *J. Biol. Chem.* **267,** 8342 (1992).
34. D. J. Takemoto, D. Morrison, L. C. Davis, and L. J. Takemoto, *Biochem. J.* **235,** 309 (1986).
35. T. Higashijima, J. Burnier, and E. M. Ross, *J. Biol. Chem.* **265,** 14176 (1990).
36. T. Voss, E. Wallner, A. P. Czernilofsky, and M. Freissmuth, *J. Biol. Chem.* **268,** 4637 (1993).
37. K. Wakamatsu, A. Okada, T. Miyazawa, M. Ohya, and T. Higashijima, *Biochemistry* **31,** 5654 (1992).
38. T. Higashijima and E. M. Ross, *J. Biol. Chem.* **266,** 12655 (1991).

39. H. R. Bourne, *Nature (London)* **366,** 628 (1993). H. R. Bourne, *Nature (London)* **369,** 611 (1994).
40. C. Oppi, T. Wagner, A. Crisari, B. Camerini, and G. P. Tocchini-Valentini, *Proc. Natl. Acad. Sci. U.S.A.* **89,** 8268 (1992).
41. T. Higashijima, S. Uzu, T. Nakajima, and E. M. Ross, *J. Biol. Chem.* **263,** 6491 (1988).
42. M. Mousli, C. Bronner, J. Bockaert, B. Rouot, and Y. Landry, *Immunol. Lett.* **25,** 355 (1990).
43. H. Mukai, E. Munekata, and T. Higashijima, *J. Biol. Chem.* **267,** 16237 (1992).
44. M. H. V. Van Regenmortel, J. P. Briand, S. Muller, and S. Plaué, *Lab. Tech. Biochem. Mol. Biol.* **19,** 41 (1989).
45. E. Atherton and R. C. Sheppard, "Solid Phase Peptide Synthesis: A Practical Approach." IRL PRESS, Oxford, 1989.
46. G. A. Grant, ed. "Synthetic Peptides: A User's Guide." Freeman, New York, 1992.
47. T. Ikezu, T. Okamoto, E. Ogata, and I. Nishimoto, *FEBS Lett.* **311,** 29 (1992).
48. T. K. Harden, L. Stephens, P. T. Hawkins, and C. P. Downes, *J. Biol. Chem.* **262,** 9057 (1987).
49. D. H. Maurice, G. L. Waldo, A. J. Morris, R. A. Nicholas, and T. K. Harden, *Biochem. J.* **290,** 765 (1993).
50. R. Dal-Toso, B. Sommer, M. Ewert, A. Herb, D. B. Pritchett, A. Bach, B. D. Shivers, and P. H. Seeburg, *EMBO J.* **8,** 4025 (1989).
51. E. Hanski, P. C. Sternweis, J. K. Northup, A. W. Dromerick, and A. G. Gilman, *J. Biol. Chem.* **256,** 12911 (1981).
52. P. C. Sternweis and J. D. Robishaw, *J. Biol. Chem.* **259,** 13806 (1984).
53. J. Codina, D. J. Carty, L. Birnbaumer, and R. Iyengar, *in* "Methods in Enzymology" (R. Johnson and J. Corbin, eds.), Vol. 195, p. 177. Academic Press, San Diego, CA, 1991.
54. M. E. Linder and A. G. Gilman, *in* "Methods in Enzymology" (R. Johnson and J. Corbin, eds.), Vol. 195, p. 202. Academic Press, San Diego, CA 1991.
55. R. A. Cerione and E. M. Ross, *in* "Methods in Enzymology" (R. Johnson and J. Corbin, eds.), Vol. 195, p. 329. Academic Press, San Diego, CA, 1991.
56. J. K. Northup, M. D. Smigel, and A. G. Gilman, *J. Biol. Chem.* **257,** 11,416 (1982).
57. D. Cassel and Z. Selinger, *Biochim. Biophys. Acta* **452,** 438 (1976).
58. D. R. Brandt, T. Asano, S. E. Pedersen, and E. M. Ross, *Biochemistry* **22,** 4357 (1983).
59. R. A. Johnson and Y. Salomon, *in* "Methods in Enzymology" (R. Johnson and J. Corbin, eds.), Vol. 195, p. 3. Academic Press, San Diego, CA, 1991.

[17] Synthesis and Expression of Synthetic Genes: Applications to Structure–Function Studies of Receptors

Cynthia J. L. Carruthers and Thomas P. Sakmar

Introduction

While a number of methods are available for site-directed mutagenesis, the use of a properly designed synthetic gene offers many advantages, particularly where extensive mutagenesis is planned. A synthetic gene used for heterologous expression codes for the correct amino acid sequence, but contains an artificial nucleotide sequence. The nucleotide sequence is generally designed to contain a large number of unique restriction endonuclease cleavage sites. The degeneracy of the genetic code and the large number of available restriction enzymes allows on the order of 40 unique restriction sites to be introduced for a synthetic gene of 1-kb pairs in length.

Mutagenesis using a synthetic gene involves the replacement of a restriction fragment within the gene by a synthetic DNA duplex counterpart that contains the desired codon alteration(s). Cassette mutagenesis can be used to introduce point mutations, or extensive changes such as deletions, insertions, and multiple nucleotide substitutions as desired. Random mutagenesis techniques such as combinatorial cassette mutagenesis are also facilitated by using a synthetic gene. Many of these site- or domain-directed mutagenesis procedures are difficult to accomplish by mismatch primer methods.

Gene synthesis has been particularly useful in the study of G protein-coupled receptors. The superfamily of G protein-coupled receptors consists of a large number of related seven-transmembrane segment plasma membrane proteins (1). In studying receptor function by heterologous expression of mutant receptors, it is often useful to create chimeric receptors in which putative functional domains are exchanged. For example, the transfer of a cytoplasmic loop sequence from one pharmacological receptor subtype to another is one approach to study the specificity of ligand-dependent G protein activation. A synthetic gene for one receptor subtype can be engineered readily to produce either a chimeric construct or the mutant gene for a functionally different receptor subtype. The use of synthetic genes allows domain exchanges without the potential limitations of naturally occurring restriction endonuclease cleavage sites.

Methods in Neurosciences, Volume 25

In this chapter, a summary of the mutagenesis methods facilitated by the use of synthetic genes is presented. The leading gene synthesis strategies are compared and contrasted. A general method for designing a synthetic gene is discussed in detail. Detailed procedures for the various steps in gene synthesis, from oligonucleotide preparation to cassette mutagenesis, are also given. Specific examples of the use of synthetic genes in the study of receptor structure and function are presented.

Mutagenesis Methods Facilitated by the Use of Synthetic Genes

The use of a synthetic gene facilitates mutagenesis methods that rely on restriction fragment replacement (cassette mutagenesis). Site-directed point mutations are easily introduced into cloned DNA by a variety of mismatch primer methods, some of which employ the polymerase chain reaction (PCR). However, site-directed cassette mutagenesis can be successfully employed to introduce extensive alterations of the nucleotide sequence within a particular gene segment. Such an approach may be useful, for example, in structure–function studies of discrete receptor domains, such as the cytoplasmic loops of a G protein-coupled receptor (2–4). A long stretch of amino acid residues easily can be replaced by a random sequence or by a homologous sequence from a related protein.

In addition, a number of random mutagenesis methods have developed based on cassette mutagenesis. For example, the classical approach to combinatorial mutagenesis involves the synthesis and cloning of a "random" DNA duplex. The utility of random mutagenesis ultimately depends on the existence of an adequate functional screening method. Various mutagenesis methods that are facilitated by the use of a synthetic gene are presented below.

Mutagenesis by Restriction Fragment Replacement (Cassette Mutagenesis)

Mutagenesis by restriction fragment replacement was first demonstrated in the naturally occurring gene of bacteriorhodopsin (5). This mutagenesis strategy was possible because of the fortuitous natural placement of unique restriction sites. Cassette mutagenesis using a synthetic gene also involves replacement of a restriction fragment by a synthetic duplex counterpart that contains the desired codon alteration(s). However, the presence of a relatively large number of unique and evenly spaced restriction sites in a carefully designed synthetic gene allows the general use of the method. Because all of the restriction sites within the gene are unique, cloning is "directional" and screening is not generally necessary for identification of a desired recombinant transformant.

Cassette mutagenesis is also of more general utility than most forms of site-directed mismatch primer mutagenesis because of the ease of producing defined mutations at multiple sites within a domain to yield deletions, extensive substitutions, domain swaps, or the construction of chimeric genes.

One beautiful example of the creative use of cassette mutagenesis was reported in a disulfide cross-linking study of the *Escherichia coli* Tar receptor (6). Individual cysteine replacements were introduced at every position in a transmembrane domain. Interactions between the four transmembrane domains that comprise the Tar receptor dimer could then be assessed by measuring the tendency of cysteine mutants to form disulfide cross-links with one another. To produce cysteine mutants at 24 separate positions with minimal synthesis, oligonucleotides were synthesized simultaneously on two columns, one producing the native sequence and the second producing repeated cysteine codons. After each codon synthesis, the resins containing the elongating chains were exchanged in a defined proportion. The resulting mutagenic duplexes were cloned and the recombinants were characterized.

Cassette mutagenesis is also the basis of a series of random mutagenesis techniques collectively known as combinatorial cassette mutagenesis as described below.

Combinatorial Cassette Mutagenesis

The approach of combinatorial cassette mutagenesis has been used to determine the "informational content" of individual amino acid residues in a protein. Previous reports using this approach suggest that proteins tolerate a range of amino acid substitutions without significant changes in biochemical properties (7–9). For example, the effects of approximately 1500 single amino acid substitutions at 142 positions in the *lac* repressor were studied (10). About one-half of all of the mutations were phenotypically silent. At some amino acid positions many different nonconservative substitutions were tolerated. These positions must play little or no role in structure and function and, therefore, were deemed to have "low informational content." At other positions, no substitutions or only conservative substitutions were tolerated without effects on function. These positions are the most important for *lac* repressor activity and were deemed to have "high informational content."

These positions were generally hydrophobic residues involved in core amino acid packing and were crucial for protein folding or stability.

In transmembrane receptor molecules, where little if any structural information is known beyond primary structure, initial mutagenesis experiments that target a particular amino acid may not prove informative. It may be appropriate to prepare large numbers of mutations in a particular putative domain if an adequate functional screening method can be devised. Combinatorial cassette mutagenesis can be employed in such situations.

The general strategy of combinatorial cassette mutagenesis is to perform restriction fragment replacement with a set of synthetic duplexes that should provide codons for each of the 20 amino acids at one or more positions within the duplex (8). This can be accomplished by synthesizing the noncoding (top) strand of the duplex with equal mixtures of all four bases in the first two positions of a codon and with equal mixtures of guanine and cytosine at the third position. Inosine is inserted at each of the randomized base positions in the bottom strand because it is able to pair with each of the four natural bases. The heterogeneous top-strand oligonucleotides and the bottom-strand oligonucleotide are annealed and the resulting duplex is ligated into an appropriate vector. Bacterial transformation essentially produces a library of mutants which can be cloned or studied batchwise depending on particular circumstances.

Another method has been described to increase the chances of sampling large numbers of amino acid changes at all positions while removing wild-type background (11). This method involves contaminating the individual strands of the mutagenic duplex to different degrees. A method has been developed for combinatorial mutagenesis that does not rely on cassette mutagenesis. This strategy uses degenerate mismatch primer mutagenesis and can be applied to a gene with a selectable phenotype (12).

Combinatorial Cassette Mutagenesis with a Limited Alphabet

Because of the large number of possibilities in a random sequence, combinatorial mutagenesis with a limited alphabet can be carried out (13). In studies of the leucine zipper domain of the *Saccharomyces cerevisiae* transcription factor GCN4, cassette mutagenesis was used to produce substitutions of four codons at eight positions. The codons of interest were those for lysine, glutamic acid, and alanine. Threonine codons were unavoidably introduced as well. Even with this limited alphabet approach, randomization of eight positions with four codons should yield 4^8 (65,536) unique sequences. A statistical analysis was carried out on a set of 121 mutants to quantify the contributions of specific sites on leucine zipper stability.

Binomial Cassette Mutagenesis

An additional method to limit the number of mutants produced while achieving the desired information has been called binomial mutagenesis (14). The method was designed to allow the study of the effect of a mutation in the context of other mutations. The goal is to identify amino acid residue pairs that may interact functionally despite being distant in primary or even tertiary structure. Alanine is substituted simultaneously at N positions with a probability of 0.5 to give a set of 2^N mutant proteins. The additivity of the effect of mutations is evaluated. The method was employed to study the helix-turn-helix of λ repressor.

Recursive Ensemble Mutagenesis

Recursive ensemble mutagenesis has been developed to allow successive rounds of combinatorial cassette mutagenesis. This method relies on an algorithm that allows information gained in one round of mutagenesis experiments to control the nucleotide composition of DNA synthesized for the next round. The result is that much more information can be gained because the relative number of "positive" mutants is greatly increased (15, 16).

Gene Synthesis Strategies

The general strategies discussed below relate to the construction of synthetic genes. Three main strategies are presented: (i) the Khorana method, (ii) the clamp method, and (iii) PCR methods. Other strategies such as stepwise synthesis with multiple cloning and amplification steps have been reported as well (17, 18). The actual design of synthetic genes, which in general involves the introduction of multiple restriction sites to facilitate mutagenesis by restriction fragment replacement, is presented later since the design process needs to be coordinated with the method chosen.

Khorana Method

The classical method of gene synthesis evolved prior to the development of automated oligonucleotide synthesis. This method involves the complete synthesis of both duplex strands. In 1976, Khorana and co-workers reported the total synthesis of the 126-bp gene for the precursor of a tyrosine suppressor transfer RNA from *E. coli* (19–21). The synthetic strategy involved the chemical synthesis of 26 oligodeoxynucleotides, the joining of several 5′ end phosphorylated oligos to form four DNA duplexes, and finally the enzymatic ligation of the four duplexes to form the full-length gene. The most important factor in this method is the high-efficiency DNA ligase-catalyzed joining of the oligonucleotides (22).

The Khorana method has evolved with automated DNA synthesis and a number of relatively long synthetic genes have been constructed by this approach. The first gene synthesis that exploited the use of solid-phase automated DNA synthesis in combination with efficient purification techniques such as reverse-phase HPLC was the synthesis of a 514-bp gene coding for human interferon-α1 (23). Nucleotide changes were introduced in this gene to eliminate complementary and repeated sequences and to optimize the sequence for *E. coli* codon usage. Subsequently, the desire to introduce restriction sites for cassette mutagenesis and to optimize codon usage for *E. coli* expression of eukaryotic genes has led to the synthesis of numerous genes based on these early methods (24–56).

The obvious advantage of the Khorana method is the high fidelity of synthesis and high general success rate since most of the synthesis is chemical and not enzymatic. The disadvantage is the time and expense involved in preparing oligonucleotides for complete upper and lower strands. However, these disadvantages can be somewhat overcome by economies of scale, the use of batchwise 5′ end phosphorylation reactions of oligonucleotides, and ligation of up to 22 oligonucleotides in a single reaction. Another minor disadvantage is that gene design involves planning the duplex joining sites as well as the restriction sites. Although not foolproof, planning the duplex joining sites is relatively straightforward.

The Khorana method is presented schematically in Fig. 1 for the synthesis of the 1464-bp rat glucagon receptor gene (57). Other long synthetic genes synthesized by this method are listed in Table I (58–69).

Clamp ("Fill-in") Method

The basic principle of the clamp method of DNA synthesis was first proposed by Khorana (70). This method involves the synthesis of relatively long oligonucleotides from alternating DNA strands. The neighboring oligos have complementary 3′ ends of 10 to 15 bp. After annealing of the oligos, the single-stranded portions are repaired by the action of DNA polymerase I (Klenow fragment or Sequenase) in the presence of excess deoxynucleotide 5′-triphosphates (dNTP's). This method is useful for short genes or fragments (200–300 bp) and may be justified by the reduction of synthesis required (58). The clamp method is presented schematically in Fig. 2.

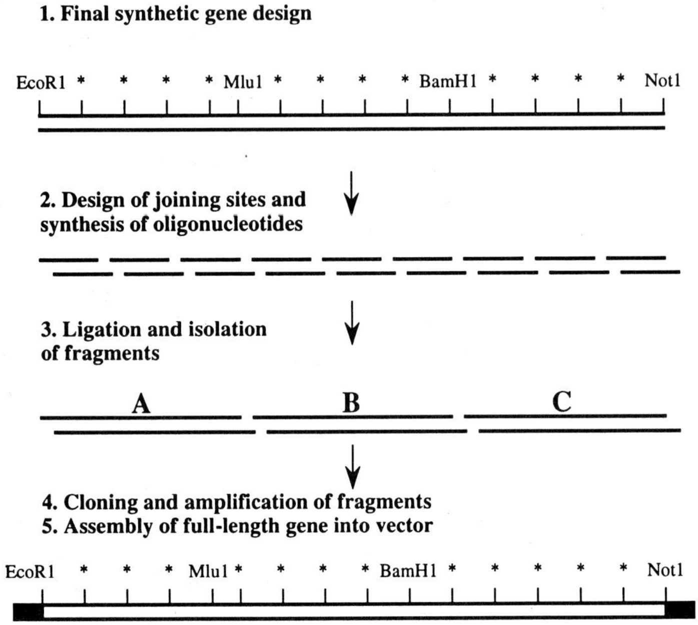

FIG. 1 Gene synthesis by the Khorana method. This method involves the total synthesis of both upper and lower DNA strands followed by ligation reactions to assemble synthetic duplexes into gene fragments. The fragments are then cloned, amplified, and subcloned to assemble the entire gene. A schematic is presented for the synthesis of the rat glucagon receptor gene as an example. The gene was designed with a large number of unique restriction sites (*). Three gene fragments were prepared *in vitro:* fragment A, *Eco*RI to *Mlu*I; fragment B, *Mlu*I to *Bam*HI; fragment C, *Bam*HI to *Not*I. Detailed procedures for gene synthesis by this method are presented in the text.

Polymerase Chain Reaction Methods

Several methods have been reported for the synthesis of genes using a combination of automated DNA synthesis and the polymerase chain reaction (66, 71, 72). One PCR method is presented schematically in Fig. 3 (66). It involves the synthesis of the full-length top and bottom DNA strands in single automated synthesizer runs. The crude oligonucleotide mixes are combined and full-length duplexes are amplified by PCR. This method was demonstrated by the synthesis of HIV-1 *rev* and *nef* genes of 393 and 655 bp, respectively.

TABLE I Selected Synthetic Genes[a]

Gene	Length (base pair)	Synthetic method	Reference
Glucagon receptor	1472	Khorana method	Carruthers *et al.* (57)
Red cone pigment	1130	Khorana method	Oprian *et al.* (59)
Green cone pigment	1130	Khorana method	Oprian *et al.* (59)
Blue cone pigment	1080	Khorana method	Oprian *et al.* (59)
Transducin α subunit	1076	Khorana method	Sakmar and Khorana (60)
Rhodopsin	1048	Khorana method	Ferretti *et al.* (61)
Thymidylate synthase	978	Khorana method	Climie and Santi (62)
Mirabilis antiviral protein	759	Khorana method	Habuka *et al.* (63)
Bacteriorhodopsin	757	Khorana method	Nassal *et al.* (64)
Sensory rhodopsin-I	720	Khorana method	Krebs *et al.* (65)
HIV-1 *nef*	655	PCR method	Ciccarelli *et al.* (66)
HBV core protein	560	Khorana method	Nassal (67)
Human interleukin-1α	484	Khorana method	Yanofsky and Zurawski (68)
HIV-1 *rev*	393	PCR method	Ciccarelli *et al.* (66)
Charybdotoxin	151	Khorana method	Park *et al.* (69)

[a] Selected genes synthesized since 1985 are listed for illustrative purposes. Genes are listed in descending order of size. A previous review contains a detailed discussion of earlier work (58).

Synthetic Gene Design

An overview of synthetic gene design is presented below that includes the choice of restriction endonuclease cleavage sites and the choice of an appropriate nucleotide sequence. The discussion applies to all methods for gene synthesis. However, the Khorana method involves the design of joining sites in addition to other considerations.

Choice of Restriction Endonuclease Cleavage Sites

The aim of gene synthesis is to prepare a DNA duplex which encodes the amino acid sequence of a particular protein while containing an optimal number of unique restriction sites. The choice of restriction endonuclease sites to be considered in the design of a synthetic gene includes the following criteria: (i) reliable availability, (ii) high activity and freedom from any exonuclease activity, (iii) a recognition sequence of five or more nucleotides, and (iv) the generation of staggered rather than blunt ends. Because of the degeneracy of the universal genetic code, a very large number of potential

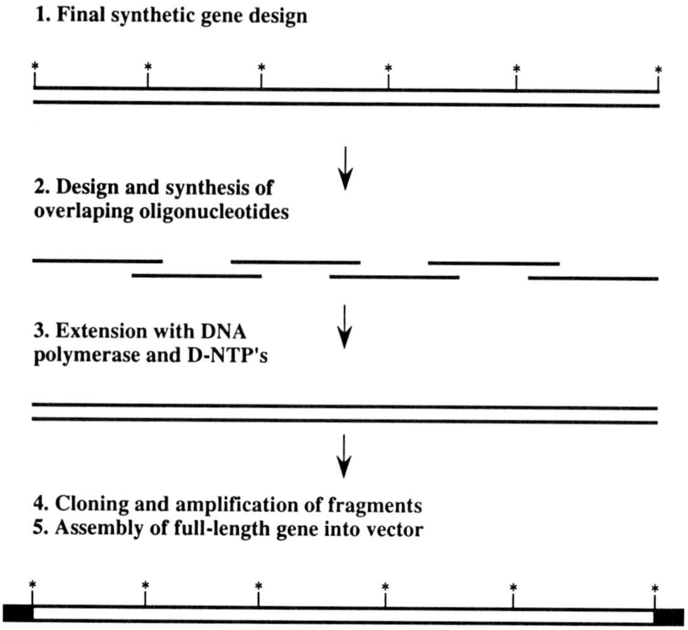

FIG. 2 The synthesis by the clamp (fill-in) method. This method involves the partial synthesis of upper and lower DNA strands followed by DNA polymerase reactions to fill in single-stranded portions. This method can be used for small- to medium-sized genes, or for fragments, which can then be assembled as in Fig. 1.

nucleotide sequences can encode a given amino acid sequence. This potential variability in nucleotide sequence generates a large number of potential restriction maps.

The traditional approach for synthetic gene design was to begin with the native DNA sequence and restriction map, retain all potentially useful restriction sites, and then attempt to add new sites in the intervening sequences (61). This approach, however, is not general. Manual approaches were used to reverse translate restriction endonuclease recognition sequences in order to consider the locations of all possible restriction sites in a particular amino acid sequence (60). More recently, this general approach has been greatly facilitated by the availability of sequence analysis software packages that allow the identification of all of the potential restriction sites within a putative gene. This can be accomplished by starting with the amino acid sequence and using a reverse translation algorithm. A fully degenerative nucleotide sequence is thus created from the initial amino acid sequence. A restriction map can then be created from the degenerate DNA sequence. Once the

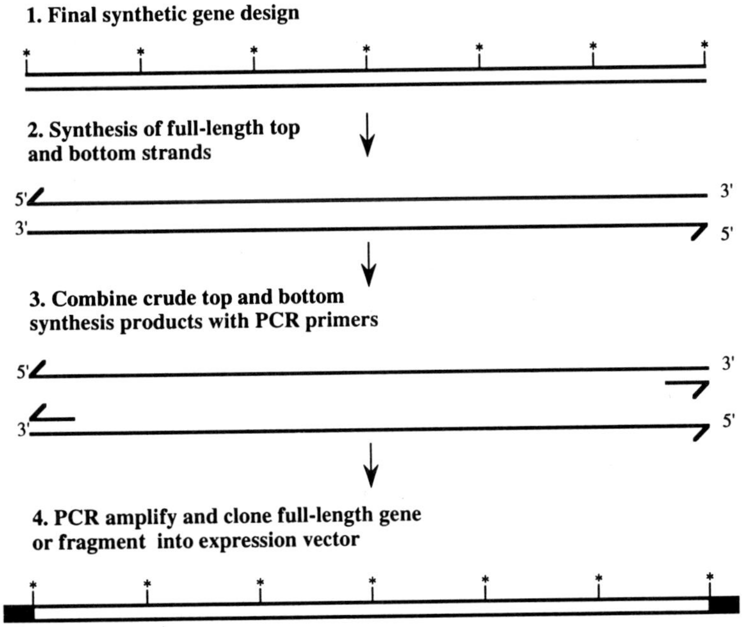

FIG. 3 Gene synthesis by very long oligonucleotide synthesis followed by polymerase chain reaction (PCR). This method involves the total synthesis of both upper and lower DNA strands as two oligonucleotides. The yield of the full-length product is exceedingly low. However, the crude oligonucleotides are combined and subjected to PCR to amplify the full-length product. This method can be used for small- to medium-sized genes, or for fragments, which can then be assembled as in Fig. 1.

restriction map of the degenerate gene sequence is available, it is straightforward to choose a maximum number of evenly spaced sites.

In order to reduce the number of restriction sites identified in the initial map, it is preferable to limit the file of restriction enzymes to those of use for cassette mutagenesis as discussed above. These restriction endonucleases would include those with unambiguous 6- or 8-base recognition sequences (palindromic or interrupted palindromic) that generate cohesive ends of two or more nucleotides. Methylase-sensitive enzymes should be avoided, but in some cases nucleotides outside of the endonuclease recognition sequence can be altered to remove the methylase recognition sequence. Sites for enzymes generating blunt ends can be used if long gaps are present after all enzymes generating staggered ends are considered. However, blunt-end cutters should not be juxtaposed in the restriction map. Convenient cloning sites should be chosen for each end of the gene. For example, a number of genes have been synthesized with an *Eco*RI site at the 5′ end and a *Not*I site at the 3′ end (57, 59, 60).

Some care must be taken so that enzymes that generate identical cohesive overhangs are not juxtaposed. For example, the juxtaposition of *Bcl*1 and *Bgl*2 would not be advisable. The restriction sites chosen define the corresponding DNA sequence and remove nucleotide ambiguity. Once the position for a particular enzyme is chosen, other potential sites for the same enzyme must be removed. Finally, the user-defined DNA sequence can be translated and the resulting amino acid sequence compared to the original to assure that the correct amino acid sequence results. This is important because the amino acid sequence translated from a degenerate codon sequence will not generally match the original amino acid sequence. This is because certain degenerate codons code for more than one amino acid. For example, serine has six codons, four of the form TCN and two of the form AGY (see Table II). The two forms reduce to the single degenerate codon WSN. If this degenerate codon is translated, it will be assigned an unknown amino acid X, since WSN can expand to any of the following: TCN (Ser), ACN (Thr), AGY (Ser), AGR (Arg), TGY (Cys), TGA (Ter), or TGG (Trp).

Creation of Appropriate Nucleotide Sequence

After defining the nucleotide sequence that corresponds to the desired restriction map, a majority of the gene sequence is still undefined. In general, the natural sequence can be retained. However, other considerations such as codon usage, potential secondary structure, and guanine-cytosine content can be considered as discussed below.

Codon Usage Considerations

In cases where a particular expression system is established, codon usage bias can be considered to attempt to optimize the level of expression (73). In some cases, codon usage can be used as well to slow translation of signal sequences so that cellular membrane translocation systems are not saturated. This was the case for the expression in *E. coli* of the gene for the light-driven proton pump bacteriorhodopsin from *Halobacterium halobium* (64). Some investigators have also placed a mammalian translation initiation consensus sequence immediately preceding the initiation methionine codon (60, 61, 74).

A dramatic example of using gene synthesis to affect expression level by altering codon bias was demonstrated in the case of the synthetic gene for human interleukin-2 (IL-2) (75). In designing the synthetic gene, the percentage of preferred codons for *E. coli* was increased from 43 to 85%. While message levels and stability were similar for the cDNA and synthetic

Table II Single-Letter Nucleotide Base Codes[a]

Code	Nucleotide base
A	Adenine
C	Cytosine
G	Guanine
T	Thymine
U	Uracil
R	(A or G)
Y	(C or T/U)
K	(G or T/U)
M	(A or C)
S	(C or G)
W	(A or T)
B	(C or G or T/U)
D	(A or G or T/U)
H	(A or C or T/U)
V	(A or C or G)
N	(A or C or G or T/U)

[a] IUPAC-IUB Commission on Biological Nomenclature, *J. Biol. Chem.* **243,** 3557 (1968).

genes expressed in *E. coli,* the synthetic gene produced up to 16 times more IL-2 than the cDNA.

Secondary Structure Considerations

Computer algorithms can also be employed to identify areas of significant secondary structure. In many cases, nucleotide sequences can be modified to remove potential hairpin loops that might affect DNA sequencing or mRNA stability.

Guanine–Cytosine Content Considerations

The synthetic gene design process allows for the reduction of G-C content if desired. Stretches of four or more guanines or cytosines can be avoided where possible to minimize potential difficulties in oligonucleotide synthesis or DNA sequencing.

Choice of Sites for Enzymatic Ligation of Synthetic Duplexes (Khorana Method)

Once a full-length synthetic gene has been designed, both strands must be assembled from individual oligonucleotides. The lengths of the oligonucleotides are designed to allow adequate purification. Generally, 70 to 90 bases are the maximum length that can be purified by HPLC or acrylamide gel electrophoresis.

One strategy of joining synthetic duplexes is based on the classical principles developed by Khorana and co-workers (20, 21). In this strategy, complementary oligonucleotides are synthesized. However, to avoid low-efficiency ambiguous blunt-end to blunt-end ligation of synthetic duplexes, the complementary oligonucleotides are designed with short 5′ overhangs. The optimal length of overhangs is 4 or 5 bases. This length is optimal for DNA ligase activity and allows enough unique sequences so that a relatively large number of duplexes can be ligated batchwise. No cohesive overhang within a gene fragment should be self-complementary, and the overhangs should obviously be unique within a gene fragment. When evaluating for self-complementarity, guanine and thymine should be considered to form stable base pairs. Restriction site overhangs are not useful for enzymatic joining of synthetic duplexes because they are generally palindromic and self-ligation will occur.

Synthetic Gene Construction (Khorana Method)

Methods are presented below for the construction of synthetic genes by the Khorana method where both upper and lower strands are synthesized. Oligonucleotides of moderate length are synthesized and purified. Phosphorylation of the 5′ ends of oligonucleotides with noncomplementary cohesive overhangs is carried out. Sets of oligonucleotides are annealed and joined with DNA ligase. The full-length gene fragment is isolated and cloned. Detailed methods relevant to other synthetic gene construction approaches are presented elsewhere as discussed above. In addition, detailed experimental methods for various mutagenesis procedures, such as combinatorial cassette mutagenesis, are presented in original sources.

The methods presented in this chapter for the design and synthesis of synthetic genes by the Khorana method were recently employed in the synthesis of a 1472-bp gene for the rat glucagon receptor (57). Glucagon is a 29 amino acid pancreatic hormone. Glucagon binds to a seven transmembrane domain G protein-coupled receptor on cells of target tissues such as liver and fat resulting in the activation of glycogenolysis and gluconeogenesis. The molecular mechanism of hormone–receptor interaction and receptor activation is not known. In order to facilitate these studies by site-directed mutagenesis, we have designed and synthetized a gene for the rat glucagon receptor (76). The gene codes for the native 485 amino acid protein, but contains 47 unique restriction sites. The full-length gene was assembled from three fragments of 362, 546, and 564 bp. The fragments consisted of 8, 14, and 14 oligonucleotides, respectively, ranging in length from 74 to 91 bases (Fig. 4).

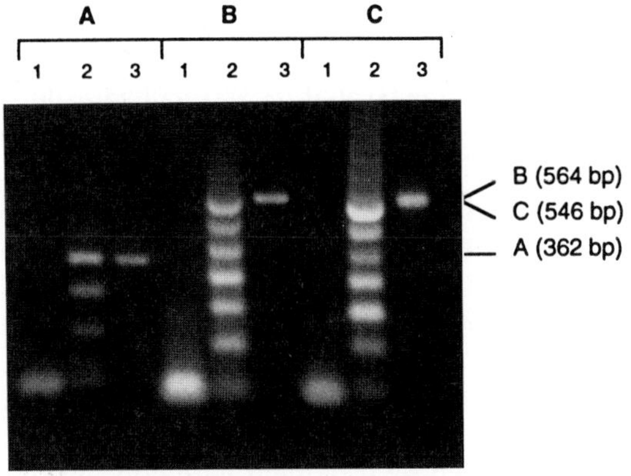

FIG. 4 Synthesis of a gene for the rat glucagon receptor. The gene is 1472 bp in length and contains 91 unique restriction sites. Agarose gel electrophoresis of annealed oligonucleotide duplexes, ligated duplexes, and purified DNA fragments was carried out for fragments A, B, and C of the synthetic glucagon receptor gene on a 1.5% agarose gel stained with ethidium bromide. Fragments A, B, and C consisted of 8, 14, and 14 oligonucleotides, respectively. Oligonucleotides ranged in length from 74 to 91 bases. All but the terminal oligonucleotides in each fragment were 5′ end-phosphorylated. Equimolar concentrations of oligonucleotides were annealed for each fragment (lanes A1, B1, C1). Ligation of the annealed oligonucleotide mixes for fragments A, B, and C resulted in a ladder of four, seven, and seven main bands, respectively (lanes A2, B2, C2). For example, in the case of fragment A, the fastest running band in lane A2 corresponds to unligated annealed duplexes. The other three bands represent the possible combinations of two, three, and four duplexes joined together. Similarly, in lanes B2 and C2, the seven main bands represent annealed duplexes and the possible combinations of two through seven duplexes joined together. The full-length synthetic fragments A, B, and C were isolated, cloned, and amplified. The correct nucleotide sequences were confirmed by dideoxy DNA sequencing on double-stranded plasmid DNA. Lanes A3, B3, and C3 show the fragments isolated from cloning vectors after restriction endonuclease cleavage and agarose gel electrophoresis. Fragment sizes are 362, 564, and 546 bp, respectively.

Experimental Procedures

Oligonucleotide Synthesis

Automated oligonucleotide synthesis can be easily carried out on oligonucleotide synthesizers with commercially available solvents and reagents. The most commonly used chemistry involves the phosphite triester approach using protected β-cyanoethyl phosphoramidite nucleosides. The fully protected 3′ terminal phosphoramidite of the oligonucleotide is coupled to a solid support such as control pore glass or polysterene (77). After protic acid treatment to remove the 5′ protecting group, the fully protected incoming phosphoramidite is activated by tetrazole so that a phosphite triester bond is formed at high efficiency. The small amount of unreacted 5′ hydroxyl of the first nucleoside is capped by a quantitative reaction with acetic anhydride in the presence of 1-methylimidazole. Finally, the newly formed internucleotide linkage is converted from a phosphite triester to a more stable phosphotriester by oxidation with idodine where water is the oxygen donor. The 5′ hydroxyl of the dinucleotide can now be deprotected with acid treatment to complete a cycle. The cycle is repeated until the full-length oligonucleotide is obtained. Thus, the oligonucleotide is elongated from 3′ to 5′. Cleavage from the support and removal of phosphate and exocyclic amine protecting groups are achieved by treatment with concentrated ammonium hydroxide.

For synthesis of the glucagon receptor gene, automated oligonucleotide synthesis is performed on an Applied Biosystems Model 392 DNA synthesizer. Phosphoramidite chemistry is employed using 0.2 μmol or 40 pmol synthesis scales and standard cycle routines. Each synthetic oligonucleotide is automatically cleaved from the solid support after removal of the terminal 5′ hydroxyl protecting group. Each oligonucleotide solution is transferred into a screw top vial. After the addition of 2 ml of fresh concentrated ammonium hydroxide, the vial is tightly capped and heated at 55°C for at least 8 hr. Each fully deprotected oligonucleotide is dried by vacuum centrifugation in a polypropylene tube and the pellets are dissolved in 50 μl of TE (10 m*M* Tris–HCl, 1 m*M* EDTA, pH 7.4). An ultraviolet spectrum is measured from 310 to 210 nm after the proper dilution in TE is made. The yield in total absorbance units at 260 nm is calculated.

Oligonucleotide Purification and Characterization

The synthetic oligonucleotides are mixed with denaturing polyacrylamide load buffer containing formamide, boiled for 3 min, quick-chilled on ice, and subjected to electrophoresis on a 40-cm-long, 3-mm-thick 6% polyacrylamide gel. The gel is run until the product band is estimated to be at least half-way down the gel. The gel is removed from the glass plates and covered with plastic wrap. The DNA is displayed by ultraviolet shadowing with a hand-held lamp, and the full-length synthetic oligonucleotide band is outlined with a marker and cut out with a razor blade. The gel fragment is transferred into a polypropylene tube, crushed thoroughly with a glass rod, and eluted from the gel into 10 ml of buffer (1 *M* triethylammonium bicarbonate, pH 7.5) by gently mixing for 12 to 24 hr at room temperature.

Triethylammonium bicarbonate (TEAB) buffer is most conveniently made by transferring 1 mol of liquid triethylamine (TEA) by weight into a 2-liter flask and adding 850 ml of distilled, deionized water. Carbon dioxide (CO_2) gas is then bubbled through the liquid (in a fume hood if possible) until the TEA goes into solution. The bubbling is continued until the pH of the buffer drops to about 7.5. The buffer can then be brought to a final volume of 1 liter. The TEAB should be stored tightly sealed at 4°C. The pH should be checked periodically and can be adjusted by additional CO_2 treatment.

Oligonucleotides are purified from gel contaminants and desalted using C18 reverse-phase cartridges (SEP-PAK, Waters) (5). Cartridges are affixed to 20-ml syringes and are primed by successive washes with acetonitrile (10 ml), 17% acetonitrile in 100 m*M* TEAB (10 ml), and 25 m*M* TEAB (10 ml). The TEAB oligonucleotide solutions are applied to the cartridges after spinning out gel fragments in a table top centrifuge. The cartridges are then washed with 25 m*M* TEAB (10 ml), and the oligonucleotides are eluted with 17% acetonitrile in 100 m*M* TEAB (4.5 ml). The eluted volume is dried by vacuum centrifugation and the pellet dissolved in TE. An ultraviolet spectrum is measured from 310 to 210 nm after the proper dilution in TE is made. The concentration of purified oligonucleotide is determined using the estimate that a solution with an absorbance at 260 nm of 1 has a concentration of 40 μg/ml.

5′ Terminal Phosphorylation

The synthetic oligonucleotides involved in joining reactions are 5′ end-phosphorylated in order to serve as substrates for DNA ligase. Phosphorylation reactions can be carried out individually (61). However, it is convenient to carry out the phosphorylations of all of the oligonucleotides within a fragment batchwise. The 5′ oligonucleotide on the upper strand and the 5′ oligonucleotide on the lower strand of a particular fragment are not phosphorylated. A batchwise reaction mixture is prepared with 100 pmol of each oligonucleotide in 50 m*M* Tris–HCl (pH 8.0). The mixture is heated for 3 min at 90°C and quick-chilled on ice. Concentrated stock solutions are added to give final concentrations of 10 m*M* $MgCl_2$, 2 m*M* spermidine, 10 m*M* DTT, 1 m*M* ATP, and 4 units of T4 polynucleotide kinase (New England Biolabs). The mixture is incubated for 45 min at 37°C, then heated 3 min at 90°C, and quick-chilled on ice. The phosphorylation reaction is repeated a second time at 37°C for 45 min with the addition of DTT and 4 units of T4 polynucleotide kinase. The reaction is stopped by adding EDTA to a final concentration of 25 m*M*. Phenol chloroform extraction is carried out followed by application to a prewashed spun-filter microconcentrator (Microcon-30, Amicon) to desalt and remove AMP and excess ATP from the sample. The concentration step is repeated three times after the addition of TE and the sample is recovered in a total volume of 50 μl of TE.

Annealing and Ligation Reaction

Complementary upper and lower strand oligonucleotides for an entire gene fragment are annealed by adding 100 pmol each of the 5′ upper strand and 5′ lower strand oligonucleotides to the phosphorylated oligonucleotide mix in a final volume of 100 μl in 50 m*M* Tris–HCl (pH 8.0), 10 m*M* $MgCl_2$. The mixture is overlayed with 50 μl of mineral oil, heated to 90°C for 3 min, and slow-cooled to room temperature over 2 to 3 hr. An aliquot is saved for analysis on an agarose gel. The remaining solution is stored at −20°C or immediately prepared for the ligation reaction.

The annealed oligonucleotides are joined by incubation for 16 hr at 14°C in 66 m*M* Tris–HCl (pH 7.5), 5 m*M* $MgCl_2$, 1 m*M* DTT, 1 m*M* ATP, and 25 units of T4 DNA ligase (Boehringer-Mannheim Biochemical) in a final volume of 100 μl. Phenol chloroform extraction followed by microconcentration on a Microcon-30 unit removes protein and desalts the solution. An aliquot is saved for agarose gel analysis.

Recovery, Cloning, and Characterization of Ligation Product

The products of the ligation reaction are separated on a preparative 1.5% agarose gel and visualized by ethidium bromide staining. The full-length synthetic DNA fragment band is cut out of the gel and purified from the agarose with the Qiaex Gel Extraction Kit (Qiagen). The DNA fragment is cloned into an appropriate cloning vector or expression vector that has been linearized by restriction endonuclease digestion to produce the two overhangs corresponding to the fragment ends. Ligation is carried out on 100 fmol of linear vector and 300 fmol of synthetic DNA fragment in a reaction mixture as previously described using 5 units of T4 DNA ligase for 16 hr at 14°C. As a control, 100 fmol of linearized vector are also subjected to the same treatment. The ligation mixtures are ethanol-precipitated and resuspended in 10 μl of TE.

Transformation of *E. coli* is carried out on 40 μl of XL1-Blue Elecro-Competent cells (Stratagene) by electroporation with 20% of the ligation reaction in a 1-mm cuvette using the Gene Pulser/*E. coli* Pulser system (Bio-Rad). After electroporation, 1 ml of chilled sterile SOC outgrowth medium (2% Bacto tryptone, 0.5% Bacto yeast extract, 10 m*M* NaCl, 2.5 m*M* KCl, 10 m*M* $MgCl_2$, 10 m*M* $MgSO_4$, 20 m*M* glucose) is immediately added to the cuvette. The cell suspension is transferred to a 10-ml culture tube and incubated for 1 hr in a 37°C shaker to allow phenotypic expression. Cells are spread on LB/ampicillin plates, which are then incubated for 16 hr at 37°C. Ampicillin-resistant colonies are streak-purified on fresh LB/ampicillin plates, and single colonies are selected for inoculating liquid LB/ampicillin (75 μg/ml) cultures.

DNA minipreps of 3-ml overnight cultures are prepared following the Qiagen-tip 20 protocol (Qiagen). Restriction analysis and subsequent dideoxy sequencing on double-stranded plasmid DNA (Sequenase, United States Biochemical) are carried out. A clone with the correct DNA sequence is selected for assembly of the full-length synthetic gene into an expression vector.

Cassette Mutagenesis

Site-directed mutagenesis of the synthetic gene is accomplished by synthesizing a pair of complementary oligonucleotides to form a duplex containing the desired codon alteration and the appropriate cohesive terminal overhangs. After purification and annealing as previously described, the 5′ end nonphosphorylated synthetic duplex is ligated into the plasmid/gene DNA fragment linearized with the appropriate restriction endonucleases. Molar ratios of synthetic duplex to linearized vector are 3 : 1. Procedures carried out after ligation are as presented above. Procedures for other types of cassette mutagenesis, such as combinatorial cassette mutagenesis, can be derived from published sources.

Expression of Synthetic Genes

As discussed above, two of the advantages of the use of synthetic genes are that they can be easily transferred among a variety of vectors and that codon usage can be optimized where relevant to achieve maximal levels of expression. Synthetic receptor genes will generally be expressed in mammalian cells in tissue culture where pharmacological and cellular physiological effects can be correlated with structural changes introduced by mutation. In the case of the synthetic gene for bovine rhodopsin, large quantities of the opsin apoprotein can be produced in monkey kidney cells by transfection where transcription is under the control of the human adenovirus major-late promotor (78). The apoprotein in the plasma membrane can be regenerated with the chromophore 11-*cis*-retinal to form rhodopsin. The recombinant rhodopsin can be solubilized with detergent treatment and purified using an affinity adsorption method (3, 78).

Conclusions

Synthetic genes have been expressed in a variety of heterologous expression systems. For example, synthetic genes have been expressed in *E. coli* (64), in monkey kidney cells in tissue culture (61), in insect Sf9 cells (79), and in yeast (80–82). Perhaps the most extensively studied synthetic genes are those of bacteriorhodopsin (64) and bovine rhodopsin (61). In visual pigment structure–function studies, synthetic receptor genes for the rhodopsin and for the human blue, green, and red cone pigment genes have been expressed in mammalian cells and purified from cell extracts after reconstitution with 11-*cis*-retinal chromophore (59). Purified site-directed mutant pigments have been studied by a variety of biochemical (83–86) and biophysical techniques (87–89). These studies have led to a greater understanding of the mechanism of wavelength regulation by visual pigments (84, 90–93) and of the mechanism of rhodopsin–transducin interaction (4, 86). Synthetic genes for the α subunit and γ subunit of the rod cell G protein, transducin, have also been expressed in a variety of systems including *E. coli* (γ subunit) (94), COS cells (α subunit) (60), and insect Sf9 cells (α subunit) (79).

In conclusion, gene synthesis should be considered when extensive long-term structure–function studies are planned. The initial investment in time and expense is offset by the ease and flexibility of DNA manipulation using a synthetic gene. Improved economical automated DNA synthesis, PCR techniques, and the availability of a large number of quality restriction endonucleases have combined to make gene synthesis possible for nearly all molecular biology laboratories.

Acknowledgment

This chapter is dedicated with respect and admiration to H. Gobind Khorana.

References

1. W. C. Probst, L. A. Snyder, D. I. Schuster, J. Brosius, and S. C. Sealfon, *DNA Cell Biol.* **11,** 1 (1992).
2. O. Moro, J. Lameh, P. Högger, and W. Sadée, *J. Biol. Chem.* **268,** 22273 (1993).
3. R. R. Franke, T. P. Sakmar, R. M. Graham, and H. G. Khorana, *J. Biol. Chem.* **267,** 14767 (1992).
4. R. R. Franke, B. König, T. P. Sakmar, H. G. Khorana, and K. P. Hofmann, *Science* **250,** 123 (1990).
5. K.-M. Lo, S. S. Jones, N. R. Hackett, and H. G. Khorana, *Proc. Natl. Acad. Sci. U.S.A.* **91,** 2285 (1984).
6. A. A. Pakula and M. I. Simon, *Proc. Natl. Acad. Sci. U.S.A.* **89,** 4144 (1992).
7. J. U. Bowie and R. T. Sauer, *Proc. Natl. Acad. Sci. U.S.A.* **86,** 2152 (1989).
8. J. F. Reidhaar-Olson and R. T. Sauer, *Science* **241,** 53 (1988).
9. J. D. York, P. Li, and S. J. Gardell, *J. Biol. Chem.* **266,** 8495 (1991).
10. J. H. Miller, C. Coulondre, M. Hofer, U. Schmeissner, H. Sommer, A. Schmitz, and P. Lu, *J. Mol. Biol.* **131,** 191 (1979).

11. K. Poindexter, R. Jerzy, and R. B. Gayle, III, *Nucleic Acids Res.* **19,** 1899 (1991).
12. W. C. Chan and T. Ferenci, *J. Bacteriol.* **175,** 858 (1993).
13. J. C. Hu, N. E. Newell, B. Tidor, and R. T. Sauer, *Protein Sci.* **2,** 1072 (1993).
14. L. M. Gregoret and R. T. Sauer, *Proc. Natl. Acad. Sci. U.S.A.* **90,** 4246 (1993).
15. A. P. Arkin and D. C. Youvan, *Proc. Natl. Acad. Sci. U.S.A.* **89,** 7811 (1992).
16. S. Delagrave, E. R. Goldman, and D. C. Youvan, *Protein Eng.* **6,** 327 (1993).
17. I. G. Ivanov, *Anal. Biochem.* **189,** 213 (1990).
18. T. K. Chang, S. A. Iverson, C. G. Rodrigues, C. N. Kiser, A. Y. Lew, J. P. Germanas, and J. H. Richards, *Proc. Natl. Acad. Sci. U.S.A.* **88,** 1325 (1991).
19. E. L. Brown, R. Belagaje, M. J. Ryan, and H. G. Khorana, *in* "Methods in Enzymology" (R. Wu, ed.), Vol. 68, p. 109. Academic Press, New York, 1979.
20. H. G. Khorana, *Bioorg. Chem.* **7,** 351 (1978).
21. H. G. Khorana, *Science* **203,** 614 (1979).
22. N. Y. Theriault, J. B. Carter, and S. P. Pulaski, *BioTechniques* **6,** 470 (1988).
23. M. D. Edge, A. R. Green, G. R. Heathcliffe, P. A. Meacock, W. Schuch, D. B. Scanlon, T. C. Atkinson, C. R. Newton, and A. F. Markham, *Nature (London)* **292,** 756 (1981).
24. S. Beck von Bodman, M. A. Schuler, D. R. Jollie, and S. G. Sligar, *Proc. Natl. Acad. Sci. U.S.A.* **83,** 9443 (1986).
25. R. Brousseau, R. Scarpulla, W. Sung, H. M. Hsiung, S. A. Narang, and R. Wu, *Gene* **17,** 279 (1982).
26. R. A. Calogero, C. L. Pon, and C. O. Gualerzi, *Mol. Gen. Genet.* **208,** 63 (1987).
27. A. Cravador, P. Jacobs, E. A. Van, C. Lacroix, B. Colau, P. van Alphen, A. Herzog, and A. Bollen, *Biochemie* **67,** 829 (1985).
28. V. Davasse and J. M. Moulis, *Biochem. Biophys. Res. Commun.* **185,** 341 (1992).
29. T. Dudler, W. Q. Chen, S. Wang, T. Schneider, R. R. Annand, R. O. Dempcy, R. Crameri, M. Gmachl, M. Suter, and M. H. Gelb, *Biochem. Biophys. Acta* **1165,** 201 (1992).
30. S. J. Eisenbeis, M. S. Nasoff, S. A. Boble, L. P. Bracco, D. R. Dodds, and M. H. Caruthers, *Proc. Natl. Acad. Sci. U.S.A.* **82,** 1084 (1985).
31. M. Eren and R. P. Swenson, *J. Biol. Chem.* **264,** 14874 (1989).
32. L. B. Farrell, D. P. Gearing, and P. Nagley, *Eur. J. Biochem.* **173,** 131 (1988).
33. F. Georges, R. Brousseau, J. Michneiwicz, G. Prefontaine, J. Stawinski, W. Sung, R. Wu, and S. A. Narang, *Gene* **27,** 201 (1984).
34. N. Groch, H. Schindelin, A. S. Scholtz, U. Hahn, and U. Heinemann, *Eur. J. Biochem.* **207,** 677 (1992).
35. G. Gross, C. Mielke, I. Hollatz, H. Blocker, and R. Frank, *J. Biol. Chem.* **265,** 17627 (1990).
36. Z. Hostomsky, K. Appelt, and R. C. Ogden, *Biochem. Biophys. Res. Commun.* **161,** 1056 (1989).
37. M. L. Howell and K. M. Blumenthal, *J. Biol. Chem.* **264,** 15268 (1989).
38. H. M. Hsiung, W. L. Sung, R. Brousseau, R. Wu, and S. A. Narang, *Nucleic Acids Res.* **8,** 5753 (1980).
39. M. Ikehara, E. Ohtsuka, T. Tokunaga, Y. Taniyama, S. Iwai, K. Kitano, S. Miyamoto, T. Ohgi, Y. Sakuragawa, K. Fujiyama, T. Ikari, M. Kobayashi, T.

Miyake, S. Shibahara, A. Ono, T. Ueda, T. Tanaka, H. Baba, T. Miki, A. Sakurai, T. Oishi, O. Chisaka, and K. Matsubara, *Proc. Natl. Acad. Sci. U.S.A.* **81,** 5956 (1984).

40. E. Jay, D. MacKnight, C. Lutze-Wallace, D. Harrison, P. Wishart, W.-Y. Liu, V. Asundi, L. Pomeroy-Cloney, J. Rommens, L. Eglington, J. Pawlak, and F. Jay, *J. Biol. Chem.* **259,** 6311 (1984).
41. E. Jay, J. Rommens, L. Pomeroy-Cloney, D. MacKnight, C. Lutze-Wallace, P. Wishart, D. Harrison, W.-Y. Liu, V. Asundi, M. Dawood, and F. Jay, *Proc. Natl. Acad. Sci. U.S.A.* **81,** 2290 (1984).
42. P. T. Jones, P. H. Dear, J. Foote, M. S. Neuberger, and G. Winter, *Nature (London)* **321,** 522 (1986).
43. H. Kaji, I. Kumagai, A. Takeda, K. Miura, and T. Samejima, *J. Biochem. (Tokyo)* **105,** 143 (1989).
44. N. Katunuma, M. Yamato, E. Kominami, and Y. Ike, *FEBS Lett.* **238,** 116 (1988).
45. I. Lindley, H. Aschauer, J. M. Seifert, C. Lam, W. Brunowsky, E. Kownatzki, M. Thelen, P. Peveri, B. Dewald, V. von Tscharner, A. Walz, and M. Baggiolini, *Proc. Natl. Acad. Sci. U.S.A.* **85,** 9199 (1988).
46. K. P. Nambiar, J. Stackhouse, S. R. Presnell, and S. A. Benner, *Eur. J. Biochem.* **163,** 67 (1987).
47. Y. A. Ovchinnikov, V. A. Efimov, I. N. Ivanova, S. V. Reverdatto, N. P. Skiba, and O. G. Chakhmakhcheva, *Gene* **31,** 65 (1984).
48. R. Quaas, Y. McKeown, P. Stanssens, R. Frank, H. Blöcker, and U. Hahn, *Eur. J. Biochem.* **173,** 617 (1988).
49. D. M. Roberts, R. Crea, M. Malecha, G. Alvarado-Urbina, R. H. Chiarello, and D. M. Watterson, *Biochemistry* **24,** 5090 (1985).
50. D. K. Smith, T. Kassam, B. Singh, and J. F. Elliott, *J. Bacteriol.* **174,** 5820 (1992).
51. M. Strauss, J. Stollwerk, B. Lenarcic, V. Turk, K. D. Jany, and H. G. Gassen, *Biol. Chem. Hoppe-Seyler* **369,** 1019 (1988).
52. M. Suzuki, S. Sumi, A. Hasegawa, T. Nishizawa, K. Miyoshi, S. Wakisaka, T. Miyake, and F. Misoka, *Proc. Natl. Acad. Sci. U.S.A.* **79,** 2475 (1982).
53. W. B. von Wilcken-Bergmann, D. Tils, J. Sartorius, E. A. Auerswald, W. Schröder, and B. Müller-Hill, *EMBO J.* **5,** 3219 (1986).
54. R. Wetzel, H. L. Heyneker, D. V. Goeddel, P. Jhurani, J. Shapiro, R. Crea, T. L. Low, J. E. McClure, G. B. Thurman, and A. L. Goldstein, *Biochemistry* **19,** 6096 (1980).
55. A. F. Worrall, C. Evans, and D. C. Wilton, *Biochem. J.* **278,** 365 (1991).
56. S. M. Zurawski, K. Pope, H. Cherwinski, and G. Zurawski, *Gene* **49,** 61 (1986).
57. C. J. L. Carruthers, C. G. Unson, H. N. Kim, and T. P. Sakmar, *J. Biol. Chem.* **269,** in press.
58. J. Engels and E. Uhlmann, *Adv. Biochem. Eng. Biotechnol.* **37,** 73 (1988).
59. D. D. Oprian, A. B. Asenjo, N. Lee, and S. L. Pelletier, *Biochemistry* **30,** 11367 (1991).
60. T. P. Sakmar and H. G. Khorana, *Nucleic Acids Res.* **16,** 6361 (1988).

61. L. Ferretti, S. S. Karnik, H. G. Khorana, M. Nassal, and D. D. Oprian, *Proc. Natl. Acad. Sci. U.S.A.* **83,** 599 (1986).
62. S. Climie and D. V. Santi, *Proc. Natl. Acad. Sci. U.S.A.* **87,** 633 (1990).
63. N. Habuka, Y. Murakami, M. Noma, T. Kudo, and K. Horikoshi, *J. Biol. Chem.* **264,** 6629 (1989).
64. M. Nassal, T. Mogi, S. S. Karnik, and H. G. Khorana, *J. Biol. Chem.* **262,** 9264 (1987).
65. M. P. Krebs, E. N. Spudich, H. G. Khorana, and J. L. Spudich, *Proc. Natl. Acad. Sci. U.S.A.* **90,** 3486 (1993).
66. R. B. Ciccarelli, P. Gunyuzlu, J. Huang, C. Scott, and F. T. Oakes, *Nucleic Acids Res.* **19,** 6007 (1991).
67. M. Nassal, *Gene* **66,** 279 (1988).
68. S. D. Yanofsky and G. Zurawski, *J. Biol. Chem.* **265,** 13000 (1990).
69. C. S. Park, S. F. Hausdorff, and C. Miller, *Proc. Natl. Acad. Sci. U.S.A.* **88,** 2046 (1991).
70. H. G. Khorana, *Pure Appl. Chem.* **17,** 349 (1968).
71. A. Di Donato, M. De Nigris, N. Russo, S. Di Biase, and G. D'Alessio, *Anal. Biochem.* **212,** 291 (1993).
72. X. C. Villarreal and G. L. Long, *Anal. Biochem.* **197,** 362 (1991).
73. P. M. Sharp, E. Cowe, D. G. Higgins, D. C. Shields, K. H. Wolfe, and F. Wright, *Nucleic Acids Res.* **16,** 8207 (1988).
74. M. Kozak, *Nucleic Acids Res.* **12,** 857 (1984).
75. D. P. Williams, D. Regier, D. Akiyoshi, F. Genbauffe, and J. R. Murphy, *Nucleic Acids Res.* **16,** 10453 (1988).
76. L. J. Jelinek, S. Lok, G. B. Rosenberg, R. A. Smith, F. J. Grant, S. Biggs, P. A. Bensch, J. L. Kuijper, P. O. Sheppard, C. A. Sprecher, P. J. O-Hara, D. Foster, K. M. Walker, L. H. J. Chen, P. A. McKernan, and W. Kindsvogel, *Science* **259,** 1614 (1993).
77. M. D. Matteucci and M. H. Caruthers, *J. Am. Chem. Soc.* **103,** 3185 (1981).
78. D. D. Oprian, R. S. Molday, R. J. Kaufman, and H. G. Khorana, *Proc. Natl. Acad. Sci. U.S.A.* **84,** 8874 (1987).
79. E. Faurobert, A. Otto-Bruc, P. Chardin, and M. Chabre, *EMBO J.* **12,** 4191 (1993).
80. M. S. Urdea, J. P. Merryweather, G. T. Mullenbach, D. Coit, U. Heberlein, P. Valenzuela, and P. J. Barr, *Proc. Natl. Acad. Sci. U.S.A.* **80,** 7461 (1983).
81. T. Tokunaga, S. Iwai, H. Gomi, K. Kodama, E. Ohtsuka, M. Ikehara, O. Chisaka, and K. Matsubara, *Gene* **39,** 117 (1985).
82. T. Tanaka, S. Kimura, and Y. Ota, *Nucleic Acids Res.* **15,** 3178 (1987).
83. G. B. Cohen, T. Yang, P. R. Robinson, and D. D. Oprian, *Biochemistry* **32,** 6111 (1993).
84. T. A. Zvyaga, K. C. Min, M. Beck, and T. P. Sakmar, *J. Biol. Chem.* **268,** 4661 (1993).
85. T. P. Sakmar, R. R. Franke, and H. G. Khorana, *Proc. Natl. Acad. Sci. U.S.A.* **86,** 8309 (1989).
86. K. Fahmy and T. P. Sakmar, *Biochemistry* **32,** 7229 (1993).

87. S. W. Lin, T. P. Sakmar, R. R. Franke, H. G. Khorana, and R. A. Mathies, *Biochemistry* **31,** 5105 (1992).
88. K. Fahmy, F. Jäger, M. Beck, T. P. Sakmar, and F. Siebert, *Proc. Natl. Acad. Sci. U.S.A.* **90,** 10206 (1993).
89. J. Resek, Z. T. Farahbakhsh, S. L. Hubbell, and H. G. Khorana, *Biochemistry* **32,** 12025 (1993).
90. T. Chan, M. Lee, and T. P. Sakmar, *J. Biol. Chem.* **267,** 9478 (1992).
91. Z. Wang, A. B. Asenjo, and D. D. Oprian, *Biochemistry* **32,** 2125 (1993).
92. E. A. Zhukovsky and D. D. Oprian, *Science* **246,** 928 (1989).
93. T. P. Sakmar, R. R. Franke, and H. G. Khorana, *Proc. Natl. Acad. Sci. U.S.A.* **88,** 3079 (1991).
94. R. L. Brown and L. Stryer, *Proc. Natl. Acad. Sci. U.S.A.* **86,** 4922 (1989).

[18] Use of Receptors Expressed in *Escherichia coli* to Study Autoimmunity against G Protein-Coupled Membrane Proteins

Johan Hoebeke, Jean-Gérard Guillet, and
A. Donny Strosberg

Introduction

Since the seminal observation of Patrick and Lindstrom (1) of an autoimmune response to the nicotinic acetylcholine receptor in patients with myasthenia gravis, attention has been focused on the role of autoantibodies against hormone or neurotransmitter receptors in several diseases of unknown origin. As a result, it has been found that the effects of autoantibodies directed against the insulin receptor can explain the phenomenology of insulin-resistent diabetes in some cases (2).

The first reports of the possible involvement of autoantibodies against G protein-coupled receptors appeared at the beginning of the 1980s. Thyroid-stimulating autoantibodies, whose target was defined as the thyrotropin receptor, were observed in patients with Graves' disease (3, 4).

The adrenergic hyporesponsiveness seen in allergic rhinitis and asthma was ascribed to autoantibodies against the β_2-adrenergic receptors (5). Borda *et al.* found autoantibodies in patients with Chagas' cardiomyopathy which modulated the activity of cardiac β-adrenoceptors (6). Based on these observations, Wallukat and Wollenberger used cultures of neonatal rat heart myocytes to confirm the existence of autoantibodies (7). These antibodies blocked the β_2-adrenergic receptor in patients with allergic asthma and stimulated the β_1-adrenergic receptor in patients with dilated cardiomyopathy. The latter finding was confirmed by Limas *et al.* (8). Finally, autoantibodies against the serotonin receptor were found in an autistic child (9).

All these observations lacked a molecular basis because no information was available on the structure of the target receptors or, more particularly, of the potential B cell epitopes involved in autoantibody recognition. Two major advances have made studies on their structures feasible. One is the successful cloning and sequencing of the β_2-adrenergic receptor of hamster lung (10). The subsequent explosion of information on the structure of the G protein-coupled membrane receptors (11) has prepared the way for a structural approach to the immunological properties of the members of this

Methods in Neurosciences, Volume 25

superfamily. The second is the successful functional expression of the β_2-adrenergic receptor in *E. coli* (12). This has made available a tool that can serve as both an immunogen and as a target in studies on the structural basis of immune recognition and regulation of G protein-coupled receptors. We have used these developments in a series of studies on the localization of immune epitopes on some of the G protein-coupled membrane receptors. We have also correlated these structural features with the functional implications of receptor interactions with induced or human autoantibodies.

Theoretical and Experimental Design Considerations

Use of Bacterial Expressed Receptors

Although receptor protein is essential for studying the mechanisms of signal transduction induced by anti-receptor autoantibodies, the difficulty of producing receptors in mammalian cells has so far limited their direct use for screening purposes or as a source of antigen. Because functional bacteriorhodopsin, a member of the receptor family, is produced by *Halobacterium halobium,* we postulated that functional G protein-coupled receptors could also be produced in *E. coli*.

The first successful attempt to express the human β_2-adrenergic receptor in *E. coli* was made using β-galactosidase as a fusion protein (12). The bacteria synthesized a considerable amount of fusion protein, but only a small fraction of the receptor generated appeared to be functional. This functional receptor was separated from its fusion partner by proteolysis and was found to be located in the bacterial inner membrane. This turned out to be the case not only when intracellular β-galactosidase was used as fusion protein, but also when the external membrane LamB or the periplasmic MalE proteins were used as fusion partners (13) (Fig. 1) (13a). In these two latter expression systems the amount of expressed proteins was evaluated at 40,000 copies while the functional receptors were estimated at 400 per bacteria.

Their ease of production, great stability, and low cost make bacteria-expressed receptors suitable for use as antigens to detect anti-receptor antibodies by immunoblots (14). They can be used to detect the functionally active receptor and also proteolytic fragments of the fusion proteins. The specificity of a typical immunoblot pattern of fusion protein and proteolytic fragments can be verified by blocking studies with receptor-derived peptide initially used for immunization. But this pattern can only be obtained with affinity-purified antibodies, since all mammalian species produce antibodies againist *E. coli*.

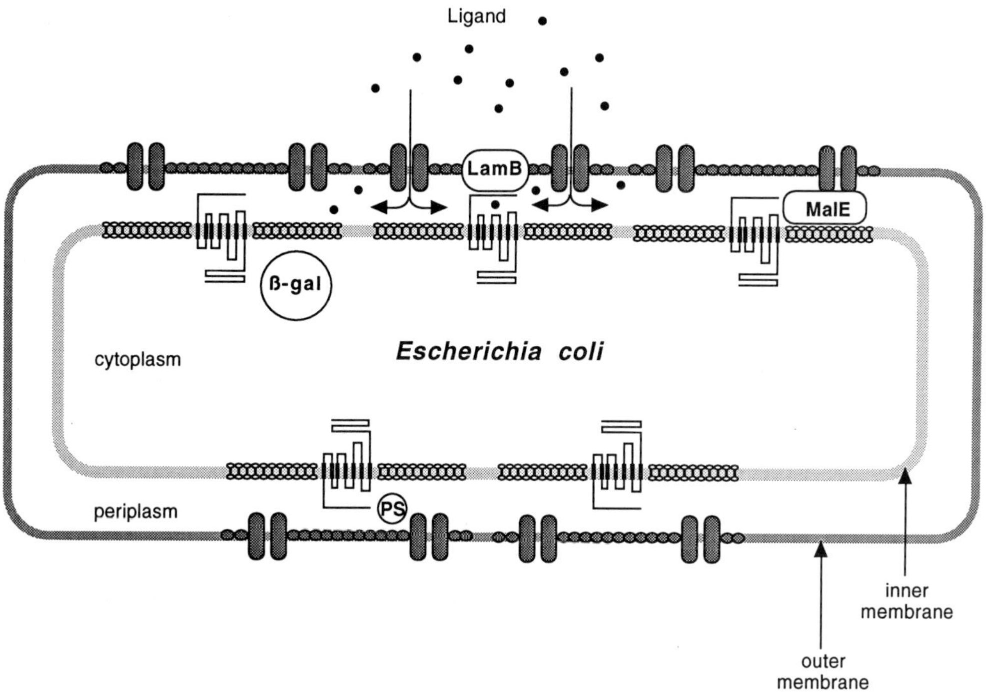

FIG. 1 Fusion partners used to express β-adrenergic receptors in *E. coli*. In the first expression system, the coding region of the β-adrenergic receptors was fused with the β-galactosidase gene coding for a cytoplasmic protein. Other fusions were performed with the bacterial genes of LamB and MalE, coding for proteins in the external membrane and the periplasm, respectively. Two other expression procedures were used. In these the coding region of the receptors was inserted at a position 60 bases after the peptide signal of MalE (PS in the figure) or directly under control of the T7 promoter. In all cases, the receptors were located at the inner membrane. Ligands can diffuse through the outer membrane to reach the receptors, but interaction with antibodies required the purification of the inner membranes. [Adapted from Marullo *et al.* (13a).]

The solubilized bacterial receptors can also be used for immunoprecipitation studies before or after forming a complex with a radioligand. Solubilization procedures are similar to those for solubilizing receptor from mammalian cells (15). In contrast, direct interaction between the receptor on the inner membrane and antibodies is not possible since the remnants of the bacterial outer wall produce a steric hindrance to the antibodies, even on spheroplasts obtained by lysozyme digestion. But it is possible to isolate inner membranes from outer membranes (13). This allows the study of receptor–antibody interaction in the absence of G protein.

Finally, the bacteria-expressing receptor can be used as antigen to study the cellular and humoral immune response.

Bacteria-expressed receptors are also useful for purposes other than immunological studies. They allow the construction of chimeric receptors for studying the importance of the different receptor domains in recognition of subclass specific ligands (16). By cotransfection of a particular receptor with different G proteins, bacteria-expressed receptors were used to study the receptor–G protein complexes (17). Using the technique of aleatory mutagenesis in certain regions of the β_2-adrenergic receptor, the importance of different amino acid residues in ligand binding could be assessed (18).

Choice of Potential Autoimmune Epitopes

The putative structure of G protein-coupled membrane receptors, based on the topology of bacteriorhodopsin, contains seven transmembrane α helices linked by intracellular and extracellular loops, with an extracellular N-terminal sequence and a cytoplasmic C-terminal sequence (Fig. 2) (18a).

There are three main factors to consider when looking at these receptors as targets for autoantibodies. First, the seven transmembrane domains and the intracellular loops are unlikely to be autoantibody recognition sites because of their inaccessibility. Second, most of the receptors in the rhodopsin superfamily are only to a very limited extent open to autoimmune attack, except for the receptors with an extended N-terminal domain (19). If the autoimmune antibodies affect the function of the receptors, which would explain their pathophysiological effects, the recognized epitopes are most likely to be located in the three extracellular loops.

Indeed, the glycosylated N-terminal sequence seems to be important only for transport to and integration into the plasma membrane for most of the monoamine receptors, but it has no effect on the subsequent ligand binding, signal transduction, or desensitization phenomena described for this group of receptors (20, 21). This may be somewhat different for peptide receptors such as that of substance P, where changes in residues of the N-terminal domain or other extracellular regions affect ligand binding (22). The first and second extracellular loops are linked together by a disulfide bridge, which seems to be essential for the functional integrity of the binding site (23–25) and interferes with coupling and densensitization functions (26–28). The first and third extracellular loops seem to be too short and too hydrophobic (29) to function as B cell epitopes, at least in the α- and β-adrenergic receptors, in the muscarinic acetylcholine receptors, and in the dopaminergic and sero-

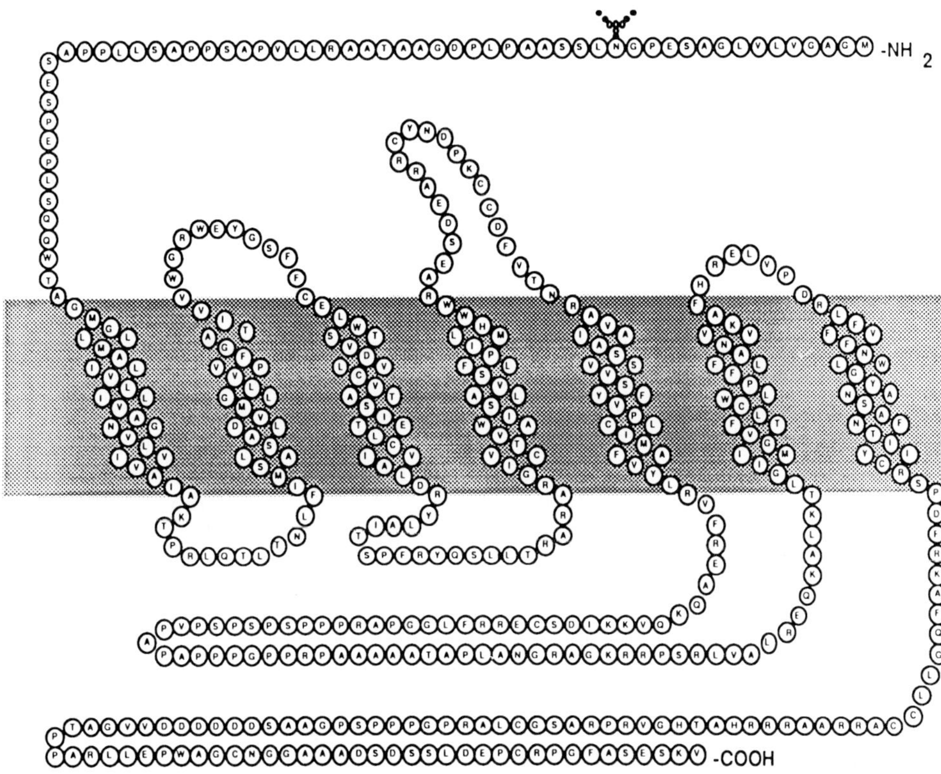

FIG. 2 Proposed topology of the β_1-adrenergic receptor in respect the plasma membrane. The transmembrane domains were chosen to include the maxima of hydrophobic regions calculated from a hydropathy plot and the minimum number of charged residues. This model is drawn by analogy to the known orientation of bacteriorhodopsin (18a).

toninergic receptors. We therefore conclude that the second extracellular loop of most of the neurotransmitter hormone receptors has the functional and immunological properties which could make it a specific target for autoantibodies with pathophysiological effects.

Choice of Immunogenic Peptides

As described elsewhere in this book (30), conjugates in which short peptides are linked to a carrier protein readily induce antibodies which recognize the receptors in immunoblots or may be used in immunocytochemistry to identify and locate receptors. The properties of synthetic peptides used to study an autoimmune response should, however, satisfy more stringent criteria than the ones used to induce anti-receptor antibodies.

Since most of the B cell epitopes recognized by "native" proteins are conformational in nature (31), longer peptides should be used as immunogens. Indeed, the longer the peptide the greater the probability is that one of its conformations will mimic the one in the native protein. Hence, longer peptides react better with anti-protein antibodies than do shorter peptides (32).

The same approach is required to induce antibodies which could mimic the functional properties of the autoantibodies. The peptides should not be coupled to a carrier protein because the coupling restricts the conformational possibilities of the peptide, even when spacer arms are used to avoid direct interaction between carrier protein and peptide. While longer peptides can yield moderate to high levels of antibody when used as immunogens (33), we have used a predictive program for MHC class II binding motifs to select free peptides yielding high titers of antibodies. Starting from sequences of model peptides known to induce a T cell helper response in a certain MHC mouse haplotype, homology scanning was used to predict immunogenic peptides from primary structures derived from cDNA sequencing (34) (The program is available from Dr. F. Borras-Cuesta, Departamento de Medecina Interna, Universidad de Navarra, Apartado 273, E-31008, Spain). The immunogenic peptides have the same T helper cell-stimulating properties as the carrier proteins and thus induce high titers of anti-peptide antibodies. Elongation of the immune peptides to include an immunogenic sequence usually results in anti-peptide antibodies with the same properties as those obtained by immunization with peptide–carrier conjugates without affecting the conformational freedom of the peptides.

Examples of Autoimmunity to G Protein-Coupled Membrane Receptors

β_1-Adrenergic Receptor

The main function of the β_1-adrenergic receptor is to stimulate the rate and the force of cardiac contraction. It occurs mainly in the left ventricle and the right atrium of the heart (70 to 90% of the β-adrenergic receptors) (35, 36). Autoantibodies against this receptor were shown in both Chagas' disease (6) and in idiopathic dilated cardiomyopathies (7), two diseases in which an autoimmune component has been suspected (37). Cloning and sequencing of the cDNA of the human β_1-adrenergic receptor (38) led to the synthesis

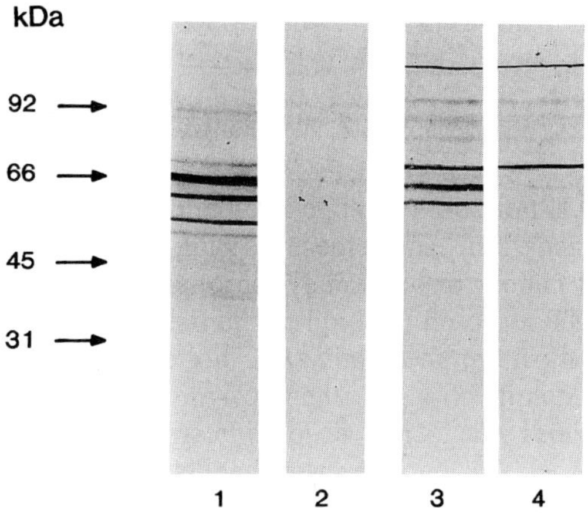

FIG. 3 Immunoblots of anti-receptor antibodies on the reduced membrane proteins of *E. coli* expressing the human β_1-adrenergic receptor. The vector used was MalE as shown in Fig. 1. (Lane 1) Proteins revealed with affinity-purified rabbit anti-peptide antibodies raised against the second extracellular loop of the receptor. (Lane 2) Affinity-purified rabbit anti-peptide antibodies preincubated with the peptide before incubation with the blotted protein. (Lane 3) Proteins revealed with human autoantibodies affinity-purified on the peptide corresponding to the second extracellular loop. (Lane 4) Affinity-purified human autoantibodies preincubated with the peptide before incubation with the blotted proteins. Three protein bands corresponding to degradation products of the fusion protein were specifically stained with both antibodies (39).

of peptides which could be used to raise antibodies and to develop a diagnostic assay.

Autoantibodies directed against the second extracellular loop of the β_1-adrenergic receptor were affinity-purified from the IgG fraction of patients and their immunological and pharmacological properties investigated (39). Immunoblots of membrane proteins from receptor produced in *E. coli* (MalE protein used as fusion partner) revealed bands from degraded fusion protein that also reacted with rabbit anti-peptide antibodies and which could be removed by preincubating the autoantibodies with the target peptide (Fig. 3). The purified antibodies also reduced noncompetitively the radioligand binding to cell membranes of rat C6 glioma cells, which possess 80% β_1- and 20% β_2-adrenergic receptors (40). The antibodies were shown to recognize the receptor independently from its coupling to the G_s protein in radioligand binding studies using inner membranes from *E. coli* expressing the receptor, since bacteria have no G protein. The antibodies completely inhibited radioligand binding to the membranes, confirming that the free receptor was

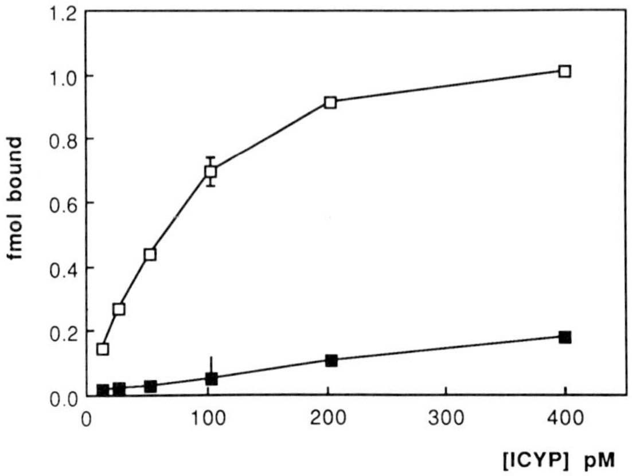

FIG. 4 Effect of antibody on the saturation curves of [^{125}I]iodocyanopindolol binding to inner membranes of *E. coli* bearing the human β_1-adrenergic receptor at 37°C. The vector used was LamB as shown in Fig. 1. Membranes were incubated with (closed symbol) or without (open symbol) 100 n*M* affinity-purified antibody. Mean and extreme values are shown for two experiments (41).

the antibody target (Fig. 4) (41). Immunoprecipitation studies also showed that the autoantibodies recognized the free receptor, but not the receptor–antagonist complex (Fig. 5) (42), suggesting that the autoimmune epitope was no more accessible in the antagonist-induced conformation. However, since the autoantibodies had only a limited effect on cAMP accumulation and did not desensitize the receptor (43), they were considered to be partial agonists (44).

Continuous stimulation of the receptor without desensitization could explain the pathophysiological role of the autoantibodies, since continuous stimulation with catecholamines is known to lead to hypertrophy of the left ventricle (45). Nonrecognition by the autoantibodies of the functional epitope in the receptor–antagonist complex could explain the beneficial effects of β-blocking agents in certain patients with idiopathic dilated cardiomyopathy (46).

The location of an autoimmune epitope on the β_1-adrenergic receptor in patients with idiopathic dilated cardiomyopathy prompted us to look for antibodies directed against the same epitope in patients with Chagas' cardiomyopathy. Preliminary results showed that 47% of the patients reacted with a peptide corresponding to the second extracellular loop in an enzyme immunoassay (47). Antibodies directed against this loop cross-reacted with an immunodominant *Trypanasoma cruzi* ribosomal protein (P0), suggesting that

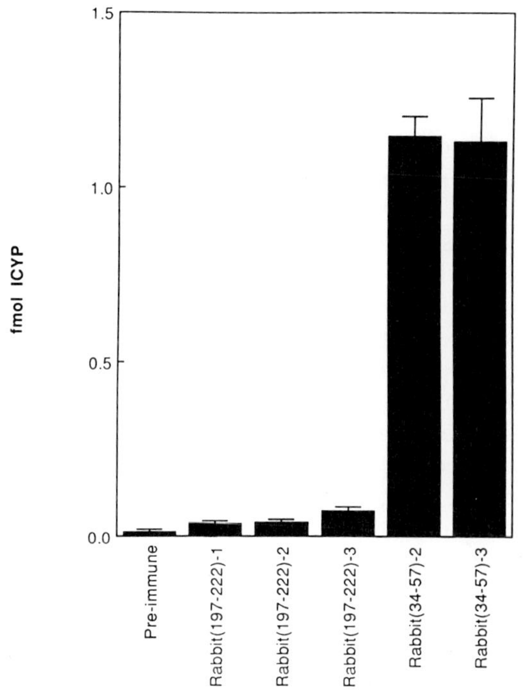

FIG. 5 Immunoprecipitation of the [^{125}I]iodocyanopindolol–β_1-adrenoceptor complex. The vector was the MalE shown in Fig. 1. Affinity-purified antibodies from rabbits immunized with the peptide corresponding to the second extracellular loop (197–222) or the peptide corresponding to a N-terminal sequence (34–57) of the human β_1-adrenoceptor were used to immunoprecipitate the radioligand–receptor complex from solubilized *E. coli* membranes bearing the receptor. The results are expressed in cpm bound ±SD (three experiments) to the protein A–Sepharose 4B IgG complexes.

the autoimmune anti-receptor response might be due to molecular mimicry (48). This observation strengthens *mutatis mutandis* the hypothesis that idiopathic dilated cardiomyopathy could result from molecular mimicry after a viral myocarditis (49).

β_2-Adrenergic Receptor

While the autoimmune response in dilated idiopathic cardiomyopathy is directed only against the β_1-adrenergic receptor subtype, autoantibodies against both the β_1- and the β_2-adrenergic receptor were found in patients with Chagas' disease (50). This autoimmune response to the β_2-adrenergic receptor is also directed against the second extracellular loop (47). The production of a monoclonal antibody against the same domain that noncompetitively inhibited radioligand binding (51) and had an agonist-like effect on cardiomyocytes in culture (G. Wallukat, personal communication) confirms that these properties are linked to the recognition of homologous receptor domains. The P0 ribosomal protein of *T. cruzi* and the β_2-adrenergic receptor had no cross-reactive epitope, suggesting that the autoantibodies were either cross-reacting with another parasite protein or arose after immune destruction of tissues bearing as primary target the β_1-adrenergic receptor.

A secondary immune response can undoubtedly explain the presence of autoantibodies against the second extracellular loop of the β_2-adrenergic receptor in 18% of patients with myasthenia gravis (52). These antibodies immunoprecipitated the solubilized free receptor, competed with radioligand binding, and immunoblotted the receptor expressed in *E. coli*. No physiological activity of these autoantibodies was tested.

In contrast to the antibodies directed against the second extracellular loop, which have agonist-like activity, the autoantibodies found in patients with asthma or atopic allergy seemed to behave as antagonists (53). This suggests that another receptor domain is the immune target. No functional monoclonal antibody could be raised against a synthetic peptide corresponding to the second transmembrane domain and containing a conserved aspartic acid residue involved in ligand binding (54). In view of the inaccessibility of this region to antibodies, this failure confirms the putative topology of the receptor. Recent data suggest that the inhibitory activity of the IgG fraction of asthmatic patients could be blocked by a peptide corresponding to the third extracellular loop of the β_2-adrenergic receptor (55). In this regard, it is worthwhile mentioning that the adjacent seventh transmembrane domain is essential for discriminating between the α_2- and β_2-adrenergic receptor antagonists (56).

Muscarinic Acetylcholine Receptor M2

As only 30% of patients with idiopathic dilated cardiomyopathy had autoantibodies against the β_1-adrenergic receptor, their sera were examined to see if other cardiovascular receptors were involved in autoimmune recognition. The muscarinic acetylcholine receptor M2, which negatively regulates the stimulating effect of catecholamines on the heart, is a potential candidate. Autoantibodies against a peptide corresponding to the second extracellular loop of the human receptor (57) were detected in 39% of the patients. Half of the positive patients also had autoantibodies against the β_1-adrenergic receptor (58). The autoantibodies blocked radioligand binding to the heart receptor by a "mixed" type of inhibition, suggesting that they did not bind to the transmembrane binding site, but at the extracellular surface. The antibody–receptor interaction probably induces a conformational change in the receptor which blocks the radioligand binding site.

Antibodies were raised against a peptide, corresponding to the second extracellular loop, which was predicted to contain a T helper cell epitope and used to determine whether the anti-receptor antibodies exerted a physiological effect on the receptor. The affinity-purified antibodies had the same functional properties as the autoantibodies. In addition, they inhibited isoproterenol-stimulated cAMP and induced a negative chronotropic effect in cardiomyocytes in culture (M. L. X. Fu *et al.*, 75). These properties mimic those of the muscarinic acetylcholinergic agonists.

The antibodies directed against the immunodominant second extracellular loop of the three receptors discussed above all display agonist activity. The interaction with the epitope, that contains a cysteine involved in a conformationally important disulfide bridge, could thus lead to a conformational change in the receptor similar to that induced by the natural hormone.

Serotoninergic Receptor 5HT1A

Ciaranello and Todd found autoantibodies against serotoninergic receptors in 7 of 13 sera from autistic children. One of these sera was directed against the 5HT1A subtype (9). These observations could not be repeated using the specific agonist 8-OH-DPAT (59). As the second extracellular loop is the main autoimmune target on G protein-coupled membrane receptors, we decided to use an enzyme immunoassay to look for autoantibodies directed against this loop in children with developmental disorders. Peptides corresponding to the same domain in the β_1-adrenergic receptor and to the first extracellular loop of the 5HT1A receptor were used as negative controls. Surprisingly, a significant amount of specific immunoreactivity against the second extracellular loop of this receptor was found in nonautistic children with development disorders who also had abnormal EEGs. Six of ten sera from this group were positive, compared to 4 of 27 sera from a control group (60).

Analysis of the 5HT1A receptor suggested that its first and second extracellular loops both contained T helper cell epitopes. This was verified when rabbits immunized with the free peptides corresponding to the two loops produced high-titer polyclonal sera. The affinity-purified antibodies from these sera were tested in immunoblots on membranes of transfected *E. coli* expressing the 5HT1A receptor (61). Only the antibodies raised against the second extracellular loop reacted with the receptor protein after electropho-

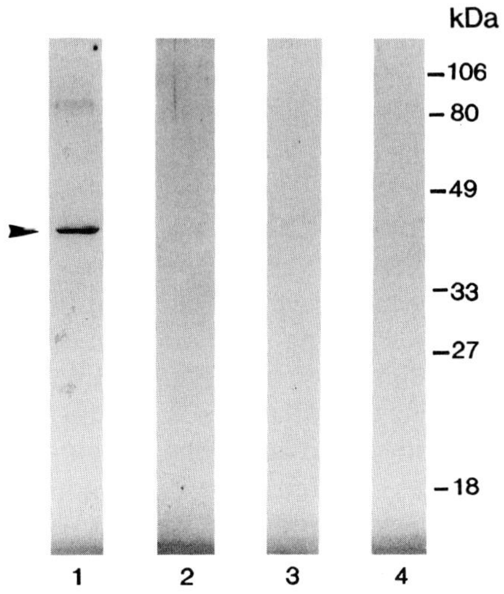

FIG. 6 Immunoblots of anti-receptor antibodies bound to the reduced membrane proteins of *E. coli* bearing the human serotoninergic 5HT1A receptor. The vector used was PS as shown in Fig. 1. (Lane 1) Proteins revealed with affinity-purified anti-peptide antibodies raised against the second extracellular loop of the receptor. Arrow indicates receptor protein. (Lane 2) Affinity-purified rabbit anti-peptide antibodies preincubated with the peptide (70 μg/ml) before incubation with the blotted proteins. (Lane 3) Proteins revealed with rabbit affinity-purified anti-peptide antibodies raised against the first extracellular receptor loop, but which cross-react with an epitope on the second extracellular loop. (Lane 4) Proteins revealed with rabbit affinity-purified anti-peptide antibodies raised against the first extracellular loop.

resis under reducing or nonreducing conditions (Fig. 6) (62). The first extracellular loop appears to be inaccessible to antibody recognition, while the second loop is recognized. This again confirms the predominance of this loop as an immune target.

The Thyrotropin (TSH) Receptor

Although the TSH receptor belongs to a family with an extended extracellular domain, it has at least one feature in common with the β_2-adrenergic receptor: autoimmune antibodies can be either blocking (antagonist) or stimulatory (agonist), depending on the disease in which they occur. Agonist-like autoantibodies are found in Graves' disease and lead to hyperthyroidism, while antagonist-like autoantibodies are found in idiopathic myxedema, where they lead to hypothyroidism (63).

Thyrotropin LH-CG receptor chimeras were used to define a domain encompassing amino acids 8–89 as necessary for interaction with agonist-like autoantibodies (64). This sequence contains peptide 32–56, which is reported to be significantly precipitated by IgG from sera of patients with Graves' disease (65). This portion of the receptor also contains a cysteine (Cys-41) which is essential for ligand binding to the receptor (28), although this region seems not to be involved, at least directly, in TSH recognition. Kosugi *et al.* located the epitope for the antagonist-like autoantibodies found in idiopathic myxedema as lying within a loop formed by a disulfide bridge at the C-terminal end of the extended extracellular domain located between cysteines 301 and 390 (66). Although this loop can be almost completely deleted, the disulfide bridge and the nearby Tyr-385 remain essential for receptor function (28). However, the deletable loop contains a polyanionic stretch (352–366) which seems to be a main antigenic epitope for autoantibodies in Graves' disease (67). Autoantibodies directed against this epitope appear to have no physiological effects.

Specific Methods

Transfection of Escherichia coli with the β_2-Adrenergic Receptor Gene(18)

Strains and Plasmids

Plasmid pRB12.65 is a derivative of the M13 plasmid pZ150 (68). It encodes ampicillin resistance and the human β_2-adrenergic receptor (Fig. 7). The receptor gene is under the transcriptional control of the bacteriophage T7ϕ10 promotor and has T7 terminator sequences downstream (69). The *Nco*I–*Bam*HI fragment containing the β_2-adrenergic receptor coding region derived from plasmid 973 (70) replaces the small *Nco*I–*Bam*HI fragment between the T7 promotor and terminator sequences of pET3d. The resulting T7/β_2-adrenergic receptor fusion uses the ATG at codon one of the β_2-adrenergic receptor gene for translation initiation. The coding region of the β_2-adrenergic receptor gene in plasmid pRB12.65 has been modified to contain several additional restriction sites not present in the wild type (71). These sites were introduced by replacing the 258-bp *Bst*EII–*Alw*NI restriction fragment of the β_2-adrenergic receptor gene with a synthetic fragment containing 17 silent mutations. These mutations result in seven new restriction sites in the β_2-adrenergic receptor gene: *Hin*dIII at codons 73/74/75, *Nhe*I at codons 83/84/85, *Bsp*HI at codons 95/96/97, *Eco*RI at codons 107/108, *Cla*I at codons 129/130, and *Spe*I at codons 136/137. These mutations were introduced to allow the introduction of aleatory mutagenesis cassettes.

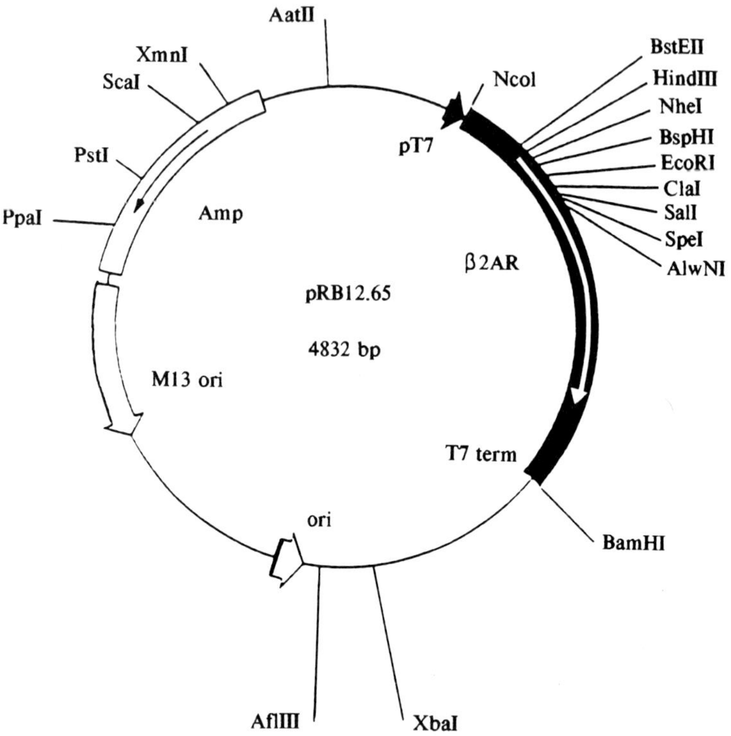

FIG. 7 Structure of the T7/β_2-adrenergic receptor plasmid. The receptor gene is under transcriptional control of the bacteriophage T7 promoter and has the T7 terminator sequence downstream. The coding region of the gene in plasmid pRB12.65 has seven new restriction sites in the receptor gene (18).

Plasmid plysS is a derivative of pACYC and encodes chloramphenicol resistance and T7 lysozyme (72).

Strain BL21(DE3)(F^- *hsdS gal* r^- m^-) contains the prophage DE3 bearing the T7 polymerase gene under the transcriptional control of the *lac-inducible* UV5 promoter (69).

Induction in Liquid Culture

BL21 (DE3)/plysS/pRB12.65 is grown in M9 medium (73) containing 0.4% glucose, 100 μg/ml ampicillin, and 25 μg/ml chloramphenicol at 37°C. When the culture reaches an A_{600} of 0.2–0.3, IPTG is added to a final concentration of 0.5 m*M* and the cultures are incubated for a further 6 hr at 23°C.

Binding Assays

Ligand binding to intact *E. coli* is measured essentially as described (12). Briefly, 1×10^7 bacteria resuspended in a 25 m*M* tris, pH 7.4, 75 m*M* $MgCl_2$ buffer are incubated for 1 hr at 37°C with increasing concentrations of [^{125}I]iodocyanopindolol (6.25 to 200 p*M*). Nonspecific binding is measured in presence of 2 μM DL-propranolol. After incubation, the bacterial suspension is filtered by suction through GF/F glass filters and washed three times with 4 ml of cold incubation buffer. Radioactivity on the filters is measured in a gamma counter (LKB, Bromma, Sweden).

In Situ [^{125}I]Iodocyanopindolol Screening Protocol

Transformants resistant to ampicillin and chloramphenicol are selected by overnight growth at 37°C, on Luria broth (LB) agar plates containing 100 μg/ml ampicillin and 25 μg/ml chloramphenicol.

To screen for ligand binding, master transformation plates are replicated onto plates prepared by overlaying 22 ml of the agar plates with 3 ml top agarose containing [^{125}I]iodocyanopindolol to a final concentration (in 25 ml) of 20 p*M* and IPTG to a final concentration of 0.5 m*M*. In some cases unlabeled dl-propranolol (final concentration 10 μM) is added to the plates as a competitor. Replica plates are incubated for 15 hr at 37°C. Colonies are then lifted from the replica plates onto nitrocellulose filter circles and exposed to X-ray films for 4–18 hr.

Colonies displaying positive [^{125}I]iodocyanopindolol binding phenotypes are located on the master plate using the [^{125}I]iodocyanopindolol probed replica filters as guides. The [^{125}I]iodocyanopindolol binding phenotypes are confirmed by patching onto master plates. Replicas are made on LB or M9 agar plates containing ampicillin, chloramphenicol, 20 p*M* [^{125}I]iodocyanopindolol, and 0.5 m*M* IPTG and reincubated for 6 hr, and filter lifts are made as above.

Alternatively, a nitrocellulose print of the master plate is made and tested for radioligand binding. The positive colonies are selected for further amplification in liquid culture and tested for radioligand binding on GF/F glass filters (Whatman, Maidstone, U.K.).

As a general rule, transfected bacteria expressing the receptor grow more slowly than untransfected bacteria. The turbidity of the culture medium of receptor-expressing bacteria after overnight incubation is around 0.7 while that of the untransfected bacteria is higher than 2.5. This property is used to select the clones to be tested for radioligand binding.

Preparation and Fractionation of Escherichia coli Membranes (13)

Inner and outer membrane fractions are isolated from bacteria grown and induced as above. Cells from a 1-liter culture are harvested and washed at 4°C with 1 liter 10 m*M* HEPES, pH 7.5. All subsequent steps are carried out on ice. The bacterial pellet is suspended in 28 ml 10 m*M* HEPES, pH 7.5, containing 20% (by mass) sucrose, 30 μg/ml DNase I, 30 μg/ml RNase A, and the following protease inhibitors: 1 m*M* phenylmethylsulfonyl fluoride, 5 μg/ml leupeptin, and 7 μg/ml pepstatin. The suspension is pressed twice through a French pressure cell at 6.9 MPa (1000 lb/in.^{2}). Unbroken cells are removed by centrifugation at 5000*g* for 10 min, and 2 ml 0.1 *M* EDTA (pH 7.5) is added to the supernatant. This supernatant is layered onto a sucrose gradient consisting of 3 ml 60% (mass/vol) sucrose (in 10 m*M* HEPES, pH 7.5, containing 5 m*M* EDTA) and centrifuged at 100,000*g* for 16 hr at 4°C in a Beckman SW-28 rotor. Two major bands are visible at the interfaces of the sucrose layers. The lower band, between the 60 and 42.5% layers, is enriched in outer membrane and the upper one, between the 42.5 and 25% layers, is enriched in inner membrane. Fractions are collected by suction with a needle and a syringe or with a glass capillary from the gradient and aliquots used to measure the binding of 4–150 p*M* [^{125}I]iodocyanopindolol. Protein concentration is determined by the method of Bradford (73).

The "outer membrane fraction" and "inner membrane fraction" are those fractions which contain the highest specific binding activity.

Preparation of Solubilized Receptor and Immunoprecipitation

Preparation of Crude Membranes (74)

A 50-ml aliquot is taken from a bacterial suspension grown in liquid medium (A_{600} 1.0–1.2) and to it is added 0.4 ml of 25 m*M* phenylmethylsulfonyl fluoride (PMSF) in ethanol. The bacteria are collected by centrifugation and resuspended in 3 ml 0.2 *M* Tris–HCl, pH 8.0, containing 0.016 ml of the PMSF solution. The suspension is further diluted in 2 ml of 1 *M* sucrose solution in the same Tris–HCl buffer. The outer bacterial wall is digested by incubating the suspension with 0.25 *M* 0.02 ml EDTA, pH 8, 0.04 ml 10 mg/ml lysozyme in water, 4 ml water at room temperature for 15 min. After addition of 0.16 ml 1 *M* $MgSO_4$, the spheroplasts are centrifuged out at 1000*g* for 10 min. The spheroplasts are lysed in 10 m*M* Tris–HCl, pH 8, EDTA 1 mM buffer containing 1 m*M* PMSF, 5 μg/ml leupeptin, and 7 μg/ml pepstatin. A sufficient volume is used to avoid gelification of DNA (10 volumes/volume of pellet). The membranes are removed by ultracentrifugation in a SW-40 Beckman rotor for 1 hr at 100,000*g*. The supernatant is discarded and the pellet resuspended in 0.025 *M* Tris–HCl, pH 8, containing 0.09 *M* NaCl and 10% (v/v) glycerol.

Solubilization of Receptors (15)

The radioligand–receptor complex is solubilized by incubating the membranes with 200 p*M* of the radioligand [^{125}I]iodocyanopindolol for 30 min at 30°C, cooling on ice for 1 hr in 2% digitonin (water soluble) in 0.01 *M*

Tris–HCl, pH 7.2, containing 100 m*M* NaCl. The free radioligand is separated from the complex by molecular sieving on a Sephadex G-25 column (PD10 ready made for desalting, Pharmacia, Uppsala, Sweden), equilibrated in 0.1% digitonin in 0.01 *M* Tris–HCl pH 7.2, 150 m*M* NaCl. The solubilized complex is stored at −80°C until used.

Free receptor is solubilized in the same way, but without preincubation with 200 p*M* [^{125}I]iodocyanopindolol.

Immunoprecipitation

The radioligand–receptor complex is incubated with an equal volume of affinity-purified antibodies for 2 hr at room temperature. Protein A-Sepharose 4B beads (Pharmacia, Bromma, Sweden) are then added and incubation continued for a further 1 hr at room temperature. The beads are extensively washed with a 0.02 *M* Tris–HCl buffer, pH 7.5, containing 1% Triton X-100, 0.5% sodium deoxycholate, 0.1% SDS, 0.5 *M* NaCl, 0.05 *M* KCl, and 1 m*M* EDTA. The radioactivity adsorbed onto the beads is counted in a LKB gamma counter.

The same methodology was is to immunoprecipitate the free receptor, but the amounts of receptor remaining in the supernatant are analyzed. The supernatants are incubated with 200 p*M* [^{125}I]iodocyanopindolol for 30 min at 37°C, precipitated with 8% polyethylene glycol 8000, filtered and washed on glass filters (Whatman GF/F), and counted in a LKB gamma counter. Nonspecific binding is measured in the presence of 5 μM propranolol.

Affinity Purification of the Anti-Receptor Antibodies

The low autoantibody titers make it necessary to take certain precautions when purifying the human autoantibodies. The IgG fraction is routinely precipitated by adding 1 volume of a saturated solution $(NH_4)_2SO_4$ (750 g/liter) to 2 volumes serum. The minimum amount of serum used is 5 ml. The precipitate is dissolved in a minimal volume (1/3 of the serum volume) of phosphate-buffered saline (10 m*M* phosphate, 150 m*M* NaCl, pH 7.4) (PBS) and dialyzed extensively against PBS.

The IgG solution is then centrifuged at 1000*g* and adsorbed onto the target peptide immobilized on CNBr-activated Sepharose 4B (Pharmacia, Sweden) columnwise. The bed volume is 2 ml in a 5-ml Bio-Rad column (Richmond, Ca). The Sepharose is washed extensively with PBS until washes have no absorbance at 280 nm, and washed once more with hypertonic PBS (NaCl 1 *M*) to remove all nonspecific IgGs.

The specific autoantibodies are desorbed under the least denaturing conditions by stepwise elution. The first eluent is 3 *M* KSCN and the desorbed fraction is immediately dialyzed against PBS. This eluent completely removes the autoantibodies to the muscarinic heart receptor (58) and the serotoninergic 5HT1A receptor (62). The autoantibodies against the B_1- and the β_2-adrenergic receptor are eluted with a 0.2 *M* glycine–HCl buffer, pH 2.8 (39, 52). The acid eluent is collected into 1 *M* Tris–HCl, pH 7.6 (1 ml for 5 ml eluent), and immediately dialyzed against PBS. The final yield of autoantibodies is 100–500 μg.

Conclusion

The studies described here clearly show that there are autoantibodies against receptors coupled to G proteins. *Escherichia coli* bacteria transfected with the genes coding for such receptors may be used in diagnostic assays to reveal the presence of such antibodies. Functional studies may also be used to define the epitopes active in the immune responses of T and B cells.

Acknowledgments

We are indebted to Drs. L. X. Fu and Y. Magnusson (Wallenberg Laboratories, Göteborg, Sweden), Dr. G. Wallukat (Max Delbrück Center, Berlin, Germany), and Mr. L. Verdot (Université François Rabelais, Tours, France) for access to unpublished data. This work was supported by grants from the C.N.R.S., I.N.S.E.R.M, and Biotechnocentre in France and from the Heart and Lung Foundation in Sweden.

References

1. J. Patrick and J. Lindstrom, *Science* **180,** 871 (1973).
2. C. Grünfeld, E. Van Obberghen, F. A. Karlsoon, and C. R. Kahn, *J. Clin. Invest.* **66,** 1224 (1980).
3. D. D. Adams,, *Vitam. Horm.* (*N.Y.*) **38,** 191 (1980).
4. W. A. Valente, P. Vitti, Z. Yavin, E. Yavin, C. M. Rotella, E. F. Grollman, R. S. Toccafondi, and L. D. Kohn, *Proc. Natl. Acad. Sci. U.S.A.* **79,** 6680 (1982).
5. J. C. Venter, C. M. Fraser, and L. C. Harrison, *Science* **207,** 1361 (1980).
6. E. Borda, J. Pascual, P. Cossio, M. de la Vega, R. Araña, and L. A. Sterin-Borda, *Clin. Exp. Immunol.* **57,** 679 (1984).
7. G. Wallukat and A. Wollenberger, *Biomed. Biochim. Acta* **46,** 634 (1987).
8. C. J. Limas, I. F. Goldenberg, and C. Limas, *Circ. Res.* **64,** 97 (1989).
9. R. D. Todd and R. D. Ciaranello, *Proc. Natl. Acad. Sci. U.S.A.* **82,** 612 (1985).
10. R. A. F. Dixon, B. K. Kobilka, D. J. Strader, J. L. Benovic, H. G. Dohlman,

T. Frielle, M. A. Bolanowski, C. D. Bennett, E. Rands, R. E. Diehl, R. A. Mumford, E. E. Slater, I. S. Sigal, M. G. Caron, R. J. Lefkowitz, and C. D. Strader, *Nature (London)* **326,** 73 (1986).
11. A. D. Strosberg, *Eur. J. Biochem.* **196,** 1 (1991).
12. S. Marullo, C. Delavier-Klutchko, Y. Eshdat, A. D. Strosberg, and L. Emorine, *Proc. Natl. Acad. Sci. U.S.A.* **85,** 7551 (1988).
13. M. P. Chapot, Y. Eshdat, S. Marullo, J. G. Guillet, A. D. Strosberg, and C. Delavier-Klutchko, *Eur. J. Biochem.* **187,** 137 (1990).
13a. S. Marullo, B. Bertin, M. Freissmuth, and A. D. Strosberg,, *Med. Sci.* **8,** 577 (1992).
14. H. Towbin, T. Staehelin, and T. Gordon, *Proc. Natl. Acad. Sci. U.S.A.* **76,** 4350 (1979).
15. B. A. Zemcik and C. D. Strader, *Biochem. J.* **251,** 333 (1988).
16. S. Marullo, L. J. Emorine, A. D. Strosberg, and C. Delavier-Klutchko, *EMBO J.* **9,** 1471 (1990).
17. M. Fressmuth, E. Selzer, S. Marullo, W. Schütz, and A. D. Strosberg, *Proc. Natl. Acad. Sci. U.S.A.* **88,** 8548 (1991).
18. R. M. Breyer, A. D. Strosberg, and J. G. Guillet, *EMBO J.* **9,** 2679 (1990).
18a. Y. Magnusson, Ph.D. Thesis, University of Göteborg (1992).
19. K. C. McFarland, R. Sprengel, H. S. Phillips, M. Köhker, N. Rosemblit, K. Nikolics, D. L. Segaloff, and P. H. Seeburg, *Science* **245,** 494 (1989).
20. P. Cervantes-Olivier, C. Delavier-Klutchko, O. Durieu-Trautmann, S. Kaveri, M. Desmandril, and A. D. Strosberg, *Biochem. J.* **250,** 133 (1988).
21. R. A. F. Dixon, I. S. Sigal, and C. D. Strader, *Cold Spring Harbor Symp. Quant. Biol.* **3,** 487 (1988).
22. T. M. Fong, R. R. Huang, and C. D. Strader, *J. Biol. Chem.* **267,** 25664 (1992).
23. R. A. F. Dixon, I. S. Sigal, M. R. Candelore, R. B. Register, W. Scattergood, E. Rands, and C. Strader, *EMBO J.* **6,** 3269 (1987).
24. H. G. Dohlman, M. G. Caron, A. DeBlasi, T. Frielle, and R. J. Lefkowitz, *Biochemistry* **29,** 2335 (1990).
25. T. M. Savarese, C. D. Wang, and C. M. Fraser, *J. Biol. Chem.* **267,** 11439 (1992).
26. C. M. Fraser, *J. Biol. Chem.* **264,** 9266 (1989).
27. S. B. Liggett, M. Bouvier, B. F. O'Dowd, M. G. Caron, R. J. Lefkowitz, and A. DeBlasi, *Biochem. Biophys. Res. Commun.* **165,** 257 (1989).
28. S. Kosugi, T. Ban, T. Akamizu, and L. D. Kohn, *Biochem. Biophys. Res. Commun.* **189,** 1754 (1992).
29. T. P. Hopp and K. R. Woods, *Proc. Natl. Acad. Sci. U.S.A.* **78,** 3824 (1981).
30. M. A. Ariano and D. R. Sibley, Chapter 21.
31. D. J. Barlow, M. J. Edwards, and J. M. Thornton, *Nature (London)* **322,** 747 (1986).
32. M. H. V. Van Regenmortel, D. Altschuh, and A. Klug, *Ciba Found. Symp.* **119,** 76 (1986).
33. S. Müller, S. Plaué, M. Couppez, and M. H. V. Van Regenmortel, *Mol. Immunol.* **23,** 593 (1986).
34. J. G. Guillet, J. Hoebeke, R. Lengagne, K. Tate, F. Borras-Herrera, A. D. Strosberg, and F. Borras-Cuestra, *J. Mol. Recognit.* **4,** 17 (1991).

35. G. L. Stiles, R. H. Strasser, T. N. Lavin, L. R. Jones, M. G. Caron, and R. J. Lefkowitz, *J. Biol. Chem.* **258,** 8443 (1983).
36. S. Golf, R. Lovstad, and V. Hansson, *Cardiovasc. Res.* **19,** 636 (1985).
37. J. F. Goodwin, *J. Mol. Cell. Cardiol.* **17,** 5 (1985).
38. T. Frielle, S. Collins, K. W. Daniel, M. G. Caron, R. J. Lefkowitz, and B. K. Kobilka, *Proc. Natl. Acad. Sci. U.S.A.* **84,** 7920 (1987).
39. Y. Magnusson, S. Marullo, S. Höyer, F. Waagstein, B. Andersson, A. Vahlne, J. G. Guillet, A. D. Strosberg, Å. Hjalmarson, and J. Hoebeke, *J. Clin. Invest.* **86,** 1658 (1990).
40. V. Homburger, M. Lucas, E. Rosenbaum, G. Vassent, and J. Bockaert, *Mol. Pharmacol.* **20,** 463 (1981).
41. Y. Magnusson, G. Wallukat, J. G. Guillet, Å. Hjalmarson, and J. Hoebeke, *J. Autoimmun.* **4,** 893 (1991).
42. K. Tate, Y. Magnusson, M. Viguier, R. Lengagne, Å. Hjalmarson, J. G. Guillet, and J. Hoebeke, *Biochimie* **76,** 159 (1994).
43. Y. Magnusson, G. Wallukat, F. Waagstein, Å. Hjalmarson, and J. Hoebeke, *Circulation* **89,** 2760 (1994).
44. A. Hedberg, E. Carlsson, E. Fellenius, and B. Lundgren, *Naunyn-Schmiedeberg's Arch. Pharmacol.* **318,** 185 (1982).
45. J. Haft, *Progr. Cardiovasc. Dis.* **17,** 73 (1974).
46. Å. Hjalmarson and F. Waagstein, *Eur. Heart J.* **12,** Suppl. F, 63 (1991).
47. P. Chiale, M. Elizari, M. Rosenbaum, M. J. Levin, L. X. Fu, Y. Magnusson, E. Angvald, K. G. Sjögren, and J. Hoebeke, *20th Annu. Meet. Basic Res. Chagas' Dis.,* Caxambu, Brazil, *1993,* p. 188 (1993).
48. I. Ferrari, H. Saunal, D. Kaplan, M. H. V. Van Regenmortel, J. Hoebeke, and M. Levin, *20th Annu. Meet. Basic Res. Chagas' Dis.* Caxambu, Brazil, *1993,* p. 152 (1993).
49. Y. Kitaura and H. Morita, *Jpn. Circ. J.* **43,** 1017 (1979).
50. L. Sterin-Borda, C. P. Leiros, M. Wald, G. Cremaschi, and E. Borda, *Clin. Exp. Immunol.* **74,** 349 (1988).
51. J. G. Guillet, R. Lengagne, Y. Magnusson, K. Tate, A. D. Strosberg, and J. Hoebeke, *Clin. Exp. Immunol.* **89,** 461 (1992).
52. H. Eng, Y. Magnusson, G. Matell, A. K. Lefvert, R. Saponja, and J. Hoebeke, *J. Autoimmun.* **5,** 213 (1992).
53. L. C. Harrison, J. Callaghan, J. C. Venter, C. K. Fraser, and M. L. Kaliner, *Ciba Found. Symp.* **90,** 248 (1982).
54. P. C. Potter, D. White, L. Lochner, F. Z. Chung, and L. Visser, *Allergologie* **12,** 9 (1989).
55. G. Wallukat and A. Wollenberger, *J. Allergy Clin. Immunol.* **88,** 581 (1991).
56. B. K. Kobilka, T. S. Kobilka, K. Daniel, J. W. Regan, M. G. Caron, and R. J. Lefkowitz, *Science* **240,** 1310 (1988).
57. E. G. Peralta, J. W. Winslow, G. L. Peterson, D. H. Smith, A. Ashkenazi, J. Ramachandran, M. I. Schimerlik, and D. J. Capon, *Science* **230,** 600 (1987).
58. L. X. Fu, Y. Magnusson, K. H. Bergh, J.Å. Liljeqvist, F. Waagstein, Å. Hjalmarson, and J. Hoebeke, *J. Clin. Invest.* **91,** 1964 (1993).
59. A. Yuwiler, J. C. Shih, C. H. Chen, E. R. Ritvo, G. Hanna, G. W. Ellison, and B. H. King, *J. Autism Dev. Disord.* **22,** 33 (1992).

60. L. Verdot, B. Garreau, C. Barthélémy, J. Martineau, M. Ferrer-di Martino, J. P. Muh, and J. Hoebeke (submitted for publication (1994).
61. B. Bertin, M. Freissmuth, R. Breyer, W. Schultz, A. D. Strosberg, and S. Marullo, *J. Biol. Chem.* **267,** 8200 (1992).
62. L. Verdot, M. Ferrer-di Martino, B. Bertin, A. D. Strosberg, and J. Hoebeke, *Biochimie* **76,** 165 (1994).
63. G. F. Fenzi, P. Vitti, C. Marcocci, L. Chiovato, and E. Macchia, *in* "Thyroid Autoimmunity" (A. Pinchera, S. H. Ingbar, J. M. McKenzie, and G. F. Fenzi, eds.), p. 83. Plenum, New York, 1987.
64. K. Tahara, T. Ban, T. Minegishi, and L. D. Kohn, *Biochem. Biophys. Res. Commun.* **179,** 70 (1991).
65. M. Murakami and M. Mori, *Biochem. Biophys. Res. Commun.* **171,** 512 (1990).
66. S. Kosugi, T. Ban, T. Akamizu, and L. D. Kohn, *Biochem. Biophys. Res. Commun.* **180,** 1118 (1991).
67. O. Takai, R. K. Desai, G. S. Seetharamaiah, C. A. Jones, G. F. Allaway, T. Akamizu, L. D. Kohn, and B. S. Prabhakar, *Biochem. Biophys. Res. Commun.* **179,** 319 (1991).
68. R. J. Zagursky and M. L. Berman, *Gene* **27,** 183 (1984).
69. F. W. Studier and B. A. Moffat, *J. Mol. Biol.* **189,** 113 (1986).
70. S. Marullo, C. Delavier-Klutchko, J. G. Guillet, A. Charbit, A. D. Strosberg, and L. J. Emorine, *Bio/Technology* **7,** 923 (1989).
71. L. J. Emorine, S. Marullo, C. Delavier-Klutchko, S. V. Kaveri, O. Durieu-Trautmann, and A. D. Strosberg, *Proc. Natl. Acad. Sci. U.S.A.* **84,** 6995 (1987).
72. F. W. Studier, A. H. Rosenberg, J. J. Dunne, and J. W. Dubendorf, *in* "Methods in Enzymology" (D. Goeddel *et al.,* eds.), Vol. 185, p. 60. Academic Press, San Diego, CA, 1990.
73. M. M. Bradford, *Anal. Biochem.* **72,** 248 (1976).
74. B. Witholt, M. Boekhout, M. Brock, J. Kingma, H. van Heerikhuizenand, and L. De Leij, *Anal. Biochem.* **74,** 160 (1976).
75. M. L. X. Fu, G. Wallukat, Å. Kjalmarson, and J. Hoebeke, *Receptors and Channels* **2,** 121 (1994).

[19] Integrated Methods for the Construction of Three-Dimensional Models and Computational Probing of Structure–Function Relations in G Protein-Coupled Receptors

Juan A. Ballesteros and Harel Weinstein

Introduction

The rapid pace of cloning and expression of G protein-coupled receptors (GPCR) offers attractive opportunities to probe the structural basis of signal transduction mechanisms at the level of these cell-surface receptors. Major insights have emerged from comparisons and classifications of the amino acid sequences of GPCRs into families defined by evolutionary developments and adapted to perform selective functions. Mutation studies and chimeric constructs of the kinds described elsewhere in this volume continue to serve in probing such insights and in relating them to an understanding of the structural underpinnings of receptor function. Yet it is quite clear that the interpretation of results from such explorations of GPCR structure–function relations depends on a level of structural understanding that is not currently available from direct experimental data. This paucity of solid structural information also hampers the development of a mechanistic understanding of the molecular process in which the ligand-binding signal is propagated to activate the receptor for subsequent interaction with G proteins.

Structural data on GPCRs, based on biochemical, immunological, and biophysical approaches, have validated a consensus architecture of GPCRs with an extracellular N-terminus, a cytoplasmic C-terminus, and a transmembrane portion comprised of seven transmembrane helical domains connected by loops. The fact that the transmembrane portion of GPCRs consists of seven helices was revealed in the projection map of the electron density of bovine rhodopsin (RH) on the membrane (1), although the structural details are not discernable in this low-resolution (9 Å) structure. To date, the models of GPCR incorporate such basic structural information, but a major impediment to their improvement is that functional data in the form of pharmacological profiles rationalized in terms of ligand structure–activity relations (SAR) appear to lack the strong discriminating power required to test the models.

Methods in Neurosciences, Volume 25

This shortcoming is most evident from comparisons of claims by various authors that agreement with such pharmacological data is achieved despite the significant differences among the models they present for similar, or even identical, receptors. In support of these claims, the modes of ligand binding in the various receptor models are compared to expectations from previous SAR studies, despite the concern that many different combinations of side-chain residues in the receptor can be viewed to correspond to the same qualitative predictions of a receptor binding pocket. Consequently, very different GPCR models can claim ligand-binding arrangements that would be compatible with a given rank order of binding affinities, especially if only a small number of compounds are being analyzed. The ambiguities are exacerbated by the limited capability of mutagenesis-based probing to distinguish between a direct, as opposed to an indirect, effect of a mutation on ligand–receptor interaction, if the sole assessment is the measurement of affinity in mutant receptors (2). Together, these difficulties have precluded a conclusive test of modeling hypotheses and of the resulting GPCR models, because a variety of molecular mechanisms can explain the observed changes in the measured properties (3).

We illustrate here approaches that can be used to resolve these apparent ambiguities that burden the pharmacological testing of GPCR models, based on the integration of structural information about the receptor, about mutants, and about the changes induced by ligand binding. In the absence of direct means for structural determination, the first component of the intergrated approach relies on experiments with structurally modified receptors (e.g., chimeric) as well as a new focus on double-revertant mutants that impose more stringent criteria for inferences on structural details such as helix–helix interactions and spatial adjacencies of various domains in the GPCR proteins (4–7). Such details can provide essential information necessary for probing GPCR models at a structural level. For the functional component of the integrated approach, novel criteria for probing CPCR models have emerged from the formulation of agonistic activity as a measurable conformational change of the receptor model, caused by ligand binding. Such considerations may prove to be a much more stringent discriminant of the ability of a GPCR model to agree with pharmacological data than the analysis of ligand-binding modes in the model, as suggested by results from simulations of a serotonin 5HT2 receptor subtype (8). At this time, the bulk of information required to interpret and integrate the results from both the structural and functional probing of receptor models is most likely to emerge from molecular modeling and computational simulation, as the structural elucidation of transmembrane protein structures is technically still a formidable task. Given the current state of the art, such applications of molecular modeling will require considerable methodological improvements and their augmentation with powerful methods of computational simulation of the dynamic, time-dependent behavior of the receptor molecules. Therefore, we have oriented the discussion of methodological approaches presented below toward an understanding of the major capabilities and possible pitfalls of a variety of modeling and computational methods used in the study of structural and functional properties of GPCRs.

Developments in the molecular modeling and computational exploration of GPCR proteins indicate a tantalizing potential to alleviate some of these difficulties. These expectations are based on the increased rate of success achieved by molecular modeling (9–12) and computational simulation methods (13–19) in providing structural insights relevant to the functions of biological molecules. A variety of such theoretical methods have already been applied to a study of GPCRs (4, 8, 20–35), as discussed in overviews (36–39). However, a coherent approach to modeling and computational probing of GPCRs has not yet emerged, as the early efforts have focused on disparate aspects in the wealth of existing information about the GPCRs and other transmembrane proteins. In contrast, the present definition and illustration of methods and approaches applicable to the elucidation of structure–function relations of GPCRs at the atomic level of detail attempt a hierarchical scheme for incorporating existing information into GPCR modeling. Thus, the sequence of steps in the modeling process is organized to reflect the hierarchical but interrelated use of information about the primary, secondary, and tertiary structural properties of GPCRs, as well as inferences from the experimental probing and biophysical analysis of transmembrane proteins.

I. Primary Structure of G Protein-Coupled Receptors: Conservation Patterns and Functional Divergence

The primary structural information about GPCRs is encoded in their amino acid (AA) sequences. As it is not feasible to translate directly the AA sequence into three-dimensional (3D) structure, sequence comparisons are used to identify the likely determinants for the structural commonality expressed in the template of seven loop-connected transmembrane helices (TMH) shared by the GPCRs. Moreover, such sequence comparisons are used to identify the basis for the different functional properties of these proteins that determine ligand binding as well as the response of the GPCR to the actions of a large variety of ligands. Consequently, a sequence alignment (SA) of GPCR is one of the first and most crucial steps in the modeling process.

Two basic assumptions underlie the extraction of structural information about GPCRs from a set of aligned sequences (10). First, they all share a structural framework, as predicted for the packing arrangement of the TMH in the superfamily of GPCRs (21, 27, 40, 41); second, highly conserved residues can be considered essential for the structural and/or functional integrity of the receptor. For AA sites with a lower degree of conservation, this second hypothesis implies that the degree of conservation observable through the alignment is proportional to the role of these residues in determining the structure and/or function of the GPCR.

The criteria guiding the construction of a sequence alignment of GPCRs are briefly reviewed in Section I,A. The alignment makes possible an analysis of the conservation pattern at individual AA sites, as discussed in Section I,B. Since the conserved residues are predicted to be molecular determinants of the structure and/or function of a GPCR, their identification should guide the modeling strategy. Identification of structurally or functionally equivalent AA positions in the great variety of GPCR sequences is facilitated by a general numbering scheme that reflects the sequence alignment and makes possible direct comparisons of loci related through this alignment. Such a general numbering scheme is presented in Section I,C.

A. *Criteria and Inferences in the Construction of Sequence Alignments*

A multiple sequence alignment is constructed so as to maximize similarities at every AA site. This is an area of great general interest in bioinformatics and an increasing variety of computational algorithms is available to perform such tasks; for a review, see Chan *et al.* (42). These computerized methods will not be discussed here. Rather, we focus on options and criteria that are useful to derive structural inferences from a sequence alignment of GPCRs. These criteria pertain to the selection of correct inputs for the alignment programs and to structural considerations applicable to checking and refining the SA generated by those programs.

An important consideration is the selection of the proper set of GPCR sequences for the SA. This selection depends on the information that is sought and is determined by the extent of homology among the compared sequences. Alignment of sequences with intermediate homologies (i.e., 30–70%) can identify continuous patterns of conservation distributed over the entire sequence. Such patterns provide structural inferences based on the relative degree of conservation among all AA sites, such as the helical periodicity in the conservation pattern described in Section II,A,3. On the other hand, alignments with maximal (i.e., above 80%) or minimal (i.e., below 30%) homologies are useful in pinpointing a few residues that provide key information about structure or function. Thus, divergent sites among highly homologous sequences (above 80%) are either responsible for functional divergence, e.g., ligand-binding specificity (43–45), or indicative of functionally nonessential positions, e.g., lipid-facing residues (40). In the comparison of highly divergent sequences (below 30% homology), conserved AA sites are candidate molecular determinants of the 3D structure and/or of receptor activation mechanisms (21, 22, 40, 41), since both ligand binding and G protein coupling have divergent specificities (40).

A quantitative measure of sequence homology identified by SA is useful in determining the relation between AA conservation and structural determinants. The "mutation matrix" provides such a measure for individual AA-to-AA substitutions. For a given set of AA substitutions identified from comparing two sequences, the integration of the pairwise measurements comprised in the mutation matrix provides a quantitative estimate of their overall homology. The AA substitutions ($AA_i \rightarrow AA_j$) are quantified by the probability of their occurrence $P(AA_i \rightarrow AA_j)$, where i and j identify a specific amino acid or a gap indicative of an insertion or deletion. The 21×21 possible mutation probabilities $P(AA_i \rightarrow AA_j)$ define a matrix called the mutation matrix (46) used in all alignment programs to quantify the extent of sequence homology between two proteins, which is proportional to the probability that one of them will convert into the other by a series of mutations. Mutation matrices sometimes contain the inverse of the mutation probabilities, the "mutation penalties," but we refer here only to mutation probabilities for the sake of consistency. It is convenient to think of mutation probabilities as a parametrization derived from a set of sequence alignments that are representative of the proteins under study, in this case the GPCR. Two complementary criteria have been presented for the selection of the most appropriate mutation matrix: one based on the degree of evolutionary divergence and the other based on structural considerations. According to the first criterion, the mutation probabilities are calculated from a set of protein sequences that are all within a similar evolutionary distance of the GPCR, such as in the mutation matrix pam200 (47). There are good reasons for using a set of sequences that are all within the same evolutionary distance, based on the argument that a given mutation will have a different probability to occur in 1 million or 200 million years of evolution. The other criterion for an appropriate selection of the mutation matrix is based on structural considerations. For instance, it has been shown for soluble proteins that mutation probabilities depend on secondary structure, i.e., that they are different for residues in an α helix compared to a β sheet or random coil (48). In addition, mutation probabilities for surface-exposed residues in TMHs are different from those in helices of soluble proteins (49), in accordance with the opposite polarity of the solvent in each case. Both the evolutionary and the structural criteria pertain to the selection of the appropriate mutation matrix for the GPCRs. From the structural standpoint, different mutation matrices should be used for the TMHs and the loops. For the loop regions connecting the TMHs, the mutation matrix could be derived from soluble proteins disregarding structural considerations, and that covers the evolutionary distances found in GPCR, e.g., as in pam200 (47). For the TMHs, both structural and evolutionary criteria could be met if the mutation matrix is derived directly from a comprehensive alignment of GPCR TMHs. However, because such a mutation matrix has not yet been presented in the literature, a reasonable alternative would be a combination of the mutation matrix for interior residues of soluble proteins in helical segments (48) and the mutation matrix for residues exposed to the lipid milieu from the PRC family (49). The combination can be achieved by averaging the two mutation matrices, or by selecting either of them for individual AA sites depending on their predicted location in interior or lipid-exposed helical faces. Those mutation matrices would fulfill the structural criteria, although not the evolutionary criteria. However, mutation probabilities are proportional to the evolutionary distances, and all GPCR within a SA share a similar evolutionary distance. Consequently, the absence of an evolutionary criterion would be expected to alter only the absolute probabilities, but not the relative degree of conservation among different AA sites from which structural inferences are derived. It should be noted that the above considerations apply for the quantification of the extent of homology or conservation within a SA of GPCRs. If the aim is to use a particular GPCR sequence to search for homologous GPCR sequences against the entire protein data base, then other considerations apply. For instance, the evolutionary criterion becomes much more important because the proteins in the data base may have evolutionary distances significantly different from those of GPCRs. Furthermore, for a data base search we cannot specify different mutation matrices for TMHs and loops with current algorithms. Because TMHs are much more homologous among GPCRs than the loops, the structural argument would favor a mutation matrix for TMHs. Thus, for a data base search against all known protein sequences, both evolutionary and structural criteria are necessary and the best alternative would be to derive a mutation matrix from GPCR TMHs fitted to that task.

In addition to their use in defining appropriate mutation matrices, structural criteria can also be useful in the refinement of SA generated for the GPCR by computer programs. Thus, the alignment can take into consideration that insertions and deletions in helices are very rare events (50) and should preferably occur in the loops. Any gaps inserted in predicted TMHs by the computerized SA can probably be relocated to loops without significant loss of homology. Furthermore, insertion of gaps in loop regions is more advisable where nonconserved Pro or Gly residues occur. The reason is that the high flexibility of Gly, or the high rigidity of Pro, are often found to have a specific structural role and thus would require conservation. For example, conserved Gly or Pro in loops is suggestive of specific structures such as turns. A nonconserved occurrence of these residues indicates a structurally permissive site where gaps can better be accommodated.

B. Analysis of the Conservation Pattern from the Sequence Alignments

Sequence comparisons and inferred degrees of conservation have been used in most attempts to model the 3D structure of GPCRs (22, 40), although few explicit methods for quantifying the degree of conservation and its pattern have been presented. It is noteworthy, however, that because the set of GPCR sequences available for SA is still limited, it remains unclear whether a certain degree of conservation observed from the alignment reflects a real property of that locus in the sequences or an artifact of the limitation in the size of the data base. For example, if a certain site contains 80% Tyr and 20% Leu, it is not clear that the real evolutionary distribution for the particular family of GPCR being modeled would not be closer to 50 : 50 or 95 : 5 ratios if a complete sample were available. Clearly, however, the model must be consistent with the information that both Tyr and Leu can be present at this locus. Therefore, the conundrum arises as to whether the conservation analyses should be based on the *nature* of the different AA present at a given site or on the *frequencies of occurrence* of each AA. In our example, the conservation analyses could be based on the existence of either Tyr or Leu at this locus, or on the 80 : 20 ratio of Tyr vs Leu appearance. As illustrated below, both approaches can be applied in a sequential fashion to draw useful inferences. The selection of GPCR sequences for a SA could stress the representativity across the evolutionary spectrum as suggested from evolutionary trees, and then the analysis could more safely be based on the frequencies of occurrence of each AA within the SA.

The first step in the conservation analysis consists of the quantification of the global degree of sequence conservation at a given AA site. This information is obtained from the set of different AAs that are found to occur at this position in the aligned sequences. For GPCRs, a variability profile V was proposed as the number of different AAs at a given position (41). The plot of V versus the AA number (naa) (Fig. 1) provides the global conservation pattern through the sequence alignment, V(naa). However, this quantification ignores the distinctions—both physical and chemical—among the AAs appearing at the given locus, and hence does not provide information about

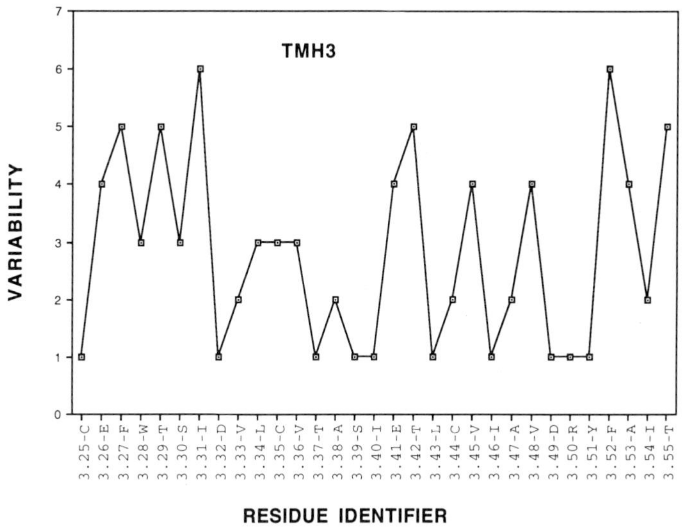

FIG. 1 Global conservation at each site of the TMH3 segment quantified through the variability profile *V*(naa) as the number of different amino acids occurring at the locus (naa). The plot is calculated from an alignment of dopaminergic, adrenergic, and serotonergic receptors. The sequence identified on the abscissa corresponds to the human β_2-adrenergic receptor, and the locus is identified by the numbering scheme described in Section I,C. Note a periodicity of three to four residues for this conservation pattern, suggestive of α-helical character as analyzed in Section II,A.

the putative physicochemical significance of the substitutions. A well-known measurement of the quantitative significance of AA substitutions is the "mutation probability" that defines a mutation matrix, as described in Section I,A, above. An alternative method to measure the global conservation at a given AA site uses mutation matrices and has been proposed for an analysis of the photoreaction center II proteins (51). It involves a similarity index defined as the average of all mutation probabilities between the AA present at a given site (51).

The conservation pattern obtained from the use of such quantification methods is essential for extracting structural inferences. For example, the analysis of the helical periodicity of a conservation pattern by Fourier transform of the *V*(naa) (41, 52) provides a powerful algorithm for predicting the boundaries of the TMH segments, as discussed below in Section II,A,3. Furthermore, Section III,A,1 describes the use of the conservation pattern as a guide for modeling the 3D architecture of GPCRs by suggesting TMH–TMH packings based on the correlation between high conservation and surface inaccessibility (49, 52, 53).

The use of the *V*(naa) defined above as guides for the 3D modeling of GPCR is based on the identification of those AA sites that are conserved beyond a given threshold. However, additional information that is important for the structural predictions resides in the detailed percentage presence of each specific AA at a given position within the conserved set, e.g., that $Tyr_{5.58}$ is 80% conserved because it occurs in 8 of 10 GPCR aligned. This information is useful because the nature of the AA present at every conserved site becomes a candidate molecular determinant of the structure and/or function of the GPCR. Since those molecular determinants exert their effects through physicochemical properties, it is important also to quantify the conservation of a given physicochemical property at a given site. This analysis should be done independently of the previous analysis of conservation of AA character since there could be positions that conserve a property such as hydrophobicity, but were not considered conserved by the previous analysis because a large number of hydrophobic AAs have appeared at this locus. Methodologically, the analysis of the conservation of physicochemical properties has been based on grouping the 20 natural AAs into subclasses according to an associated property. For example, Baldwin (40) grouped the 20 AA into three classes according to their size: small, medium, and large. Amino acid sites containing residues belonging to only one size class or two neighboring classes were considered conserved, and thus predicted to lie in helix–helix interfaces. A clearly significant physicochemical property for membrane proteins is the hydrophobicity/hydrophilicity. The method used for the conservation analysis based on this criterion classifies the 20. AAs into polar or hydrophobic, with the AA sites that conserve a polar character predicted to be at helix–helix interfaces while positions that conserve an apolar are considered more likely to face the lipids (27, 40, 54). Other physicochemical properties that can be identified by grouping AAs are the electrostatic charge (R,K vs D,E), analogous H-bonding characteristics (S,T), or aromaticity (F,Y,W,H), and these could serve in similar analyses leading to structural predictions.

C. Numbering Scheme for G Protein-Coupled Receptors

To relate site-defined properties to the sequences of the many different classes of GPCR proteins, we have developed a common numbering scheme that is informative of the relative position of each AA, the AA present at that position, and the real AA number in a particular GPCR. We refer to these three generalized numbers associated with each AA position in a GPCR sequence as *identifiers,* derived as follows: Every AA identifier starts with the TMH number, e.g., 4, for TMH4, and is followed by the position relative to a reference residue among the most conserved AA in that TMH. That reference residue is arbitrarily assigned the number 50. For example, the most conserved AA in TMH4 is a tryptophan whose identifier would be 4.50, i.e., $W_{4.50}$. A serine residue located five AA after $W_{4.50}$ will be $S_{4.55}$. This general numbering scheme is illustrated in Fig. 2 on the helical net representation of the human β_2-adrenergic GPCR. To relate the identifier to the AA sequence of a particular GPCR protein, each identifier defined above can be followed by the numbering in the particular sequence. For example, $W_{4.50}$ is number 158 in the human β_2-adrenergic GPCR and would be referred to as $W_{4.50(158)}$ for that specific receptor.

Mutations are identified in this numbering scheme in the usual manner, with the wild-type identifier followed by the mutant AA. For example, $W_{4.50(158)}F$ defines the Phe mutation of the Trp wild type. This identification scheme allows for a systematic comparison of mutations done in different receptors at the same loci.

Table I lists selected reference AAs in each TMH and illustrates the new identifiers for TMHs of the β_2-adrenergic receptor. The relative position of each reference AA in the GPCR TMHs can be seen in the helical net representation of the human β_2-adrenergic GPCR in Fig. 2.

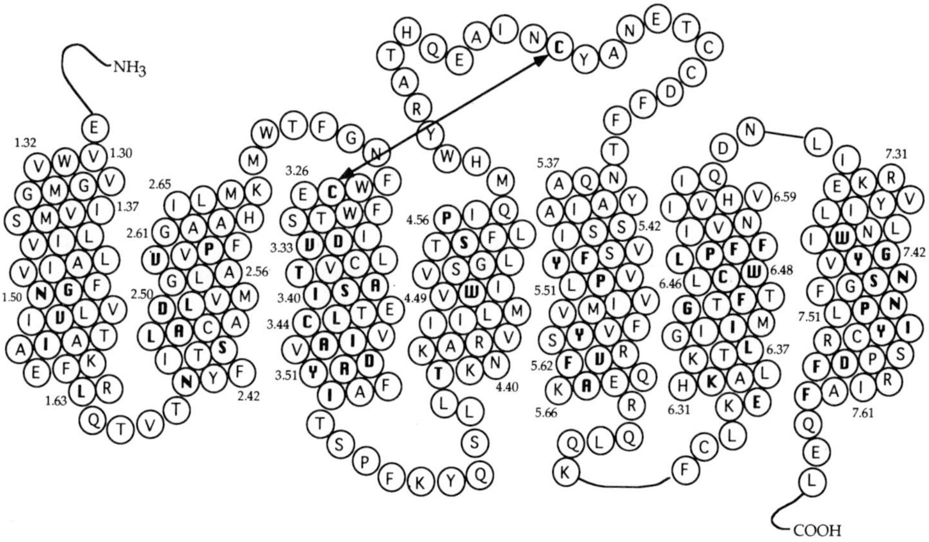

FIG. 2 Generalized numbering scheme illustrated on a helical net representation for the human β_2-adrenergic receptor. Highly conserved residues are shown in bold letters. See Fig. 9 for an explanation of helical net representations. The TMH ends are taken from Donnelly *et al.* (30).

TABLE I Generalized Numbering Scheme for GPCR Sequences

TMH	100% conserved in neurotransmitter GPCR	AA identifier	AA No. in β_2-ADR	AA identifier in β_2-ADR
1	Asn	N 1.50	51	N 1.50 (51)
2	Asp	D 2.50	79	D 2.50 (79)
3	Arg	R 3.50	131	R 3.50 (131)
4	Trp	W4.50	158	W4.50 (158)
5	Pro	P 5.50	211	P 5.50 (211)
6	Pro	P 6.50	288	P 6.50 (288)
7	Pro	P 7.50	323	P 7.50 (323)

If the AA is in a loop, this numbering scheme could provide it with two different identifiers, e.g., $L_{1.63} = L_{2.34}$. For a specific AA site, the proximity to a TMH boundary determines which identifier is actually used, although both define uniquely the same position.

II. Secondary Structure of G Protein-Coupled Receptors: Prediction and Modeling of Individual Transmembrane Helices

The helical nature of the transmembrane domains of GPCRs was confirmed by the 9 Å resolution structure of bovine RH (1), validating the prediction of seven TMHs for rhodopsin based on the hydrophobicity profile (55). Consequently, we assume the general validity of the parsing of GPCRs sequences into segments representing the seven TMHs and concentrate on methods to predict the TMH ends, presented in Section II,A. With the TMH ends defined, the essential considerations for the 3D modeling of each TMH at atomic detail are presented in Section II,B. Finally, some approaches for secondary structure predictions for the TMH-connecting loops as well as for the N- and C-terminal domains are discussed in Section II,C.

A. The Prediction of Transmembrane Helix Boundaries

The most common approach for predicting the transmembrane helical domains from GPCR sequences rests on the hydrophobicity profile (56), but at least four other useful approaches are available to achieve such parsing of the primary protein structure. The five methods presented below are qualitatively different, thus providing alternative paths to the prediction of the TMHs. Slightly conflicting predictions are likely to occur due to the inconclusive nature of each of these methods when applied individually. Consequently, the ability to quantify and compare predictions from each method to pursue convergence among their results becomes a desirable aim in the selection of the precicted TMH boundaries. As described below, the application of these methods leads to a definition of the TMH domain that goes beyond the limits provided by hydrophobicity criteria and extends the TMHs into other physicochemical environments such as the phospholipid head groups.

The predictions from the hydrophobicity profile presented in Section II,A,1 are challenged in Section II,A,2 by the hypothesis that Arg and Lys adjacent to the predicted cytoplasmic ends belong to the TMHs, where they supposedly interact with phospholipid head groups. A combination of these predictions leads to the redefinition of some TMHs that is supported by the analysis of the α-helical periodicity in the AA sequence. The α-helical periodicity of the sequence has been suggested as yet another criterion for the prediction of TMH ends, as discussed in Sections II,A,3 and II,A,4, below. Other approaches to predict the TMH boundaries rely on structural and evolutionary considerations, such as the position of nonconserved Pro residues (Section II,A,5). Finally, statistical methods based on the probability of each AA belonging to a TMH are presented in Section II,A,6. All these prediction methods use the conservation analysis of the GPCR sequence alignment described in Section I, based on the assumption that the aligned GPCRs share a common structure, in this case similar TMH boundaries. This assumption is verified by the comparison of the L versus M subunit of the photosynthetic reaction center (PRC) in the available crystal structure (57).

1. The Hydrophobicity Profile as Parsing Criterion

In this approach, a sequence is scanned with a fixed window size and the average hydrophobicity is calculated within that window (56). The plot of the average hydrophobicity against the sequence number identifies regions with high overall hydrophobic character such as expected of the TMH domains (56). A hydrophobic region identified in the plot is predicted as a TMH if it spans at least 18 AA, a length considered to represent the minimum number of residues needed to traverse the cell membrane in an α-helical conformation. This method has been extensively applied to sequences suspected to encode GPCR proteins, although its focus has rarely been to define the exact boundaries of the seven TMHs. Nonetheless, TMH ends have been inferred as the AA at which the sequence returns to residues with high hydrophilic character, as shown in Fig. 3, based on the assumption that the hydrophobic environment of the membrane excludes the hydrophilic residues. This assumption is likely to be more correct for the region of the protein facing the lipid chains of the membrane than for the ends of the TMH

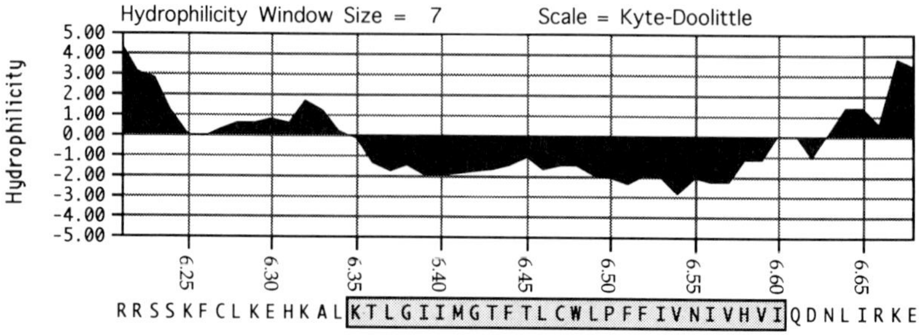

FIG. 3 Hydrophobicity profile for TMH6 of the human β_2-adrenergic receptor used to predict TMH boundaries. Note that the predicted helical ends by this criterion would be residues 6.35 and 6.60 for the N- and C-terminus, respectively. The locus is identified by the numbering scheme described in Section I,C.

domains that are likely to be adjacent to the polar lipid head group regions of the membrane. Consequently, TMH boundaries predicted by the hydrophobicity profile should represent minimal TMH boundaries, since the helices could actually extend into the more polar lipid head group region of the membrane, as discussed in the following section. Although the hydrophobicity profile was originally derived for a single sequence, it can also be applied through the sequence alignment of GPCRs in the form of average hydrophobicities (20). The advantage of the "evolutionary averaged" hydrophobicity plots is that they may overcome irrelevant deviations from the hydrophobic pattern of TMHs present in particular GPCRs. The assumption of common TMH boundaries for the GPCR within the SA allows for the analysis of the evolutionary consistency of their prediction.

The optimal choices of a hydrophobicity scale from which the averaged hydrophobicity is calculated, as well as the optimal window size and the threshold values for the prediction of TMHs using the hydrophobicity profile, have been the subject of several studies (58–60). The authors suggested the use of different hydrophobicity scales, pursuing convergence among their results as the guiding predictive principle. Notably, scales based on partition coefficients of peptides and/or small molecules between aqueous and nonpolar solvents yielded larger errors than scales based on residue solvent accessibility in protein structures, or on the statistical occurrence of each natural AA within TMH segments. The evaluation of the optimal choice of the window size was also inconclusive; small windows (e.g., seven AAs) offered better resolution but produced more false positives (i.e., TMH-like regions), whereas larger windows reduced false positives but also merged closely spaced TMHs. Since the aim is to predict TMH ends with the total number of helices assumed to be fixed at seven, a shorter window of seven AAs is expected to perform better due to its better resolution.

2. The Use of Arg and Lys Positions to Define the Cytoplasmic Ends

Predicted to lie outside the TMH by the hydrophobicity profile, the Arg and Lys residues, positioned predominantly at the cytoplasmic ends of BR and PRC, were actually found to belong to the TMH domains (61) in the available structures of these proteins (57, 62, 63), as shown in Fig. 4 for BR. It is thus very likely that when such residues appear in GPCRs next to the hydrophobic domains of the putative TMHs, they actually belong to these helices and can be used to define their cytoplasmic end (61). The mechanistic explanation given for the observation that Arg/Lys residues are part of the TMH region in BR and PRC is that these residues anchor the TMH to the membrane through ionic interactions with phospholipid head groups, as suggested by NMR studies on other membrane proteins (64–66). This would explain the cytoplasmic localization of the Arg/Lys since it is known that the inner leaflet of cell membranes is richer in negatively charged phospholipids than in the outer one. For use as a criterion, it is reasonable to predict that any Arg/Lys at the cytoplasmic ends lie inside a TMH if they appear in the sequence with an α-helical periodicity ($i,i+3$, or $i+4$). An example is seen in a helical net representation of TMH6 of the β_2-adrenergic GPCR (see Section II,A,4 for definition), shown in Fig. 5. In that helix, the patch formed by $\{K_{6.29}, K_{6.32}, K_{6.35}\}$ is predicted to be part of the TMH and to face the membrane environment. Such helical patches of Arg/Lys are often observed in the representation of GPCR sequences continuing the hydrophobic patch of the TMH core (see Section III,A). This continuity of the helical domain, shown by the analysis of the helical periodicity in Sections II,A,3 and II,A,4, integrates the predictions of TMH boundaries provided by the hydrophobicity profile with the Arg/Lys criterion. Since the phospholipid head groups are considered the limit of the membrane in the intercellular face, the ends predicted by this motif constitute maximal cytoplasmic TMH ends, as opposed to the minimal ends predicted from the hydrophobicity profiles.

The extension of the single TMH of M13 filamentous bacteriophage from the hydrophobic core into the Arg/Lys region, in accordance with the considerations described above, was confirmed with NMR measurements (67). Notably, the three Arg/Lys residues present in M13 at the cytoplasmic boundary were shown to be the energetic determinant of M13–lipid interactions because mutation of these Arg/Lys to Glu/Asp converted M13 into a soluble protein (68). The authors concluded from the properties of these mutants that the M13 protein anchors to the membrane through electrostatic interactions, consistent with the proposed role of the Arg/Lys in anchoring the TMH to the membrane through ionic pairs with the phospholipid head

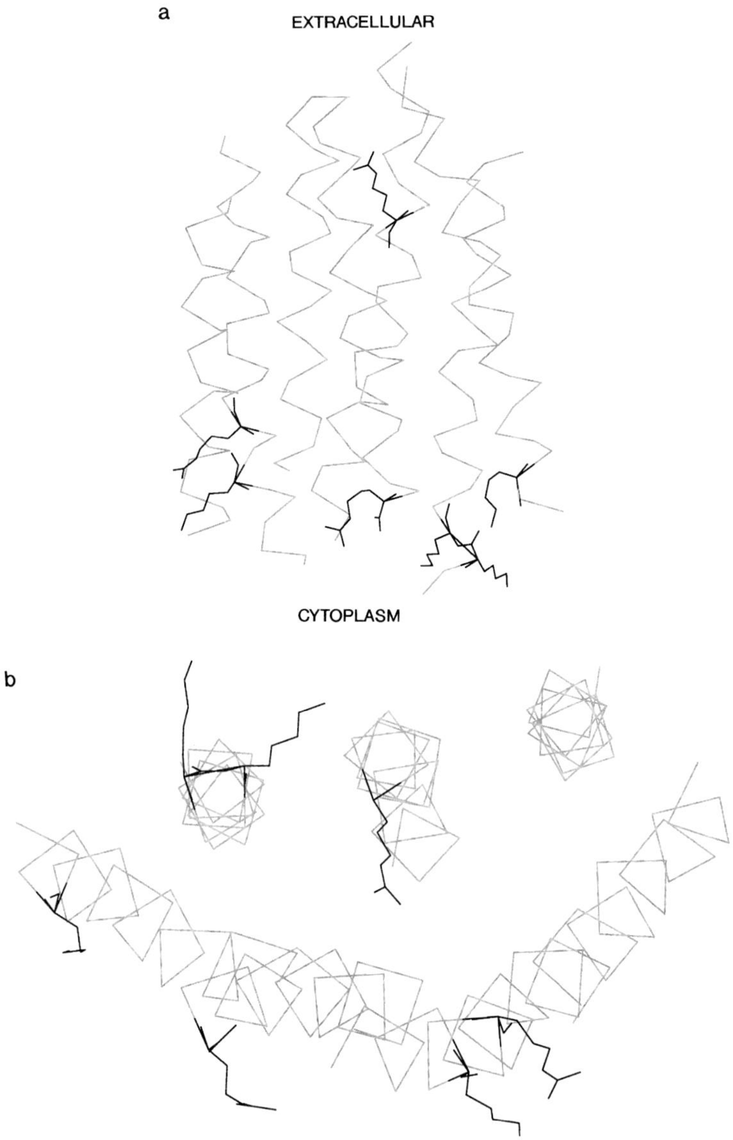

FIG. 4 The seven TMH bundle of BR is shown by the C_α trace plus the Arg and Lys residues. (a) View parallel to the membrane plane. Note that six of seven Arg/Lys residues are concentrated at the cytoplasmic boundaries. (b) View along the membrane axis from the cytoplasmic side. All six Arg/Lys at the cytoplasmic boundaries shown in (a) belong to the TMH and are protruding into the lipid phase, presumably to interact with the phospholipid head groups through ionic bonds.

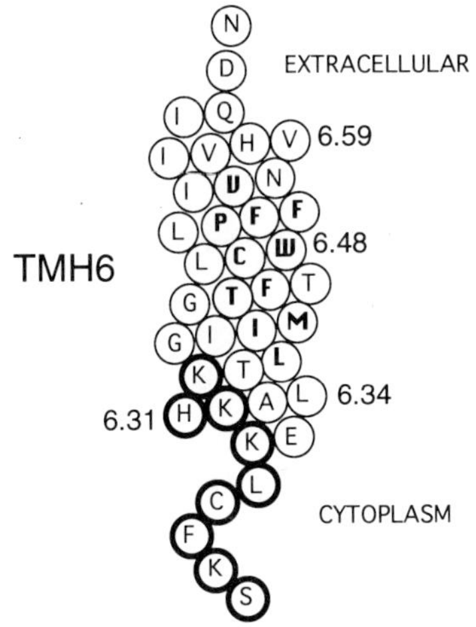

FIG. 5 Prediction of TMH6 boundaries by the Arg/Lys cytoplasmic motif. The sequence corresponds to the human β_2-adrenergic receptor. Residues highlighted by a thicker circle represent positions where there are Arg or Lys residues in any sequence within an alignment of neurotransmitter GPCRs. Note that beyond the predicted TMH6 cytoplasmic end Arg/Lys occur at consecutive positions, thus no longer consistent with an α-helical periodicity. Highly conserved amino acids in an SA of adrenergic, dopaminergic, and serotonergic GPCRs are shown in bold letters.

groups. The direct interaction between the Arg/Lys side chains and phospholipid head groups has been measured by ^{31}P-NMR for a number of TMH-containing proteins using the signature of the phosphorus group which displays an immobilized component due to the presence of the protein [e.g. glycophorin (66)], an effect that can be mimicked by polylysine (64).

The picture of a typical TMH that emerges from this analysis is thus of a hydrophobic core with polar ends facing the phospholipid head groups, especially in the cytoplasmic extension which is rich in basic residues, as shown in Fig. 6.

The extent of the refinement in the definition of cytoplasmic ends of TMHs that results from the inclusion of these Arg/Lys in addition to the predictions from the hydrophobic profile is dependent on the TMH topology, i.e., whether the N-terminus of the TMH is at the cytoplasmic side (N_{in}-to-C_{out}) or at the extracellular side (C_{in}-to-N_{out}). Transmembrane helices with N_{in}-to-C_{out} topology would have a larger error in the cytoplasmic ends predicted with the hydrophobicity profile than the TMHs with opposite topology (C_{in}-to-N_{out}). The reason, illustrated in Fig. 7, is that in an α helix the $C_\alpha \rightarrow C_\beta$ bond is always oriented toward the N-terminus. Therefore, an Arg or Lys

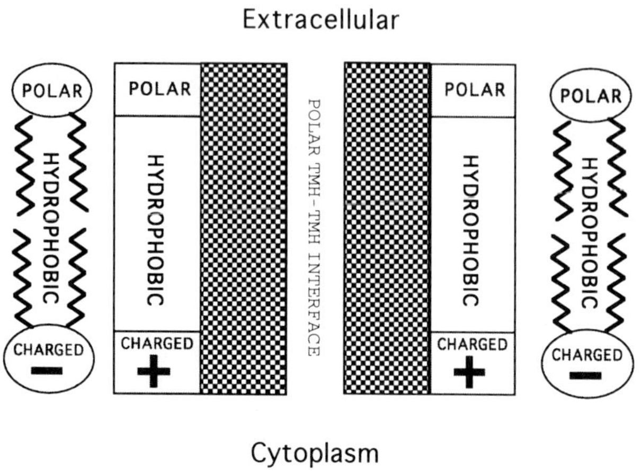

FIG. 6 Emerging picture of the polarity pattern for a prototype TMH from the inclusion of the phospholipid head group regions. Note the complementarity in the polarity distribution pattern at the TMH–lipid interface, especially the ionic complementarity of the Arg/Lys at cytoplasmic boundaries with the negatively charged character of inner monolayers in biological membranes. (Note that the extracellular polar extension of the TMH is drawn by analogy to the cytoplasmic end, although no clear example has been presented for this portion of the TMH.) Residues on interior faces of the TMH bundle have polarity properties similar to those found for soluble proteins.

residue in an $N_{in} \rightarrow C_{out}$ TMH (TMH6 in Fig. 7) can extend over two turns of the helix (seven AAs) toward the cytoplasmic side to interact with the phospholipid head groups, while the hydrophobic portion of their side chains can interact with the lipid chains. On the other hand, Arg/Lys residues in $C_{in} \rightarrow N_{out}$ TMHs (TMH7 in Fig. 7) can barely extend one-half turn (two AAs) toward the cytoplasmic side to reach the phosphate groups. This observation explains why the errors committed in predicting TMH ends are greater for the N_{in}-to-C_{out} helices, as illustrated by tests for proteins for which experimental structural information has become available. (e.g., BR in Table II). The different magnitudes of the errors in the predicted cytoplasmic ends have a topological dependence that not only agrees with, but can be satisfactorily explained by, the observed presence of the Arg/Lys in the TMHs. The magnitude of the error in the prediction of TMH boundaries using the hydrophobicity profile can now be estimated from Table II. For TMH with an N_{in}-to-C_{out} topology the error is around seven AAs (TMH4 is not well defined in BR), i.e., 25–30% (7/22) of an average TMH length. For TMH with an N_{out}-to-C_{in} topology the average error is insignificant. As discussed in Section

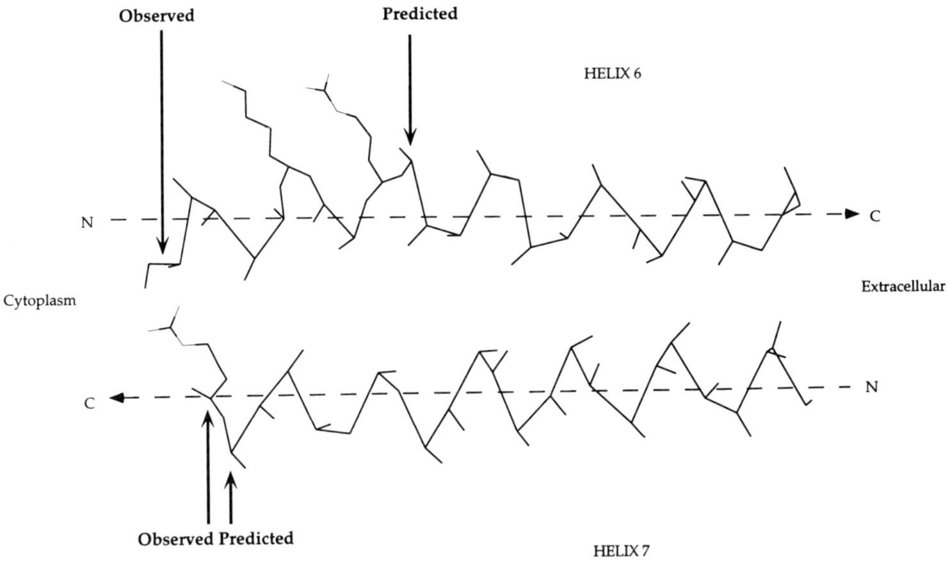

FIG. 7 The TMH6 and 7 of BR showing the C_α–C_β trace plus the Arg/Lys residues, in a view parallel to the membrane. The TMH6 has the topology $N_{in}\rightarrow C_{out}$ while TMH7 has $C_{in}\rightarrow N_{out}$. The topological dependence of the error in the predicted cytoplasmic ends using the hydrophobicity profile can be observed: eight AAs for TMH6 and two AAs for TMH7. This topological dependence can be explained by the Arg/Lys motif. The $C_\alpha\rightarrow C_\beta$ vector is always oriented toward the N-terminus. Therefore, Arg/Lys residues for $N_{in}\rightarrow C_{out}$ TMHs (e.g., TMH6) can extend over two turns of the helix (seven AA) toward the cytoplasmic side to interact with the phospholipid head groups. On the other hand, Arg/Lys residues for $C_{in}\rightarrow N_{out}$ TMHs (e.g., TMH7) can barely extend half turn (three AA) toward the cytoplasmic side to reach the phosphate groups. Note that the actual helix–helix interface is offset from the cytoplasmic ends of the hydrophobic segment by seven AA.

III,B,2 below, the topological dependence of this error (i.e., even-numbered helices in GPCRs are affected more than the odd-numbered ones) could seriously affect the validity of helix–helix packing models if TMH boundaries are taken from the hydrophobicity profile.

Although the Arg/Lys criterion was originally derived for a single sequence, it can also be applied through the sequence alignment of GPCRs. The assumption of common TMH boundaries for the GPCR within the SA allows for the analysis of the evolutionary consistency of these predictions. For example, the positions where those Arg/Lys occurred within an alignment of neurotransmitter GPCR are shown in Fig. 5 on the helical net representation of TMH6 in the β_2-adrenergic receptor. Note the consistent helical

TABLE II Predicted and Observed TMH of BR Using the Hydrophobicity Profile

Helix	Predicted Kyte–Doolittle	Observed	Cytoplasmic difference
1	10–34	10–32	2
2	44–68	38–62	6
3	78–102	80–101	1
4	106–130	108–127	−2
5	134–158	136–157	1
6	175–199	167–193	8
7	200–224	203–227	−3

periodicity in the evolutionary disposition of the Arg/Lys in the predicted TMH; they occur on a single face of TMH6 predicted to be oriented toward the lipids (see Section III,A). Beyond the predicted cytoplasmic end of TMH6 there are six Arg/Lys residues occurring at consecutive AA sites, inconsistent with helical periodicity.

3. Prediction of TMH Segments from the α-Helix Periodicity of Property Profiles Measured by Fourier Transform Analysis

The regularity of α-helical structures, and more specifically their characteristic periodicity of 3.6 AA/turn, can be used for prediction purposes based on the assumption that if the profile of a given property, e.g., hydrophobicity, within a segment of the sequence alignment displays an α-helical periodicity, this segment can be predicted to be a TMH. Properties whose value for every residue along the sequence reflect a degree of surface exposure, such as hydrophobicity (52, 69) or conservation (52, 70), would exhibit an amphipathic profile in a TMH and can thus serve in the detection of the α-helical periodicity.

In the computational algorithm developed by Komiya *et al.* (71) and Donnelly and Cogdell (49, 72) to take advantage of these criteria for TMH modeling, a property profile U is calculated for a sequence from the individual properties $U(j)$ assigned at each position j in a sequence alignment. $U(j)$ can be the hydrophobicity $H(j)$ or the conservation as measured by the variability $V(j)$ described in Section I,B. Two methods are used to analyze the α-helical periodicity of a property profile $U(j)$: a Fourier transform analysis presented in this section and the visual inspection of helical net representations in search of continuous patches along the TMH surface, presented in Section II,A,4 below. The Fourier transform analyzes the overall pattern throughout a sequence segment, e.g., the amphiphatic character, whereas the analysis of continuous patches of AA with a common property in a helical net representation identifies residues that fulfill the TMH periodicity criterion if they arc separated by three to four AAs. Both approaches can be used to predict TMH boundaries, or to define TMH–TMH versus TMH–lipid interfaces as described below in Section III,A.

In the Fourier transform analysis of the property profile $U(j)$ over a window size N (69, 71), the power spectrum $P(\omega)$ reveals all existing periodicities ω. If the sequence contained within the window N adopts an α-helical conformation, a peak should appear around $\omega = 105°$, the angle between adjacent side chains for an α helix viewed down its axis. The power spectrum $P(\omega)$ is calculated as shown in Eq. (1).

$$P(\omega) = \left[\sum_{j=1}^{N} (U(j) - \overline{U})\sin(j\omega)\right]^2 + \left[\sum_{j=1}^{N} (U(j) - \overline{U})\cos(j\omega)\right]^2, \quad (1)$$

where $\overline{U}$ is the mean value of $U(j)$ over the window N. If $U(j)$ is the hydrophobicity $H(j)$, the Fourier transform leads to the commonly used hydrophobicity moment (69). An α-helix periodicity index (AP) calculated from $P(\omega)$ describes the extent of the periodicity in the helical region compared to that over the entire spectrum and is defined as shown in Eq. (2).

$$\mathrm{AP} = \frac{(1/30)\int_{90°}^{120°} P(\omega)d\omega}{(1/180)\int_{0°}^{180°} P(\omega)d\omega}. \quad (2)$$

A value of AP>2 was suggested as a significant indication of α-helical character based on an analysis of the known TMH structure of the PRC (52), although parameters such as the window size and number of sequences within the alignment can affect the AP value. Since AP measures the extent of α-helix periodicity, a plot of AP versus the residue number can be used to identify the TMHs (52) and to infer their boundaries (72). The plotted values are calculated by scanning the sequence or sequence alignment with a constant window size N. The sequence segment at which the AP value drops significantly indicates the N- or C-terminus of the TMH.

For an illustrative example of the application of this procedure to use the variability profile in order to define the putative ends of TMHs, we have scanned a sequence alignment of dopaminergic, adrenergic, and serotonergic GPCRs calculating the AP values for each consecutive segment of 12 AA. The plots obtained for each of the seven TMHs (Fig. 8) clearly identify each of them as a region of increased AP values within the AP plot, indicating the presence of significant helix periodicity within these regions. The TMH

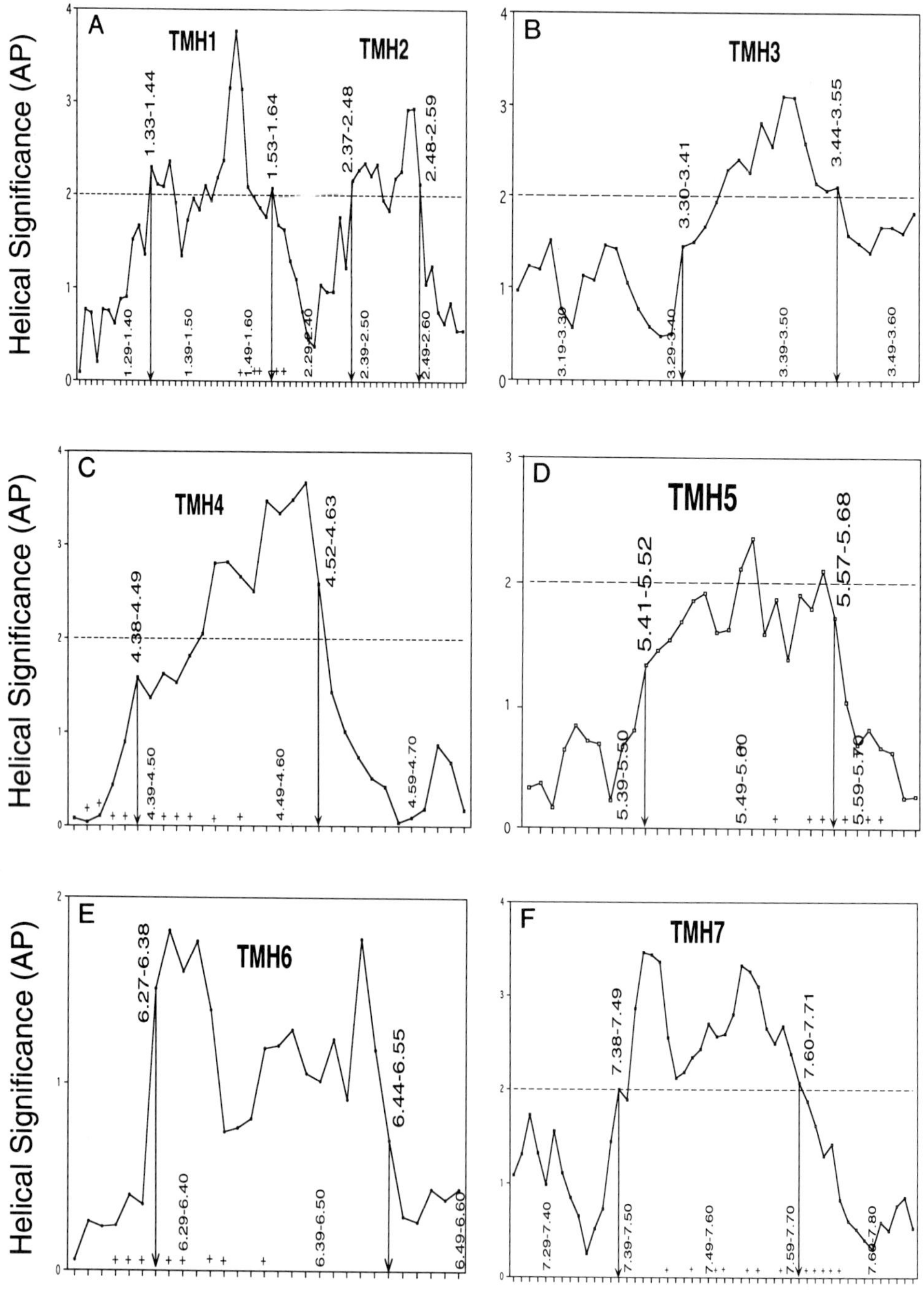
Helical Significance (AP)
A
TMH1
TMH2
1.33-1.44
1.53-1.64
2.37-2.48
2.48-2.59
1.29-1.40
1.39-1.50
1.49-1.60
2.29-2.40
2.39-2.50
2.49-2.60
B
TMH3
3.30-3.41
3.44-3.55
3.19-3.30
3.29-3.40
3.39-3.50
3.49-3.60
Helical Significance (AP)
C
TMH4
4.38-4.49
4.52-4.63
4.39-4.50
4.49-4.60
4.59-4.70
D
TMH5
5.41-5.52
5.57-5.68
5.39-5.50
5.49-5.60
5.59-5.70
Helical Significance (AP)
E
TMH6
6.27-6.38
6.44-6.55
6.29-6.40
6.39-6.50
6.49-6.60
F
TMH7
7.38-7.49
7.60-7.71
7.29-7.40
7.39-7.50
7.49-7.60
7.59-7.70
7.69-7.80

boundaries predicted by these plots are determined from the first 12 AA segment at which the AP changes from a basal value to one indicating significant helix periodicity and then back again, as indicated by arrows in Fig. 8. If this change occurs in the 12-AA window at the N-terminus of the TMH, the predicted beginning of this TMH is the first AA within this window, e.g., residue 6.27 in TMH6. Similarly, if the drop in AP value is in the 12-AA window at the C-terminus of the TMH, the predicted end of this TMH is the last AA within this window, e.g., residue 6.55 in TMH6. Note, however, that the predicted C-terminus of TMH6 could be either 6.55 or 6.54 since the drop in the AP plot is not very sharp.

Some additional considerations are necessary in the interpretation of the TMH boundaries predicted by the Fourier transform of the conservation pattern (Fig. 8). *First,* the assumption that lipid-exposed residues are less conserved than residues facing the interior of the TMH bundle is based on structural premises. But interior residues that are functionally divergent, e.g., determining ligand-binding specificity, may induce a similar variability on both helix faces thus leading to a loss of helical periodicity. This seems to be the case for the C-terminus of TMH2 that shows a subtype-specific AA distribution at positions 2.60 and 2.64. *Second,* the distortion of α-helix character induced by a Pro residue includes a face-twist of the helical portions N- and C-terminal from the Pro (61, 73) (see also Section II,B,2 and Fig. 10). The periodicities are offset by this Pro-induced shift, thus diminishing the AP value for window segments that include the Pro-kink region as defined in Section II,B,2 below. Another artifact in the AP plot produced by a Pro residue occurs when it is facing the lipids and yet it is highly conserved. Such an apparent departure from the general expectation that lipid-facing residues are not conserved applies specifically to Pro because this AA can exert its structural effects without facing the protein interior, e.g., $Pro_{6.50}$ in TMH6 according to Baldwin (40). To correct for this occurrence, Donnelly *et al.* (49) proposed to alter the Pro variability to an average value in order to increase the detection of the helical periodicity. A consequence of the

FIG. 8 α-Helix periodicity of the variability profile measured by Fourier transform analysis used to predict TMH boundaries for TMH1–2 (A) to TMH7 (F). The plots were calculated with a scanning window of 12 AAs shown at the abscissa. The TMH ends predicted by this criterion are indicated by arrows, the occurrence of Arg/Lys residues are indicated by positive signs, and the suggested significance threshold for α-helix periodicity is marked by a dotted line. Note the presence of Arg/Lys residues inside predicted TMH segments at the cytoplasmic side, i.e., C-termini for odd TMHs and N-terminal for even TMHs. The degree of α-helical periodicity was calculated from an alignment of adrenergic, dopaminergic, and serotonergic GPCRs.

presence of Pro in TMHs 2, 4, 5, and 6 in the GPCR aligned for Fig. 8 is that the extracellular segments of the helix relative to the Pro residues are shorter than the 12-AA window chosen for the Fourier analysis. This is responsible for decreasing the AP values (e.g., see Fig. 8) and prevents the accurate prediction of the extracellular TMH boundaries for these helices. Donnelly *et al.* argued that the presence of AA sites whose variability was significantly larger (or smaller) than the average within a window can dominate the Fourier transform and thus induce inaccuracies. They proposed a smoother profile (either conservation or hydrophobicity) by modifying the $U(j)$ values (49).

Based on the special considerations and possible artifacts in the calculation of AP plots, the prediction of TMH boundaries by the Fourier transform method may be incomplete and should be augmented by predictions using other methods presented in this chapter. Convergence of results from several different methods can be sought to secure the prediction of TMH ends. This is illustrated by a comparison of the TMHs predicted from the AP plots in Fig. 8 (e.g., 6.27 for TMH6 N-terminus) with the predictions shown in Fig. 3 obtained using the hydrophobicity profile (6.36 for TMH6 N-terminus). As discussed in Section II,A,2 above, the TMH boundaries predicted from AP plots in the cytoplasmic side of TMHs can extend beyond the boundaries predicted by the hydrophobicity profile, due to the Arg/Lys motif. The implication from this comparison is that the hydrophobic core of the TMH predicted by the hydrophobicity profile is an α helix that is continuous with the cytoplasmic extensions predicted by the Arg/Lys motif since they have the same helical periodicity as measured by Fourier transform. If a break existed in the helix between the hydrophobic core facing the lipid chains and the Arg/Lys region facing the phospholipid head groups, the corresponding AP values of this nonhelical portion would be significantly smaller. Such a conclusion was reached by Donnelly *et al.* in their analysis of the TMHs in the antenna complexes (72). Thus, it appears that the redefinition of a TMH as suggested in Fig. 6 is validated by the analysis of the helical periodicity. Interestingly, the TMH boundaries on the extracellular side show the opposite pattern whereby the AP plots suggest shorter TMHs (e.g., 6.55 for TMH6 C-terminus) than the hydrophobicity plots (6.59 for TMH6 C-terminus). The reason here seems to be the above-mentioned inaccuracy in the prediction of extracellular ends by the Fourier transform of TMH 2, 4, 5, and 6 due to the presence of Pro residues close to the end of the helix; in this case predictions made using the hydrophobicity profile would be more reliable than the Fourier transform.

In addition to the conservation and hydrophobicity profiles for the analysis of α-helical periodicity with Fourier transform methods, Donnelly *et al.* proposed the use of another property, estimating the extent to which a position is exposed to the solvent as evaluated from environment-dependent substitution tables calculated from either lipid-facing or aqueous-facing residues (49, 72). However, no significant improvement was reported for the prediction of TMH boundaries using this set of parameters. Nevertheless, these properties are useful in the prediction of the lipid and interior faces of the TMHs, as described below in Section III,A.

4. Prediction of Transmembrane Helix Segments from the α-Helix Periodicity Identified from Surface Patches on a Helical Net Representation

Since α-helices have a periodicity of 3.6 AA/turn, individual residues separated by 3 or 4 AAs and sharing a given property such as hydrophobicity or conservation would define a continuous patch on the helical surface. Since any such patch is characteristic of an α helix, the limits of these patches in the sequence are predicted to be the TMH boundaries. For example, Donnelly *et al.* (30) predicted TMH ends by the limit of helical patches consisting of AA sites where hydrophilic residues are excluded. This method is analogous to the Fourier transform method described in the previous section, since both measure the same property, i.e., α-helical periodicity of a given property. The difference between these two methods is that the Fourier transform is applied to a property profile over the entire sequence, rather than to individual residues. This difference allows for a better prediction using helix patches than the Fourier transform for TMH ends close to Pro-kinks. The Fourier transform method has difficulties with such structures for reasons described in the previous section.

The analysis of helical patches is best undertaken on helical nets, which are a correct two-dimensional representation of positions on a three-dimensional cylinder defined by the C_α atoms of an α helix. Due to their value as analytic tools of helical periodicity as described in this chapter, Fig. 9 depicts in detail this mode of representation in terms of helical nets and its translation to 3D.

5. Analysis of the Occurrence of Pro Residues within Putative Transmembrane Helix Segments as an Indication of Helix Ends

Due to the characteristic distortion produced by a Pro residue on an α helix (see Section II,B,2 below), single Pro residues within TMHs are generally highly conserved (74). Single Pro residues that are not conserved are common at the ends of helices and can be used for prediction of TMH ends. It has been shown that such single and nonconserved Pro residues can occupy the first three positions at the N-terminus and at most the last position at the C-terminus without significantly distorting the helix (75, 76). The reasons for

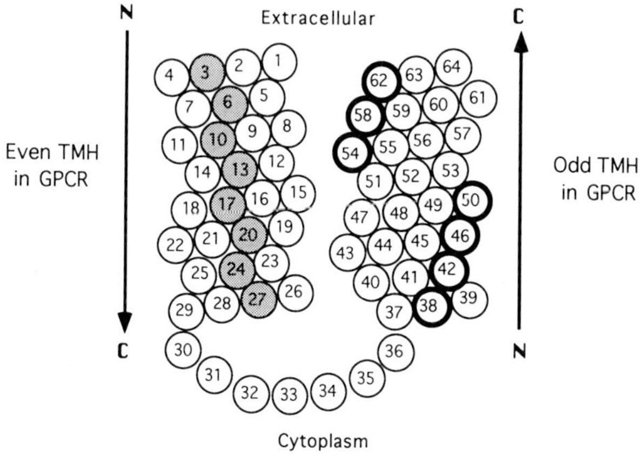

FIG. 9 The helical net representation, illustrated by two consecutive TMHs with opposite topologies. Each circle represents the position of a C_α of an ideal α-helix on the surface of the cylinder enveloping a helix with the same axis. When this cylinder is unwrapped, it provides a two-dimensional surface of the envelope of an ideal α helix. The positions of the C_α on this envelope constitute the helical net. The path described by consecutive residues can be followed in this figure from the residue numbers. Note that consecutive residues on the continuous cylinder surface can appear at the opposite margins of the helical net, e.g., see residues 4 and 5. Thus, residue 5 (AA_5) should be considered for all purposes as if it was positioned next to AA_4. Note the handedness of the helical net progressing from the N- to the C-termini. This reflects the right-handed symmetry of the α helix that is not reversed for an even- relative to an odd-numbered TMH for the GPCRs. The nearer neighbor's helix periodicity that positions on the same "face" of the helix those residues separated by three to four other AAs is reflected in the helical net by the fact that the C_α of such pairs are adjacent in position, above and below, respectively. These representations allow the exploration of α-helical periodicity by visual inspection, as described in Section III,A,2. For example, note the continuous helical patch defined by residues 3–6–10–13–17–20–24–27 shown shaded, which runs parallel to the TMH axis and thus appears on a single face of the TMH. Helical net representations also identify continuous patches that run at a certain angle from the TMH axis, as the one defined by residues 38–42–46–50–54–58–62 shown by thicker circles. This later patch would result from a tilted and/or supercoiled TMH relative to the membrane axis. Note that by tilting the helical net, the direction of this patch would become parallel to the membrane axis defined by the first helix.

these position preferences are (i) that an amide hydrogen at the N-terminus in an α helix is not required to H-bond, and (ii) that the Pro ring within an α helix produces a steric clash with the AA in the preceding turn, but not with residues following the Pro. Within an α helix, Pro–Pro or Pro–X–Pro motifs never occur at the C-terminus and are severely restricted at the N-terminus where they could adopt a poly(Pro) helix conformation (75). In fact, the four Pro–Pro motifs that appear in the PRC structure are all in an extended conformation located in extracellular loops. Such a motif appears toward the C-terminus of TMH4 in many neurotransmitter GPCRs.

6. Methods for Defining Transmembrane Helix Ends Based on the Statistical Probability of Occurrence of Specific Amino Acids in Transmembrane Helices

A statistical method based on conformational preference functions has been proposed to predict TMHs (77), in which the probability for each AA in a sequence to belong to a TMH is calculated and used for prediction. The accuracy of this method was around 79% for five membrane proteins. Granatir *et al.* (78) combined this method with hydrophobicity-based methods of the type described in Section II,A,1. The authors provided a consensus algorithm where different methods are applied and their respective predictions of TMH ends are compared pursuing a convergent or consensus prediction among them. The statistical combination of different approaches is expected to improve the results due to the inconclusive nature of each of these methods applied individually. A similar conclusion was reached by Edelman (60) from a more complex mathematical formulation applied to derive TMH predictions based on known structural data. Although all these methods represent algorithmic improvements relative to the hydrophobicity profile, their predictive power does not seem substantially improved. The reason may be that all these methods are restricted to revealing the hydrophobic core of the TMHs either by their definition, e.g., based on calculated hydrophobicity, or through the data base from which the frequencies of residue occurrence in TMHs were derived. Because ignoring the Arg/Lys motif can lead to the major errors described above for BR in the prediction of the TMH boundaries by the hydrophobicity profile (Table II), the absence of this additional criterion in all these methods may preclude a significant predictive improvement. For instance, the test of Edelman's method on the BR and PRC structures (29 TMHs) yielded a total error of 4 residues at the extracellular boundaries, compared to 53 residues at the cytoplasmic side where the Arg/Lys motif is located. It is likely that the addition of the Arg/Lys motif criterion to this type of methods could significantly improve their predictive power.

B. Modeling of the Three-Dimensional Structure of Individual Transmembrane Helices

Modeling the three-dimensional structure of the helix bundle is greatly facilitated by the availability of structural information about preferred backbone and side-chain dihedral angles in α helices (79–81), giving rise to side-chain rotamer libraries (82, 83) that can be used in GPCR modeling (25). The allocation of appropriate torsional angles to residues in the TMHs is important because inappropriate rotamers may hinder helix packing interactions. The (ϕ,ψ) backbone dihedral angles for a TMH are the same as for those of a soluble protein α helix, judged by the PRC crystal structure (57), i.e., $-60°$, $-40°$. Side-chain dihedral angles can be classified broadly according to their relative energy minima: $\{-60°, 180°, 60°\}$ for all side-chain angles, except when preceding a bond with π-electron character such as for {F,Y,W,H,D,N,E,Q,R} in which case the values are $+/-90°$. An α-helix environment significantly restricts the values of the X_1 angle for all AAs to values around $-60°$ and 180° (80). The reason is the steric overlap between the side-chain atoms in position gamma of residue *i* and the carbonyl oxygen of the third preceding residue (N-terminal to *i*), denoted $C{=}O_{i\text{-}3}$. The statistical analysis of X_1 distributions in helical environments in high-resolution crystal structures clearly supports this generalization, particularly for aromatic side chains (80). Two sets of residues have additional specific interactions between their side-chains and the helix backbone that give rise to preferential X_1 rotamers:

1. Ser/Thr/Cys

The ability of these residues to H-bond their side chain to the $C{=}O_{i\text{-}4}$ of the preceding turn, and the significant preference observed in the PDB for structures with such conformations, constrains the starting rotamers for the modeling purpose to $-60°$ (79, 80). Since this criterion arises from favorable interactions rather than steric hindrances, these residues could also adopt other X_1 rotamers, e.g., $X_1 = 180°$ for Ser and Cys, although such values appear with much lower frequency. Thr residues are generally restricted to $X_1 = 60°$ due to the C_β-branched character of the side chain, as explained below.

2. Val/Ile/Thr

C_β-branched AAs in an α helix are restricted to one rotamer, as suggested by the relative rotamer populations observed in the PDB structures, as well as by results from computational energy minimizations [see Table III in Piela *et al.* (81)]. The reason for this preference is the above-mentioned steric clash of a C_γ atom at $X_1 = 60$ with the $C{=}O_{i\text{-}3}$. Because C_β-branched AAs have two atoms at position γ, the only rotamer that avoids the steric clash at 60° is the one that places the two γ atoms at X_1 values of $-60°$ and 180°. For reasons of nomenclature, this rotamer corresponds to $X_1 = 60°$ for Ile and Thr, and $X_1 = 180°$ for Val.

The choices of torsional angles mentioned above as preferred starting values for the construction of TMHs assume standard helical conformations. However, this is not true for THMs including internal Pro, due to the disruption of the helical character by the presence of Pro-kinks (PK) as found in TMHs 2, 5, 6, and 7 of the GPCRs. The geometry of the PK regions within α helices is characterized by backbone dihedral angles that deviate from α-helical values for the residues comprised between the Pro_i and the amino acid at position *i*-4 from the Pro, termed $AA_{i\text{-}4}$ (84). These local distortions produce a highly exposed, H-bond free $C{=}O_{i\text{-}3}$ and $C{=}O_{i\text{-}4}$ (85). The loss of two helical H-bonds in one turn suggests a significant intrinsic flexibility of the TMH at the PK region. This flexibility has been demonstrated by NMR (86, 87) and computational molecular dynamics simulations (88, 89). There are also nonlocal deviations from an ideal α helix induced by a PK, characterized by a bending (73, 84, 90) and twisting (61, 73) of the regular α helix halves both N- and C-terminal from the PK region, as illustrated in Fig. 10. The disruption of α-helical character occurs due to steric clashes of Pro rings with the preceding turn, and the loss of a helical backbone H-bond due to the lack of the amide H in Pro. These reasons, and the intrinsic flexibility of a PK, confer to these positions in a TMH some crucial structural, and possibly also functional roles (74, 91–93). However, *de novo* structural modeling of PK regions is currently problematic due to three interrelated factors:

1. Their intrinsic flexibility at the level of secondary structure increases significantly the exploration of the conformational space necessary to find the preferred packing.
2. In known structures, PK regions can adopt a variety of (ϕ,ψ) angles, making it difficult to model PK *de novo* in TMH structures.
3. The nonlocal disruption of the TMH pattern by PK implies large deviations of the relative 3D localization of the helical portions before and after the PK. The nonlocal perturbation of helical character can be separated into a bend and a twist of the helical portions N- and C-terminal to the PK, as shown schematically in Fig. 10. These distortions produce a twisting of the helix "face" (61, 73), as illustrated in Fig. 10, which may hinder the analysis of helical periodicity as mentioned in Section II,A,3 for the Fourier transform method. In known structures, the bending and twisting angles vary between $\{9°,40°\}$ and $\{-44°,87°\}$, respectively (73).

Sankararamakrishnan *et al.* suggested specific backbone dihedral angles for modeling PKs (94). Use of these initial angles for the PK is preferable over modeling PK-containing TMHs as ideal α helices and relying on energy minimizations to produce the appropriate conformations, because the latter approach may produce structures that differ from known PK geometries (26). Due to the intrinsic flexibility of the PKs, those starting values will

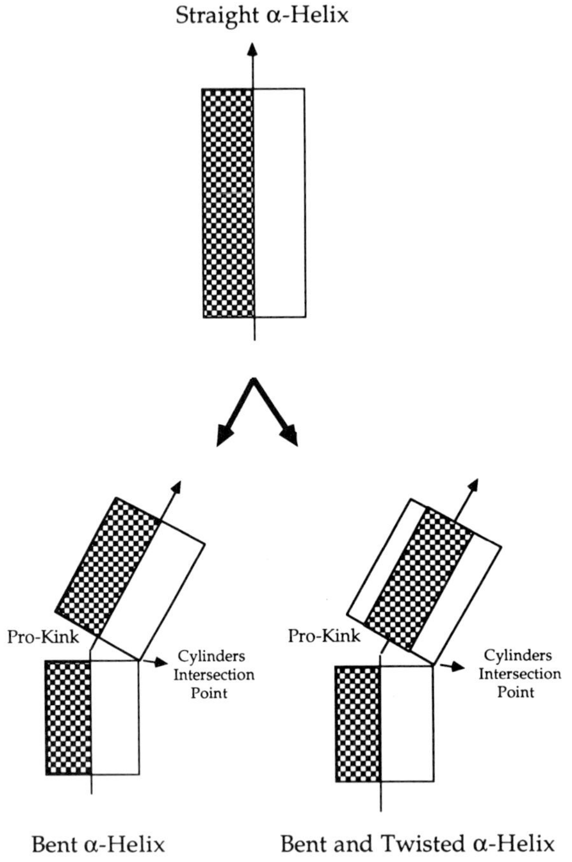

FIG. 10 Nonlocal deviations from an ideal α helix induced by a Pro-kink, characterized by a bending and twisting of the regular α-helix halves both N- and C-terminal from the Pro-kink region. The TMH is schematically represented by a cylinder with two distinct faces. Note that the twist induced by the Pro-kink on the square face would be expected to offset the periodicities of these two halves, thus hindering the analysis of helical periodicity that assumes a continuous helix such as the Fourier transform analysis described in Section II,A,3.

usually be modified through the modeling process so as to optimize tertiary structure interactions. For example, a specific H-bond between $Asn_{7.49}$ and the $C{=}O_{i\text{-}4}$ from $Pro_{7.50}$ (i.e., $C{=}O_{7.46}$) could dictate the conformation of the PK of TMH7.

The last step in modeling individual TMHs is the conformational search among feasible side-chain rotamers within each TMH, pursuing both the absence of steric conflicts and the presence of specific stabilizing intrahelical side-chain to side-chain interactions. For example, Donnelly *et al.* (30) found that $N_{7.49}$ can H-bond the backbone $C{=}O_{7.46}$ of the Pro-kink in TMH7.

C. Secondary Structure Prediction for the Loops and N- and C-Terminal Domains

The secondary prediction methods applied to extramembranal portions of GPCRs (20, 25) are those commonly used for prediction in soluble proteins (12, 95) and will not be reviewed here. For predicting the conformation of short loop structures connecting TMHs, additional constraints are imposed by the TMHs and these can be used for prediction purposes (see Section III,C, below). For example, the identification of sequence motifs favoring specific conformationally defined turns, or the use of conformational selection based on statistical trends, can be combined with energy minimization to model backbone conformations for these regions (96, 97). A note of caution is necessary in the application of secondary structure prediction methods derived from soluble proteins to extramembranal portions of GPCRs, because interactions between these protein regions and the adjacent membrane environment may lead to physicochemical situations not considered in the development of such methods for soluble proteins, thus vitiating their predictive powers. This seems to be the case for the cytoplasmic portions of GPCRs, due to the tendency of Arg/Lys residues to be predominantly localized in these regions where they can interact with negatively charged phospholipid head-groups, as discussed above (Section II,A,2).

III. Modeling the Tertiary Structure of the Three-Dimensional Model of a G Protein-Coupled Receptor

Modeling the interactions that determine the tertiary structure of the GPCRs can be divided into three procedural steps, discussed below in Sections III,A–C. The first involves the prediction of the orientation of each TMH relative to the lipid and the protein interior. This entails predicting for each TMH the residues that lie in the helix–helix or the helix–lipid interface. Throughout this chapter, the terms helix–helix or TMH–TMH interfaces are used to refer to the protein interior of the transmembrane helix bundle of GPCRs. The second step involves the packing of the seven TMHs into a bundle that fulfills the predictions from step 1 and represents the transmembrane domain of the GPCR. The third step proceeds to the completion of the 3D model with protein portions that lie outside the membrane environment, such as the loops and the N- and C-termini.

A. Prediction of TMH–TMH versus TMH–Lipid Interfaces

All methods for predicting the orientation of TMHs toward the lipid or toward other TMH interfaces rely on the same conservation or polarity criteria that served in the definition of TMH segments (see Sections II,A,3 and II,A,4, above). Based on the conservation criterion, the face of the TMH containing more conserved residues is predicted to face the protein interior. The rationale for this criterion is the common observation that residues exposed to the surrounding environment are less conserved than interior residues, a prevalent assumption that was corroborated for TMHs in an analysis of the PRC (52). The second criterion for defining the orientation of each TMH in the helix bundle is based on the complementarity of the surface polarity between a TMH and the membrane lipids. Such complementarity considerations must differ in character for the hydrophobic region of the TMHs compared to the cytoplasmic extensions of these TMHs that contain the Arg/Lys patches: in the hydrophobic core, the more apolar face of the TMH is predicted to be oriented toward the membrane lipids, whereas, for the Arg/Lys region, the more hydrophilic face (containing the patch of basic residues) of the TMH cytoplasmic extension is oriented toward the membrane at the level of the phospholipid head groups (Fig. 6). Since all these considerations rely on α-helix periodicity, the methods available for the analysis of the conservation and hydrophobic profile presented above for the detection of helix periodicity (Sections II,A,3 and II,A,4) can be used for prediction of TMH–TMH versus TMH–lipid interfaces. While the Fourier transform procedure would thus predict the orientation of a TMH, the identification of TMH patches formed by AAs located at the same interface is very useful in predicting the *extent* to which each TMH is exposed to the lipids or buried among other TMHs.

1. The Hydrophobic or Conservation Moment Predicts the Relative Orientation of Helices in the Transmembrane Bundle

The formulation of the Fourier transform analysis of a given property profile, e.g., the conservation $U(j)$, was described in Section II,A,3. Donnelly *et al.* (41, 49) showed that the same methodology can be applied to predict the TMH orientation. The orientation of the internal, protein-facing side of a TMH can be estimated from the direction of the moment $(P(\omega))^{1/2}$ when $\omega = 100°$. This is the moment produced by the property profile $U(j)$ when the sequences form an ideal α helix. The angle ω describes the direction of the moment relative to the first residue ($j = 1$) and can be calculated as shown in Eqs (3) and (4).

$$\theta = \begin{cases} \gamma, & \text{if } \left(\sum_{j=1}^{N} (U(j) - \overline{U}) \sin(j\omega) \right) > 0 \\ 360 - \gamma, & \text{if } \left(\sum_{j=1}^{N} (U(j) - \overline{U}) \sin(j\omega) \right) < 0 \end{cases} \tag{3}$$

$$\gamma = \arccos \left[\frac{\sum_{j=1}^{N} (U(j) - \overline{U}) \cos(j\omega)}{\sqrt{P(\omega)}} \right] \quad \text{where } 0° \leq \gamma \leq 180°. \tag{4}$$

When the conservation profile is used, the vector of the property moment points toward the protein interior. Note that the variability moment will point in the opposite direction, toward the lipid. When the hydrophobicity profile $H(j)$ is used, the moment points toward the more hydrophobic face of the TMH (98). However, this direction is not always facing the TMH–lipids interface. Although the hydrophobicity moment of the hydrophobic core of the TMH is predicted to point toward the membrane lipid, the Arg/Lys region on the cytoplasmic side inverts the direction of the hydrophobic moment 180° away from the lipid interface due to the high hydrophilic character of these basic residues. These conflicting orientations predicted for a continuous TMH face explain the surprising observation about the seemingly chaotic direction of the hydrophobic moments calculated for the TMHs in the BR structure (34). The reason was that the Arg/Lys region of the TMHs was included, as it is present in the BR structure (see Fig. 4) and therefore the calculated moments were a composite of two different moments of opposite directions as discussed above.

Most authors have applied the hydrophobic moment method to the TMH boundaries as predicted by the hydrophobicity profile, which do not include the Arg/Lys region, resulting in predictions that were consistent with expectations. Thus the surprising situation arises that the hydrophobic moment as originally presented by Eisenberg *et al.* (98) has yielded useful predictions although in its test with a real TMH structure it failed to predict correctly the TMH orientations (34) for the reasons explained above. The proper application of this methodology was outlined by Donnelly and Cogdell (72), with the Arg/Lys extensions of the TMH taken into account, and used for prediction purposes. An analysis of the antenna complexes showed the opposite directions of the hydrophobic moment depending on which segment of a TMH is considered. The angle between the moment for the hydrophobic core of the helix and the Arg/Lys extension was indeed close to 180° (72). The authors suggested this 180° difference as a useful tool for prediction purposes. A schematic representation of the relative direction of each property moment is shown in Fig. 11.

Properties other than the variability and/or hydrophobicity profiles can by used in a similar manner to predict TMH orientation. In particular, Donnelly *et al.* (49, 72) used the estimate of the extent to which a position is exposed to solvent as predicted from environment-dependent substitution tables. These tables were calculated from either lipid-facing (49) or water-facing (48) residues. These two properties are directly related to the prediction of the orientation of TMH with respect to their TMH–TMH versus TMH–lipid interfaces. In fact, the direction of moments calculated for these

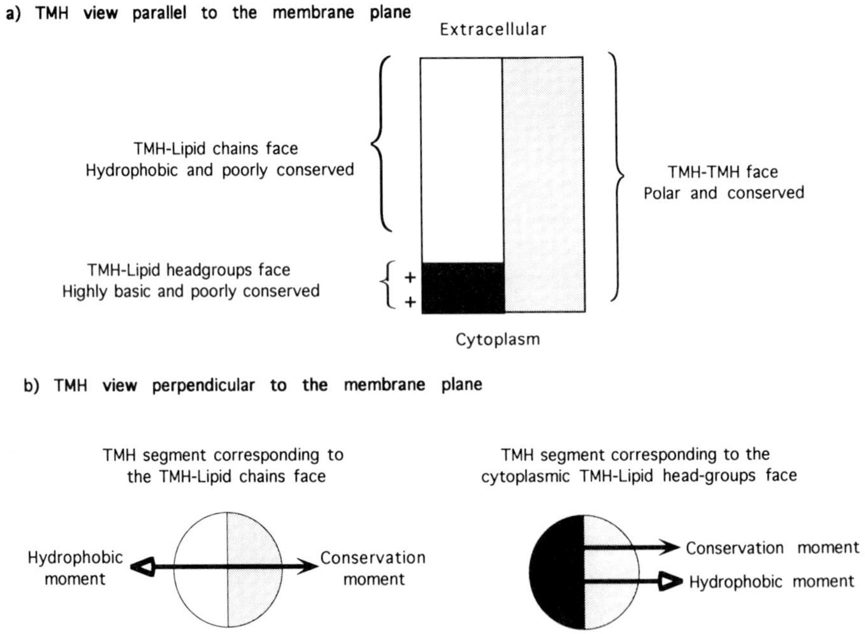

FIG. 11 Directionality of the hydrophobic (open arrows) and conservation (closed arrows) moments relative to helix–helix or helix–lipid interfaces. While the conservation moment consistently points toward the protein interior, note the opposite orientation (i.e., 180°) of the hydrophobic moment depending on the helical portion of the TMH. For the hydrophobic core, the hydrophobic moment points toward the TMH–lipid interface. For the Arg/Lys cytoplasmic extensions facing the phospholipid head groups, the hydrophobic moment point toward TMH–TMH interfaces.

properties was shown to predict correctly the orientation of the seven TMHs in BR (49).

2. Patches on the Transmembrane Helix Surface Identify TMH–TMH vs TMH–Lipid Interfaces

Similar to their use in identifying helix ends (see Section II,A,4) "patches" of AAS identified on helical net representations, and sharing a particular property, can be useful as an alternative method to predict the face of each TMH that is in contact with either the lipid or the other TMHs. Individual residues separated by three to four AAs and predicted from their properties to be likely to face a given environment, e.g., the lipids, will appear as continuous property-sharing patches on a helical net representation. The orientation of such patches defines the predicted interface of a TMH according to the shared property, e.g., the lipid interface identified from a hydrophobic patch. An alternative representation to helical nets for the identification of such patches is the helical wheel representation in which the helix is projected along its axis, as used by Baldwin (40). To summarize the use of AA patches to identify the nature of TMH interfaces we present first the criteria for predicting a lipid or interior environment for individual AA sites. We then show how these positions can be identified as continuous patches on a helical net representation to become the predicted TMH–TMH or TMH–lipid interfaces (40). Because these predictions are done at the level of single residues, as opposed to a global orientation for a TMH segment provided by the Fourier transform method described above, we show how this method can be used to estimate the extent to which each TMH is exposed to the lipid phase. This degree of surface exposure will become a constraint used in the subsequent packing of the TMH into a model of the transmembrane domain of the GPCR, as illustrated in the subsequent Section III,B.

The two major criteria for predicting whether a given residue is facing the lipid milieu or other TMHs are based on the same two AA properties discussed in previous sections: the conservation and the degree of polarity. Consequently, the same general guidelines apply for the use of these two properties as described in the previous sections. In evolutionary terms, conserved AAs are predicted to be at TMH–TMH interfaces while nonconserved sites are likely to face the lipid environment. The quantification of the degree of conservation as presented in Section I leaves unspecified the criterion of a conservation threshold for prediction purposes, and different authors have used different thresholds. In the most comprehensive conservation analysis presented to date, Baldwin (40) considered the extent of conservation at different levels beyond a conservative threshold, i.e., 66% presence of a given AA. Selecting a conservative threshold avoids the exclusion of residues that are significantly, but not absolutely, conserved. The subdivision of these selected AA sites into different levels of conservation is useful for the comparison of the respective types of predictions that can be expected; thus, predictions derived from more-conserved positions are stronger than predictions derived from less-conserved positions, so that if contradictory predictions arise from any two AA sites, the prediction derived from the most conserved position prevails. The polar character of an AA site can also serve as a criterion (54), because polar sites are predicted to be at TMH–TMH interfaces, while hydrophobic AA are predicted to face the lipid. These expectations are validated by data from known structures of transmembrane proteins (54, 70). As discussed in the previous sections, this polarity pattern is reversed in the Arg/Lys regions toward the cytoplasmic boundaries of TMHs where these basic residues are predicted to face the membrane environment at the level of the phospholipid head groups. Although similar conservation or polarity criteria are used in the helical patches method and in the Fourier transform method (see Section III,A,1), the former has the advantage that the criteria can be applied to only a few AA sites within TMHs, because a continuous profile over a sequence segment is not required. This not only allows the modification of criteria based on conservation or polarity properties, but also makes possible the use of qualitatively different criteria such as experimental data on ligand or lipid interaction sites for GPCR. These three sets of criteria are described below.

1. The conservation analysis as presented in Section I identifies specific residues that can be predicted to lie either at lipid- or interior-facing positions. For example, Baldwin (40) predicted that divergent residues among highly homologous and functionally equivalent GPCR would face the lipid environment, because these sites are likely to occur in functionally nonimportant positions. Baldwin also proposed the conservation analysis at the level of physicochemical properties rather than chemical identity. Thus, she identified residues that conserve a similar size during evolution and predicted them to face TMH–TMH interfaces (40). Zhang and Weinstein (54) analyzed the conservation of polar character at positions where more than 90% of the GPCR sequences considered retained polar residues. These positions, called "polarity-conserved positions" (PCPs) were predicted to be at TMH–TMH interfaces, as verified by the structure of BR family (54).

There are notable exceptions to the validity of predictions using the conservation analysis. An accepted exception to the rule that conserved residues point toward the protein interior is the case of Pro residues because they can exert their effects from a lipid-facing position (40, 49), as discussed in Section II,A,3. Similarly, an exception to the rule that a poorly conserved AA site is facing the lipid would be the case of a residue responsible for functional divergence, such as ligand binding; such a residue is expected to face the interior of the TMH bundle in the ligand recognition pocket.

2. The criterion of a polar complementarity between the TMH and the lipid phase can also identify specific residues predicted to face the lipid or protein interior phase. For example, Donnelly *et al.* (30) identified a continuous patch of AA sites not containing polar residues and predicted that they face the lipid. Similarly, Arg and Lys residues at the cytoplasmic boundaries are predicted to face the phospholipid head groups (61). A note of caution is necessary in defining the polar character of each AA in a TMH. The polar character of the 20 natural AAs that is relevant to the predictions discussed here may differ from the polarity assigned by hydrophobicity scales, due to specific interactions within TMHs. Such is the case of Ser, Thr, Cys, Pro, and Tyr. In the first group, {Ser, Thr, Cys} residues are generally found to H-bond back to the carbonyl backbone of the preceding turn, i.e., C_β-$O(S)_\gamma$-H...O=C_{i-4} (79). Such an arrangement decreases the polarity of the face they present for interaction, so that this AA site can be exposed to the lipid milieu (79). For example, 45% of the Ser residues in BR are exposed to the lipid phase. In the second group, Pro residues induce a characteristic kink in TMH, as described in Section II,B, that results in the two carbonyls of the preceding turn being highly exposed and H-bond free. This arrangement enhances the polar effect of a Pro residue relative to the hydrophobicity scale. This indirect polar character of Pro residues would suggest that they are likely to be positioned facing the protein interior, as suggested by Woolfson (91) based on an analysis of the hydrophobic moments of Pro-containing α helices. On the other hand, Tyr residues have been observed facing the lipid head groups in the structures of BR (62) and bacterial porins (99, 100).

3. Some experimental data on GPCR are available concerning the exposure of AA sites to the membrane lipid or the protein interior. Such information includes (i) the identification of covalent attachment sites for activated probes, (ii) the identification of residues proposed from mutagenesis experiments to be in direct contact with the ligand, and (iii) data on specific TMH–TMH interactions implying that the residues involved should face the protein interior. In particular, experiments on adrenergic GPCR identified positions $W_{7.40}$ (101) and {$S_{2.63}$ and/or $H_{2.64}$} (102) as ligand attachment sites from alkylation experiments, thus identifying AAs lying at TMH–TMH interfaces. Covalent attachment AA sites identified for the bovine rhodopsin GPCR belong to two classes, from hydrophilic (55) and hydrophobic (103, 104) probes, respectively. The hydrophilic probes were predicted to label AA sites exposed to the aqueous phase or at least lying in the phospholipid head group region. The hydrophobic probes, on the other hand, were predicted to label AA sites exposed to the lipid milieu. The alkylation experiments thus provided a powerful discriminant of TMH–TMH versus TMH–lipid interfaces. Using similar arguments, AA

sites proposed to lie in direct contact with receptor ligands based on mutagenesis experiments were also useful to predict TMH–TMH interfaces [e.g., Maloney-Huss and Lybrand (25), Hibert *et al.* (22), and Baldwin (40)]. However, the strength of those predictions is less than that of those obtained from covalent attachment sites, because they are not supported by evidence of direct contact. Finally, some considerations were proposed for predicting specific interactions among TMHs based on interactions of specific AAs. Such is the case of the interaction between the conserved $D_{2.50}$ and $N_{7.49}$ loci. A double-revertant mutant of the gonadotropin-releasing hormone receptor (GnRHR) (4) showed that the pair $N_{2.50}/D_{7.49}$ in the GnRHR has the same ligand-binding properties as the $D_{2.50}/N_{7.49}$ pair found in most GPCRs, thus identifying a TMH2–TMH7 interaction. These inferences are discussed below in the context of modeling the packing of TMHs in the transmembrane bundle.

Once the prediction of a tertiary environment has been achieved for individual AA sites, the regularity of an α helix implies that residues predicted in a common environment—either TMH–TMH or TMH–lipid—should lie on a common face of the TMH. The consistency of the predicted TMH–TMH or TMH–lipid environment at the level of individual AA sites was shown by Baldwin using helical wheel representations for the seven TMHs (40). As an illustration, her resulting predictions are shown for TMH6 in Fig. 12a. Note that when the Arg/Lys criterion is added to the Baldwin predictions, the resulting helix–lipid interface is continuous. This consistency is also illustrated in Fig. 13b for all seven helices in our model of the GnRHR (4), where the Arg/Lys at cytoplasmic boundaries are facing the phospholipid head groups.

From these predictions of TMH interfaces, one can estimate the extent to which each TMH is exposed or buried within the TMH bundle. A helical wheel projection of the individual AA predictions, as shown in Fig. 12b for TMH6, serves to characterize the degree of surface exposure. Because these predictions should be consistent with the final 3D model, the predicted extent of lipid exposure (or TMH exposure) of any given helix becomes a constraint for modeling the helix packing arrangement of the seven TMHs (see section below). An extensive analysis of a sequence alignment of 204 GPCRs leading to comprehensive predictions of the degree of surface exposure for each TMH has been presented by Baldwin (40); the degree of lipid exposure in decreasing order was TMH1, 4, 5, followed by TMH6, then TMH2, 7, and finally TMH3, which is predicted to have very little lipid-facing area, mostly at the extracellular end. The author used these predictions (combined with other criteria) to identify each TMH on the experimentally determined electron density projection map of rhodopsin at 9 Å resolution. It should be noted that the degree of surface exposure for a TMH may vary along the membrane height, i.e., a TMH could be more exposed on the extracellular than on the intracellular side of the membrane.

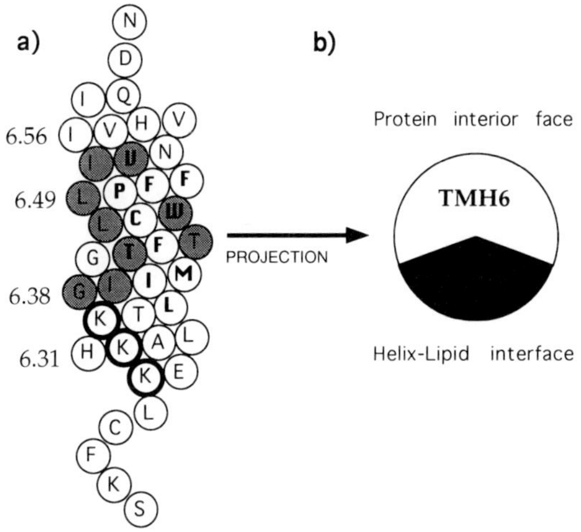

FIG. 12 Prediction of TMH–TMH versus TMH–lipid interfaces shown on a helical net representation (a) and its projection on a helical wheel representation (b) of TMH6 of the human β_2-adrenergic GPCR. (a) Highlighted are the predictions made of AA facing the lipids (shaded) or the interior of the TMH bundle (bold letters) by Baldwin (40), as well as the Arg/Lys motif toward the cytoplasmic boundaries (thicker circles) predicted by Ballesteros and Weinstein (61). Note the remarkable consistency among different prediction methods, so that a continuous TMH between the hydrophobic core and the Arg/Lys cytoplasmic extension is supported by the continuity in the helical periodicity suggested by a common TMH–lipid interface. (b) A helical wheel projection of the individual AA predictions shown in (a) serves to characterize the degree of surface exposure in Section III,B,1.

No single packing arrangement is expected to fulfill all the predictions for each AA site on a helix for which a TMH–TMH or TMH–lipid environment has been defined, because the approach is integrating a set of predictions emerging from different criteria and considerations. Therefore, as for the predictions at the secondary structure level, conclusions are based on the comparison and mutal reinforcement of different methods. Convergent or contradictory structural inferences from such an analysis can be integrated and explored in a 3D model at atomic detail (see below).

B. Modeling the Tertiary Structure of G Protein-Coupled Receptor: Packing the Transmembrane Helices in the Transmembrane Bundle

Modeling helix–helix packing interactions among the seven TMHs leads to a 3D model of the transmembrane domain of a GPCR, a major objective of any receptor modeling strategy. Until recently, this modeling step has been

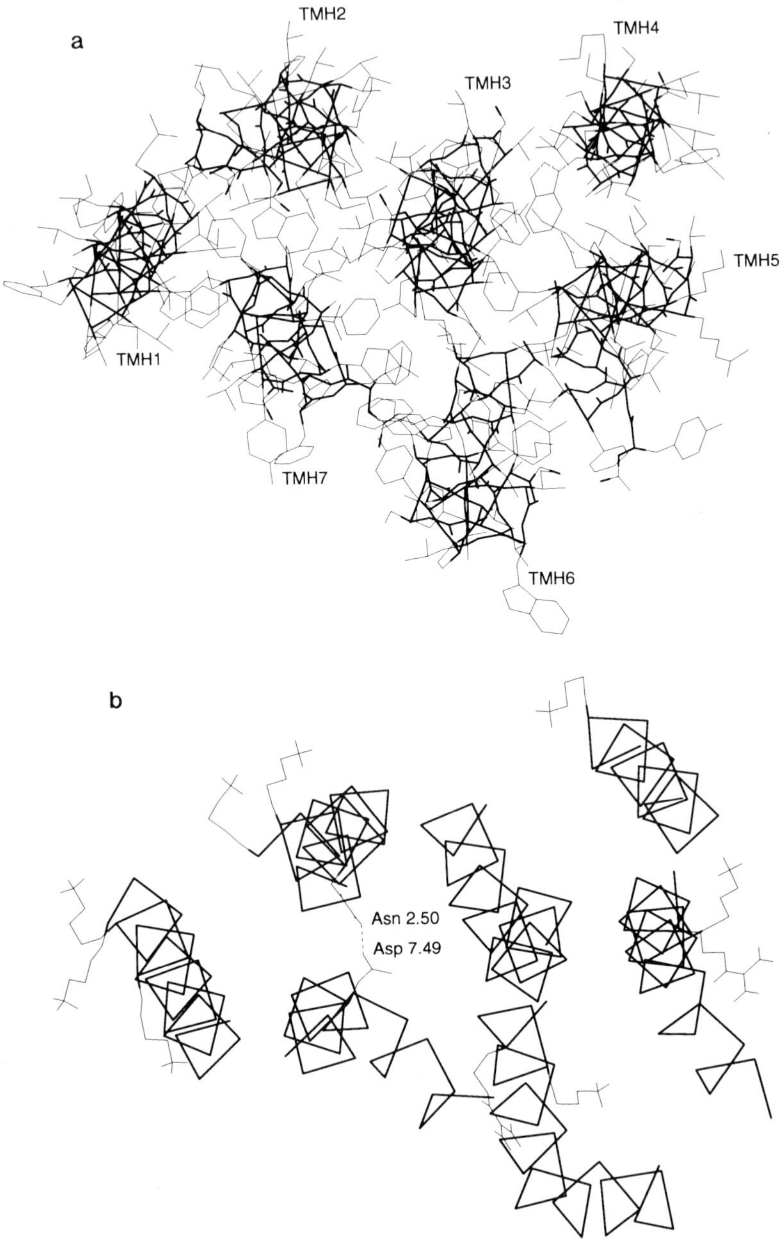
a
TMH2
TMH4
TMH3
TMH5
TMH1
TMH7
TMH6
b
Asn 2.50
Asp 7.49

the subject of controversy regarding the use of the known structure of BR as a suitable template for GPCR (40, 54, 105, 106). The use of the structure of BR as a template for GPCR modeling requires the assumption of a common 3D-fold between these two distant families of transmembrane proteins. Its attractivity is that it significantly simplifies the modeling process, since the seven TMHs of a GPCR can be superimposed onto the BR backbone following an alignment of their respective TMH sequences (20–22, 24, 26, 27, 31–34). The arguments supporting the structural similarity are the following: BR and the opsin GPCR seem to be evolutionarily related, as reflected by a similar architecture based on seven TMHs surrounding a retinal chromophore attached through a Schiff base to a Lys residue on TMH7. This functional similarity in the "ligand" and the conformational changes associated with it during activation by light do not extend to the signal transduction mechanism; BR pumps protons out of the cell while rhodopsin (RH) activates the G-protein transducin. Although there is no detectable sequence homology between the BR and GPCRs (107), detectable homologies have been found if evolutionary events such as exon shuffling (105) and/or gene duplication (108) are hypothesized. However, these similarities do not in themselves suggest a specific alternative 3D-fold for the GPCR because only few TMHs are involved. The possibility remained that the AA sequences of BR and GPCR would be similar at the level of physicochemical properties in the absence of direct AA homology, but a report indicates that the pattern of polarity-conserved positions through their sequences is clearly divergent (54). Since helix–helix interactions are significantly driven by complementarity of their polar–apolar surfaces, this result suggests that the TMH packing arrangement is likely to be significantly different between BR and the GPCR.

The projection map of the electron density of bovine RH confirmed the suspected divergence in the packing arrangement of the seven TMHs in BR

FIG. 13 (a) Our model of the gonadotropin-releasing hormone receptor. This model was constructed so as to comply with the electron density projection map of RH and was based on criteria described in this chapter. (b) The criterion that highly conserved residues interact among themselves is illustrated for a proposed interaction between $N_{2.50}$ and $D_{7.49}$ which received experimental support through the double-revertant mutant character observed experimentally (4). This criterion focuses attention to clusters of conserved AA in the 3D model that are very useful in predicting the detailed helix packing among the TMHs involved. Note the presence of Arg/Lys at the cytoplasmic side facing the lipid head groups, as discussed in Sections II,A,2 and III,A.

compared to GPCRs (1). However, the low resolution (9 Å) of the RH projection map precluded the experimental identification of the individual TMHs as well as the identification of the view as cytoplasmic or extracellular. Although the helix packing arrangements appeared to be significantly different in RH compared to BR, the low resolution prevented a more detailed comparison between the structures. A comprehensive attempt to identify the view side and the individual TMHs of RH in the projection map has been undertaken by Baldwin (40); the results share with most GPCR models and BR with the sequential and anti-clockwise order (viewed from the extracellular side) of the seven TMHs. However, the details of the helix arrangement and packing are different, as discussed by Baldwin (40).

Currently, a GPCR model is expected to comply with the projection map of RH, although due to the short time since this information has been published only two such models have been presented to date; Baldwin's (40) and ours (4) (Fig. 13). The projection map of the seven TMHs for RH thus becomes a starting point for modeling helix–helix interactions. Having modeled the seven individual TMHs as described above, methods can be used to position the individual TMHs on the projection map of RH as described in Section II,B,1, as a point of departure for modeling the transmembrane bundle. *De novo* modeling methods that can be used to infer the relative axial displacement among the seven TMHs to produce a specific 3D model at atomic detail are described in Section III,B,2. Methods used to model GPCRs based on the structure of BR as a template are described in Section III,B,3, since a similar strategy would be useful once a comparable 3D structure of RH becomes available and because of the interest these models have evinced as hypotheses generators in mutagenesis studies.

In developing a general method for helix–helix packing in the transmembrane helix bundle of GPCR, a note of caution is necessary regarding the assumption that all GPCRs share a common template. The ability to predict structural similarity is actually modulated by two factors: First, it would depend on the extent of sequence homology across members of the superfamily, 50% having been proposed as a threshold for structurally homologous proteins (109). Second, it would depend on the extent to which the presence of AA substitutions can be expected to entail significant structural modifications, such as the nonlocal structural effects of Pro in TMHs discussed above or new anchors for TMH–TMH interactions. As GPCR subfamilies differ in the degrees of homology among them, as well as in the presence and relative positions of Pro residues in TMHs 2, 4, and 5, significant differences in the TMH bundle packing may occur, and these may have been exploited by evolution for functional differentiation.

1. Inferences for Transmembrane Helix Packing from Methods Used to Identify the Individual Transmembrane Helices on the Projection Map of Rhodopsin

Baldwin (40) identified the TMHs on the RH projection map (1) based on three criteria: *First,* the minimum length of all interhelical loops over the whole GPCR family (6 to 13 AA) was considered short enough to suggest a sequential packing arrangement where each TMH is in contact with the two adjacent TMHs in the sequence. *Second,* the extent of lipid exposure predicted for each TMH (see Section III,A,2) should be consistent with the observed lipid exposure for each TMH on the RH map. Thus TMH3 would be most buried, TMH2, 6, and 7 would have intermediate exposure, and TMH1, 4, and 5 would be most exposed. Application of the surface exposure criteria assuming a sequential helix packing arrangement led to two alternative assignments of the seven TMHs in the RH map. These alternatives differ in having either clockwise or anti-clockwise connectivity when viewed from the intracellular side of the membrane, as shown in Fig. 14a for the clockwise arrangement prepared after Fig. 3 of Baldwin (40). Note that the solvent-accessible surface for each helix around the TMH bundle agrees fairly well with the predicted degree of lipid exposure, perhaps with the exception of TMH2. *Third,* criteria based on proposed ligand–GPCR interactions favored the clockwise connectivity (Fig. 14a) because the anti-clockwise could not explain a variety of proposed sites of interaction for the retinal (in RH) and adrenergic ligands based on mutagenesis experiments. The required consistency between the predicted (GPCR) and the observed (RH) degree of lipid exposure for each TMH provides yet another criterion for testing the suitability of BR as a template for GPCR (40). The predicted degree of surface exposure for each GPCR TMH is not consistent with the observed degree of lipid exposure for each TMH in BR, as shown in Fig. 14b [Figure 5 in Baldwin (40)]; note that the solvent accessibility surface around the TMH bundle indicates that according to the GPCR properties, TMH2, 3, and 7 in BR are overexposed while TMH4 is underexposed.

2. Modeling TMH–TMH Packing on the Rhodopsin Template

Throughout this chapter several methods incorporating a variety of criteria have been presented at every structural level, up to the level of TMH–TMH packing in three dimensions in an atomic resolution model of the TMH bundle in GPCR. Having predicted the relative positions of the seven TMHs in a two-dimensional projection on the membrane plane from the RH map, modeling considerations now face the problem of the axial displacement of each TMH in the TMH bundle and the detailed residue-to-residue interactions among them. An initial axial positioning of each TMH can be accomplished by aligning the respective TMH end positions so as to maximize TMH–TMH

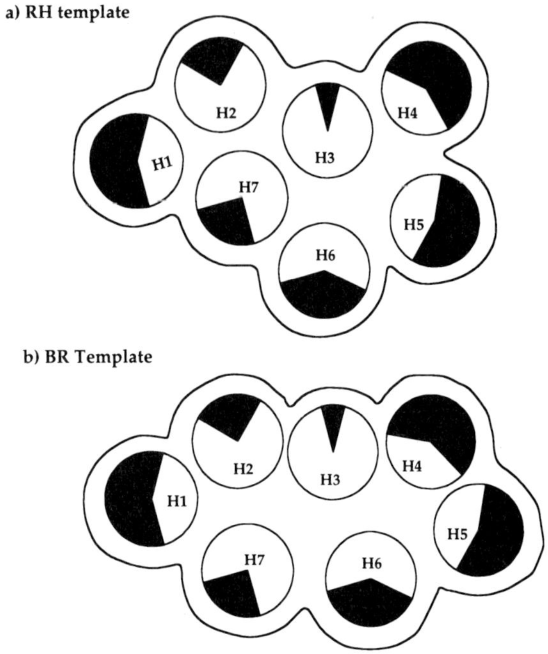

FIG. 14 Predicted degree of surface exposure for each GPCR TMH (40) superimposed on a cytoplasmic view of the projection map of RH (a) and BR (b), where it can be compared with the extent of solvent accessibility for each TMH, marked with a continuous line surrounding the TMH bundle. (a) The identification of the RH seven TMHs follows that of Baldwin (40) for a sequential and clockwise helix packing arrangement. Note that the solvent-accessible surface for each helix around the TMH bundle agrees fairly well with the predicted degree of lipid exposure (see text). (b) The predicted degree of surface exposure for each GPCR TMH is not consistent with the observed degree of lipid exposure for each TMH in BR, thus suggesting the inadequacy of using BR as a template for GPCR modeling (see text).

contacts. If the TMH ends have been predicted from the hydrophobicity profile, the resulting axial displacements of the TMHs relative to each other correspond to their hydrophobic cores, thus missing the helical portions facing the phospholipid head groups defined by the Arg/Lys motif (see Section II,A,2). The error in this initial relative axial displacement among adjacent TMH has been estimated to be three to four residues (40). However, larger errors can be expected due to the topological dependence of the cytoplasmic extensions of TMH based on the Arg/Lys motif as discussed in Section II,A,2. This was shown for BR in Table II, where the average error in the predicted TMH end at the cytoplasmic side with an $N_{in} \rightarrow C_{out}$

topology is seven AAs, while for TMHs with an $N_{out} \rightarrow C_{in}$ topology the average error is minimal. As a result, because any two TMHs adjacent in the GPCR sequence have an opposite topology, their respective cytoplasmic helix ends (predicted from the hydrophobicity profile) would be on average offset by seven residues or two turns of an α helix, thus inducing a similar error in the subsequent helix–helix packing. *Notably, if a length of 22–28 AAs is assumed for a TMH, the error in the modeled axial displacement between adjacent TMHs predicted only from the hydrophobicity profile could amount to 25–30%.* This error induced by the Arg/Lys motif would be predicted to affect current GPCR models whose TMH boundaries were derived from hydrophobicity criteria (8, 21–28, 31–34, 38, 110). In a most illustrative example of such a possibility, this error was shown to vitiate models of the BR TMH bundle in which TMH boundaries were predicted by hydrophobicity criteria before the structural elucidation of BR (111). In an interesting molecular dynamics approach to packing the seven TMHs of BR based on the electron density projection map of the helices on the membrane—an approach that is very similar to the current situation with RH—the authors (111) found that the best model could be improved both in energy and in the prediction of residues involved in retinal binding by shifting TMH6 *two turns upward.* Figure 7 illustrates that these two turns in TMH6 actually correspond to the helical portion of TMH6 missed due to the Arg/Lys motif. Furthermore, the TMH6–TMH7 packing in the structure of BR shown in Fig. 7 illustrates that the actual helix–helix interface is offset from the cytoplasmic ends of the hydrophobic segment by seven AAs—see predicted vs observed TMH boundaries for these. When the same authors repeated these calculations with the same predicted TMH boundaries after the structure of BR was elucidated, they reached similar conclusions (18). Interestingly, in this case they could modify the cytoplasmic end of TMH6 to include the previously missed Arg/Lys portion, thus improving the accuracy of their models (18). Although we have used the Arg/Lys redefinition of the TMH boundaries as an illustrative case, the general conclusion is that inaccuracies in the prediction of TMH boundaries could seriously affect the modeling of TMH packing arrangements.

The problems described above for the modeling of TMH–TMH packing can be circumvented with methods to predict interhelical residue-to-residue interactions that could guide the TMH packing independent of the selected TMH boundaries. This is a most difficult step in the modeling process, as illustrated by the fact that few approaches have been offered to determine specific residue-to-residue contacts among the seven TMHs. Even in cases where specific residues are predicted to interact, or to bind a ligand, the resulting constraints imposed by such information on the possible helix packings at atomic level are rather weak. The ambiguity arises from the intrinsic conformational degrees of freedom of the residue side chain and/or the ligand, as well as the different possible orientations of the TMHs in a 3D space that can satisfy the proposed interactions. As a consequence, a similar set of criteria considered by several authors has yielded significantly different 3D models of highly homologous GPCR. For example, significant differences are present in the details of models of GPCR that have used a common 3D template, such as BR, to infer the interhelical contacts (see Section III,B,3 below) even within the same laboratory (22, 27). The lack of appropriate theoretical methodologies is echoed in the lack of appropriate experimental data to derive information on interhelical contacts. Notably, attempts to find double-revertant mutants in GPCR have shed some light on what promises to become the most valuable source of experimental testing of GPCR models. The assumption of spatial proximity between residues shown to display a double-revertant mutant character is based on the hypothesis of direct interaction among them (3). Thus, we present below first those methods that are useful to predict residue-to-residue interhelical contacts and then a description of the two reported double-revertant mutants for the GPCR family. The methodology used to predict specific interhelical contacts for GPCRs is based on the proposed ligand interaction sites, as well as polarity or conservation criteria.

Proposed ligand–GPCR contacts have been widely used in modeling TMH–TMH packing arrangements. The most solid case is the use of covalently bound ligands where the attachment site has been identified experimentally. The retinal attachment site ($K_{7.43}$) in rhodopsin is an example in which the combination with the proposed counterion to the Schiff base ($E_{3.28}$) leads to constraints in the relative position of TMH7 and TMH3 [for a review, see Oprian (112)]. Similarly, Maloney-Huss and Lybrand (25) used structural inferences from the binding of a pindolol derivative that labels $Ser_{2.63}$ and/or $His_{2.64}$ of the β_2-adrenergic receptor to infer the relative positioning of TMH2 and TMH3; this ligand could be accommodated in their model structure with the alkylation group pointing toward residues 2.63–2.64 and the protonated amine making an ionic bond with the $Asp_{3.32}$ in TMH3. The proposed ionic interaction between the protonated amine moiety of neurotransmitters and the corresponding conserved $Asp_{3.32}$ in the cationic neurotransmitter GPCRs is the most widely accepted ligand–receptor interaction site inferred from mutagenesis studies.

Another set of widely accepted ligand interaction sites from mutagenesis experiments involve Ser and Thr residues occurring at positions 5.42–5.43–5.46 in a variety of neurotransmitter GPCRs (43, 45, 113–117). These three positions, which define a common patch on the TMH surface (see Fig. 2), are differentially used in homologous receptors for ligand binding and/or activation. The strongest support for a direct interaction of some ligands at this site came from systematic double-modification of the hydroxyl groups in the adrenergic catecholamine ligands and the β_2-adrenergic receptor (117).

The other AA site, extensively studied by mutagenesis experiments, that has been proposed to be in direct contact with the ligand for a variety of receptors (35) is position 7.39. Interestingly, the RH projection map with the TMHs identified by Baldwin (40) brings close in space the key residues $D_{3.32}$, $S/T_{5.42\text{-}5.43\text{-}5.46}$ and $AA_{7.39}$, consistent with a large body of pharmacological and structure–activity data pertinent to the nature of the ligand recognition sites on GPCRs.

Molecular models of the recognition site obtained earlier from structure–activity considerations and computational probing have also served as a source for predicted ligand–receptor interaction sites. For example, following a precise definition of the residues forming the binding site of the 5TH2 GPCR, Zhang and Weinstein (8) used this approach as a primary criterion to model the TMH bundle.

The criterion of residue polarity predicts a preferential interaction of residues with similar polarity, i.e., polar–polar and apolar–apolar contacts. Yet this criterion in itself is rather general in the context of putative interactions among the many residues of each kind in the TMHs. Additional constraints are needed to predict specific residue–residue contacts. The constraints established so far for the packing arrangement of the seven TMHs significantly restrict the possibilities of residues interacting based on their polar character. A clear example is provided by GPCR modeling based on the BR template (see below) where the superposition of the GPCR TMHs onto the BR backbone severely restricts the possible combinations of polar–polar interactions within the TMH bundle. The same criterion can become a test of the model: Maloney-Huss and Lybrand (25) found a glutamic acid in position 3.39 that had no polar counterpart in their model and was significantly exposed to the lipids. An explanation is provided by the projection map of RH which indicates that TMH3 at the level of $E_{3.39}$ is significantly more buried among the other TMHs than initially predicted by these authors, calling for a refinement of the packing in this region of the model. Zhang and Weinstein (54) proposed to combine the polarity criterion with the conservation analysis by identifying AA sites that conserved their polarity across more than 90% of the aligned GPCR and called these loci "polarity-conserved positions" (PCPs). When applied to the BR structure, these PCPs were shown to preferentially interact among themselves compared to other polar residues within the sequence (54).

The last set of criteria for predicting residue-to-residue interactions to be discussed here as guides of the packing of the seven TMHs is based on the degree of evolutionary conservation. The final 3D model is expected to explain the observed degrees of conservation for each AA site, which becomes a constraint for modeling the detailed packing of the seven TMHs. Because highly conserved residues are candidate molecular determinants of the structural and/or functional integrity of the GPCR, the final 3D model should encompass specific interactions involving these residues that are responsible for their apparent structural and/or functional requirement through evolution. These molecular determinants exert their effects through physicochemical properties such as charge, aromaticity, volume, and H-bonding ability, so that the characteristic properties of each conserved residue should be engaged in specific interactions within the model. For example, a residue that conserves its volume through evolution should be significantly buried, surrounded by a closely packed environment. Similarly, the acidic character of $D_{2.50}$ should have a counterpart within the TMH bundle that stabilizes a charge in a protein interior environment. Because the counterpart of a conserved residue should be equally preserved through evolution, highly conserved residues are predicted to interact among themselves. This criterion focuses attention to clusters of conserved AA in the 3D model that are very useful in predicting the detailed helix packing among the TMHs involved. An example of such clusters involves TMH1, 2, and 7 at the level of the conserved residues [$N_{1.50}$, $D_{2.50}$, $N_{7.49}$, $P_{7.50}$, $Y_{7.53}$] as proposed by several authors (4, 8, 25, 27, 30). an illustration of such a predictive criterion is given in Fig. 13 for a model of the GnRHR where a proposed interaction between $N_{2.50}$ and $D_{7.49}$ received experimental support through the double-revertant mutant character observed experimentally (4), as outlined briefly below in the context of coordinated conservation patterns.

A variation of the conservation analysis seems especially well suited for the prediction of tertiary structure interactions. Thus, Benner and Gerloff (11) and Altschuh *et al.* (118) have proposed to identify for this purpose the AA sites where mutations during evolution show a coordinated or parallel pattern in which mutations at one site are accompanied by mutations at a distant position. For example, the conserved AA sites 2.50 and 7.49 appear to be an example of such evolutionary correlation, because when $D_{2.50}$ changes to $N_{2.50}$ in the GnRHR, then residue $N_{7.49}$ changes to $D_{7.49}$. Residues that display a coordinated AA substitution pattern are thus predicted to interact among themselves, through functional coupling and/or through direct contact (3). The initial translation of the steric constraints offered by these coordinated AA sites into 3D structures is based on the simple assumption of direct interactions, because this hypothesis was found to be almost always correct for five different protein families with known structures (118–120). Benner and Gerloff have applied the same criteria for structural predictions, achieving remarkable success in the secondary structure prediction of the catalytic domain of protein kinases (11). From an evolutionary standpoint, step-by-step AA substitutions imply successive adaptations of the structure, the driving force being to optimize function. Assuming a common framework for the GPCR within the alignment means that the structural changes during this adaptation process are local in character, without gross alterations of TMH–TMH packing arrangements. This hypothesis allows the evaluation of which structural frameworks, i.e., which TMH packing arrangements, are consistent with the set of identified correlated AA sites. It should be noted that the initial criterion that conserved AA interact among themselves can now be seen as a particular case of coordinated AA substitution patterns. Although the structural hypothesis of spatial proximity between coordinated AA sites has been applied to infer tertiary structure contacts, the correlated mutation pattern may arise also from functional coupling among noninteracting residues. The identification of such related sites may provide a molecular basis for understanding how functional diversity is achieved across a protein family (11) such as the GPCRs.

Experimentally, tertiary structure interactions can be probed by the construction of double-revertant mutants. If the phenotype of a single mutant can be reverted back to wild type by a second mutation at another site, it is likely that these two residues interact with one another, the simplest manner being a direct contact. Kobilka's laboratory (5) proposed an interaction between position 7.39 (TMH7) and TMH1–TMH2 based on chimeric constructs between adrenergic receptors; the deleterious phenotype of a N7.39F mutation in the β_2-adrenergic receptor was suppressed by chimeras containing the first two TMHs of the α_2 subtype. Applying the same logic, Pittel and Wess (6) proposed the adjacency of TMHs 1 and 7 in the muscarinic GPCR. More recently, double-revertant mutagenesis studies on RH by Rao *et al.* (7) suggested that the retinal attachment site, $K_{7.43(296)}$, is in proximity to $G_{2.57(90)}$ in addition to its wild-type counterion $E_{3.28(113)}$, thus positioning TMH7 adjacent to TMHs 2 and 3. Notably, double-revertant mutants can be predicted as a special case of coordinated AA substitutions, when two AA sites predicted to affect function significantly have a coordinated AA substitution pattern that is complementary in nature and is restricted to two residues within the sequence (see above). The simplest case would be a reciprocal exchange between two conserved AA sites, as the above-mentioned $D_{2.50} \rightarrow N_{2.50}$ and $N_{7.49} \rightarrow D_{7.49}$ occurring in the GnRHR. Zhou *et al.* (4) showed the double-revertant mutant character between these two loci in the GnRHR, thus providing a bridge between theoretical prediction methods and the equivalent experimental counterpart, used to validate the models. Note that distinct double-revertant mutant experiments have involved different residues within the same TMHs (i.e., 7.39, 7.43–2.57–3.28, 7.49–2.50), so that when combined these experiments provide significant structural constraints for modeling the relative placement and mutual interactions of the TMHs involved.

A seemingly powerful and stringent test of a 3D model of the transmembrane domain of a GPCR is to require a packing density for interior residues similar to that commonly observed for known protein structures (25, 110), including membrane proteins as shown for the PRC (52). These techniques, as implemented in the program QPACK (121), have been applied in the Lybrand laboratory to provide one of the most stringent and helpful tools for the modeling process (25, 110). While it is always possible to pack seven TMHs into a model of the transmembrane domain of a GPCR, it is quite difficult to attain a closely packed interior of the seven-helix bundle that resembles the packing density found for known structures (122). The application of this method to probe and improve current GPCR models is likely to increase significantly in the near future, since it provides not only a valuable modeling tool, but also a quantitative measurement of the "correctness" of the modeled structures that constitutes an objective measure for comparisons among different models.

3. Assembly of the Transmembrane Helix Bundle According to a BR Template

As discussed at the beginning of Section III,B and illustrated throughout, the use of the helix bundle packing observed for BR (62) as a direct template for GPCR modeling is likely to be inappropriate. Nevertheless, the procedural steps and the inferences that can be drawn from the application of a known structural template to the specific modeling of GPCR merit scrutiny because structural information at the same level as currently available for BR is likely to emerge shortly for RH—a member of the GPCR family.

Modeling helix–helix interactions based on the BR template involves three sequential steps, as illustrated in many of the reported modeling studies that have utilized this approach (20–22, 24, 26, 27, 31–35): (i) alignment of TMH sequences for proteins in the BR and GPCR families, (ii) superposition of the modeled GPCR TMH backbone onto the structurally known BR backbone, and (iii) refinement of the side-chain dihedral angles in order to relieve steric clashes and to improve interresidue interactions. Although the process seems straightforward, several caveats and considerations apply. First, a meaningful alignment of the GPCR sequences with BR is hindered by the lack of significant sequence homology among them. Solutions to this problem have involved a "structural" alignment whereby residues predicted to face the lipids (or the protein interior) in the GPCR are aligned with residues known to face the lipids (or protein interior) in the BR structure. However, because the degree of predicted lipid exposure for the GPCR sequences does not match the observed exposure in the BR structure (see discussion above and Fig. 14b), this "structural" alignment is not unique so that different alignments have been proposed in the literature. The published TMH alignments are offset from each other by one to four residues [compare (22) vs (123) vs (26)] depending on the particular helix. The structural implications of these differences in the alignments are very significant and can be estimated from the parameters of an ideal α helix: 1.5 Å pitch and 105° per residue. When the same GPCR TMH is superimposed on its BR counterpart based on two different alignments, the resulting positioning of the same TMH on the respective GPCR models can deviate up to 6 Å axially, and their faces could be displaced by a rotation of up to 150°. As a result, GPCR models derived from different alignments would position residues critical for ligand binding quite differently. However, because these problems are due to an ambiguous alignment of the modeled and template sequences, they should not be expected when the RH template becomes available for use in the modeling of the highly homologous GPCRs.

It is important to note that the superposition of a GPCR TMH backbone onto the BR backbone is even more ambiguous when the position of Pro residues in these two proteins is not conserved, as occurs for all TMHs except TMH1. The kink introduced in the TMH by a Pro residue (see Section II,B) renders the two helical portions of the TMH that flank the Pro discontinuous both in the directions of their helical axis and in the orientations of their "faces" (see Fig. 10). Whether kinked TMHs are superimposed onto straight TMHs or vice versa, the superposition could be done for either helical portion, or for a combination of both. Nordvall and Hacksell (26) found that superimposing the GPCR helices on only one of the two helical portions of kinked helices led to unacceptable steric clashes or large openings between TMHs, thus suggesting an averaged full backbone superposition. Notably, the nonconserved character of Pro-kinks applies also among members of the GPCR family. For example, the Pro in TMH2 occurs at different positions within the alignment and is even absent in some receptors. Since the cytoplasmic half of the TMHs in GPCRs is much more conserved than the extracellular half (40), and sequence conservation suggests structural preservation, these helices should probably be modeled superimposing their helical portions closer to the cytoplasmic side. The consequence is thus that the extracellular portion of TMH2 and others (e.g., TMH5) is structurally more divergent, and significant alterations of the helix–helix packing interactions are likely to occur at these loci. This hypothesis is also consistent with functional considerations, the extracellular half being more likely to have to accommodate chemically divergent ligands while the intracellular half is responsible for the activation mechanism that could be shared by all members of the GPCR family.

C. Modeling Extramembrane Segments of the G Protein-Coupled Receptors

Modeling the interhelical linkers, termed loops, has become increasingly important due to information obtained from mutagenesis experiments that have identified some of these protein segments as binding sites in peptide receptors (124) or G protein-coupling domains (125). Modeling the N- or C-terminal domains of the GPCR is a very different task from that concerned with interhelical loops and is not discussed here. For loop modeling, a number of loop searching algorithms have been used, some of them implemented in standard modeling software packages [e.g., as used by Trumpp-Kallmeyer *et al.* (28) to model the neurokinin-1 GPCR]. These computational algorithms search structural data bases for protein fragments that have AA sequences homologous to portions of the loops. The yield of such searches is usually low, and the validity of structural predictions for such short, inherently flexible fragments remains unclear. Maloney-Huss and Lybrand (25) generated the structure of interhelical loops using a *de novo* procedure based on molecular dynamics (MD) simulations carried out at a low temperature and applying a series of constraints. Each loop segment was first assigned a completely extended structure (all $\phi,\psi = 180°$) and then attached to the end of one target helix via a transpeptide linkage, before a short constant-temperature (10–20K) MD run was initiated. Weak harmonic constraints (5–10 kcal/mol) were used to pull the free end of the loop into position to form a transpeptide bond with the end of the second target helix. The final conformation defined for loops from such a procedure is probably one among many possible structures, due to the predicted flexibility and high degree of dynamical motion observed in such segments of soluble proteins. As a consequence, no reliable methods are currently available to suggest the specific modes of interaction between these loops and the TMH bundle. An exception is a disulfide bridge between the top of TMH3 and the TMH4–TMH5 linker shown in adrenergic (126) and muscarinic (127) receptors, as well as in opsins (21), and by analogy assumed to exist in most other GPCRs.

Of the six interhelical loop segments, the one that has attracted the most attention is intracellular loop III, whose N- and C-terminal ends have been implicated in the interaction of GPCR with the G proteins (128). These portions of loop III, adjacent to the cytoplasmic ends of TMH5 and TMH6, were predicted to be in an α-helical conformation by Maloney-Huss and Lybrand (25). This prediction was substantiated by extensive mutagenesis studies (125), with the authors concluding that the N- and C-terminal ends of the loop were a continuous helix extending from TMH5 and TMH6 into the cytoplasm where they could interact with the G protein. Notably, these cytoplasmic extensions of TMH5 and 6 agree with the predictions of TMH boundaries made by the Arg/Lys motif discussed in Section II,A,2.

IV. Probing of G Protein-Coupled Receptor Models Based on Molecular Energy Criteria

Minimum energy-based criteria can be used throughout the modeling process at every structural level, but it is only when a 3D model of the GPCR is nearly complete that a full energetic refinement of the model can be undertaken. Computational methods ranging from quantum mechanical calculations (129–131) to energy minimizations by molecular mechanics (MM) and the search for structures with the lowest free energy with MD simulations are available for this purpose (13–19).

The energy-based probing of the GPCR model requires an exploration of the conformational space around the 3D model to search for structures with increasingly lower energies. This procedure is continued until convergence is reached, i.e., when the lowest energy structure has been identified within the limitations of the computational algorithms used. The two major computational algorithms that have been used for this task, MM minimizations and MD simulations, are applicable in order of increasing complexity. While MM minimization performs an optimization of the structure based on gradients of the potential energy surface, MD simulates the dynamic behavior of the protein where the incorporation of kinetic energy allows the algorithm to cross higher barriers of the potential energy surface exploring a wider portion of the conformational space available to the 3D model toward an "average structure" that corresponds at least to a local free energy minimum. As the computational requirements for MD simulations make this procedure much more costly than MM calculations, most published reports of models of GPCR have used only the former. However, for the purpose of energetic probing of the 3D model, MD simulations are much more stringent than MM: while MM will always reach a local minimum providing a minimized structure, in MD simulations the 3D model may fall apart if it lacks energetic consistency whereby the attractive forces do not prevail over dynamic fluctuations at room temperature. It is noteworthy, however, that Maloney-Huss and Lybrand found that excessive use of energy minimization leads to structures that are too densely packed (25), thus hindering the model-building process due to the difficulty in loosening the model structure for further modeling explorations. The same effect of excessive compactness was observed for energy minimizations *in vacuo* of the entire model (25,

39), although in this case the problem could be overcome by a treatment of the protein environment (110), as described below.

Methodological improvements and a probing of the effects of the environment, both lipid and aqueous, are required for further validation and quantitative evaluation of the GPCR models. Progress in that area is hindered by the lack of structural information at atomic detail about the protein–lipid interface and by the necessary simultaneous treatment of a lipid and aqueous environments for a complete model of the receptor structure. An exploration of the effects of the environment on a GPCR model has been presented by Kontoyianni and Lybrand (110). Molecular dynamics simulations of a complete model of the β_2-adrenergic receptor were carried out with three different considerations of the environments: (i) simulation of the GPCR model *in vacuo,* (ii) a simulation *in vacuo* with an explicit model of the membrane consisting of 60 DPPC lipid molecules surrounding the receptor, and (iii) a continuum solvent model for the aqueous phase combined with the explicit treatment of the DPPC bilayer as in point 2. Although the last and more sophisticated representation of the environment produced the most physically reasonable behavior for the receptor models, the authors report that no major qualitative differences were observed among these three simulations except in the loop regions and TMH bundle surface residues. Residues of the TMH portions that would be exposed to lipids were found to be packed against the TMH backbone when explicit lipids were not included. On the other hand, the loop regions collapsed onto the lipid head group regions when the explicit bilayer model was included without a model for the aqueous phase, producing an unlikely physical situation. As a result, the authors have suggested the application of MD methods with an *in vacuo* explicit bilayer model where only the TMH bundle is included, in order to optimize TMH–TMH packing contracts. The resulting transmembrane models could then be completed by a subsequent addition of the loop regions. Although other authors have used different approaches, such as restricting the model to the transmembrane portion where a low dielectric environment can be assumed (8, 27) or simulating the entire receptor model with a homogenous dielectric environment (20), in the future MD simulation of the form described by Kontoyianni and Lybrand (110) is likely to provide GPCR models that are more reliable in their description of the extramembranal and juxtamembranal regions.

An alternative treatment of the membrane environment was proposed by Jahnig and Edholm (18, 111) in a MD simulation study of alternative TMH packing arrangements of BR. To simulate the membrane environment, the authors added to the potential energy function a hydrophobic potential term that applies only to surface atoms. These MD simulations for BR bear special significance for the study of GPCR models because their aims and procedures parellel those applicable to GPCR models (111), but the availability of the BR structure makes possible a test of the adequacy of the structures resulting from the computations (18). Using the electron density projection map of BR (62) as a starting point for modeling helix–helix packings, the ability of energy-based criteria to distinguish the correct helix packing arrangement among alternative models was tested by MD simulations. Although the sequential and anti-clockwise packing arrangement of the BR structure was found to be lower in energy than that of any other structural alternative considered by the authors, the energy differences were not significant enough to substantiate the ability of MD simulation to guide the packing of the seven TMHs of BR (18). A particularly significant conclusion of these studies is that differences in the relative *axial* displacements among adjacent helices discussed in Section III,B,2 were not overcome by the MD simulations (111), thus stressing the importance of the approaches described in Section III of this chapter to model TMH–TMH packing at atomic detail. However, a caveat of these studies is that the authors did not use the experimentally known TMH boundaries, but rather those predicted by hydrophobicity criteria [see Table 4 of Jahnig and Edholm (18)], which can be in error by up to seven or eight AA as described in Section II,A,2. Interestingly, these criteria for the selection of TMH ends parallel those used for most published GPCR models (8, 21–28, 31–34, 38, 110), thus providing an indication of the consequences of the expected inaccuracies in the prediction of TMH boundaries. As described in the discussion of the results obtained by Jahnig and Edholm in Section III,B,2, errors in the predicted TMH boundaries seriously vitiated the BR models constructed, so that the correct ligand–receptor interactions were obtained only after shifting TMH6 two turns toward the extracellular side (111). Note that this correction is precisely what would be expected to be required according to the criterion that calls for including Arg/Lys patches at the cytoplasmic ends of the GPCR helices, as illustrated in Fig. 7. In summary, it appears that at the present stage of modeling, MD simulations are best suited to refine carefully built 3D models, but are not likely to provide a good approach to helix packing from a crude starting structure.

V. Computational Approaches to Probe Structural Details and Functional Mechanisms of G Protein-Coupled Receptors

Pharmacological profiles, in the form of ligand-binding affinities, or data on the efficacy of a ligand in eliciting a response (i.e., degree of agonism or antagonism), are available for probing 3D models of GPCRs with respect to functional data. The approaches required to relate the models of ligand–receptor complexes to the binding properties are far less elaborate and turn out to be much more ambiguous than the equivalent methodological approaches required to relate the receptor models to the experimental data on receptor response. Probing the constructed models of GPCRs against rank orders of binding affinity for various ligands is discussed in Section V,A. The apparent lack of discriminatory power of these criteria in probing the results of modeling approaches makes necessary the use of the receptor models in the prediction of relative ranks of pharmacological efficacies, as described below in Section V,B, in order to achieve more stringent tests of the structural and functional details.

A. Ligand Binding Properties as Criteria for Probing G Protein-Coupled Receptor

The most widely used approaches for probing GPCR models by criteria related to their pharmacological properties relate to the ability of the models to rationalize ligand-binding affinities. Models of ligand–receptor complexes have been proposed for each of the modeled receptors, and validation of the models was sought on the basis of a proposed agreement between the ligand-binding scheme in the receptor model and the experimentally determined rank order of affinities and stereoselectivities of various ligands (8, 21, 24–27, 30–33, 35, 132, 133). To translate binding affinities into ligand–receptor interactions at the atomic level, structure activity relationships (SAR) of the ligands were used to infer the nature of the chemical groups of the receptor that make direct contact with the ligands. For example, an H-bond acceptor or an aromatic ring may be proposed to interact with specific moieties of the ligands under consideration, based on the identified environment of the ligand-binding pocket in the receptor model. However, such structural requirements in the ligand-binding pocket can be fulfilled by different combinations of side chains from residues in different TMHs of the receptor model, thus producing a high degree of ambiguity in the fulfillment of structural criteria. This lack of stringency in the criteria applied to receptor probing based on the mere fit of ligands in a predetermined binding pocket may be responsible for the apparent agreement of dissimilar GPCR models with the same affinity data and SAR considerations [cf. claims made for structurally very different models for the β_2-adrenergic receptor (21, 25, 27, 32) or for the different models for the 5TH2 receptor (8, 27, 133, 134)]. Consequently, the application of more stringent criteria related to the functional response of the receptor to the binding of agonist appears necessary for the validation of GPCR models. These criteria require the application of a more complex set of computational approaches in which the effect of agonist binding on the time-dependent structural rearrangements of the receptor model is compared to experimental data about the pharmacological efficacies of the ligands and the structural elements involved in receptor–effector coupling, as described below.

B. Criteria Based on Computational Simulation of Receptor Activation Mechanisms

The most challenging task in computational modeling in support of the efforts to understand structure–function relations of GPCR and their complexes with ligands remains the incorporation of the various available experimental results connecting structural information to functional data for the receptor. New and experimentally testable hypotheses must be derived from such approaches, based on the behavior of models for the receptors and their ligands and on the computational exploration of their interactions. For this exploration, methods of static energy minimizations and MD simulations have been presented in the literature (13–15). A detailed discussion of the computational methods providing functional inferences from the simulation of molecular mechanisms such as ligand–receptor interactions is beyond the scope of this chapter. However, the significant methodological and practical differences between molecular modeling of receptor structure and the computational simulation of receptor dynamics required to probe receptor function merit some emphasis. As discussed by Weinstein (19) and summarized in Fig. 15, the computational simulation of the time-dependent behavior of the molecular model of a GPCR in the presence and absence of bound ligand can provide essential clues about the relation between structural details and the properties of the receptor. Consequently, such simulations provide a much more reliable basis for understanding the effects of mutations and chimeric constructs than mere inferences from static models. This is because subtle structural rearrangements could redistribute local interactions at mutated sites so as to compensate for modifications intended to be introduced by the mutations. It follows that modified proteins may maintain structural and functional integrity through rearrangements that can be predicted from computational simulations, but not from the analysis of the static models. Similarly, computational simulations of the signal transduction mechanism triggered in a model of the 5HT2 subtype of the serotonin GPCR by the binding of a ligand have revealed a characteristic series of structural rearrangements produced by the binding of agonists that differ from the structural effects of antagonist binding (8, 133). Consequently, it is to be expected that mutations affecting the structural rearrangements, not the actual interactions between ligands and the receptor, would affect the measured affinities of agonists and antagonists. Such mutations would thus be difficult to discern

Molecular Modeling Approaches

Offer: 3-D constructs of *structural hypotheses.*

Best application: To identify and refine *specific questions* to be answered in new experiments designed according to the structural hypotheses.

Computational Simulation Approaches

Offer: A formal framework for computational explorations of the structure, dynamics and functional properties of specific molecular systems.

Features include:

- rigorous definitions of the explored systems (e.g., boundary conditions; nature of the environment);
- systematic explorations of specific hypotheses;
- control experiments.

Best applications: To predict the behavior of the systems in *new experiments* designed to test specific hypotheses resulting from the analysis of their dynamic behavior.

FIG. 15 Differences in the definitions and most appropriate uses of computational approaches to molecular modeling and computational simulation.

from mutations affecting the actual binding sites if only affinities are compared, making it difficult to interpret the results of mutagenesis probing of receptor structure (2). The combined use of ligand affinity and functional probing of the effects of structural modifications of GPCR can overcome such difficulties if the results are interpreted with the help of insight obtained from simulations of the consequences of these modification for both the structural and the dynamic properties of the GPCR.

The computational simulation studies that can provide dynamic information about the receptor proteins are complex and fraught with methodological pitfalls and shortcomings related to the special properties of the membrane-bound species (8, 18, 39). Nevertheless, the validation of some of the inferences obtained from such studies by comparison to the known properties of the simulated system can help substantiate their use in probing the properties of the system. For example, the simulation of the complexes between a model of the transmembrane portion of the 5HT2 receptor with ligands chosen to represent full agonists, partial agonists, and an antagonist (8, 133) produced results that were in full agreement with the pharmacological properties of the ligands and with experimental data concerning the structural rearrangements expected from receptor activation (8). The detailed mechanistic insights provided by these simulations at the atomic level of structure identified the steps in the propagation of the signal produced by ligand binding to the region of the receptor molecule that governs the interaction with G proteins. Key residues responsible for these interactions were identified (135) in a manner not feasible with structural inferences alone. It is important to note that experimental probing (e.g., by mutagenesis) of these specific structural and functional inferences should provide valuable information not only about the validity of the specific 3D model of the 5HT2 GPCR, but also about the detailed nature of a process of intramolecular signal transduction that may be common to many, if not all, GPCR families.

VI. Concluding Remarks

By focusing this description of a methodological approach to GPCR modeling and probing on the hierarchical use and intrinsic caveats of each available method, we have attempted to illustrate the extent to which each method is inconclusive taken individually, but can gain great predictive value if applied appropriately in conjunction with complementary approaches. Therefore, major inferences and conclusions guiding the construction and functional probing of 3D models of GPCR should be reached by comparing predictions from different methods with valid experimental data, while pursuing convergence among the results of the various prediction methods. This procedure requires an understanding of the basis and limitations of each available method, as sought from the juxtaposition and discussion of these approaches in the present overview. It is clear, however, that a deeper and more rewarding understanding of the methods, approaches, and inferences presented here will be achieved when a correct three-dimensional structure of a GPCR (e.g., a refined structure of RH) is available at atomic detail for use in the exploration of convergent or contradictory structural inferences. This exploration should provide the required preparation for a complete structural and functional modeling of individual proteins from the great variety of GPCRs that are likely to exhibit tantalizing differences related to their specific functions and localizations.

Acknowledgments

Critical reading of the manuscript and many helpful suggestions by Daniel Strahs and Juan Luis Pascual-Ahuir are gratefully acknowledged. The work was supported in part by NIH Grants DA-06620 and DA-00060 (to HW) and DK-46943 (to Stuart C. Sealfon), and by a Fulbright/MEC (Spain) fellowship (to JAB). Computations for the illustrative examples were performed on the supercomputer systems at the Pittsburgh Supercomputer Center (sponsored by the National Science Foundation), Cornell National Supercomputer Facility (sponsored by the National Science Foundation and IBM), and the Frederick Biomedical Supercomputing Center of the NIH (NCI-Laboratory for Mathematical Biology).

References

1. G. F. Schertler, C. Villa, and R. Henderson, *Nature* (*London*) **362,** 770 (1993).
2. D. Colquhoun and M. Farrant, *Nature* (*London*) **366,** 510 (1993).
3. W. H. Ward, D. Timms, and A. R. Fersht, *Trents Pharmacol. Sci.* **11,** 280 (1990).
4. W. Zhou, C. Flanagan, J. A. Ballesteros, K. Konvicka, J. S. Davidson, H. Weinstein, R. P. Millar, and S. C. Sealfon, *Mol. Pharmacol.* **45,** 165 (1994).
5. S. Suryanarayana, M. von Zastrow, and B. K. Kobilka, *J. Biol. Chem.* **267,** 21991 (1992).
6. Z. Pittel and J. Wess, *Mol. Pharmacol.* **45,** 61 (1994).
7. V. R. Rao, G. B. Cohen, and D. D. Oprian, *Nature* (*London*) **367,** 639 (1994).
8. D. Zhang and H. Weinstein, *J. Med. Chem.* **36,** 934 (1993).
9. J. M. Thornton, T. P. Flores, D. T. Jones, and M. B. Swindells, *Nature* (*London*) **354,** 105 (1991).
10. A. M. Lesk and D. R. Boswell, *Curr. Opin. Struct. Biol.* **2,** 242 (1992).
11. S. A. Benner and D. Gerloff, *Adv. Enzyme Regul.* **31,** 121 (1991).
12. B. Rost and C. Sander, *J. Mol. Biol.* **232,** 584 (1993).
13. C. L. Brooks, M. Karplus, and B. M. Pettitt, "Proteins: A Theoretical Perspective of Dynamics, Structure, and their Thermodynamics." Wiley, New York, 1988.
14. J. A. McCammon, *Science* **238,** 486 (1987).
15. W. F. van Gunsteren and P. K. Weiner, "Computer Simulations of Biomolecular Systems." ESCOM, Leiden, The Netherlands, 1989.
16. W. F. van Gunsteren, R. M. Brunne, A. E. Mark, and S. P. van Helden, *in* "Molecular Aspects of Biotechnology: Computational Models and Theories" (J. Bertran, ed.), p. 105. Kluwer Academic Publishers, Dordrecht, The Netherlands, 1992.
17. H. Weinstein and E. L. Mehler, *in* "Molecular Aspects of Biotechnology: Computational Models and Theories" (J. Bertran, ed.), p. 153. Kluwer Academic Publishers, Dordrecht, The Netherlands, 1992.
18. F. Jahnig and O. Edholm, *J. Mol. Biol.* **226,** 837 (1992).

19. H. Weinsterin, *in* "Trends in Ecological Physical Chemistry" (L. Bonati, U. Cosentino, M. Lasagni, G. Moro, D. Pitea, and A. Schiraldi, eds.), p. 1. Elsevier, Milan, Italy, 1992.
20. S. G. Dahl, O. Edvardsen, and I. Sylte, *Proc. Natl. Acad. Sci. U.S.A.* **88,** 8111 (1991).
21. J. Findlay and E. Eliopoulos, *Trends Pharmacol. Sci.* **11,** 492 (1990).
22. M. F. Hibert, S. Trumpp-Kallmeyer, A. Bruinvels, and J. Hoflack, *Mol. Pharmacol.* **40,** 8 (1991).
23. J. P. Huggins, S. Trumpp-Kallmeyer, M. F. Hilbert, J. M. Hoflack, B. O. Fanger, and C. R. Jones, *Eur. J. Pharmacol.* **245,** 203 (1993).
24. C. D. Livingstone, P. G. Strange, and L. H. Naylor, *Biochem. J.* **287,** 277 (1992).
25. K. Maloney-Huss and T. P. Lybrand, *J. Mol. Biol.* **225,** 859 (1992).
26. G. Nordvall and U. Hacksell, *J. Med. Chem.* **36,** 967 (1993).
27. S. Trumpp-Kallmeyer, J. Hoflack, A. Bruinvels, and M. Hibert, *J. Med. Chem.* **35,** 3448 (1992).
28. S. Trumpp-Kallmeyer, J. Hoflack, and M. Hibert, *in* "The Tachykinin Receptors" (S. H. Buck, ed.). Humana Press, New York, 1993 (in press).
29. R. B. Westkaemper and R. A. Glennon, *Pharmacol. Biochem. Behav.* **40,** 1019 (1991).
30. D. Donnelly, J. B. C. Findlay, and T. L. Blundell, *Receptors & Channels* (1994) in press.
31. A. P. Ijzerman, G. P. J. Van, and K. A. Jacobson, *Drug Des. Discov.* **9,** 49 (1992).
32. X. Q. Lewell, *Drug Des. Discov.* **9,** 29 (1992).
33. Y. Yamamoto, K. Kamiya, and S. Terao, *J. Med. Chem.* **36,** 820 (1993).
34. P. Cronet, C. Sander, and G. Vriend, *Protein Eng.* **6,** 59 (1993).
35. A. Smolyar and R. Osman, *Mol. Pharmacol.* **44,** 882 (1993).
36. C. Humblet and T. Mirzadegan *Annu. Rep. Med. Chem.* **27,** 291 (1992).
37. J. B. C. Findlay and D. Donnelly, *in* "GTPases in Biology II" (B. Dickey and L. Birnbaumer, eds). Springer-Verlag, Heidelberg, 1994, pp. 17–32.
38. J. Hoflack, S. Trumpp-Kallmeyer, and M. Hibert, *in* "3D QSAR and Drug Design: Theory, Methods, and Applications" (H. Kubinyi, ed.). ESCOM, New York, 1993 (in press).
39. M. Kontoyianni and T. P. Lybrand, *Perspect. Drug Discov. Des.* **1,** 291 (1993).
40. J. M. Baldwin, *EMBO J.* **12,** 1693 (1993).
41. D. Donnelly, M. S. Johnson, T. L. Blundell, and J. Saunders, *FEBS Lett.* **251,** 109 (1989).
42. S. C. Chan, A. K. Wong, and D. K. Chiu, *Bull. Math. Biol.* **54,** 563 (1992).
43. R. Link, D. Daunt, G. Barsh, A. Chruscinski, and B. Kobilka, *Mol. Pharmacol.* **42,** 16 (1992).
44. D. Oksenberg, S. A. Marsters, B. F. O'Dowd, S. Havlik, S. J. Peroutka, and A. Ashkenazi, *Nature* (*London*) **360,** 161 (1992).
45. H. T. Kao, N. Adham, M. A. Olsen, R. L. Weinshank, T. A. Branchek, and P. R. Hartig, *FEBS Lett.* **307,** 324 (1992).
46. M. O. Dayhoff, W. C. Barker, and L. T. Hunt, *in* "Methods in Enzymology"

(C. Hirs and S. Timashe'ff, eds.). Vol. 91, p. 524. Academic Press, New York, 1983.
47. S. F. Altschul, *J. Mol. Biol.* **219,** 555 (1991).
48. J. Overington, D. Donnelly, M. S. Johnson, A. Sali, and T. L. Blundell, *Protein Sci.* **1,** 216 (1992).
49. D. Donnelly, J. P. Overington, S. V. Ruffle, J. H. A. Nugent, and T. L. Blundell, *Protein Sci.* **2,** 55 (1993).
50. S. Pascarella and P. Argos, *J. Mol. Biol.* **224,** 461 (1992).
51. B. Svensson, I. Vass, and S. Styring, *Z. Naturforsch. C: Biosci.* **46,** 765 (1991).
52. D. C. Rees, H. Komiya, T. O. Yeates, J. P. Allen, and G. Feher, *Annu. Rev. Biochem.* **58,** 607 (1989).
53. J. Overington, M. S. Johnson, A. Sali, and T. L. Blundell, *Proc. R. Soc. London, Ser. B* **241,** 132 (1990).
54. D. Zhang and H. Weinstein, *FEBS Lett.* **337,** 207 (1994).
55. J. B. Findlay, P. L. Barclay, M. Brett, M. Davison, D. J. Pappin, and P. Thompson, *Vision Res.* **24,** 1501 (1984).
56. J. Kyte and R. F. Doolittle, *J. Mol. Biol.* **157,** 105 (1982).
57. J. Deisenhofer and H. Michel, *EMBO J.* **8,** 2149 (1989).
58. M. Degli Esposti, M. Crimi, and G. Venturoli, *Eur. J. Biochem.* **190,** 207 (1990).
59. J. Edelman and S. H. White, *J. Mol. Biol.* **210,** 195 (1989).
60. J. Edelman, *J. Mol. Biol.* **232,** 165 (1993).
61. J. A. Ballesteros and H. Weinstein, *Biophys. J.* **62,** 107 (1992).
62. R. Henderson, J. M. Baldwin, T. A. Ceska, F. Zemlin, E. Beckmann, and K. H. Downing, *J. Mol. Biol.* **213,** 899 (1990).
63. J. Deisenhofer, H. Michel, T. O. Yeates, H. Komiya, D. C. Rees, J. P. Allen, and G. Feher, *Proc. Natl. Acad. Sci. U.S.A.* **84,** 6438 (1987).
64. C. M. Deber, D. W. Hughes, P. E. Fraser, A. B. Pawagi, and M. A. Moscarello, *Arch. Biochem. Biophys.* **245,** 455 (1986).
65. M. Roux, J. M. Neumann, R. S. Hodges, P. F. Devaux, and M. Bloom, *Biochemistry* **28,** 2313 (1989).
66. P. L. Yeagle, *Biophys. J.* **37,** 227 (1982).
67. G. D. Henry and B. D. Sykes, *Biochem. Cell Biol.* **68,** 318 (1990).
68. A. Gallusser and A. Kuhn, *EMBO J.* **9,** 2723 (1990).
69. D. Eisenberg, R. M. Weiss, and T. C. Terwilliger, *Proc. Natl. Acad. Sci. U.S.A.* **81,** 140 (1984).
70. D. C. Rees, L. DeAnatonio, and D. Eisenberg, *Science* **245,** 510 (1989).
71. H. Komiya, T. O. Yeates, D. C. Rees, J. P. Allen, and G. Feher, *Proc. Natl. Acad. Sci. U.S.A.* **85,** 9012 (1988).
72. D. Donnelly and R. J. Cogdell, *Protein Eng.* **6,** 629 (1993).
73. R. Sankararamakrishnan and S. Vishveshwara, *Int. J. Pept. Protein Res.* **39,** 356 (1992).
74. K. A. Williams and C. M. Deber, *Biochemistry* **30,** 8919 (1991).
75. M. W. MacArthur and J. M. Thornton, *J. Mol. Biol.* **218,** 397 (1991).
76. K. G. Strehlow, A. D. Robertson, and R. I. Baldwin, *Biochemistry* **30,** 5810 (1991).

77. D. Juretic, B. Lee, N. Trinajstic, and R. W. Williams, *Biopolymers* **33,** 255 (1993).
78. C. A. Granatir, L. A. Parodi, and G. M. Maggiora, *Protein Sci.* (1994), in press.
79. T. M. Gray and B. W. Matthews, *J. Mol. Biol.* **175,** 75 (1984).
80. M. J. McGregor, S. A. Islam, and M. J. Sternberg, *J. Mol. Biol.* **198,** 295 (1987).
81. L. Piela, G. Nemethy, and H. A. Scheraga, *Biopolymers* **26,** 1273 (1987).
82. J. W. Ponder and F. M. Richards, *J. Mol. Biol.* **193,** 775 (1987).
83. R. L. J. Dunbrack and M. Karplus, *J. Mol. Biol.* **230,** 543 (1993).
84. D. J. Barlow and J. M. Thornton, *J. Mol. Biol.* **201,** 601 (1988).
85. D. N. Woolfson and D. H. Williams, *FEBS Lett.* **277,** 185 (1990).
86. R. Bazzo, M. J. Tappin, A. Pastore, T. S. Harvey, J. A. Carver, and I. D. Campbell, *Eur. J. Biochem.* **173,** 139 (1988).
87. A. Pastore, T. S. Harvey, C. E. Dempsey, and I. D. Campbell, *Eur. Biophys. J.* **16,** 363 (1989).
88. R. H. Yun, A. Anderson, and J. Hermans, *Proteins* **10,** 219 (1991).
89. R. Sankararamakrishnan and S. Vishveshwara, *Proteins: Struct. Funct. Genet.* **15,** 26 (1993).
90. L. Piela, G. Nemethy, and H. A. Scheraga, *Biopolymers* **26,** 1587 (1987).
91. D. N. Woolfson, S. R. J. Mortishire, and D. H. Williams, *Biochem. Biophys. Res. Commun.* **175,** 733 (1991).
92. M. S. Sansom, *Protein Eng.* **5,** 53 (1992).
93. J. A. Ballesteros and H. Weinstein, *Biophys. J.* **62,** 110 (1992).
94. R. Sankararamakrishnan and S. Vishveshwara, *Biopolymers* **30,** 287 (1990).
95. G. D. Fasman, "Prediction of Protein Structure and the Principles of Protein Conformation." Plenum, New York, 1989.
96. N. L. Summers and M. Karplus, *in* "Methods in Enzymology" (J. Langone, ed.). Vol. 202, p. 156. Academic Press, San Diego, CA, 1991.
97. M. Hilbert, G. Bohm, and R. Jaenicke, *Proteins: Struct., Funct. Genet.* **17,** 138 (1993).
98. D. Eisenberg, E. Schwarz, M. Komaromy, and R. Wall, *J. Mol. Biol.* **179,** 125 (1984).
99. M. S. Weiss, U. Abele, J. Weckesser, W. Welte, E. Schiltz, and G. E. Schulz, *Science* **254,** 1627 (1991).
100. S. W. Cowan, T. Schirmer, G. Rummel, M. Steiert, R. Ghosh, R. A. Pauptit, J. N. Jansonius, and J. P. Rosenbusch, *Nature (London)* **358,** 727 (1992).
101. S. K. Wong, C. Slaughter, A. E. Ruoho, and E. M. Ross, *J. Biol. Chem.* **263,** 7925 (1988).
102. H. G. Dohlman, M. G. Caron, C. D. Strader, N. Amlaiky, and R. J. Lefkowitz, *Biochemistry* **27,** 1813 (1988).
103. M. D. Davison and J. B. Findlay, *Biochem. J.* **236,** 389 (1986).
104. M. D. Davison and J. B. Findlay, *Biochem. J.* **234,** 413 (1986).
105. L. Pardo, J. A. Ballesteros, R. Osman, and H. Weinstein, *Proc. Natl. Acad. Sci. U.S.A.* **89,** 4009 (1992).
106. J. Hoflack, S. Trumpp-Kallmeyer, and M. Hibert, *Trends Pharmacol. Sci.* **15,** 7 (1994).
107. R. Henderson and G. F. Schertler, *Philos. Trans. R. Soc. London, Ser. B* **326,** 379 (1990).

108. E. W. Taylor and A. Agarwal, *FEBS Lett.* **325,** 161 (1993).
109. C. Chothia and A. M. Lesk, *EMBO J.* **5,** 823 (1986).
110. M. Kontoyianni and T. P. Lybrand, *Med. Chem. Res.* **3,** 407 (1993).
111. F. Jahnig and O. Edholm, *Z. Phys. B* **78,** 137 (1990).
112. D. D. Oprian, *J. Bioenerg. Biomembr.* **24,** 211 (1992).
113. I. Gantz, J. DelValle, L. D. Wang, T. Tashiro, G. Munzert, Y. J. Guo, Y. Konda, and T. Yamada, *J. Biol. Chem.* **267,** 20840 (1992).
114. B. Y. Ho, A. Karschin, T. Branchek, N. Davidson, and H. A. Lester, *FEBS Lett.* **312,** 259 (1992).
115. A. Mansour, F. Meng, W. J. H. Meador, L. P. Taylor, O. Civelli, and H. Akil, *Eur. J. Pharmacol.* **227,** 205 (1992).
116. N. J. Pollock, A. M. Manelli, C. W. Hutchins, M. E. Steffey, R. G. MacKenzie, and D. E. Frail, *J. Biol. Chem.* **267,** 17780 (1992).
117. C. D. Strader, M. R. Candelore, W. S. Hill, I. S. Sigal, and R. A. Dixon, *J. Biol. Chem.* **264,** 13572 (1989).
118. D. Altschuh, A. M. Lesk, A. C. Bloomer, and A. Klug, *J. Mol. Biol.* **193,** 693 (1987).
119. D. Altschuh, T. Vernet, P. Berti, D. Moras, and K. Nagai, *Protein Eng.* **2,** 193 (1988).
120. N. Pastor, D. Pinero, A. M. Valdes, and X. Soberon, *Mol. Microbiol.* **4,** 1957 (1990).
121. L. M. Gregoret and F. E. Cohen, *J. Mol. Biol.* **211,** 959 (1990).
122. C. Chothia, M. Levitt, and D. Richardson, *J. Mol. Biol.* **145,** 215 (1981).
123. M. F. Hibert, S. Trumpp-Kallmeyer, J. Hoflack, and A. Bruinvels, *Trends Pharmacol. Sci.* **14,** 7 (1993).
124. T. M. Fong, R. R. Huang, and C. D. Strader, *J. Biol. Chem.* **267,** 25664 (1992).
125. A. H. Cheung, R. R. Huang, and C. D. Strader, *Mol. Pharmacol.* **41,** 1061 (1992).
126. J. R. Raymond, M. Hnatowich, R. J. Lefkowitz, and M. G. Caron, *Hypertension (Dallas)* **15,** 119 (1990).
127. T. M. Savarese, C. D. Wang, and C. M. Fraser, *J. Biol. Chem.* **267,** 11439 (1992).
128. T. M. Savarese and C. M. Fraser, *Biochem. J.* **283,** 1 (1992).
129. H. Weinstein and J. P. Green, eds. "Quantum Chemistry in Biomedical Sciences" *Ann. N.Y. Acad. Sci.* **367** (1981).
130. D. M. Hirst, "A Computational Approach to Chemistry." Blackwell, London, 1990.
131. E. Clementi, "Modern Techniques in Computional Chemistry: MOTECC-91." ESCOM, Leiden, 1991.
132. J. Hoflack, M. F. Hibert, T.-K. S., and J. M. Bidart, *Drug Des. Discov.* **10,** 157 (1993).
133. D. Zhang and H. Weinstein, *Med. Chem. Res.* **3,** 357 (1993).
134. O. Edvardsen, I. Sylte, and S. G. Dahl, *Brain Res. Mol. Brain Res.* **14,** 166 (1992).
135. X. Luo, D. Zhang, and H. Weinstein, *Protein Sci.* **2,** Suppl. 1, 158 (1993); X. Luo, D. Zhang, and H. Weinstein, *Protein Eng.* **7** (1994) in press.

Section IV

Localization and Regulation

[20] Anti-fusion Protein Antibodies Specific for Receptor Subtypes

Brian J. Ciliax, Craig Heilman, Sharon Edmunds,
Steven M. Hersch, and Allan I. Levey

Introduction

Immunocytochemistry of specific neurochemical markers has been very useful for the study of functional neuroanatomy and has been employed for the localization of presynaptic markers (such as neuropeptides, neurotransmitters, and their synthetic enzymes) of chemically defined neurons. For example, dopaminergic neurons have been visualized using antibodies against tyrosine hydroxylase (1–3) and more recently dopamine itself (4). However, other molecules important in synaptic transmission, e.g., receptors and transporters, are intriguing candidates for immunocytochemical localization, but present several challenges. These molecules are hydrophobic intergral membrane proteins, which are more difficult to study and also frequently less antigenic. Furthermore, most receptors are much more diverse with large numbers of structurally related subtypes. Through newer immunological and molecular methods, we have localized genetic receptor subtypes in the central nervous system using subtype-specific antibodies (5, 6). This chapter focuses on the methodology for raising and characterizing antifusion protein antibodies specific for the genetic D1 and D2 dopamine receptors and their application for localization of the subtypes in basal ganglia.

Until recently, based on biochemical, anatomical, and pharmacological data, dopaminergic receptor subtypes were only discriminated into two classes (7), now considered to represent receptor subfamilies. Receptors of the D1 subfamily bind benzazepines with high affinity and increase adenylate cyclase activity when stimulated. Receptors of the D2 subfamily bind butyrophenones and benzamides with high affinity and either reduce or have no effect on adenylate cyclase activity when stimulated. D1 and D2 receptors, determined with radioligand-binding autoradiography, are enriched in all dopaminergic regions, especially basal ganglia.

In contrast to this simple dichotomy, molecular biological research has demonstrated a broader diversity of dopamine receptor subtypes. To date, five different dopamine receptor genes (D1–D5) have been cloned (8, 9), each falling into either of the pharmacological D1 or D2 subfamilies. All have seven transmembrane regions that share extensive homology across

Methods in Neurosciences, Volume 25

subtypes. These transmembrane regions are believed to comprise the ligand-binding site and contain a conserved aspartate residue (which presumably interacts with the amine group of dopamine) and two conserved serine residues (which presumably interact with the catechol groups). The high degree of homology between receptor subtypes in these regions may explain the difficulty in distinguishing between them with pharmacological agents. *In situ* hydridization studies have shown that the mRNA distributions for these various dopamine receptor subtypes differ from one another. D1 and D2 mRNAs are concentrated in basal ganglia (10–12), whereas D3–D5 mRNAs have more discrete distributions (11, 13). These varied dopamine receptor subtypes likely play important individual roles in dopaminergic neurotransmission, since there are differences in constitutive activities, affinities for dopamine, regulatory phosphorylation sites, G protein coupling and second messengers, and mRNA distributions among select subtypes. Some propose that each subtype serves a unique function within individual dopaminergic circuits. One critical aspect of subtype function is where it is located within dopaminergic circuitry, e.g., among select neuronal populations or at pre- vs postsynaptic sites. However, localization of the individual receptor proteins has been difficult to achieve by radioligand-binding autoradiography, since even the most selective ligands cannot always distinguish between all members of the same pharmacological subfamily. Although there is considerable homology between members, antibodies can be raised against unique regions, thus producing subtype-selective probes. We detected individual receptor subtypes by combining immunological and molecular biological techniques (see below). These techniques can generate antibodies which target unique regions of the receptor protein (5, 14, 15). Such antibodies demonstrate high specificity and can be affinity-purified to minimize background signal, particularly advantageous properties for immunocytochemical probes.

Several additional aspects of immunochemical detection provide distinct advantages over radioligand-binding autoradiography for the localization of neurochemical markers. First, whereas binding assays depend on a high-affinity radioligands, antibodies can be generated against many different types of molecules (e.g., receptors, transporters, synthetic enzymes, neurotransmitters, or neuropeptides), regardless of their "ligand-binding" properties. Second, spatial resolution at the light and electron microscopic levels is superior with immunocytochemical techniques, permitting more precise neuroanatomical localization and allowing more specific questions to be addressed. Third, immunocytochemical methods are readily adapted to multiple-label experiments, using immunofluorescent, silver-intensified immunogold, or immunoperoxidase methods. In addition to one label marking an antibody, a second label can identify a second antibody or a neuroanatomical tracer, which would provide combined neurochemical and connectional information. Multiple-label immunocytochemical assays can also be conducted at the electron microscopic level using dual-peroxidase, peroxidase–immunogold, or peroxidase–ferritin double-label methodologies.

In addition to radioligand-binding assays, another prominent methodology for studying neurochemical anatomy is *in situ* hybridization. Although *in situ* hybridization has many desirable features (e.g., high specificity and sensitivity, ability for double-labeling, and cellular resolution), immunocytochemistry has critical advantages, because it directly detects the actual, functional molecule (receptor, transporter, enzyme, etc.), instead of the corresponding messenger RNA. Since the functional molecule mediates a physiological effect, it is important to determine its distribution and relative concentration. These parameters do not necessarily coincide with those of the corrsponding mRNA because translation rates, transport to distinct subcellular compartments, and turnover rates may all vary from one molecule to the next. Therefore, direct detection of receptor proteins by immunochemical means provides information unattainable from, but complementary to, other methodologies.

Fusion Proteins

Background

Molecular biological techniques can be used to raise antibodies against select regions of a receptor subtype (5, 14). This process avoids several problems with antigen purification, as used in conventional immunological schemes, and produces greater antibody specificity. Purifying a single receptor subtype would be difficult since affinity ligands cannot generally distinguish between the different genetic subtypes. Furthermore, purification of membrane proteins generally requires treatment with detergents, which makes the retention of a functional binding site more difficult. Finally, even if the receptor protein subtypes could be purified and isolated from each other, immunization against an entire protein would generate many antibodies cross-reactive with related subtypes containing homologous regions. By immunizing against either a synthetic peptide or a fusion protein containing a sequence identical to a unique receptor region, individual receptor subtypes can be targeted, generating subtype-specific antibodies.

Recombinant proteins containing unique, antigenic receptor segments can be made using genes of individual dopamine receptor subtypes. Specific regions of interest of the receptor gene can be subcloned in a variety of vectors and expressed in transfected bacterial or mammalian cells. The receptor gene segments can be fused with the genes of other useful proteins. For our purposes, a binding protein is advantageous for affinity purification. Expression of this hybrid creates a fusion protein of a single polypeptide chain containing the receptor protein segment on one-half and the binding protein on the other. Once the fusion protein has been expressed in transfected bacteria, it can be affinity-purified from bacterial proteins via the binding protein domain. If desired, the binding protein may be cleaved from the receptor segment by a specific peptidase, the consensus sequence of which is located at the junction between the two halves of the fusion protein. Large (milligram) amounts of select regions of receptors (i.e., receptor subtype-specific epitopes) can be synthesized and purified in this manner. Purified fusion protein can then be used for immunization, affinity purification of the resulting antibodies, and their immunological characterization.

Selection of Target Region

When designing immunochemical probes against receptor subtypes, several criteria are useful in the selection of the optimum region. First, the region should have a unique amino acid sequence. Dopamine receptors belong to the superfamily of G protein-coupled receptors (8, 9), the members of which have considerable structural similarity. All contain seven transmembrane segments which are highly conserved. Monoaminergic receptors contain a conserved aspartate residue and two conserved serine residues, thought to interact with the amine and aromatic hydroxyl groups on the various neurotransmitters (DA, NE, 5HT). Short segments of the third intracellular loop adjacent to the membrane probably interact with specific G protein species to affect second-messenger production. Because of the high degree of homology of these segments within and across receptor families (16), they should be excluded. The selection of a unique region should be cross-checked with other family members and other families with a computer program. Second, the region should be hydrophilic. This is important because hydrophobic regions tend to produce insoluble fusion proteins, which require significant effort to isolate and purify. Furthermore, any epitopes contained in the hydrophobic region could be buried in a membrane compartment, making them inaccessible to the antibodies during immunocytochemistry. Fortunately, for members of the G protein-coupled receptor superfamily, the least-conserved regions also tend to be the hydrophilic regions outside of the membrane-spanning segments (16). Moreover, highly charged, hydrophilic sequences also tend to be the most immunogenic (15, 17, 18). Third, when designing fusion proteins for antibody generation, a large region is preferable, a condition not practical when synthesizing peptides. Selecting a large region maximizes the number of potential epitopes and may impart a conformation to the protein segment similar to that seen in the native receptor (19). This latter point can be critical when attempting to use the resulting antibodies to recognize the target receptor in its native conformation, e.g., for immunoprecipitation and immunocytochemistry. This may explain why many antipeptide antibodies are excellent when used for probing denatured G protein-coupled receptors on Western blots or the small synthetic peptide in ELISA assays, yet are not as sensitive when used for immunocytochemistry (20). Moreover, because many different antibody clones are generated against the larger region in the fusion protein, there is a better chance of having a few of the corresponding epitopes survive the tissue fixation treatment. However, anti-peptide antibodies have worked extremely well for many ionotropic receptor subunit proteins. Finally, if possible, it is best to avoid regions containing consensus sites for glycosylation,e.g., the amino termini of G protein-coupled receptors, since antibodies specific for the peptide sequence can be obstructed by the carbohydrate tree. An example of this was observed with a panel of antibodies raised against muscarinic receptor subtypes. These anti-peptide antibodies blotted the homologous synthetic peptides with great sensitivity yet did not recognize the native receptors (20).

We have successfully generated subtype-specific antibodies for dopaminergic receptors by targeting the carboxy terminus (D1) or the third intracellular loop (D2). Either region, for the respective subtypes, is large (97 amino acids for D1, and 120 for D2), hydrophilic, antigenic, and unique (5). The combination of these characteristics has led to the production of highly sensitive and specific antibodies which recognize native and denatured receptors in several different immunochemical assays. The structures of D1 and D2 receptor subtypes are shown in a schematic (Fig. 1) as examples of each subfamily. A high degree of homology can be seen in the transmembrane regions, whereas the C-terminus and the third intracellular loop, for D1 and D2, respectively, are relatively heterogeneous.

Plasmid Construction

We have used the pGEX expression system (21) to produce subtype-specific fusion proteins for m1–m5 muscarinic receptors and D1, D2, and D5 dopamine receptors. The pGEX plasmid encodes for glutathione *S*-transferase (GST), a 26-kDa enzyme that binds to immobilized glutathione affinity resin and can be eluted under mild conditions. Furthermore, fusion proteins containing GST are usually soluble and easy to manipulate. The plasmid is now

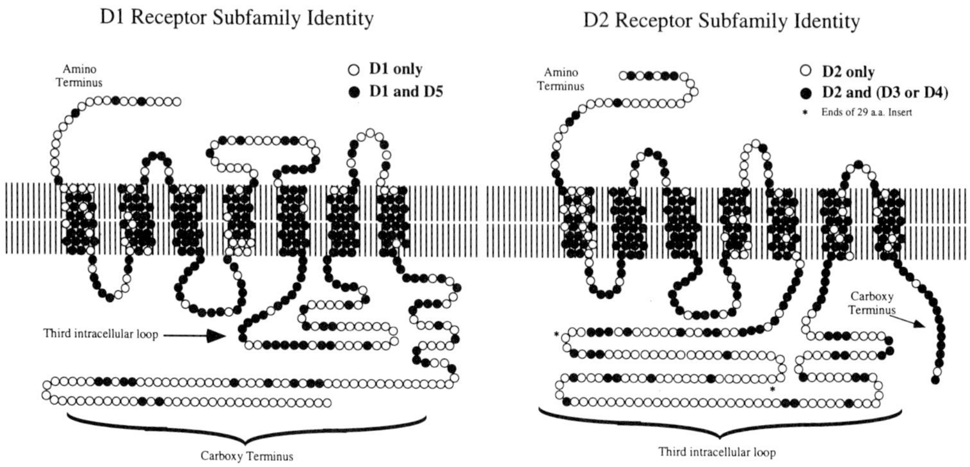

FIG. 1 Schematic of D1 and D2 receptor structure and divergence. Putative structures and divergent amino acids of the dopamine receptors were constructed from the published sequences and hydropathy analysis for the human genes for the D1 and the D2 receptor. Closed circles represent amino acids in the receptor which are also present in another subtype. Open circles represent amino acids unique to either D1 or D2 receptors. Note the contrasting sizes of the carboxy termini and the third intracellular loops. These regions are also the most divergent for their respective subfamilies.

commercially available in several versions, including pGEX-2T and pGEX-3X, encoding the consensus sequences for cleavage of the fusion protein by thrombin or factor Xa, respectively. The consensus sequence codes are positioned after the GST gene and before the insertion site. Stop codons, aligned in each of the three reading frames, positioned after the insertion site, terminate synthesis of the polypeptide chain. Expression of the fusion protein is regulated by a lac I^q gene, which represses a tac promoter until transcription is induced with isopropylthiogalactopyranoside (IPTG). Ampicillin resistance conferred by the pGEX plasmids is used to select for transfected bacteria. The steps involved in plasmid construction and fusion protein production are outlined in a schematic figure (Fig. 2).

Recombinant proteins which contain only portions of a particular receptor can serve as antigens in the generation of subtype-specific antibodies. Using standard molecular biological techniques (22), the genetic code of a selected region of receptor protein is amplified by polymerase chain reaction (PCR) and fused with the gene of a binding protein. Polymerase chain reaction amplification is accomplished using oligonucleotide primers containing the

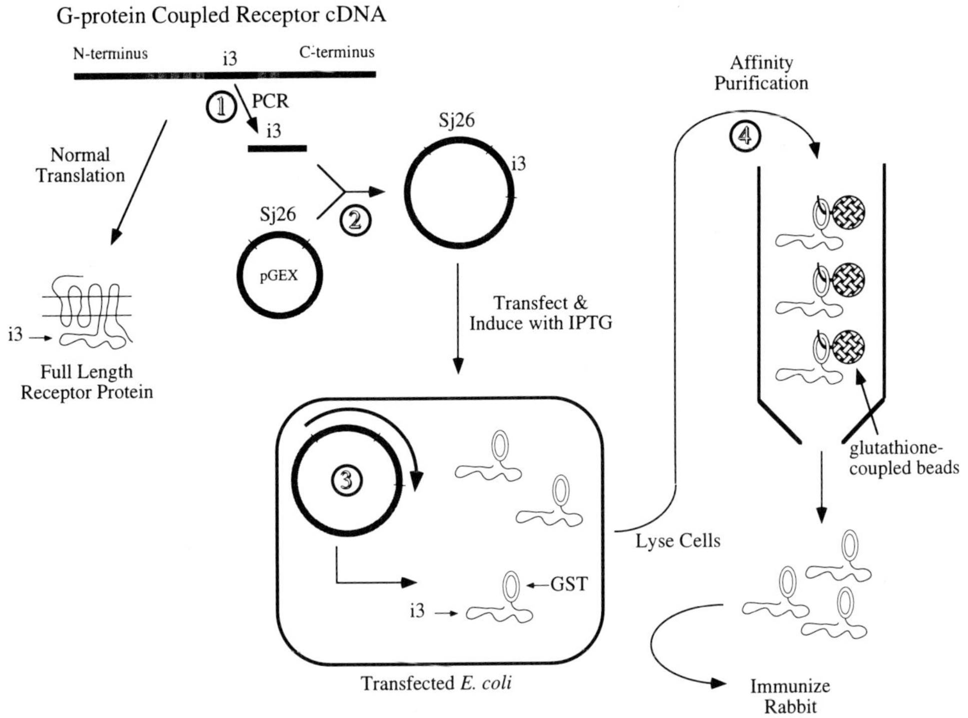

FIG. 2 Schematic of fusion protein production. Illustration of (1) the subcloning of the heterogeneous third intracellular loop of a G protein-coupled receptor (e.g., D2), (2) the insertion of the segment into pGEX plasmid, (3) the expression of the fused "genes" in transfected *E. coli,* and (4) the affinity purification of the resulting fusion protein. Normal translation of the full-length receptor is represented on the left; the structure of the receptor's third intracellular loop is shown to be retained in the fusion protein. The GST moiety of the fusion protein binds to the glutathione-coupled beads during affinity purification.

same restriction sites as those selected in the insertion site of the plasmid as well as sequences complementary to the 5′ and 3′ ends of the receptor segment. Avoid using restriction sites present within the receptor DNA fragment to be amplified. The amplified sequence is digested with restriction enzymes and then ligated into the pGEX plasmid. Finally, the sequence of the purified plasmid is confirmed by the double-stranded dideoxynucleotide method (23). The recombinant plasmid is transfected into RR1 bacteria for plasmid amplification. Transfected RR1 bacteria are simply cultured in a large volume and lysed, and the plasmid DNA is purified by ion-exchange chromatography (Qiagen).

Fusion Protein Expression and Purification

The amplified plasmid is used to transfect BL21 bacteria for fusion protein expression. BL21 bacteria are optimal for expression because basal synthesis of endogenous proteins is suppressed with a pLysS plasmid (encoding lysozyme, which acts as a repressor). Also, the strain is deficient in several proteases which otherwise could degrade the expressed protein during purification. Fusion protein expression is induced with IPTG and pre- and post-induction samples of bacterial cultures are compared by sodium dodecyl sulfate (SDS)–polyacrylamide gel electrophoresis (PAGE). Among the constitutive bacterial proteins seen in the preinduction sample, at least one additional band should be conspicuous in the postinduction sample. The mobility of fusion protein should be roughly consistent with the predicted molecular mass based on the amino acid sequence of GST plus the receptor segment.

For antibody production and other studies, the fusion protein should be purified to remove contaminating bacterial proteins. Fusion protein-laden bacteria are lysed in the presence of peptidase inhibitors and the DNA is digested enzymatically to reduce the viscosity of the lysate. The lysate is passed over affinity chromatography beads containing immobilized glutathione (S-linked), to which GST binds. After several washes, the fusion protein is eluted with excess free reduced glutathione. The purified fusion protein is then subjected to SDS–PAGE to determine its purity and assess its size as above. Additional bands are usually of lower molecular weight than the parent band and may represent degradation products or incompletely synthesized copies. Following dialysis to remove excess free ligand, the fusion protein is used for immunization of the host animal and for purification of the resulting sera via affinity methods. Alternately, GST can be proteolytically cleaved at this point and reabsorbed on immobilized glutathione to purify the receptor peptide.

The carboxy terminus of the D1 receptor and the third intracellular loop of the D2 receptor were used to construct fusion proteins using the pGEX-2T system. An example of induction of D1 fusion protein synthesis is shown in Fig. 3. After 4 hr of IPTG induction of BL21 bacteria transfected with D1-pGEX plasmid, a dense band appeared. Lysates of D1- or D2-expressing BL21 cultures were affinity-purified over immobilized reduced glutathione. Affinity purification removed essentially all background proteins, yielding pure D1 or D2 fusion proteins, which have parent bands that are larger than GST (Fig. 4).

Immunization and Purification

The optimal conditions for immunizing animals with novel fusion protein antigens are still poorly understood. Different conformations of the immunogen can generate antibodies with reactivities that differ depending on the type

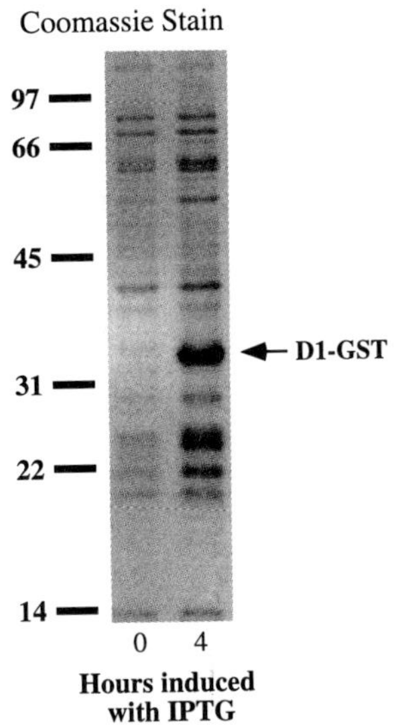

FIG. 3 Sodium dodecyl sulfate–polyacrylamide gel electrophoresis of BL21 bacterial lysates expressing D1 fusion protein. BL21 bacteria transfected with D1-pGEX plasmid were grown to confluency and then induced for 4 hr with IPTG. A pre-induction sample on the left reveals many constitutive bacterial proteins. A post-induction sample on the right reveals an additional dense parent band (arrow) and several other bands below. Positions of molecular weight markers are shown on the left.

of immunoassay employed. For example, detergent-treated immunogens are denatured and may generate antibodies with different conformational specificity than nondenatured immunogens. One can exploit this phenomenon to generate antibodies optimal for a particular immunoassay. If the primary goal is to develop an antibody reactive with the receptors on immunoblots, it may be useful to SDS-denature soluble fusion proteins prior to immunization. In contrast, we have found that antibodies raised to soluble, nondenatured fusion proteins are better for both immunocytochemistry and immunoprecipitation (6, 14, 24).

Animals are immunized for polyclonal antibody production with purified, subtype-specific fusion protein. We give rabbits monthly injections of approximately 200 μg soluble purified fusion protein emulsified in Freund's adjuvant. Bleeds are obtained at 3 and 4 weeks post-boost, clotted overnight at 4°C, and clarified, and the sera are aliquotted and frozen at −80°C. For repeated use, 1- to 2-ml aliquots are stored at 4°C for several months using aseptic techniques. Alternatively, the immunizations may be performed by commercial institutions.

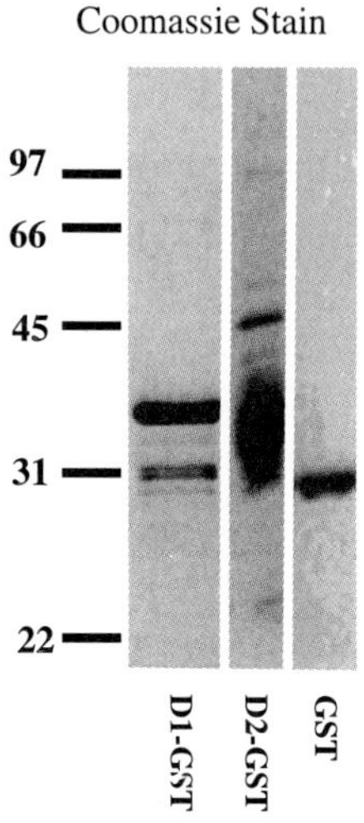

FIG. 4 Sodium dodecyl sulfate–polyacrylamide gel electrophoresis of purified D1, D2, and GST fusion proteins. Induced D1, D2, and GST fusion proteins were affinity-purified over glutathione agarose beads and eluted with excess free glutathione. Virtually all bacterial proteins were removed. Positions of molecular weight markers are shown on the left.

High-titer antisera still contain many antibodies that can cross-react with unrelated antigens and increase the background of immunochemical assays. Some of the contaminating antibodies are specific for the GST moiety of the fusion protein or any bacterial proteins still present in the purified fusion protein sample. Both of these classes of antibodies may be extracted by first passing the antisera over an affinity column made from lysates of BL21 bacteria expressing GST. In addition, serum contains many different antibodies directed against a variety of antigens to which the host animal had been exposed. The antifusion protein antibodies may be purified from these unwanted constitutive antibodies by affinity chromatography of the antisera (25) against the same fusion protein used to immunize the animal. Affinity columns are made by covalently conjugating purified fusion protein to activated chromatography resin. After antisera are passed over the affinity column, bound antibodies can be washed free of most contaminating antibodies and eluted with acidic or basic buffers. The eluted antibodies are then characterized as described below.

Characterization

In order to properly interpret immunoassay results, the affinity-purified antibody must first be characterized. Three different methods (immunoblotting, immunoprecipitation, and immunocytochemistry) offer independent confirmation of specificity by measuring different aspects of the antibody–receptor interaction. Immunoblotting determines the molecular weight(s) of immunoreactive material and is critical for ruling out cross-reactions with other proteins. Immunoprecipitation determines antibody recognition of functional binding sites labeled with radioligand, thus linking antibody binding with an operational measure of the desired antigen. Immunocytochemistry determines the anatomical distribution of the immunoreactive material. Although an antibody may perform well in one of these assays, it may not perform well in the others. This is partially due to the fact that different forms of the antigen are being tested. For immunoblotting, the receptor is denatured with detergent and adsorbed to a membrane surface. For immunoprecipitation, the receptor is solubilized in mild detergent, but still in a functional form, close to the native state. For immunocytochemistry, the receptor is cross-linked with aldehyde fixative to other proteins and lipids via its primary amino groups. Modification of the epitopes by the tissue treatments in each of these techniques may disrupt antibody–antigen interactions. For example, if the antibody's recognition site is dependent on a specific conformation of the protein, then denatured versions may not be immunoreactive. Moreover, if the recognition site is dependent on a free amino group in the epitope, then aldehyde-fixed versions may not be immunoreactive. Polycloncal antibodies may hold an advantage over monoclonal antibodies in this regard since many diverse antibody clones are present, increasing the odds that one will work if others do not. This also applies to situations where the tissue fixation parameters for immunocytochemistry are less than ideal and many antigens could be lost, e.g., immunocytochemistry using human brain tissue with long postmortem delay prior to tissue fixation. Furthermore, a single polyclonal antiserum may contain entirely different sets of antibody clones that react in immunocytochemistry, immunoblotting, or immunoprecipitation assays, since the epitopes surviving the respective tissue treatments may be mutually exclusive. Thus, although each method has limitations, all provide valuable information in the characterization of subtype-specific antibodies.

Immunoblot Analysis

Immunoblotting (Western blotting) is useful for characterizing antibody specificity and for determination of the molecular weight of the immunoreactive material. Antigen proteins are separated using SDS–PAGE and electrophoretically transblotted to membrane using wet-cell or semi-dry methods. Soluble fusion proteins transblot readily and minute amounts (nanogram) are detectable. Native receptors are more difficult to transblot and require the omission of methanol from the transblotting buffer for successful transfer. This may be due to the displacement of detergent of the lipophilic regions of the receptor by the alcohol causing the membrane protein to precipitate in the gel. Our GST-coupled fusion proteins on the other hand do not have lipophilic regions, and so they are soluble to begin with and can still exit the gel.

As a first screening step, an antibody is routinely blotted against the individual fusion protein used as immunogen, and its reactivity is compared to other fusion proteins derived from all other members of the receptor family. The antibody must be specific for the homologous fusion protein if the antibody will be valuable for immunoanalysis of subtype proteins. Since receptors within the same family are the molecules with the greatest homology to the subtype of interest, their respective fusion proteins should be the most likely to cross-react with the antibody. This is actually an advantage when designing fusion proteins from dopamine receptors, since G_s-coupled receptors (D1 subfamily) tend to have short third intracellular loops but long carboxy tails, and G_i-coupled receptors (D2 subfamily) tend to have long third intracellular loops but short carboxy tails. Thus, fusion proteins derived from the long carboxy tails of receptors from the D1 subfamily or from the long third intracellular loops of receptors from the D2 subfamily have no comparable regions in the short carboxy tails of the D2 subfamily or in the short third intracellular loops of the D1 subfamily, respectively. However, comparable regions do still exist across members of the same subfamily.

In the next step, the antibody is blotted against homologous, full-length receptor protein as well as all other receptor proteins in its family. For this purpose, the cDNAs encoding the individual receptor subtypes are expressed either transiently or in stable cell lines. The antibody should be specific for those cells transfected with the homologous receptor gene and the immunoreactive band should have the appropriate molecular weight. However, aggregation of membrane proteins, including receptors and transporters, is common, which could increase the apparent molecular weight of the antigen or even hinder its solubilization or entry into the gel.

Finally, the antibody is blotted against native receptor protein from brain tissue. Multiple receptor subtypes may be present in the brain samples, and so immunodetection does not necessarily indicate subtype specificity. However, the molecular weight of the immunoreactive band should be generally consistent with the size of the native protein as predicted by its primary structure (plus glycosylation, usually) and gel exclusion chromatography. Moreover, the regional distribution of the antigen should coincide with the known distribution of the receptor protein (determined or inferred from radioligand binding autoradiography or *in situ* hybridization data). For example, D1 and D2 receptor binding and mRNA are very high in striatum but low in cerebellum, hippocampus, and thalamus. Additional experiments can be conducted for further characterization, e.g., proteolytic and deglycosylation treatments of the antigen can be used to determine whether the molecular weights of the immunoreactive bands match those of the expected digestion products. For any of these immunoblotting experiments, whether against fusion protein, cloned receptor, or brain tissue region, preabsorption of the antibodies on homologous and heterologous fusion proteins is an important control experiment to verify the immunological specificity of the reaction.

Anti-fusion protein antibodies directed against D1 or D2 receptor subtypes were specific for their respective antigens in immunoblot experiments. The antibodies were sensitive for their respective fusion proteins and did not recognize the other dopamine receptor fusion proteins, GST, or bacterial proteins. Affinity-purified D1- or D2-specific antibodies were specific for their respective fusion proteins, without cross-reactivity to other subtypes or GST (Fig. 5, left). Moreover, the antibodies also recognized their respective cloned receptors expressed in transfected mammalian cells, without cross-reactivity to the other subtype (Fig. 5, right). The molecular weights of the detected cloned receptor bands were in agreement with the calculated molecular weight of the full-length receptors. Furthermore, the antibodies detected bands in canine brain tissue with molecular weights corresponding to those of the cloned receptors (5). The immunoreactive bands were enriched in striatal membranes but undetected in cerebral cortex and hippocampus. These data confirm the specificity of these antibodies for their respective receptor subtypes.

Immunoprecipitation

Immunoprecipitation measures the antibody's recognition of a functional receptor binding site. We use two variations of the experiment to measure antibody binding either directly or indirectly. In the first, the receptor proteins are solubilized with a mild detergent, labeled with radioligand, and then immunoprecipitated with the purified antibody. The precipitated material is washed and then counted for radioactivity. The second version is subtractive, in which the solubilized receptors are first removed by precipitation and then the remaining material is labeled with radioligand and counted. Receptor binding sites may be solubilized from cells transfected with an individual receptor cDNA or from fresh brain tissue. Cloned receptors should only be recognized by the homologous antibody. Since brain tissue may express

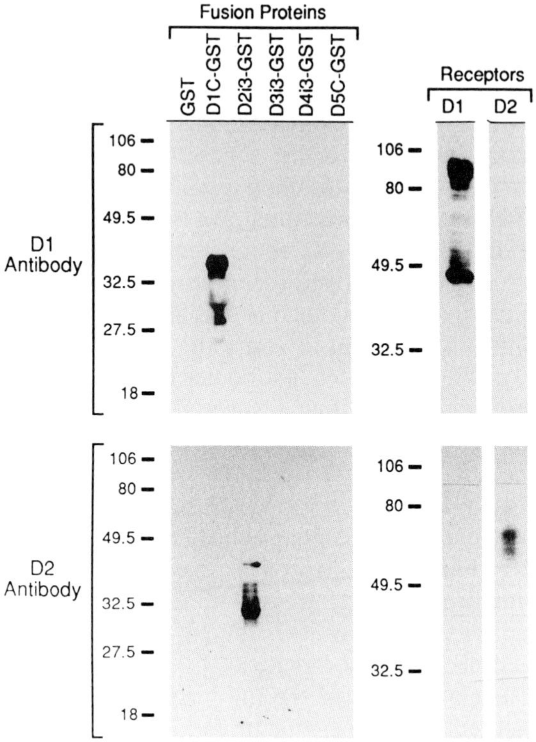

FIG. 5 Western blots of affinity-purified antibodies specific for D1 and D2 receptors. (Top) D1 antibody was used to probe membranes blotted with GST and D1–D5 fusion proteins (left) or membranes from COS cells transfected with either D1 or D2 receptor cDNAs (right). (Bottom) D2 antibody was used to probe the same blots after they were stripped of the previous antibodies. Note that each antibody recognizes a single fusion protein and full-length cloned receptor with complete subtype specificity. The two bands in D1-transfected cells correspond to the size of the unmodified gene product (approximately 49 kDa) and a high-molecular-weight glycosylated product. Positions of molecular weight markers are shown on the left. Reprinted from Ref. 30, by courtesy of Marcel Dekker, Inc.

multiple receptor subtypes, antibodies of differing specificities may all detect immunoreactivity from a single region but the sum of all precipitated sites should not exceed the total number of binding sites in that region (6). As with immunoblotting, preabsorption of the antibodies with fusion protein is again an important control experiment. In addition, the regional distribution of immunoreactivity should match the known distribution of the receptor protein.

Using the subtractive method, D1 and D2 receptor antibodies have been shown to immunoprecipitate striatal binding sites radiolabeled with the corresponding selective ligands (Fig. 6) (5). The immunoprecipitation was dependent on the concentration of antibody present. Neither antibody demonstrated cross-reactivity to the other subtype. Control experiments omitting primary antibody or substituting with preimmune sera were negative. These data demonstrate that the antifusion protein antibodies are subtype specific and bind to functional receptor protein.

Immunocytochemistry

Immunocytochemistry determines the anatomical distribution of receptor-like immunoreactivity in perfusion-fixed brain tissue. Sections from cryoprotected, aldehyde-fixed brain are incubated with primary antibody and detected by immunoperoxidase methods, using diaminobenzidine as chromagen. If the distribution can be predicted based on receptor binding or *in situ* hybridization studies, a matching distribution of immunoreactivity can be

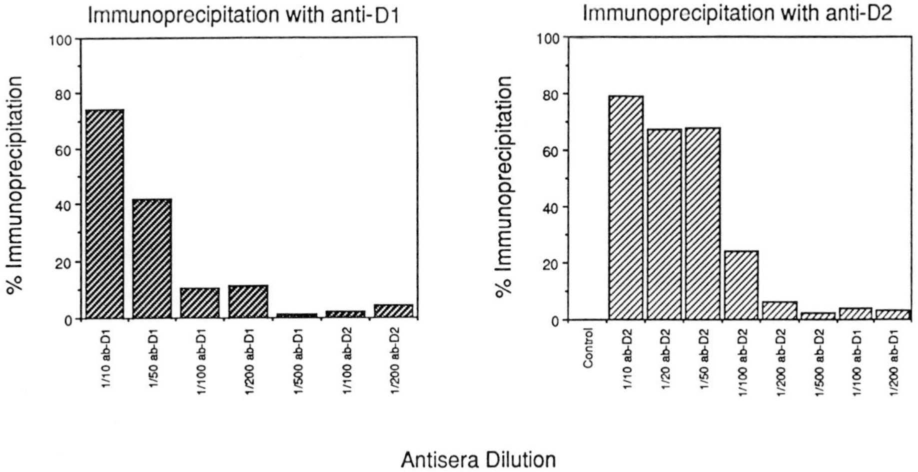

FIG. 6 Immunoprecipitation of D1 and D2 receptors with subtype-specific antibodies. Solubilized striatal membranes were incubated with various dilutions of D1- or D2-specific antibodies, followed by precipitation with protein A-Sepharose. Remaining receptors were assayed by [^{3}H]SCH 23390 or [^{3}H]spiperone binding. Each antibody precipitated the corresponding cloned receptor in a concentration-dependent manner. Antibodies did not immunoprecipitate the heterologous receptor. [Data taken from Levey *et al.* (6).]

strong evidence that the antibody is specifically labeling the target receptor. However, caution must also be employed when interpreting "novel" receptor distributions for they may differ from classical patterns and of course still be valid. For instance, the D3–D5 receptor mRNAs are concentrated in limbic regions (26–29) and do not coincide with classical D1 or D2 binding distributions. It is important to realize that the autoradiographic radioligand-binding distributions represent composites of the distributions of each family member. In addition, receptor protein distributions determined by immunocytochemistry can also be divergent from that predicted by mRNA distributions because of protein transport.

For D1 and D2 receptor proteins, both radioligand-binding and *in situ* hybridization data indicate that they should be concentrated in neostriatum, a brain region rich in dopaminergic input. In addition, substantia nigra labeling differentiates D1 and D2 radioligand binding, with D1 binding concentrated in pars reticulata and D2 binding concentrated in pars compacta. Immunocytochemical distributions using D1- and D2-specific antibodies agree with these reports. D1 and D2 immunoreactivities are both high in the neostriatum, but are segregated to pars reticulata and pars compacta, respectively (Fig. 7). Therefore, this characterization supports the conclusion that the antibodies are specific for the D1 and D2 dopamine receptor genetic subtypes.

Generation and Characterization of D1- and D2-Specific Antibodies

We have had success with the pGEX system in developing antibodies to the muscarinic receptor subtypes (m1–m5) and D1 and D2 receptor subtypes. The following methods use the pGEX system as an example for developing D1 and D2 fusion proteins. Other established systems should provide similar fusion proteins. The basic steps for generating antifusion protein antibodies are as follows:

1. Construct plasmid by inserting PCR-amplified receptor segment.
2. Express fusion protein in transfected BL21 colony and affinity-purify.
3. Immunize rabbits and affinity purify antisera using purified fusion protein.
4. Characterize purified antibody by immunoblot analysis, immunoprecipitation, and immunocytochemistry.

Protocols

Specific protocols are listed below.

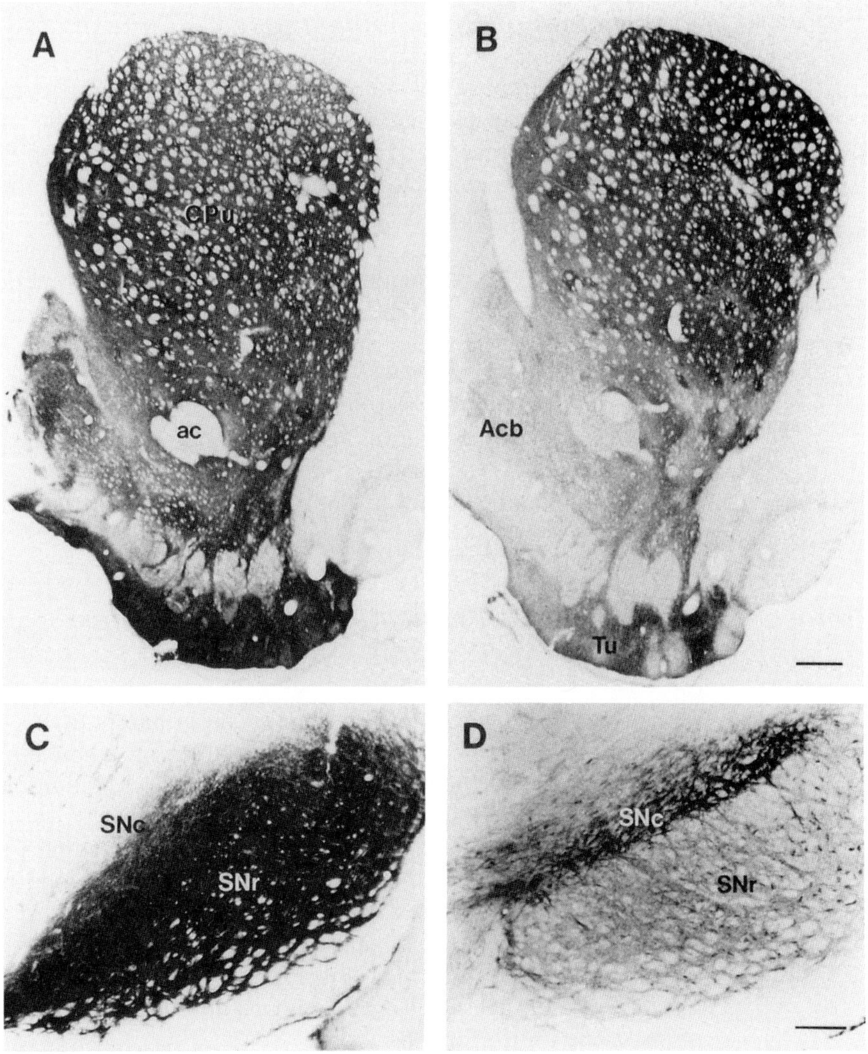

FIG. 7 Immunocytochemistry of D1 and D2 receptors with subtype-specific antibodies. D1 (A and C) and D2 (B and D) are both enriched in the striatum (A and B) and substantia nigra (C and D). The subtypes have overlapping but heterogeneous distributions in the neuropil in caudatoputamen (CPu; neostriatum), nucleus accumbens (Acb), and olfactory tubercle (Tu). The subtypes are complementary in the substantia nigra, with D1 in the neuropil of the pars reticulata (SNr) and D2 in perikarya densely packed in the compacta (SNc). ac, anterior commissure. (Bars: A and B, 250 μm; C and D, 200 μm.)

Protocol 1: Construction of Recombinant pGEX-2T Plasmids

1. Synthesize or purchase oligonucleotide primers (48 mers) complementary to sense and antisense strands of the cDNA sequence to be amplified by polymerase chain reaction. Incorporate restriction sites for *Bam*HI and *Eco*RI or other available cloning sites, into the primers at the 5′ and 3′ ends of the gene fragments, respectively (make sure these sites are not present in the cDNA).
2. Amplify receptor DNA fragments by polymerase chain reaction.
3. Digest receptor DNA and pGEX-2T with *Bam*Hi and *Eco*RI endonucleases.
4. Gel-purify the digested DNAs.
5. Ligate the receptor DNA and pGEX-2T and transform *Escherichia coli.*
6. Analyze transformants by miniprep DNA purification and digestion with *Bam*HI and *Eco*RI.
7. Verify recombinant plasmids by double-stranded dideoxy sequencing (23).

Protocol 2: Expression and Purification of GST-Receptor Fusion Proteins

1. Transform *E. coli* with either a recombinant or parent pGEX-2T vector, grow in 400 ml Luria broth containing 100 μg/ml ampicillin to optical density of 0.4, and then induce fusion proteins with addition of 400 μl of 1.0 *M* IPTG stock for 3–4 hr.
2. Check induction by SDS–PAGE analysis of 25-μl samples of cultures taken immediately before and 4 hr after IPTG. With good results, the fusion protein is the most abundant protein in culture.
3. Harvest cultures by centrifugation at 5000*g* for 10 min, wash cell pellets with 10 ml of buffer A (50 m*M* Tris, pH 8.0, 10 m*M* EDTA, 25% sucrose), resuspend, and digest cell walls with 12 ml buffer A containing 40 mg of lysozyme on ice for 1 hr.
4. Wash cells and resuspend in 10 ml buffer B [10 m*M* Tris, 1.0 m*M* EDTA, 1.0 m*M* dithiothreitol (DTT)] containing a protease inhibitor mixture (1.0 m*M* PMSF, 5.0 μg/ml aprotinin, 1.0 μg/ml leupeptin, and 1 μg/ml pepstatin).
5. Freeze/thaw three to four times on dry ice (may freeze overnight at this step).
6. Add DNase (20 μg/ml) or sonicate gently on ice to reduce viscosity.
7. Add 1/10 volume 10% Triton X-100.
8. Pellet at 5000*g*, remove, and save both supernatant and pellet. (Check

both fractions by SDS–PAGE for fusion protein; only soluble protein in the supernatant can be further purified. Can repeat the above steps on the pellet if a significant amount of protein is in this fraction.)

9. To the supernatant add 15 ml buffer C (20 mM HEPES, pH 7.6, 100 mM KCl, 0.2 mM EDTA, 20% glycerol) containing fresh 1 mM DTT and protease inhibitor mixture.
10. Add 4 ml of 1 : 1 slurry of glutathione agarose beads (Sigma Chemicals); preabsorb beads in buffer C overnight.
11. Shake at 4°C for 1–2 hr.
12. Pellet at 1000g for 5 min, and batch wash beads twice with 40 ml buffer C.
13. Elute protein with 6 × 2 ml of 50 mM reduced glutathione in buffer C.
14. Collect elutions and centrifuge at 2500g for 2 min to remove contaminating beads.
15. Analyze fractions by SDS–PAGE, pool, and dialyze into phosphate-buffered saline using Centriprep chambers (Millipore) or tubing. Check protein concentration (glutathione interferes with most assays if not completely dialyzed), aliquot, and freeze at −80°C.

Protocol 3: Animal Immunization and Antisera Affinity Purification

Immunization

1. Inject rabbits monthly with 200 μg of purified fusion protein emulsified in Freund's adjuvant.
2. Collect bleeds 3–4 weeks post-boost.
3. Clot blood overnight at 4°C, clarify, and store aliquots at −80°C.
4. Aliquots, thawed and stored at 4°C, may be used repeatedly over several months using aseptic techniques.

Antisera Affinity Purification

1. Conjugate purified fusion protein and the lysate of BL21 bacteria expressing GST to individual batches of activated affinity resin. Place each in separate mini-chromatography columns.
2. Add 100 units of heparin to 1 ml of antisera.
3. Pass heparinized antisera over GST column to absorb anti-GST and anti-bacterial protein antibodies.
4. Pass effluent from "BL21/GST" column over fusion protein column.
5. Wash fusion protein column with excess buffer to remove unbound protein.
6. Elute bound proteins in fractions with acidic and/or basic buffers. Collect

elutions in tubes containing 1/10th volume of 1 *M* Tris–HCl, pH 8.0, in order to neutralize the solution immediately.

7. Measure OD_{280} to estimate protein concentration of elution fractions.
8. Pool fractions containing the peak of the elution curve and store at a concentration of 0.1 mg/ml in 50% glycerol at −20°C. Alternatively, dilute samples can be treated with 0.01% sodium azide and stored at 4°C.
9. Columns can be stripped with 10 bed volumes of acidic or basic buffer and regenerated with excess 0.1 *M* Tris–HCl, pH 8.0.

Protocol 4a: Immunoblot Analysis

1. Prepare crude membranes by homogenizing (in the presence of protease inhibitors: 1.0 m*M* PMSF, 5.0 μg/ml aprotinin, 1.0 μg/ml leupeptin, and 1 μg/ml pepstatin) brain tissue or cells transfected with receptor cDNA.
2. Separate membrane proteins by SDS–polyacrylamide gel electrophoresis using prestained molecular weight markers.
3. Transblot overnight onto PVDF membrane.
4. Block membrane with 5% dry milk in Tris-buffered saline pH 7.4 (TBS) at room temperature for 30 min.
5. Incubate with primary antibody overnight at 4°C and wash with TBS.
6. Incubate with secondary antibody–horse radish peroxidase conjugate at room temperature for 1 hr and wash with TBS.
7. Visualize immunoreactive bands with enhance chemiluminescence (ECL, Amersham).

Protocol 4b: Immunoprecipitation Assay for Soluble Dopamine Receptors

Direct Detection Method

1. Harvest transfected cells expressing dopamine receptor cDNAs with rubber policeman, wash, resuspend in cold TE buffer (10.0 m*M* Tris/1.0 m*M* EDTA, pH 7.4) containing a mixture of protease inhibitors (aprotinin, 5 μg/ml; leupeptin, 1–2 μg/ml; PMSF, 100 μg/ml), and homogenize on ice with polytron. Animal tissues can be substituted for determining subtype composition.
2. Centrifuge at 500*g* to remove nuclei and cell debris and then recentrifuge supernatant at 30,000*g* for 20 min. Wash membrane pellet by repeating centrifugation. Resuspend in TE (plus inhibitors) and determine protein concentration. Save aliquot for membrane filtration binding and protein assays to assess total number of receptors/milligram membrane protein.

3. Resuspend membranes in TE containing 1.0% digitonin (Wako) and 0.1% cholate (Sigma) to a protein concentration of 1 mg/ml. The optimal concentration of detergents varies with source of digitonin and should be titered separately. Concentrations as low as 0.4% digitonin are typically effective.
4. Polytron and leave on ice for 60 min to solubilize the receptors.
5. Centrifuge at 30,000*g* for 30 min at 4°C.
6. Transfer supernatant to new tube and add [^{3}H]ligand to final concentration of $10 \times K_d$.
7. Pipet 200 μl of receptor–[^{3}H]ligand mixture into 96-well microtiter plates. Save aliquots of the mixture for gel filtration on G25 Sephadex to determine total specific binding to soluble receptors (calculated from the difference in binding in presence and absence of $100 \times K_i$ of D1- or D2-selective antagonist).
8. Add 5 μl of rabbit antisera to each well and incubate for 4 hr at 4°C. Control experiments substitute nonimmune sera and sera preabsorbed with fusion proteins.
9. Add 50 μl of goat anti-rabbit sera (diluted 1 : 1 in TE) to each well and incubate overnight at 4°C. The optimal concentration of goat anti-rabbit sera varies depending on the batch.
10. Centrifuge microtiter plate at 2400*g* for 5 min at 4°C. Rapidly aspirate supernatants, resuspend pellets in 250 μl ice-cold TE containing 0.1% digitonin/cholate, and recentrifuge.
11. Repeat aspiration, resuspension, and pelleting.
12. Resuspend pellets with 100 μl of 10% SDS and count radioactivity in each sample.
13. Percentage bound by each antibody is calculated by formula:

$$\%\text{Bound} = \frac{\text{Specific binding in immunoprecipitates}}{\text{Total specific binding of soluble receptors (gel filtration)}}$$

$$= \frac{([^3\text{H}])\text{ligand in immune pellets}) - ([^3\text{H}]\text{ligand in nonimmune pellets})}{[^3\text{H}]\text{ligand-labeled soluble receptors} - \text{antagonist}}.$$

Indirect Detection (Subtractive) Method

1. Prepare membranes and solubilize receptors as described in steps 1–5 above.
2. Incubate solubilized receptors (0.2–0.4 pmol) with preimmune sera or antisera for 12 hr at 4°C in Tris-HCl buffer, pH 7.4, containing 1 m*M* EDTA, 5 m*M* KCl, 1.5 m*M* $CaCl_2$, 4 mM $MgCl_2$, and 120 m*M* NaCl.
3. Remove bound receptors by absorption of antibodies onto 100 μl of

protein A-Sepharose CL-4B, containing 0.1% digitonin, for 90 min at 4°C. Pellet beads by centrifugation.

4. Incubate receptors remaining in supernatant with 5 n*M* [^{3}H]SCH 23390 or 1 n*M* [^{3}H]spiperone for 90–120 min, respectively, at 22°C in the same buffer as in step (2).
5. Terminate binding assay by Sephadex G-50 chromatography.
6. Count radiolabeled receptors present in void volume.
7. Calculate percentage bound by each antibody from the following formula:

$$\%\text{Bound} = \frac{\text{Specific binding in immunoprecipitates}}{\text{Total specific binding of soluble receptors}}$$

$$= \frac{([^3\text{H}]\text{ligand in void volume after preimmune precipitation}) - ([^3\text{H}]\text{ligand in void volume after immune precipitation})}{([^3\text{H}]\text{ligand-labeled soluble receptors (no antibody treatment)} - \text{antagonist})}.$$

Protocol 4c: Immunocytochemical Localization of Receptor Proteins

Perfusion Fixation of Rats with Paraformaldehyde

1. Deeply anesthetize rats with 50 mg/kg pentobarbital administered intraperitoneally.
2. Incise abdomen and expose thoracic cavity widely after incision of diaphragm and retraction of thoracic cage.
3. Insert 16-gauge intravenous catheter into left ventricle, remove stylette, and thread catheter sleeve into aortic root. Incise right atrium for outflow.
4. Perfuse 20–30 ml of ice-cold 0.9% saline at 20 ml/min flow rate with peristaltic pump.
5. Clamp descending aorta with hemostat immediately after starting saline rinse.
6. Perfuse 200 ml of freshly prepared 3% paraformaldehyde in 0.1 *M* phosphate buffer, pH 7.4.
7. Perfuse 200 ml of ice-cold 10% sucrose in 0.1 *M* phosphate buffer, pH 7.4.
8. Remove brain and let sink overnight in 30% sucrose in 0.1 *M* phosphate buffer, pH 7.4.

Receptor Immunocytochemistry

1. Section brains at 40 μm on a freezing sliding microtome and collect tissue in TBS on ice.
2. Treat sections in 4% normal goat serum (NGS)/0.4% Triton X-100/TBS for at least 1 hr at 4°C on shaker table.
3. Incubate sections in 2% NGS/0.1% Triton X-100/TBS containing affinity-

purified antibody (final concentration of ~0.5 μg/ml) or control for 12–48 hr at 4°C.

4. Rinse sections 3 × 10 min in cold TBS.
5. Incubate sections in 2% NGS/0.1% Triton X-100/TBS containing 1% goat anti-rabbit sera for 1 hr at 4°C.
6. Rinse sections 3 × 10 min in cold TBS.
7. Incubate sections in 2% NGS/TBS (no Triton) containing 0.5% rabbit peroxidase–anti-peroxidase complexes for 1 hr at 4°C.

Note. Steps 5–7 can be performed with avidin–biotin methods (e.g., ABC Elite) with similar results.

8. Rinse sections 3 × 10 min in cold TBS.
9. Develop peroxidase reaction by incubating sections in 0.05% diaminobenzidine/0.01% hydrogen peroxide for 5–10 min (by visual inspection).
10. Rinse sections 3 × 10 min in cold TBS.
11. Mount sections onto gel-alum subbed slides, air-dry, dehydrate, and coverslip.

References

1. T. F. Freund, J. R. Powell, and A. D. Smith, *Neuroscience* **13,** 1189 (1984).
2. A. M. Graybiel, E. C. Hirsch, and Y. A. Agid, *Proc. Natl. Acad. Sci. U.S.A.* **84,** 303 (1987).
3. B. E. Jones and A. Beaudet, *J. Comp. Neurol.* **261,** 15 (1987).
4. P. Voorn, C. R. Gerfen, and H. J. Groenewegen, *J. Comp. Neurol.* **289,** 189 (1989).
5. A. I. Levey, S. M. Hersch, D. B. Rye, R. Sunahara, H. Niznik, C. A. Kitt, D. L. Price, R. Maggio, M. R. Brann, and B. J. Ciliax, *Proc. Natl. Acad. Sci. U.S.A.* **90,** 8861 (1993).
6. A. I. Levey, C. A. Kitt, W. F. Simonds, D. L. Price, and M. R. Brann, *J. Neurosci.* **11,** 3218 (1991).
7. P. H. Anderson, J. A. Gingrich, M. D. Bates, A. Dearry, P. Falardeau, S. E. Senogles, and M. G. Caron, *Trends Pharmacol. Sci.* **11,** 231 (1990).
8. D. R. Sibley and F. J. Monsma, Jr., *Trends Pharmacol. Sci.* **13,** 61 (1992).
9. J. A. Gingrich and M. G. Caron, *Annu. Rev. Neurosci* **16,** 299 (1993).
10. A. Mansour, J. H. Meador-Woodruff, Q. Zhou, O. Civelli, H. Akil, and S. J. Watson, *Neuroscience* **46,** 959 (1992).
11. D. M. Weiner, A. I. Levey, R. K. Sunahara, H. B. Niznik, B. F. O'Dowd, P. S. Seeman, and M. R. Brann, *Proc. Natl. Acad. Sci. U.S.A.* **88,** 1859 (1991).
12. R. T. Fremeau, Jr., G. E. Duncan, M. G. Fornaretto, A. Dearry, J. A. Gingrich, G. R. Breese, and M. G. Caron, *Proc. Natl. Acad. Sci. U.S.A.* **88,** 3772 (1991).

13. A. Mansour, J. H. Meador-Woodruff, J. R. Bunzow, O. Civelli, H. Akil, and S. J. Watson, *J. Neurosci.* **10,** 2587 (1990).
14. A. I. Levey, T. M. Stormann, and M. R. Brann, *FEBS Lett.* **275,** 65 (1990).
15. P. Stern, *Trends Biotechnol.* **9,** 163 (1991).
16. Z. W. Hall, *Trends Neurosci.* **10**(3), 99 (1987).
17. G. Walter, *J. Immunol. Methods* **88,** 149 (1986).
18. R. A. Lerner, *Adv. Immunol.* **36,** 1 (1984).
19. M. Freissmuth, E. Selzer, S. Marullo, W. Schutz, and A. D. Strosberg, *Proc. Natl. Acad. Sci. U.S.A.* **88,** 8548 (1991).
20. A. I. Levey, W. F. Simonds, A. M. Spiegel, and M. R. Brann, *Soc. Neurosci. Abstr.* **15,** 64 (1989).
21. D. B. Smith and K. S. Johnson, *Gene* **67,** 31 (1988).
22. J. Sambrook, E. F. Fritsch, T. Maniatis, "Molecular Cloning: A Laboratory Manual." Cold Spring Harbor Press, Cold Spring Harbor, NY, 1989.
23. R. J. Kraft, K. S. Tardoff, K. S. Krauter, and L. A. Leinwand, *BioTechniques* **6,** 544 (1988).
24. F. Dorje, A. I. Levey, and M. R. Brann, *Mol. Pharmacol.* **40,** 459 (1991).
25. D. Harlow and D. Lane, "Antibodies: A Laboratory Manual." Cold Spring Harbor Lab., Cold Spring Harbor, NY, 1988.
26. M.-L. Bouthenet, E. Souil, M.-P. Martres, P. Sokoloff, B. Giros, and J.-C. Schwartz, *Brain Res.* **564,** 203 (1991).
27. P. Sokoloff, B. Giros, M. P. Martres, M. Andrieux, R. Besancon, C. Pilon, M. L. Bouthenet, E. Souil, and J. C. Schwartz, *Drug Res.* **42,** 224 (1992).
28. H. H. M. Van Tol, J. R. Bunzow, H.-C. Guan, R. K. Sunahara, P. S. Seeman, H. B. Niznik, and O. Civelli, *Nature (London)* **350,** 610 (1991).
29. R. K. Sunahara, H.-C. Guan, B. F. O'Dowd, P. S. Seeman, L. G. Lauier, G. Ng, S. R. George, J. Torchia, H. H. M. Van Tol, and H. B. Niznik, *Nature (London)* **350,** 614 (1991).
30. B. J. Ciliax, S. M. Hersch, and A. I. Levey, "Dopamine Receptors and Transporters," H. B. Niznik (ed.), p. 385. Marcel Dekker, Inc., New York, 1994.

[21] Development of Antireceptor Antibodies Using Synthetic Peptides

Marjorie A. Ariano and David R. Sibley

Introduction

Drug manipulations of dopaminergic systems in the central nervous system are usual therapies given to alleviate psychomotor disturbances such as Parkinson's disease, Huntington's disease, Tardive dyskinesia, epilepsy, schizophrenia, and cocaine addiction. As the site of drug action, the precise cellular distribution patterns for the dopamine receptors have become an important experimental focus due to the critical role for this neurotransmitter in these prevalent neurological dysfunctions. Pharmacology can distinguish two functional classes of dopamine receptors: the D1 and D2 subfamilies. These are defined by their unique ligand recognition characteristics and by the selective activation of different transduction systems following agonist binding to the receptor. The morphological detection of these two dopamine receptor families can be achieved through analysis of *in vitro* ligand binding using fresh frozen tissue slices (1–4). However the cloning of at least five unique dopamine receptor subtypes (D1A, D1B(rat) = D5(human), D2, D3, D4), two of which contain functional RNA splice variants, has demonstrated a greater response diversity and functional complexity in the effects mediated by the dopamine systems of the central nervous system (5).

The receptor subtypes can be distinguished individually by anatomical techniques like *in situ* hybridization histochemistry, which can localize cells containing the dopamine receptor messenger RNAs (6–9). An alternative technique which examines the expression of the receptor protein, and can detect the five different receptor subtypes, is the generation of polyclonal antisera to defined sequences of the native receptor protein based on the cDNA sequence. In this chapter, we describe the methods we have used to produce selective immunological reagents to the D1A, D2, and D3 subtypes of the rat dopamine receptors. The utility of this approach is apparent from the fact that anti-peptide polyclonal antibodies have been used successfully to distinguish a number of G protein-coupled receptors, i.e., the D1 dopamine receptor (9); the β-adrenergic receptor (10, 11) m1, m2, m3, m4 muscarinic receptors (12–15); and the GluR1 and AMPA-selective glutamate receptors (16, 17).

Methods in Neurosciences, Volume 25

Experimental Methods

Choosing the Antigenic Peptide Sequence

The amino acid sequences for each dopamine receptor subtype were analyzed using IBI-Pustell MacVector DNA/Protein sequence analysis software for areas of predicted antigenicity. Special care was taken to restrict peptide sequences to nonconserved regions of the receptors within the cytoplasmic and extracellular loops, based on the two-dimensional cDNA sequence maps (18). Based on this information, epitopes were chosen from both extracellular and intracellular portions of the receptor proteins with a length of 8–16 residues. Sequences should be produced against both external and internal faces of the receptor protein to minimize the potential for false-positive localization of a homologous peptide sequence in a protein(s) with an enriched neuronal distribution. At least one peptide sequence is generated against an extracellular peptide fragment, since we have found that external epitopes produce better immunochemical staining reagents. The different peptide sequences used as antigenic determinants for the D1A, D2, and D3 dopamine receptor subtypes are illustrated in Fig. 1.

Immunogen Preparation

Peptide sequences are produced by solid-phase methods using an automated synthesizer. The purity and authenticity of the peptide sequences should be determined using HPLC analysis. The peptides which we employed were synthesized by Peninsula Laboratories (Belmont, CA). A terminal cysteine residue is placed at the carboxyl end of each peptide fragment in order to provide a conjugation point for the larger, carrier protein molecule. Conjugation is performed on an equal weight basis using *M*-maleimidobenzoyl-*N*-hydroxysuccinimide to either bovine thyroglobulin or keyhole limpet hemocyanin (both reagents purchased from Sigma Chemical Co., St. Louis, MO).

Immunization Schedule

Prebled New Zealand White rabbits are initially immunized, for each peptide antigen, with 0.5 mg of the peptide conjugate emulsified in Freund's complete adjuvent. After 3 weeks, the rabbits are given booster immunizations every

---

FIG. 1 Peptide sequences and corresponding anti-peptide antisera generated against fragments of the predicted amino acid sequence of (A) D1A, (B) D2, and (C) D3 dopamine receptor subtypes.

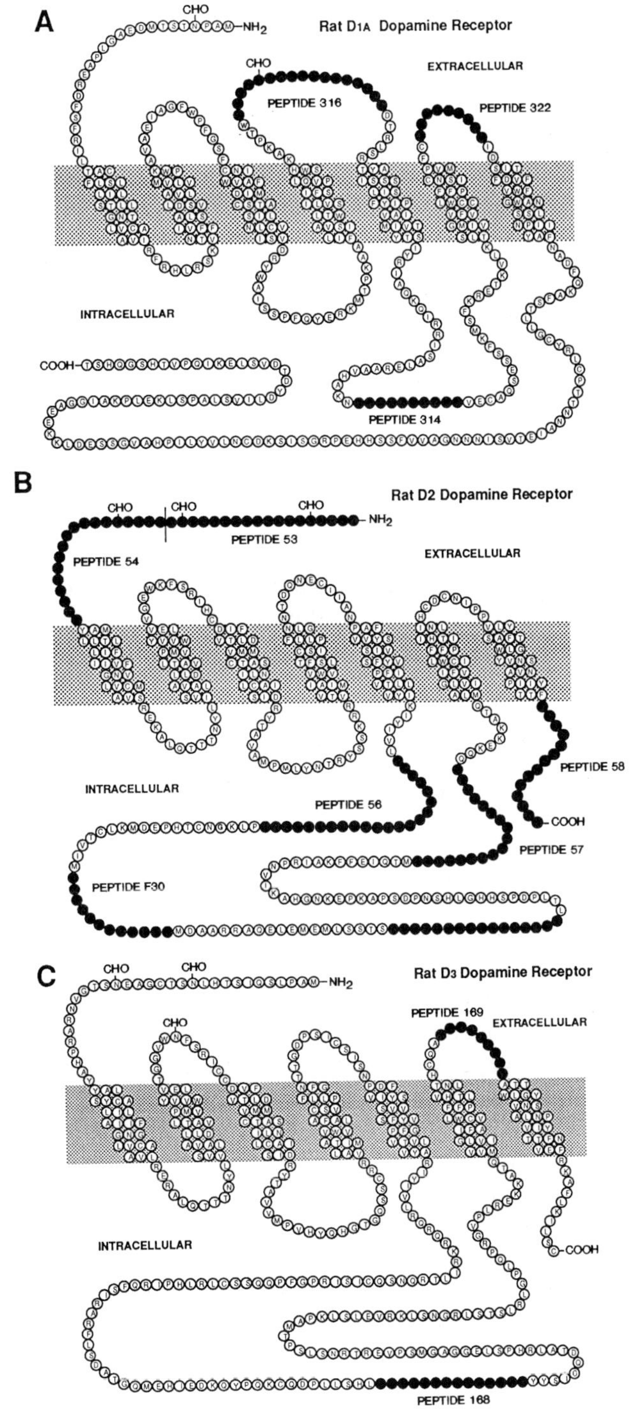
A
Rat D1A Dopamine Receptor
CHO
NH2
CHO
EXTRACELLULAR
PEPTIDE 316
PEPTIDE 322
INTRACELLULAR
COOH
PEPTIDE 314
B
Rat D2 Dopamine Receptor
CHO
CHO
CHO
NH2
PEPTIDE 53
PEPTIDE 54
EXTRACELLULAR
PEPTIDE 58
INTRACELLULAR
PEPTIDE 56
COOH
PEPTIDE 57
PEPTIDE F30
C
CHO
CHO
Rat D3 Dopamine Receptor
NH2
PEPTIDE 169
CHO
EXTRACELLULAR
INTRACELLULAR
COOH
PEPTIDE 168

other week with 0.5 mg of peptide conjugate suspended in incomplete adjuvant until maximum titers of 10^4 to 10^6 are attained; typically 7–13 weeks after the primary inoculation. The rabbit with the highest titer after 13 weeks, which demonstrates immunofluorescent specificity using genetically engineered Chinese hamster ovary (CHO) cells (see below), is maintained for further boosts. The antibody production is monitored by solid-phase ELISA using the synthetic peptides and *p*-nitrophenyl phosphate as the chromagen (19).

Antisera Specificity

Permanently transfected CHO cells are genetically engineered to express the rat D1A, D2, or D3 dopamine receptor subtypes (20). The CHO cell lines contain about 1 pmol/mg protein of receptor expression, as assessed using radiolabeled ligand binding. The CHO cells are cultured onto poly-L-lysine-coated 22-mm circular glass coverslips, grown to confluence, and then used for assessment of the specificity of the different immunological reagents via immunofluorescent processing. The coverslips are placed cell-side up into Costar six-well plastic culture dishes for the immunohistochemical processing. The CHO cells are initially permeabilized to ensure that anti-peptide antisera directed against internal epitopes of the dopamine receptor proteins are accessible to the large immunoglobulin molecules. Omission of this step is a useful control experiment; no immunoreactivity should be detectible for an intracellular epitope without initial CHO permeabilization. Permeabilization of the CHO cells (e.g., fixation in 4% paraformaldehyde with 0.001% saponin for 15 min at 22°C, or freezing at −75°C for 30 min) is followed by three 5-min washes in phosphate-buffered saline (PBS), pH 7.2. Then primary antipeptide antibody is applied overnight at 4°C diluted 1 : 1000 in PBS. Unbound anti-peptide antiserum is washed off using three 15-min washes in PBS at 22°C the following day.

The receptor expression is visualized using commercially available fluorescent-labeled secondary antisera (Jackson ImmunoResearch Labs, West Grove, PA), conjugated to a number of different fluorophores. This secondary antiserum is incubated for 1 hr at 4°C and then rinsed off in 3 × 15 min PBS at 22°C, and the glass coverslips are Superglued to standard glass microscope slides (cell-side up) and covered with glycerine–PBS (9 : 1, v : v) for subsequent examination using the fluorescence photomicroscope. Controls for determining antisera specificity include (i) incubation of untransfected CHO cells with the primary anti-peptide antisera, (ii) use of a number of different anti-peptide antisera, directed against extracellular and intracellular amino acid sequence epitopes of the native receptor protein (iii) substitution of preimmune serum at the experimental dilution (e.g., 1 : 1000 in PBS), (iv) adsorption challenge of the anti-peptide antiserum with its corresponding peptide antigen at 1 mg/ml antiserum followed by incubation at a final experimental dilution of 1 : 1000, (v) omission of the primary antiserum, and (vi) incubation with CHO cells transfected with alternative dopamine receptors, e.g., using a D1A-specific anti-peptide antiserum on a D2-transfected CHO cell line. Immunofluorescent staining following each of these procedures should be absent compared to the primary anti-peptide antiserum incubation on appropriately transfected CHO cells.

Our initial experiments on the D2 dopamine receptor generated eight different anti-receptor antisera to peptide sequences of the native protein. We have subsequently determined that a minimum of one extracellular and one intracellular receptor peptide sequence is sufficient to generate immunologic agents that recognize authentic receptor protein. The deciding factor is principally a financial consideration; the more antigenic determinants which are probed, the more confident one can be with the localization data. The assumption made in generating anti-receptor antisera to both an extracellular and an intracellular epitope is that an unknown peptide sequence within the CNS which has homologous antibody recognition to two unique amino acid sequences of a dopamine receptor would have an extremely low probability of occurrence.

Tissue Immunofluorescence Procedure

Animals are killed by guillotine without prior sedation; the brain is rapidly extracted from the calvaria, frozen, on dry ice and sectioned at 10 μm using a cryostat. Tissues are thaw-mounted onto chrom alum-coated glass microscope slides and allowed to briefly air-dry. Sections are rehydrated in PBS, 3 × 5 min at 22°C and then the primary anti-peptide antiserum is applied at various empirically determined experimental dilutions and incubated overnight at 4°C in a moist environment. Unbound anti-peptide antiserum is rinsed off using 3 × 15 min each in PBS at 22°C the following day. Visualization of the dopamine receptor-expressing elements is accomplished by using the commercially available secondary fluorescent-labeled antiserum, as noted above. A final PBS rinse is performed after the 1-hr incubation at 4°C in a moist environment, then the tissue slices are covered with glycerine–PBS. Routine controls were as follows: (i) the primary antiserum is omitted, (ii) preimmune serum is substituted for the primary antiserum at the experimental dilution, (iii) the primary antiserum is challenged with its peptide antigen and then applied to the tissue section at the experimental dilution, and (iv) CNS areas known to be devoid of dopamine receptor binding sites are examined. No immunofluorescent staining was detected in these control studies.

Perfused Tissue Localization Procedure

Animals are overdosed with sodium pentobarbital and transcardially perfused using cold 4% paraformaldehyde in PBS. The brain is removed from the calvaria, postfixed for an additional hour at 4°C, and then cut at 100 μm thickness using a vibratome. The tissue sections are incubated as free-floating pieces overnight at 22°C with different anti-peptide antisera at various experimental dilutions. Antigen-binding sites can be visualized using fluorescence or peroxidase indirect immunohistochemical methods (21). Alternatively, the perfused, postfixed tissue can be cryoprotected using 20% sucrose, frozen and sectioned at 10 μm. Perfused–frozen tissue slices can be processed for either bright-field (peroxidase) or fluorescence immunohistochemistry of the dopamine receptor sites, as described above.

Photomicrography Techniques

Fluorescent-labeled tissues are examined by epifluorescence using a photomicroscope equipped with a mercury vapor ultraviolet light source and dichroic filters of 545/590 nm for rhodamine fluorophores, 485/520 nm for fluorescein conjugates, and 360/420 nm for AMCA derivatives. Typically, the section is scanned using a 25× glycerine-immersion objective. The experimental results are documented using HP-5(+) black and white film (Ilford, Ciba-Geigy, Paramus, NJ) pushed to 800 to 3200 ASA, depending on the fluorochrome employed; red wavelengths need higher ASA values, while yellow-green- and blue-emitting compounds should be exposed at slower

---

FIG. 2 Dopamine receptor anti-peptide antisera specificity determined using permanently transfected CHO cells. All CHO cultures were processed identically; all photomicrographs were printed using identical exposure and printing parameters. (a) D1A anti-peptide antisera (peptide 322) incubated on CHO cells stably transfected with the full-length cDNA for the D1A dopamine receptor subtype. (b) Corresponding D1A preimmune sera incubated on D1A-transfected CHO cells. (c) D1A anti-peptide antisera incubated on D2-transfected CHO cells. (d) D2 anti-peptide antisera (peptide F30) incubated on CHO cells stably transfected with the full-length cDNA for the D2 dopamine receptor subtype. (e) Corresponding D2 preimmune sera incubated on D2-transfected CHO cells. (f) D2 anti-peptide antisera incubated on untransfected CHO cells. (g) D3 anti-peptide antisera (peptide 169) incubated on CHO cells stably transfected with the full-length cDNA for the D3-dopamine receptor subtype. (h) Corresponding D3 preimmune sera incubated on D3-transfected CHO cells. (i) D3 anti-peptide antisera incubated on untransfected CHO cells.

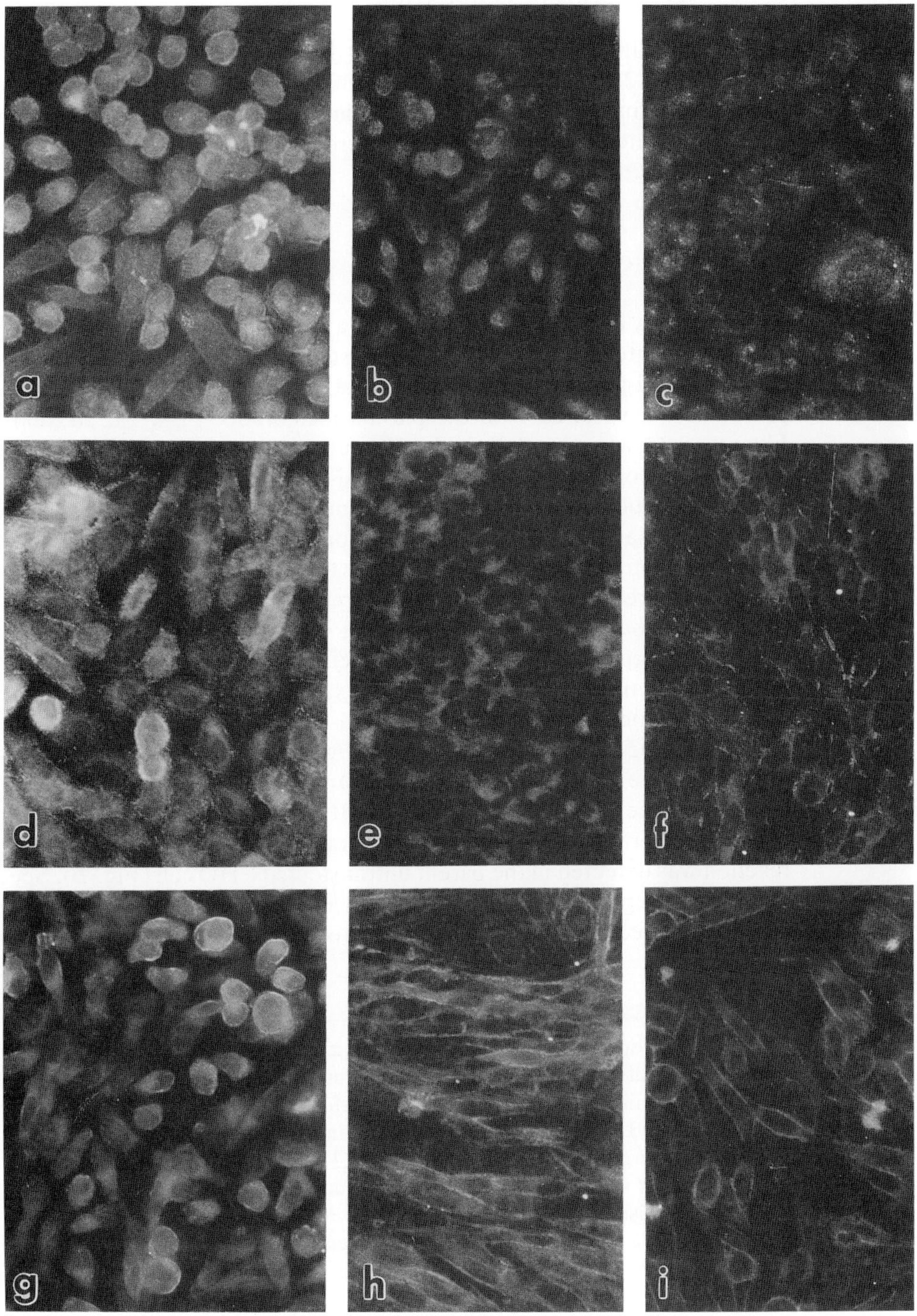
a
b
c
d
e
f
g
h
i

speeds. The film is developed using D-19 developer (Kodak, Rochester, NY) for 3 min at 1 : 2 dilution, in order to increase the contrast of the film negatives. Bright-field immunohistochemistry is documented using a medium-speed black and white film (FP-4, Ilford) and processed according to the manufacturers' instructions.

Results and Discussion

The specificity of the anti-peptide antisera using genetically engineered CHO cells is presented in Fig. 2. Fluorescent staining for the dopamine receptors using the anti-peptide antisera was most robust at the CHO cell surface, presumably associated with the plasma membrane (Figs. 2a, 2d, and 2g). Fluorescence was also diffuse throughout the cytoplasm, corresponding to putative sites of receptor protein synthesis. The immunofluorescent staining pattern was equivalent for each of the anti-peptide antisera generated for a particular dopamine receptor subtype. For the purposes of illustration, only one peptide antiserum from each of the three different dopamine receptor subtypes is presented. Substitution of preimmune serum greatly attenuated the immunofluorescent receptor staining in the cell cultures (Figs. 2b, 2e, and 2h). Specific dopamine receptor detection was only obtained in the appropriately transfected cell type. As examples, (i) D1A anti-peptide antisera only recognize the D1A-transfected CHO cells (Fig. 2a); no reaction was seen in D2-transfected CHO cells (Fig. 2c), (ii) D2 anti-peptide antisera only recognize the D2-transfected CHO cells (Fig. 2d); no reaction was detected in the parent untransfected CHO cell (Fig. 2f), and (iii) D3 anti-peptide antisera only recognize the D3-transfected CHO cells (Fig. 2g); no reaction was detected in the parent untransfected CHO cell (Fig. 2i). Specific dopamine receptor binding in the striatum is about 200–600 fmol/mg protein, and thus the receptors would be considered rare protein components of this

---

FIG. 3 D1A dopamine receptor expression in rat neostriatal tissue slices. Arrow denotes a typical medium-diameter neuron expressing the receptor protein. Asterisk is within a myelinated fiber bundle which perforates the substance of the neuropil. (a) Anti-peptide 322 antisera at 1 : 1000 dilution in 10 μm fresh frozen section processed for immunofluorescence. (b) Anti-peptide 322 antisera at 1 : 500 dilution in 6 μm perfused frozen section processed for peroxidase immunohistochemistry. (c) Anti-peptide 322 antisera at 1 : 1000 dilution, following adsorption challenge with the 322 peptide in 10 μm fresh frozen section processed for immunofluorescence. (d) Preimmune sera from the rabbit used to generate anti-peptide 322 antisera at 1 : 1000 dilution. Calibration bars, 50 μm.

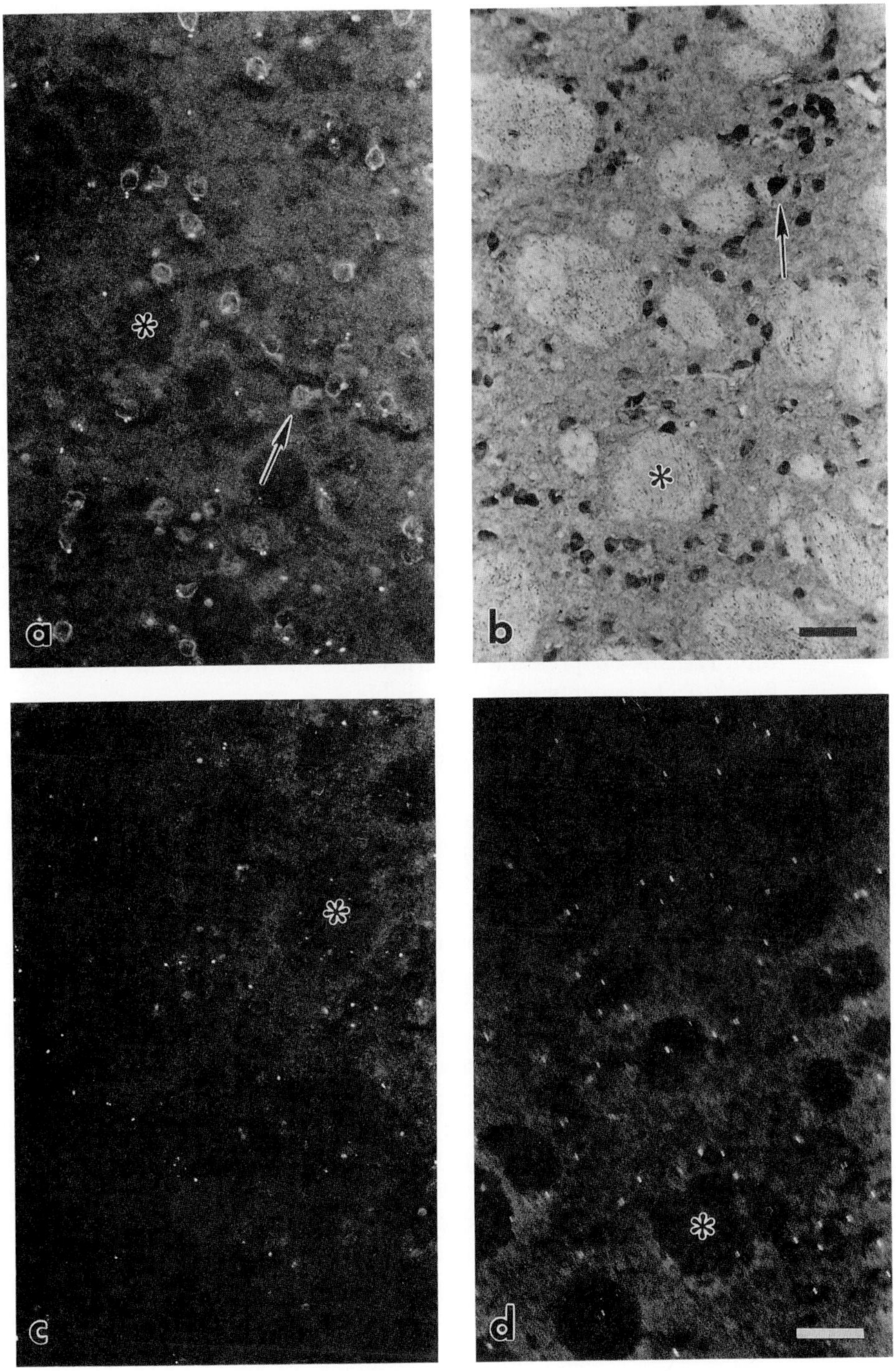
a
b
c
d

basal ganglia structure. Consequently, it is difficult to assess the biochemical selectivity of the antisera in tissue homogenates derived from the neostriatum using Western blot analyses. However, the CHO cell results demonstrate the selectivity of the antisera reagents for their specific dopamine receptor subtype.

The expression of the D1A dopamine receptor subtype in tissue slices is demonstrated in Fig. 3. All three D1A anti-peptide antisera produced homologous staining patterns; however, the extracellular epitopes were more robust in their immunofluorescent reaction. Neostriatal D1A-like expression was visible within medium-diameter neurons (Fig. 3a) and was especially robust at the peripheral border of the cell body. The neuropil was also reactive for this dopamine receptor subtype, while the myelinated fiber bundles (asterisk) which penetrate through the nucleus were not reactive. Adsorption challenge of the anti-peptide antiserum with its corresponding 322 peptide virtually abolished the immunofluorescent staining (Fig. 3c). The occasional punctate element was still visible and would therefore be considered an artifact of the whole sera employed in the experimental paradigm. Incubation with the preimmune sera from the rabbit used to generate the anti-peptide 322 antisera (Fig. 3d), likewise, attenuated the cellular and neuropil D1A-like expression pattern within the neostriatum, but the punctate fluorescent elements were still visible.

A number of experimental advantages can be achieved using immunochemical detection of receptor protein expression in brain slices. The tissues can be cross-linked with aldehyde fixatives and still retain antigenic sites (Fig. 3b). The peroxidase immunohistochemical detection of D1A-like receptor expression demonstrates reaction product within an analogous population of medium-diameter neurons to those visualized in fresh frozen immunofluorescent experiments. Moreover, the resolution is improved enough to distinguish fine punctate grains within the myelinated fiber bundles at this magnification. Electron microscopic observations demonstrate the presence of D1A receptor protein within axons of the striatal projection system (9), and other experimental approaches have determined that anterograde transport of the D1A receptor from the neostriatum to the substantia nigra and globus pallidus occurs (7, 22).

D2-like receptor expression is demonstrated in Fig. 4. Anti-peptide F30, directed against an intracellular epitope unique to the long form of this subtype, shows medium-sized neurons reactive for this dopamine receptor (arrows, Fig. 4a). The fluorescence is especially prevalent at the cell membrane. Myelinated fiber bundles (askerisk) do not exhibit staining, and the neuropil reaction within the neostriatum is less than that seen for D1A expression. Substitution of preimmune sera (Fig. 4b) abolishes the D2-like

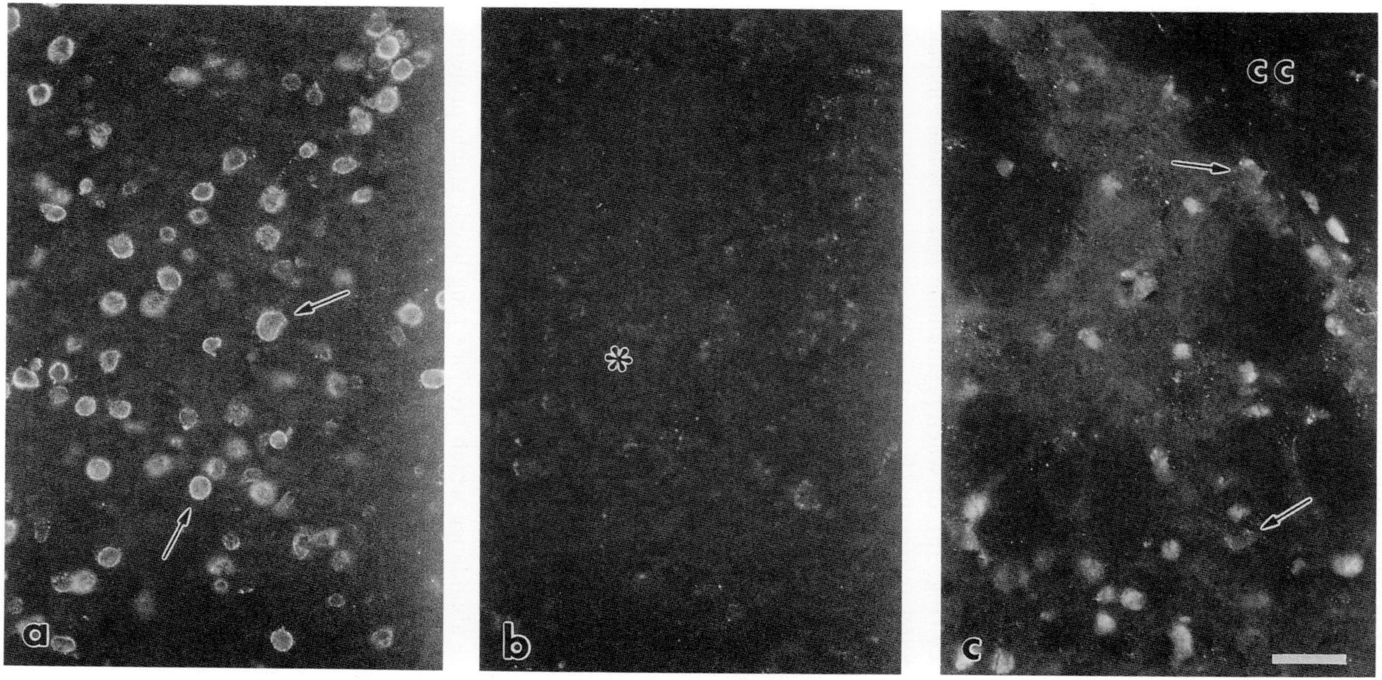

FIG. 4 D2 dopamine receptor expression in rat neostriatal tissue slices. Arrows denote typical medium-diameter neurons expressing the receptor protein. Asterisk is within a myelinated fiber bundle which perforates the substance of the neuropil. (a) Anti-peptide F30 antisera at 1 : 10,000 dilution in 10 μm fresh frozen section processed for immunofluorescence. (b) Preimmune sera from the rabbit used to generate anti-peptide F30 antisera at 1 : 10,000 dilution. (c) Anti-peptide 53 antisera at 1 : 3000 dilution in 6 μm perfused frozen tissue section, processed using immunofluorescence. CC is the corpus callosum. Calibration bar, 50 μm.

receptor labeling. Fixation of the tissue using paraformaldehyde, followed by immunofluorescent detection of the D2 receptor, illustrates medium-diameter neurons (arrow, Fig. 4c), as well as large-sized cells with reactive initial processes (cf. 23, 24). We have also determined that D2-like receptor expression is present in retrogradely identified striatonigral neurons using anti-peptide antisera directed against both intracellular and extracellular portions of the receptor protein (25).

D3-like receptor expression is presented in Fig. 5. Anti-peptide 169, directed toward a portion of the third extracellular loop, demonstrates immunofluorescent staining at the cellular membranes of medium-sized neurons in the neostriatum (Fig. 5a). This selective staining pattern is abolished when the anti-peptide antiserum is challenged with the appropriate peptide 169 sequence (Fig. 5b), or when preimmune serum is used in place of the anti-peptide 169 antiserum (Fig. 5c).

A very important advantage of immunohistochemical detection of receptor protein is the ability to examine subtype-selective forms of the receptor. This cannot be accomplished using *in vitro* autoradiography due to the lack of subtype-selective ligands. Moreover, cellular resolution for the dopamine receptors using radioligands is not as precise as immunohistochemistry, nor amenable to fixation followed by electron microscopic analyses. Alternative experimental methods have employed the cellular detection of the mRNA for the different dopamine receptor subtypes (6–9); however, this does not elucidate whether the mRNA transcript is translated into functional protein within the neuron. *In situ* hybridization histochemistry cannot determine the full extent of the receptors' cellular expression into the dendritic arborizations nor transport of the protein to the terminal fields. A further confound is that *in situ* hybridization may not detect neural populations with low message abundance which produce long-lived, stable proteins.

Receptor immunohistochemistry, using anti-peptide antisera, alleviates many of this experimental difficulties. There are of course caveats associated with this paradigm which must be recognized. These immunologic reagents could elucidate unknown proteins with homologous structure(s) to the receptor peptide sequences in brain slices. This potential limitation can be minimized by using multiple anti-peptide antisera, directed against unique polypeptide fragments in both the extracellular and the intracellular portions of the dopamine receptor protein. In addition, using antisera as dilute as possible in the initial incubation helps to limit false-positive staining reactions.

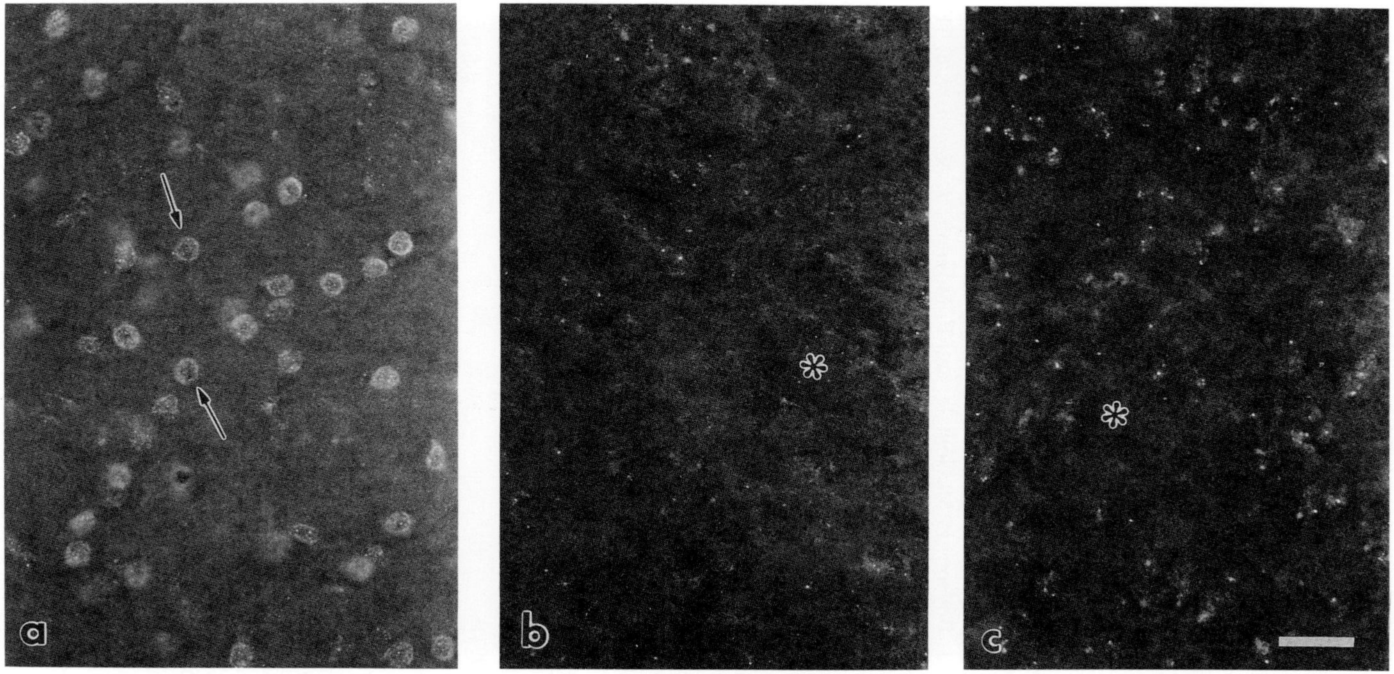

FIG. 5 D3 dopamine receptor expression in rat neostriatal tissue slices. Arrows denote typical medium-diameter neurons expressing the receptor protein. Asterisk is within a myelinated fiber bundle which perforates the substance of the neuropil. (a) Anti-peptide 169 antisera at 1 : 3000 dilution in 10 μm fresh frozen section processed for immunofluorescence. (b) Anti-peptide 169 antisera at 1 : 3000 dilution, following adsorption challenge with the 169 peptide in 10 μm fresh frozen section processed for immunofluorescence. (c) Preimmune sera from the rabbit used to generate anti-peptide 169 antisera at 1 : 3000 dilution. Calibration bar, 50 μm.

Acknowledgments

The technical assistance of Mr. Eric R. Larson is greatly appreciated. Dr. Sara Fuchs generously provided reagents for our screening. This work was supported in part by USPHS NS 23079 (MAA).

References

1. M. A. Ariano, *Brain Res.* **421,** 245 (1987).
2. M. A. Ariano, H. C. Kang, R. P. Haugland, and D. R. Sibley, *Brain Res.* **547,** 20 (1991).
3. S. J. Boyson, P. McGonigle, and P. Molinoff, *J. Neurosci.* **6,** 3177 (1986).
4. S. L. Vincent, Y. Khan, and F. Benes, *J. Neurosci.* **13,** 2551 (1993).
5. D. R. Sibley, F. J. Monsma, Jr., and Y. Shen, *Int. Rev. Neurobiol.* **35,** 391 (1993).
6. C. LeMoine, E. Normand, and B. Bloch, *Proc. Natl. Acad. Sci. U.S.A.* **88,** 4205 (1991).
7. J. H. Meador-Woodruff, A. Mansour, D. K. Grandy, S. P. Damask, O. Civelli, and S. J. Watson, Jr., *Neurosci. Lett.* **145,** 209 (1992).
8. J. H. Meador-Woodruff, A. Mansour, H. H. M. Van Tol, S. J. Watson, Jr., and O. Civelli, *Proc. Natl. Acad. Sci. U.S.A.* **86,** 7625 (1989).
9. Q. Huang, D. Zhou, K. Chase, J. F. Gusella, N. Aronin, and M. DiFiglia, *Proc. Natl. Acad. Sci. U.S.A.* **89,** 11988 (1992).
10. A. Luxembourg, M. Herman, and E. M. Ross, *FEBS Lett.* **283,** 155 (1991).
11. H.-Y. Wang, L. Lipfert, C. C. Malbon, and S. Bahouth, *J. Biol. Chem.* **264,** 14424 (1989).
12. M. I. Fonesca, J. S. Aguilar, A. Skorupa, and W. L. Klein, *Brain Res.* **563,** 163 (1991).
13. G. R. Luthin, J. Harkness, R. P. Artymyshyn, and B. B. Wolfe, *Mol. Pharmacol.* **34,** 327 (1988).
14. S. J. Wall, R. P. Yasuda, F. Hory, S. Flagg, B. M. Martin, E. Ginns, and B. B. Wolfe,, *Mol. Pharmacol.* **39,** 643 (1991).
15. S. J. Wall, R. P. Yasuda, M. J. Li, and B. B. Wolfe, *Mol. Pharmacol.* **40,** 783 (1991).
16. E. Molnár, A. Baude, S. A. Richmond, P. B. Patel, P. Somogyi, and R. A. J. McIlhinney, *Neuroscience* **53,** 307 (1993).
17. R. S. Petralia and R. J. Wenthold, *J. Comp. Neurol.* **318,** 329 (1992).
18. D. R. Sibley and F. J. Monsma, Jr., *Trends Pharmacol. Sci.* **13,** 61 (1992).
19. A. Voller, D. E. Bidwell, and A. Bartlett, "The Enzyme Linked Immunosorbent Assay (ELISA)." Boraugh House, Guernsey, England, 1979.
20. F. J. Monsma, Jr., L. C. Mahan, L. D. McVittie, C. R. Gerfen, and D. R. Sibley, *Proc. Natl. Acad. Sci. U.S.A.* **87,** 6723 (1990).
21. L. A. Sternberger, "Immunocytochemistry." Wiley, New York, 1979.

22. D. Van der Kooy, P. Weinrich, and J. I. Nagy, *Neuroscience* **19,** 139 (1986).
23. L. D. McVittie, M. A. Ariano, and D. R. Sibley, *Proc. Natl. Acad. Sci. U.S.A.* **88,** 1441 (1991).
24. M. A. Ariano, R. S. Fisher, E. Smyk-Randall, D. R. Sibley, and M. S. Levine, *Brain Res.* **609,** 71 (1993).
25. M. A. Ariano, C. J. Stromski, E. Smyk-Randall, and D. R. Sibley, *Neurosci. Lett.* **144,** 215 (1992).

[22] Receptor mRNA Measurement by Multiplex Nuclease Protection Assay

Moshe Jakubowski

Overview

The methods most commonly employed in mRNA analysis can be divided into four groups: (i) filter hybridization techniques, such as Northern blot and dot blot; (ii) solution hybridization-nuclease protection; (iii) reverse transcriptase-polymerase chain raction (RT-PCR); and (iv) *in situ* hybridization histochemistry. These techniques differ in a number of factors: type of probe used, method of signal detection, type and amount of information obtained, specificity, background, sensitivity, quantitative accuracy, and sample handling.

Type of Probe

The types of probe most commonly used in hybridization techniques include DNA, RNA, and deoxyoligonucleotide probes. DNA and RNA probes, usually 100–1000 bases in length, are typically generated using a labeled nucleotide that is incorporated uniformly along the sequence. With deoxyoligonucleotide probes (usually 20–50 bases in length) the labeled nucleotide is added at the end of the molecule. Double-stranded DNA probes are commonly used in dot blot and Northern blot analyses. Single-stranded RNA probes are typically used in nuclease protection and *in situ* hybridization. Deoxyoligomers are required as primers in RT-PCR analysis and are also popular as probes in *in situ* hybridization as well as in dot and Northern blot analyses.

Method of Signal Detection

Solution hybridization-nuclease protection analysis consists of three steps: (i) hybridization in solution of the RNA probe and sample RNA; (ii) nuclease digestion that degrades any nonhybridized single-stranded RNA in the mixture; (iii) gel electrophoresis to resolve the specific target : probe hybrids protected from the nuclease action. In Northern blot analysis, the RNA is first fractionated by gel electrophoresis and transferred onto filter. Target

Methods in Neurosciences, Volume 25

sequences in the filter are then hybridized with the probe, and the background is minimized by posthybridization washing of the filter. In dot blot analysis, the sample RNA is spotted directly on filter, and the filter is then hybridized with probe and washed. The resulting signal and background remain inseparable in the spot of application. *In situ* hybridization employs tissue sections or cell preparations and is carried out under the same principles as filter hybridization. The signal is localized in specific cells that express the mRNA of interest. In RT-PCR, the mRNA sequence is reversed-transcribed to first stand cDNA which is then amplified by PCR between two given points using specific primers. The PCR product is then resolved by gel electrophoresis.

Type and Amount of Information

Solution hybridization-nuclease protection analysis shows the size(s) of the protected region(s) of the probe(s). It is therefore suitable for determination of several target mRNAs simultaneously using multiple probes predesigned to generate protected bands of separable sizes. A RT-PCR analysis shows the size of the target region chosen to be amplified and is also suitable for determination of several target RNAs simultaneously. Northern blot analysis shows the original size of the target RNA, including its poly(A) tail, and is typically performed using a single probe at a time; simultaneous multiple probe analysis is possible only if the target RNAs have distinct native sizes. Dot blot analysis provides no specific characterization of the signal and can be used only with one probe at a time. *In situ* hybridization indicates the cellular localization of the signal. It is suitable for simultaneous localization of two or more RNA sequences, using probes that are labeled and detected differentially (e.g., radioisotopic, colorimetric, fluorometric). *In situ* hybridization allows one to study heterogeneity among populations of cells expressing a given gene in the context of their function; e.g., co-expression of two genes in individual cells.

Specificity

The specificity of the signal depends on the method of signal detection and the specificity of the probe sequence, particularly when directed against a receptor mRNA that shows homology to related receptor mRNAs. With solution hydridization-nuclease protection assay, gel electrophoresis is employed to resolve the protected hybrids, which represent regions of perfect complement between the probe and target RNA. Protected bands generated by related RNAs will be smaller than the specific protected band, due to the nuclease ability to cleave at points of mismatches inside the duplex. This method, therefore, offers absolute specificity even when the probe is partially complementary to related target sequences. On the other hand, filter hybridization, *in situ* hybridization, and RT-PCR require a unique probe sequence in order to maximize signal specificity.

Background

In addition to the obvious effect of probe (or primer) specificity, the control of background level depends on the method of analysis employed. With dot blot, Northern blot, and *in situ* hybridization, the level of background depends ultimately on the stringency of the posthybridization washes. Increased wash stringency reduces the background, but may also decrease the signal. With solution hybridization, the level of background is effectively minimized by nuclease digestion. However, the resistance of the target : probe hybrid to nuclease action varies with the type of nuclease used (see below). With RT-PCR, the background level is primarily a function of the reaction conditions.

Sensitivity

The level of sensitivity increases in the following order: Northern blot, dot blot, nuclease protection, *in situ* hybridization, RT-PCR. With RT-PCR, one can theoretically detect a single copy of target RNA. *In situ* hybridization has a single-cell resolution and can detect about 400 copies of target RNA (1). The supersensitivity of *in situ* hybridization is essentially due to high signal density, namely, the signal is concentrated in individual cells and then magnified using the microscope. Solution hybridization-nuclease protection assay has a limit of detection of about 20 fg target RNA, an equivalent of about 100,000 molecules (2). The sensitivity of Northern blot analysis is considered to be 10–50% of that of solution hybridization-nuclease protection (3). Dot blot may have a sensitivity similar to or slightly better than that of Northern blot.

Quantitative Accuracy

The best quantitative accuracy is offered by solution hybridization-nuclease protection assay. Hybridization in solution is ideal for complete association of probe and target RNA, and the RNase protection step removes background selectively without impinging on the signal. The signal is linearly proportional to the amount of target RNA as long as hybridization is carried out to completion using adequate amount of probe (>10-fold excess of amount of target RNA) and ample time (typically overnight incubation). Reverse transcriptase-polymerase chain reaction analysis is also based on hybridization in solution and can provide high quantitative accuracy depending on the reaction conditions employed. The major drawbacks of this method are that (i) the linear increment in signal may be skewed due to the exponential amplification by PCR and (ii) the resulting signal depends on the polymerase activity which may vary among assays. In filter hybridization, the target RNA is fixed on a solid phase which sets a limit on its accessibility to the probe. Both Northern and dot blot analyses often result in scattered background spots on the filter that confound the signal. With *in situ* hybridization, the target mRNA is embedded in the tissue matrix and is partly blocked by the ribosomes with which it is engaged. This method should be considered semi-quantitative at best.

Sample Handling

One final factor to consider in choosing between solution and filter hybridization methods is the amenability of the technique to processing a large number of samples in the same assay. A small-scale solution hybridization-RNase protection assay (<30 tubes) can be completed leisurely by one person in 2 days: Day 1, synthesis of probe(s) and overnight hybridization; Day 2, all steps from RNase digestion through gel electrophoresis and commencement of overnight autoradiography. A large-scale assay (50–100 samples) can be completed comfortably over 3 days by performing the gel electrophoresis on Day 3. Northern blot and dot blot analyses can be performed in a similar timetable but typically with only one probe at a time, whereas a single multiplex nuclease protection assay can generate data on several different mRNAs simultaneously.

Choosing a Method

It is apparent from the above discussion, that solution hybridization-nuclease protection assay offers the highest specificity and quantitative accuracy and is the most suitable method for multiplex mRNA analysis and utilization of a large number of samples. The sensitivity of the technique is superior to that of filter hybridization methods, though it cannot match the sensitivity of *in situ* hybridization or RT-PCR. The method is suitable for quantitative determination of mRNAs encoding dopamine D2 receptor and gonadotropin-releasing hormone (GnRH) receptors in the rat anterior pituitary gland, using only a small fraction (2–5 μg) of the total RNA isolated from a single gland. On the other hand, the levels of GnRH receptor mRNA in the human ovary and placenta barely reach the detection limit of solution hybridization-nuclease protection, even with 100 μg total RNA analyzed. In such a case, one should opt for RT-PCR or *in situ* hybridization.

Theoretical and Experimental Design Considerations

Tissue Homogenization

The protocol given in this chapter utilizes a double-detergent lysis buffer that ruptures the cytoplasmic compartment of the cell, while rendering the cell nuclei intact. The cell nuclei are removed by low-speed centrifugation, and cytoplasmic RNA is purified from the supernatant. The lysis buffer contains the nonionic detergent Nonidet-P40 which disrupts the cell plasma membrane, and the relatively mild anionic detergent sodium deoxycholate which helps dissociating proteins from nucleic acids, promotes dissolution of lipid membrane structures (4), and facilitates the removal of outer nuclear membrane together with the attached fragments of rough endoplasmic reticulum, thereby maximizing the separation of cytoplasmic RNA from the nuclear fraction (5, 6). This subcellular fractionation provides a simple and a rapid means of removing chromosomal DNA with only minor loss of the cellular mRNA, given that ~95% of the cellular mRNA is present in the cytoplasm. It also allows one to purify and analyze nuclear RNA separately. The method provides high yield of intact RNA from a variety of tissue, such as rat brain and pituitary gland, and does not require RNase inhibitors.

For tissues rich in RNase activity, one should opt for lysis buffers containing >4 *M* guanidinium, which disrupts all cellular compartments, resulting in liberation of all cellular RNA together with chromosomal DNA (7). This approach provides full protection from degradation by RNase (8), but requires extra cumbersome steps to remove the chromosomal DNA in the process of purification of the RNA (7). The more popular and rapid method, using a single phenol/chloroform extraction at low pH, was reported to yield pure RNA with no trace of DNA (9). In our hands, however, this method may not always eliminate DNA contamination entirely and can significantly bias the spectrophotometric determination of total RNA.

DNA Determination

The choice between the two methods of DNA determination described below depends on whether the DNA is intact or degraded, the available amount of DNA, availability of the required spectrometer, and the chemical hazards involved. The colorimetric assay involves hazardous chemicals, it requires a spectrophotometer, and its limit of detection is 3 μg DNA/ml (10). The dye employed in the fluorometric assay is a potential mutagen, the method requires a fluorometer, and it detects as little as 10 ng DNA/ml (11). Both assays are performed with a standard curve using purified reference DNA (free of RNA and protein) isolated from the same species as the samples and quantified by A_{260}.

Template DNA for in Vitro Transcription

The length of the protected fragment is usually designed to be 100–500 bp, in order to accommodate for the separation range of polyacrylamide gel electrophoresis. A probe of corresponding length is transcribed from an insert of template DNA cloned in a plasmid vector between two distinct RNA polymerase promoters. Although each RNA polymerase recognizes its specific promoter sequence, it may also initiate transcription at low level from the opposite promoter, resulting in cosynthesis of two complementary transcripts (12, 13). Therefore, template DNA for antisense RNA is linearized at the 5′ end of the insert and transcribed from the RNA polymerase promoter proximal to the 3′ end of the insert. Conversely, template for sense RNA is linearized at the 3′ end of the insert and transcribed from the promoter proximal to the 5′ end. We strongly recommend electrophoresing the digested DNA through agarose gel and purifying the linearized DNA band from the gel in order to remove traces of uncut plasmid. This effectively minimizes generation of RNase-resistant background bands that result from transcripts complementary to the probe. The problem of cosynthesis of two complementary RNA strands also occurs when the linear template DNA has 3′ overhanging ends (14). Therefore, template DNA should be cut only with a restriction enzyme that produce blunt ends or 5′ protruding ends.

Hybridization Solution

Solution hybridization is carried out at about 25°C below the melting temperature of the expected hybrid, in the presence of an agent that reduces the effective melting point. Hybridization is typically carried out at 45°C in the presence of 80% formamide, 0.4 *M* NaCl (15). We prefer the protocol developed by Thompson and Gillepsie (16) which permits hybridization at 25–30°C in the presence of 4 *M* guanidinium thiocyanate (GuSCN). Although the GuSCN and the formamide protocols are essentially equivalent in terms of sensitivity of the assay (17), hybridization in 4 *M* GuSCN provides the advantage of protecting both the target and probe RNAs from potential degradation by RNase.

Nuclease Protection

Nuclease digestion removes the excess bulk of the radioactive probe molecules that remained unhybridized and the bulk of total RNA from the sample. Nuclease digestion also removes the flanking, noncomplementary single-stranded ends of both the probe and the target RNAs, yielding blunt-end double-stranded fragments. The nuclease of choice for protecting RNA : RNA hybrids is a cocktail of RNase A/RNase T1; it gives optimal signal/noise ratio that remains constant under a wide range of enzyme concentrations, incubation temperatures, and durations of digestion. Using the standard concentrations of RNases A and T1 (15), we have never experienced any problem of underdigestion (high background) or overdigestion (reduced signal due to trimming of the RNA : RNA hybrid). Other nucleases used are S1 nuclease and RNase T2. Protection with these nucleases, however, can result in either underdigestion or overdigestion, depending on enzyme concentration, temperature, and duration of digestion.

Gel Electrophoresis

The protected RNA : RNA hybrids are resolved through nondenaturing polyacrylamide gel electrophoresis. Under these conditions, the target (nonisotopic) strand preserves the uniformly-labeled probe in the face of ongoing radiolysis. Dissociation of the protected RNA : RNA duplex into single-stranded fragments using denaturing gel electrophoresis is not necessary nor is it recommended. Under nondenaturing conditions, the relative electrophoretic mobility of RNA : RNA duplex is almost the same as that of double-stranded DNA fragments. The single-stranded RNA probe typically migrates slower in a nondenaturing gel, compared to the corresponding RNA : RNA fragment.

Amount of RNA Analyzed

The amount of total RNA taken into the analysis depends on the relative abundance of the target transcript in the specimen. When target transcript level is low, it may be necessary to perform the analysis utilizing the entire amount of RNA available from each sample (e.g., GnRH mRNA in the hypothalamus). In other cases it may even be necessary to pool RNA from two or more samples (e.g., proopiomelanocortin pre-mRNA in the mediobasal hypothalamus). Conversely, when target transcript level is high, RNA analysis can be performed utilizing fixed aliquots from each sample (e.g., GnRH receptor mRNA in the anterior pituitary gland). In cases of highly abundant transcripts, the amount of probe synthesized may become limiting to the number of samples that can be analyzed (e.g., prolactin mRNA in the anterior pituitary gland). In such cases, the amount of RNA taken from each sample is minimized according to the available amount of probe, so that the probe is still added in adequate excess of the target RNA to drive hybridization to completion. The decision on the amount of RNA analyzed becomes further compounded when multiple probes are employed simultaneously against transcripts of different abundance. As shown in Fig. 1, each probe is added in adequate excess of its respective target in the tube, and the specific activity of each probe is adjusted in an inverse relationship to its relative amount in the tube, resulting in comparable signals for target transcripts whose levels are as much as 1000-fold different from each other.

Quantitative Analysis of Signal

Phosphorimaging is the most sensitive, most amenable technology available today for quantitative autoradiography. Unfortunately, the system is not available in every institution. An alternative approach is to cut out the protected bands from the gel and measure the radioactivity using β-scintillation counting. This method is cumbersome and time consuming, requiring accurate alignment of visible radioactive landmarks on the gel with their counterpart landmarks on the film and precise excision of the individual bands from the gel (2). The method is also limited by the sensitivity of the β-counter which is inferior to that of phosphorimaging. A third approach involves densitometric analysis of the X-ray image of the gel. This method, however, suffers from a narrow linear range of signal intensity and is by far inferior to phosphorimaging technology, even when several exposure times are employed.

Data Expression

The measured values can be expressed as mass of the protected portion of the target mRNA, as mass of the full-length target mRNA protected, or in terms of moles of the protected mRNA. The latter option is more meaningful in terms of the relative levels of several mRNAs in the same tissue. The

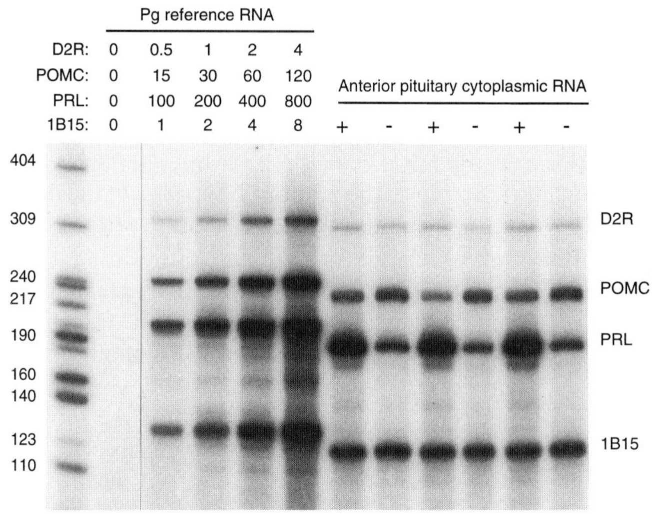

FIG. 1 Multiplex RNase protection assay. Autoradiogram of protected dopamine D2 receptor (D2R), proopiomelanocortin (POMC), prolactin (PRL), and cyclophilin (1B15) mRNA fragments in the cytoplasmic fraction from anterior pituitary gland of ovariectomized rats (−) and ovariectomized rats receiving estradiol 17β replacement (+). Note the differential changes in D2R, POMC, and PRL mRNAs in the face of constant 1B15 mRNA level across the lanes. Three micrograms of total cytoplasmic RNA was hybridized with four RNA probes simultaneously: 300 pg of D2R probe (~1100 cpm/pg), 1000 pg of POMC probe (~50 cpm/pg), 4000 pg of prolactin probe (~10 cpm/pg), and 300 pg of cyclophilin probe (~1100 cpm/pg). Note that PRL and POMC mRNAs were assayed with probes whose specific activities were, respectively, only ~5 and 1% of that of the D2R and 1B15 probes in order to accommodate for the marked difference in the abundance of the four mRNAs in the sample. Transcript levels (pg protected mRNA/μg total RNA) were approximately 0.2 (D2R), 0.8 (1B15), 9 (POMC), and 30–200 (PRL). The sizes of the resulting protected fragments, electrophoresed through 5% polyacrylamide nondenaturing gel, are ~320, ~220, 181, and 111 bp for D2R, POMC, PRL, and 1B15 mRNAs, respectively. Usually, the probe and reference RNAs contain at their ends short complementary sequences derived from the plasmid polylinker and/or synthetic linkers attached to the cloned template DNA, so that the portion of the probe protected by the reference RNA is only slightly longer than that protected by the sample RNA. Double-stranded DNA marker, *Msp* I-digested pBR322 DNA. The X-ray film was exposed to the dried gel for 20 hr at room temperature.

measured mRNA values can then be expressed in terms of content per tissue specimen or normalized relative to total RNA or total DNA, depending on the nature of the tissue and experimental conditions. Tissue such as anterior pituitary gland, ovary, and the developing brain can show significant dynamic changes in cell number and/or cell size as a function of the hormonal milieu, resulting in corresponding changes in total DNA and RNA content in the tissue. In addition, we always include a probe for the mRNA encoding cyclophilin, a housekeeping protein, to validate the specific changes in levels of the mRNA(s) of interest (Fig. 1). The steady-state levels of cyclophilin mRNA in a given tissue remain generally unchanged under a variety of experimental paradigms, but can also be regulated in other situations. Thus, cyclophilin mRNA may or may not show the ideal behavior of an internal *standard* control, but it is always useful to include it in the assay as an internal *reference*.

Specific Methodologies

Chemicals and Enzymes

Magnesium chloride, sodium chloride, calcium chloride, ammonium acetate, EDTA, diethyl pyrocarbonate (DEPC), bisbenzimidazole (Hoechst 33258), diphenylamine, Ficoll, bromophenol blue, xylene cyanole FF, RNase A and RNase T1 were from Sigma. Tris, sucrose, *N,N,N',N'*-tetramethylethylene diamine (TEMED), ammonium persulfate, urea were from ICN. Nonidet-P40 was from Polysciences, deionized formamide from IBI, and double-distilled phenol from Aldrich. Chloroform, isoamyl alcohol, isopropanol, 8-hydroxyquinoline, sodium deoxycholate, acetaldehyde, sulfuric acid, glacial acetic acid, hydrochloric acid, agarose (low EEO), guanidinium thiocyanate (GuSCN), and 40% acrylamide (19 : 1 acrylamide : bisacrylamide) stock solution were from Fisher. Perchloric acid was from Anachemia, ethanol from AAPER Alcohol Chemical, and Freon 11 from Miller-Stephenson. Proteinase K, RNase-free DNase I, and RNase-free yeast tRNA were from Boehringer-Mannheim. RNasin was from Promega. T3, T7, and SP6 bacteriophage RNA polymerases, and *in vitro* transcription kit were from Strategene or Promega. NEN supplied 800 Ci/mmol [α-^{32}P]UTP.

Reagents

Double-detergent lysis solution: 10 m*M* Tris–HCl, pH 7.4, 1.5 m*M* $MgCl_2$, 0.3 *M* sucrose, 0.5% Nonidet-P40, 0.25% sodium deoxycholate. *Cushion solution:* 10 m*M* Tris–HCl, pH 7.4, 1.5 m*M* $MgCl_2$, 0.4 *M* sucrose. Lysis and cushion solutions are used either fresh or within 1 week of preparation (stored at 4°C). *Concentrated (10×)proteinase K buffer:* 100 m*M* Tris–HCl, pH 7.6, 50 m*M* EDTA, 10% SDS (stored at room temperature). *High-salt*

DNase buffer: 10 m*M* Tris–HCl, pH 7.4, 500 m*M* NaCl, 5 m*M* $MgCl_2$, 0.1 m*M* $CaCl_2$ (stored at 4°C). *Hybridization solution:* 4 *M* guanidinium thiocyanate, 0.1 *M* EDTA, pH 7.5 (stored at room temperature). *Concentrated (5×) TBE solution:* 0.45 *M* Tris, 0.45 *M* boric acid, 0.01 *M* EDTA (filtered and stored at room temperature). *Concentrated type II gel-loading buffer (10×):* 25% Ficoll, 0.2 *M* EDTA (pH 8), 0.25% bromophenol blue, 0.25% xylene cyanole FF (stored at room temperature). *Concentrated denaturing gel-loading buffer (100%):* 98% deionized formamide, 10 m*M* EDTA (pH 8), 0.25% bromophenol blue, 0.25% xylene cyanole FF (stored at −20°C). *Working solution of RNase:* 10 m*M* Tris–HCl, pH 7.5, 300 m*M* NaCl, 40 μg/ml RNase A, 2 μg/ml RNase T1 (stored at 4°C). *RNase A and proteinase K stock solutions:* 10 mg/ml in water (stored at −20°C). *Molten double-distilled phenol* containing 0.1% 8-hydroxyquinoline is divided into ~30-ml aliquots in 50-ml Falcon tubes wrapped with aluminum foil and stored at −20°C.

The lysis buffer should be prepared as follows in order to prevent precipitation of magnesium deoxycholate salt. First, dissolve sucrose in water. Next, add crystalline sodium deoxycholate and mix gently to dissolve the detergent. Place the tube in ice for 10 min. Add the appropriate volumes of 1 *M* Tris and 1 *M* $MgCl_2$, mix immediately, and put tube back in ice. Finally, add the appropriate volume of concentrated Nonidet-P40 and vortex well.

Frozen aliquots of phenol are melted at 65°C. For RNA extraction, add ~20 ml TE, pH 8, to ~30 ml molten phenol and shake 15 min at room temperature. This TE-saturated phenol has a final pH ~6.7. For DNA extraction, add 0.5 *M* Tris–HCl, pH 8, and shake 15 min, discard upper aqueous phase, add 0.1 *M* Tris–HCl, pH 8, and shake again. This Tris-saturated phenol has a final pH ~7.8 (it is also suitable for RNA extraction). Store the TE- or Tris-saturated phenol at 4°C and use within 1 month of preparation. Chloroform is mixed with isoamylalcohol at a 24 : 1 ratio and stored at room temperature. To prepare a 1 : 1 phenol : chloroform mixture, add a volume of 24 : 1 chloroform : isoamylalcohol to an equal volume of phenol and remove the aqueous phase formed at the top. This mixture is made at the required volume and used fresh.

All solutions used for extraction of RNA must be RNase free. These solutions are prepared and stored in sterile plastic bottles (Costar) or 50-ml plastic tubes (Falcon). Stock solutions are prepared using molecular-grade reagents and ultrapure deionized (Type 1) water. Alternatively, one can use distilled (Type 2) water that was treated overnight with 0.1% DEPC at 37°C and then autoclaved. Stock solutions are sterilized using 0.2-μm filter (Nalgene or Costar).

Gel Electrophoresis

Nondenaturing, polyacrylamide slab gel is used for analysis of double-stranded RNA protected from RNase digestion. The gel solution is prepared fresh using 40% acrylamide stock solution (usually 5% final concentration), concentrated (5×) TBE solution (1× final concentration) and adjusting the final volume with water. Mix gently and add 60 μl 10% ammonium persulfate and 20 μl TEMED per 10 ml of gel solution, gently mix again, and pour.

Denaturing polyacrylamide gel electrophoresis is used for analysis of single-stranded RNA, i.e., reference and probe RNA. The gel solution is prepared using crystalline urea (7 *M* final concentration), 40% acrylamide stock solution (usually 4% final concentration), and 5× TBE (1× final concentration) and adjusting the final volume with water. Warm up the mixture in a microwave oven for a few seconds (do not heat up), mix to allow the urea to dissolve, filter, and store at 4°C up to 3 months. To cast a gel, take the required volume from the acrylamide/urea stock solution, add 100 μl 10% ammonium persulfate and 10 μl TEMED per 10 ml of gel solution, mix, and pour.

The integrity of total RNA isolated from the tissue can be analyzed using nondenaturing 1% agarose gel electrophoresis, provided the gel and running buffer are not contaminated with RNase. Agarose gel electrophoresis is also used for purifying linearized template DNA (see below).

The type of gel-loading buffer used depends on the type of gel. For nondenaturing gels (polyacrylamide or agarose), use type II loading buffer at a final concentration of 1.5×. For denaturing polyacrylamide gels, use formamide (denaturing) loading buffer (100%) at a final concentration >50% after mixing with RNA sample; denature sample at 80°C for 1 min, quickly spin down, and immediately place in ice prior to loading on denaturing gel.

Collection and Storage of Specimens

Tissue specimens are snap-frozen and stored at −80°C until further processing. Specimens can be kept on ice or ice-chilled glass plate until they are frozen. Snap-freezing is achieved by submerging the tissue for several seconds in liquid Freon prechilled in small beaker on dry ice. Excess of Freon is blotted off the specimen onto tissue paper, and the frozen specimen is placed in a 1.5-ml (or larger) tube prechilled on dry ice. The protocols given below for tissue processing and RNA and DNA purification refer to specimens smaller than ~60 mg wet weight. These protocols should be scaled up for larger specimens.

Tissue Homogenization

The frozen specimen is thawed and homogenized in ice-cold, double-detergent lysis buffer, using >10 μl of lysis buffer per mg tissue. In order to process an entire sample in one 1.5-ml tube through the subsequent steps of RNA isolation, the volume of the homogenate should not exceed 600 μl (up to ~60 mg wet tissue). The tissue is homogenized using disposable 1-ml tuberculin syringes attached to 22-G hypodermic needles (for larger specimens use larger syringes). Remove the syringe plunger, place the frozen specimen inside the syringe barrel, and push it down with the plunger to the bottom of the barrel. Allow the tissue to thaw completely, and then withdraw the appropriate volume of cold lysis buffer through the needle into the syringe. Begin homogenization by expelling the tissue through the needle into a 1.5-ml tube placed on ice. Withdraw the tissue fragments emerging from the needle back into the syringe and expel again into the tube repeatedly about 10 times to bring homogenization to completion. Keep homogenates in the tubes on ice until completing processing of a series of samples (usually 12–18). This method is comparable to homogenization using glass tube/Teflon pestle tissue grinders (2).

Specimens rich with connective tissue (e.g., placenta, gonads) are pulverized using a pestle and mortar on dry ice. The frozen powder is transferred into the barrel of a syringe, thawed, and homogenized in the above lysis buffer, rendering the cell nuclei intact. Alternatively, one can homogenize such specimens using a homogenizer with motor-driven rotating blade, but this results in disruption of the cell nuclei and liberation of DNA and nonfractionated total cellular RNA.

Isolation and Quantitation of Total Cytoplasmic RNA

The homogenate, stored on ice in a 1.5-ml tube, is withdrawn back into the respective syringe homogenizer and transferred to a new tube that contains ~400 μl of ice-cold cushion buffer. Slowly push the plunger of the syringe and gently layer the homogenate on top of the cushion buffer, resulting in two phases due to the difference in sucrose concentrations. Centrifuge for 10 min at ~800*g*, 4°C, using a variable-speed microcentrifuge, to form a nuclear pellet at the bottom of the tube and a cytoplasmic fraction above the cushion phase. Carefully withdraw the cytoplasmic fraction (a volume 10% greater than that of the original homogenate) and transfer to a new 1.5-ml tube placed on ice (keep the pipette tip on top of the upper phase throughout its collection). Add 0.1 vol of 10× proteinase K buffer, 0.02 volumes of 10 mg/ml proteinase K (200 μg/ml final concentration), and 0.03

volumes of 5 *M* NaCl (150 m*M* final concentration). Vortex the mixture and incubate at 45°C for 1 hr. Add 1 volume of 1 : 1 TE-saturated phenol : chloroform mixture, shake 5 min, centrifuge 3 min at room temperature, and transfer upper aqueous phase to a new tube (discard lower organic phase). Add 1 volume of chloroform, shake, centrifuge, and transfer upper phase to a new tube as described above (discard lower organic phase). Add either 2.5 volumes of cold ethanol or 1.5 volumes of cold isopropanol (whichever fits in the tube), vortex thoroughly, and store at −20°C for 2 hr or longer.

Following centrifugation, the RNA pellet is washed with cold 70% ethanol, dried, and resuspended in 50 μl of TE. A 2-μl aliquot (usually 1–2 μg) is diluted in 1 ml 10 m*M* Tris-HCl, pH 8, and total RNA content is determined spectrophotometrically using absorbance at 260 nm (A_{260}). Typically, the A_{260}/A_{280} ratio ranges between 2.0 and 2.2. At this point, aliquots of ~1 μg total RNA can be analyzed for RNA integrity using agarose gel electrophoresis. The amount of total RNA taken into the analysis is precipitated again using salt and ethanol in order to be resuspended in 20 μl hybridization solution; alternatively, the sample can be lyophilized in the SpeedVac Concentrator (Savant) and redissolved in 20 μl hybridization solution.

Isolation of Total Nuclear RNA

Following removal of the cytoplasmic phase, withdraw and discard the remaining cushion phase (keep the pipet tip on top of the cushion phase throughout its collection, so that remaining traces of the cytoplasmic fraction above are collected first, thereby minimizing cross-contamination of the nuclear pellet). The nuclear pellet can be kept on ice or stored at −80°C until subsequent processing. Total RNA can be isolated from the nuclear pellet using a protocol that involves partial DNase digestion or another protocol in which DNA is rendered intact. Total RNA content in the nuclear fraction cannot be measured using A_{260} due to the presence of DNA and/or degradation products resulting from the DNase digestion. Partial DNase digestion minimizes the viscosity of the sample and is the procedure of choice when DNA determination is not required, or when DNA is determined using the diphenylamine-colorimetric method. Isolation of intact DNA is necessary for DNA determination using the bisbenzimidazole (Hoechst 33258)-fluorometric method.

Isolation of Total Nuclear RNA with Partially-Degraded DNA

Take up the nuclear pellet in high-salt DNase buffer (usually 120–150 μl) and vortex thoroughly to disrupt the cell nuclei. Add 40–60 U RNase-free DNase and incubate at 37°C for 10 min. Increase the volume with ultrapure water to adjust NaCl final concentration to 150 m*M*, and add 0.1 volume of 10× proteinase K buffer and 0.03 volumes of 10 mg/ml proteinase K (300 μg/ml final concentration). Vortex the mixture and incubate at 45°C for 2 hr. If required, save an aliquot containing about 5–10 μg DNA and store at −20°C for colorimetric DNA assay. The remaining sample is extracted, precipitated, dried, and resuspended in 20 μl hybridization solution, as per the cytoplasmic fraction.

Isolation of Total Nuclear RNA with Intact DNA

Take up the nuclear pellet in high-salt DNase buffer to disrupt the cell nuclei as described above. However, do not add DNase. Increase the volume and adjust conditions for proteinase K digestion as described above. Extract with 1 volume of 1 : 1 Tris-saturated phenol : chloroform and again with 1 volume of chloroform. However, instead of transferring the upper aqueous phase to a new tube, aspirate and discard the lower organic phase, leaving the interface and upper phase in the tube. This allows maximal recovery of the intact DNA which is trapped in the turbid interface. Following precipitation, the pellet is washed with 70% ethanol, lyophilized, and resuspended in a large volume of TE (e.g., 500 μl) to ensure the DNA is dissolved completely. An aliquot containing about 1–5 μg DNA is then carefully removed and stored at −20°C for fluorometric DNA determination. If nuclear RNA is required, adjust conditions to 5 m*M* $MgCl_2$ and 150 m*M* NaCl, add 40–60 U RNase-free DNase, and incubate at 37°C for 30 min. Extraction, precipitation, and resuspension of nuclear RNA in 20 μl hybridization solution are performed as described above.

Quantitation of Nuclear DNA

Diphenylamine-Colorimetric Method

Each sample and reference DNA (0–60 μg) is adjusted to 5 m*M* NaOH, 0.5 *N* perchloric acid in a final volume of 0.5 ml, incubated at 70°C for 15 min, and cooled down to room temperature. Add 1 ml of diphenylamine solution (see below), vortex, and incubate at 50°C for 5 hr. The resulting blue color is very stable for at least 24 hr at room temperature, and DNA concentration is determined by absorbance at 600 nm, using disposable plastic cuvettes. Diphenylamine solution is prepared by dissolving 1.5 g diphenylamine in 100 ml glacial acetic acid and adding 1.5 ml concentrated sulfuric acid. This reagent is stored at 4°C (where it solidifies). Prior to use, the reagent is melted at room temperature, and 0.5 ml of 1.6% aqueous acetaldehyde is added per 100 ml. The concentrated acetaldehyde (100%) should be stored in an explosion-proof refrigerator.

Bisbenzimidazole (Hoechst 33258)-Fluorometric Method

To each aliquot of sample and reference DNA (0–2 μg), add 1.6 ml of 1 μg/ml Hoechst 33258 in TNE solution (10 m*M* Tris–HCl, pH 7.4, 100 m*M* NaCl, 4 m*M* EDTA). Vortex and measure emission at 470 nm in response to excitation with a 350-nm beam. Fluorescence is stable for at least 24 hr when tubes are kept covered with aluminum foil. The dye solution is prepared when required using 1 mg/ml Hoechst 33258 stock solution (stored at −20°C) and TNE (stored at room temperature).

Preparation of Template DNA for Transcription of RNA in Vitro

Probe (antisense) RNA and reference (sense) RNA are transcribed *in vitro* from template DNA cloned in pBluescript (Stratagene) or pGEM (Promega) vectors. The construct is linearized at the 5′ end of the insert to generate a template for probe RNA, or at the 3′ end of the insert to generate a template for reference RNA. Linearization of the template can be performed only with restriction enzymes that produce either blunt ends or 5′ overhanging ends. Restriction enzymes that produce 3′ overhanging ends should not be used. The restriction reaction is electrophoresed through 0.8% agarose gel to separate the linear molecules from traces of uncut plasmid. The reaction should be loaded at a concentration that does not exceed 1 μg DNA per 20 μl per 0.5-cm wide well (overloading will cause the band of linear plasmid to smear over uncut material). The linearized DNA band is purified from the gel and stored at −20°C.

Synthesis and Quantitation of Reference RNA

For the most consistent results, use a kit of transcription reagents from a commercial source (e.g., Stratagene or Promega). For synthesis of low-specific-activity reference RNA (~1 cpm/pg) set up a 50-μl reaction using 10 μl 5X transcription buffer; 5 μl 100 m*M* DTT; 10 μl 2.5 m*M* ATP/CTP/GTP/UTP mixture; 1–2 μl [α-^{32}P]UTP; 1 μg template DNA; 1 μl 40 U/μl RNasin; 20–50 U of T3, T7, or SP6 RNA polymerase; and ultrapure water. Following incubation at 37°C for 1 hr, add 1 μl of 20 U/μl RNase-free DNase, mix, and incubate for another 15 min at 37°C. Take 3–5 μl for denaturing polyacrylamide gel electrophoresis in order to determine the fraction of full-length transcripts in the preparation (see below). Increase the reaction volume to 200 μl with TE. The incorporation of the nucleotides into RNA is monitored by the [α-^{32}P]UTP tracer using aliquots spotted on DE-81 filters (see below); with 20–40% incorporation, about 6–12 μg of reference RNA is synthesized under the above conditions. Extract reaction once with 1 volume (200 μl) of 1 : 1 phenol : chloroform mixture and again with 1 volume of chloroform. Add 0.5 volume (100 μl) of 7.5 *M* ammonium acetate (2.5 *M* final concentration) plus 2.5 volumes (750 μl) ethanol and incubate at −20°C for 2 hr or longer. Centrifuge at 4°C, wash the tube two times with 1 ml of 70% ethanol, dry the pellet, and resuspend it in 50 μl TE. This procedure precipitates the RNA while removing the free dNMPs resulting from digestion of the template DNA, as well as the unincorporated rNTPs, as verified with DE-81 filters (see below).

Take a 5-μl aliquot from the purified reference RNA preparation to measure RNA content using A_{260} reading (the A_{260}/A_{280} ratio is typically 2.0–2.2). Correct this measurement for the amount of the full-length transcripts in the preparation, as determined by the proportion of radioactivity contained in the full-length product and in the shorter fragments resolved by the gel electrophoresis. The reference RNA is serially diluted in hybridization solution to the desired stock concentration (usually 10–20 times more concentrated than the top concentration in the standard curve) and stored in aliquots at −80°C (do not store at −20°C). For each assay, a new aliquot is thawed and serially diluted to establish the standard curve.

Synthesis of Probe RNA

For synthesis of high-specific-activity probe (~1500 cpm/pg), 15 μl (~0.19 nmol) 800 Ci/mmol, 10 mCi/ml [α-^{32}P]UTP is dried down in the SpeedVac, and the transcription reaction is set up by adding 2 μl 5X transcription buffer, 1 μl 100 m*M* DTT, 2 μl 2.5 m*M* ATP/CTP/GTP mixture, 1–2 μl 0.1 μg/μl template DNA, 0.5 μl 40 U/μl RNasin, 0.5-1 μl containing 5–25 U of the appropriate RNA polymerase, and ultrapure water to a final volume of 10 μl. Following incubation at 37°C for 15–30 min, increase the reaction volume to 150 μl with TE, add 0.8 μl 1 *M* $MgCl_2$ and 1 μl of 20 U/μl RNase-free DNase, mix, spin down, and incubate at 37°C for 15 min. The fraction of the radioactive UTP incorporated into the riboprobe is determined using aliquots spotted on DE-81 filters, as described below. Another aliquot (1 μl) of the newly synthesized probe is resolved through 4% polyacrylamide/7 *M* urea denaturing gel to verify that most of the product consists of full-length transcripts.

To remove the unincorporated [α-^{32}P]UTP, add 20 μg yeast tRNA, 0.5 vol (75 μl) 7.5 *M* ammonium acetate, plus 2.5 volumes (~600 μl) of ethanol. Vortex and incubate at −20°C for 30 min or longer. Centrifuge at 4°C for 10 min, collect the supernatant with a pipet tip, wash the tube with 1 ml cold 70% ethanol, centrifuge again at 4°C for 5 min, and collect supernatant with a pipet tip (both supernatants can be added to an absorbent material and discarded as solid radioactive waste). Finally, dry the RNA pellet and resuspend in 50 μl of hybridization solution. The amount of probe recovered is determined using an aliquot spotted on a DE-81 filter (see below).

Monitoring Incorporation and Removal of Free Nucleotides

Incorporation of radiolabeled nucleotides into nucleic acids is estimated using ~5 × 5-mm DE-81 filters (Whatman). Dilute the transcription reaction with TE to 150–200 μl, vortex, and spin down. Spot 1.5 μl on a DE-81 filter and allow to dry for 5 min. Wash the filter twice for ~5 min in ~5 ml 0.5 *M* sodium phosphate buffer. This filter indicates the amount of radioactivity incorporated. Spot another 1.5-μl aliquot on a second filter to be rendered unwashed (= total radioactivity). Using β-scintillation counting, divide the counts per minute (cpm) in the washed filter by the cpm in the unwashed filter. The resulting quotient is the proportion of [α-^{32}P]UTP incorporated in the RNA.

Estimation of the amount of probe synthesized is important in order to ensure that excess amount of probe is used in the subsequent hybridization reaction. This amount is estimated from the input amount of the [α-^{32}P]UTP (in nmol) multiplied by incorporation quotient (determined with DE-81 filters) and divided by the fraction content of the UTP in the transcribed sequence (assume 0.25 if the sequence is unknown). The result (given in nmol rNMPs incorporated) is then multiplied by 323 (average M.W. of rNMPs), resulting in nanograms of RNA synthesized. The nanomole amount of the input [α-^{32}P]UTP should be calculated based on (a) its specific activity corrected for the day of the reaction and (b) the actual total cpm present in the transcription reaction. This is calculated as follows: (i) 800 Ci/mmol is 0.8 mCi/nmol or, conversely, 1 nmol/0.8 mCi, hence 1.25 nmol/mCi; (ii) 1 mCi = 2.22×10^9 dpm as per the date of calibration indicated by manufacturer; this number is corrected for the date of usage using the decay factor (DF); (iii) nmol input [α-^{32}P]UTP = total cpm in reaction divided by ($2.22 \times 10^9 \times$ DF) and multiplied by 1.25 nmol. In this calculation, the measured cpm is compared to the theoretical dpm because the β-counter typically detects >98% of the disintegrations of ^{32}P. With 50% incorporation, about 100 ng of probe RNA is synthesized under the above conditions. For synthesis of probe of lower specific activity, unlabeled UTP at the appropriate amount is included in the reaction, so that the eventual specific activity is adjusted according to the ratio of [α-^{32}P]UTP to cold UTP.

After precipitating and resuspending the purified probe, spot 1.5 μl on a

DE-81 filter and wash the filter as described above, thus removing traces of unincorporated [α-^{32}P]UTP that might have coprecipitated with the probe. This filter indicates how many of the cpm incorporated originally in the transcription reaction have been recovered (usually 70–100%).

The quantitation of reference RNA is based on A_{260} reading, and it is therefore crucial that the measurement is not biased by the presence of free nucleotides in the preparation. Relying on the [α-^{32}P]UTP tracer included in the transcription reaction, spot two 1.5-μl aliquots of the purified reference RNA on DE-81 filter; wash one filter as described above, and leave the other filter unwashed. The two filters should yield the same amount of radioactivity, indicating the complete removal of the free nucleotides from the reference RNA preparation.

Solution Hybridization–RNase Protection Assay

Depending on the amount of the target RNA, mix 0.2–2 ng of each probe (usually a 20 to 200-fold excess amount) with sample RNA in a 1.5-ml tube at a final volume of 25 μl of hybridization solution. Usually, the sample RNA is dissolved in 20 μl, and the probe (or probe mixture) is added at a volume of 5 μl/tube. For the standard curve, add 5 μl containing the same amount of probe(s) to 20 μl containing varying amounts of the respective reference RNAs, covering the range of the target RNA levels in the samples. In addition, a fixed amount of yeast total RNA comparable to the amount of total RNA per sample is dried down before the reference RNA and the probe(s) are added. Vortex the hybridization mixture, spin down, place the tubes in a floating rack, and incubate for 5 min in a water bath at 55°C [= melting point for duplex RNA at 4 *M* GuSCN (18)]. Following this initial denaturation of probe and target RNAs, transfer the rack with tubes to a 30°C water bath for hybridization overnight.

After overnight hybridization, add 300 μl of RNase solution, vortex thoroughly, and incubate for 1 hr at 30°C (the addition of the RNase solution to the hybridization reaction results in final concentrations of 0.3 *M* GuSCN and 8 m*M* EDTA). After RNase digestion, add 20 μl of 10% SDS and 5 μl of 10 mg/ml proteinase K, vortex thoroughly, and spin down; this will be associated with a white precipitate of SDS (due to the presence of GuSCN). Place tubes in a floating rack and incubate for 1 min at 65°C to allow the precipitate to go back into solution, then at 45°C for 15–30 min. After proteinase K digestion, extract with 1 volume (350 μl) of 1 : 1 TE-saturated phenol : chloroform mixture. Add 12 μg yeast tRNA and 2.5 volumes (900 μl) ethanol, vortex thoroughly, and incubate at −20°C for 1–2 hr, or overnight. Following centrifugation for 10 min at 4°C, decant the supernatant, wash the tube with 0.7 ml cold 70% ethanol, centrifuge 5 min at 4°C, decant the supernatant, dry the pellet, and resuspend it in 7 μl 1.5X gel-loading buffer type II. To ensure that the pellet dissolves completely in the gel-loading solution, keep the samples at room temperature for 10 min, mix well, spin down, incubate at 65°C for 5 min, mix well, and spin down again. Keep tubes at room temperature until gel loading.

The samples are electrophoresed through a nondenaturing polyacrylamide gel (14–20 cm tall, 0.75 mm thick) at 200–250 V for about 2 hr. The gel is mounted on 3M paper (Whatman) and dried using a gel drier. For the purpose of image documentation and reproduction, the gel is exposed to an X-ray film for 1–3 days (usually overnight) at room temperature. Quantitative determination of the signal in the protected bands is performed using phosphorimaging technology and ImageQuant software (Molecular Dynamics). The quantified signals from the protected bands are plotted against the known amounts of reference RNA. This should yield a linear curve from which the respective amounts of target RNA in the samples are computed.

Troubleshooting Common Problems

Homogenate mixes with cushion solution: The lysis and cushion buffers have the same concentration of sucrose (prepare again).

Presence of DNA in cytoplasmic RNA fraction: This can be caused by a number of different factors: (i) the tissue was not allowed to thaw completely before initiation of homogenization; (ii) needle bore is too small (do not use needles of a gauge number higher than 22); (iii) detergent concentrations are too high (do not exceed 0.5% Nonidet-P40 and 0.25% sodium deoxycholate); (iv) slow freezing of tissue, resulting in ice-crystal formation and rupture of nuclear membranes (make sure Freon is prechilled on dry ice before usage).

Total RNA appears to be degraded on agarose gel: Always include a lane of control RNA previously confirmed to be intact. If the control RNA appears degraded too, then the problem may arise from RNase contamination in the gel box, the gel-loading buffer, the running buffer, or the gel itself. If degradation is specific to your samples, then your specimen is likely to contain a high level of RNase activity. In this case, resort to RNA extraction using a GuSCN protocol.

Fluorometric DNA assay yields a saturation curve: This is an indication that there is not enough Hoechst 33258 to detect the DNA quantitatively. At a dye concentration of 1 μg/ml, the signal should remain linear up to ~15 μg DNA.

Probe or reference RNA shows little proportion of full-length transcripts: This may be due to RNA degradation (i.e., RNase contamination) or premature termination of transcription. In the case of high-specific-activity probe reaction, where UTP concentration is restricted (15–25 μM) and nucleotide incorporation is high (greater than 50%), the problem of premature termination of transcription may be alleviated by slowing down the reaction rate. This may be achieved using one or any combination of the following modifications: (i) reducing the concentration of the RNA polymerase; (ii) shortening the duration of the reaction; (iii) increasing the concentration of the limiting NTP; (iv) incubating at 30°C rather than 37°C. Frequently, however, the problem of premature termination of transcription arises from the inability of a given bacteriophage RNA polymerase to transcribe a given mammalian sequence processively. This is particularly evident when the problem occurs in the face of optimal reaction conditions, such as with low-specific-activity probe or reference RNA synthesis. In such a case, one should try to subclone the insert in a different orientation, such that the same sequence can be transcribed using another RNA polymerase. With the availability of different vectors and three different RNA polymerases (SP6, T3, T7), one can find the optimal insert/polymerase combination for any given insert.

Background protected band(s) following RNase digestion: This is an indication that the probe preparation contains transcripts complementary to the probe itself. Such a problem is typically due to either (i) presence of uncut template DNA or (ii) linearizing the template DNA using a restriction enzyme that generates 3′ overhanging ends.

Erratic signal in standard curve: This may be an indication of erroneous preparation of standard curve. However, it may also suggest that sample processing through the course of the assay was not uniform.

Signal becomes saturated at the higher end of standard curve: An indication that the probe was not added at adequate excess amount over target RNA. Try to increase amount of probe, or reduce amount of input RNA. Of course, the signal arising from the samples must fall within the linear range of the standard curve.

Autoradiogram of protected RNA shows high level of radioactivity in wells of the gel: (i) wells were not rinsed adequately prior to sample loading; (ii) RNA pellets were not dissolved thoroughly prior to loading.

References

1. R. T. Fremeau, Jr., D. J. Autelitano, M. Blum, J. Wilcox, and J. L. Roberts, *Mol. Brain Res.* **6,** 197 (1989).
2. M. Jakubowski and J. L. Roberts, *J. Neuroendocrinol.* **4,** 79 (1992).
3. J. E. Krause, J. D. Cremins, M. S., Carter, E. R. Brown, and M. R. MacDonald, *in* "Methods in Enzymology" (P. Conn, ed.), Vol. 168, p. 634. Academic Press, San Diego, CA, 1989.

4. D. M. Wallace, *in* "Methods in Enzymology" (S. L. Berger and A. R. Kimmel, eds.), Vol. 152, p. 33. Academic Press Orlando, FL, 1987.
5. S. Penman, *J. Mol. Biol.* **17,** 117 (1966).
6. E. Holtzman, I. Smith, and S. Penman, *J. Mol. Biol.* **17,** 131 (1966).
7. J. M. Chirgwin, A. E. Przybyla, R. J. MacDonald, and W. J. Rutter, *Biochemistry* **18,** 5294 (1979).
8. M. Sela, C. B. Anfinsen, and W. F. Harrington, *Biochim. Biophys. Acta* **26,** 502 (1957).
9. P. Chomczynski and N. Sacchi, *Anal. Biochem.* **162,** 156 (1987).
10. G. J. Gendimenico, P. L. Bouquin, and K. M. Tramposch, *Anal. Biochem.* **173,** 45 (1988).
11. C. Labarca and K. Paigen, *Anal. Biochem.* **102,** 344 (1980).
12. P. R. Chakraborty, P. Sarkar, H. H. Huang, and U. Maitra, *J. Biol. Chem.* **248,** 6637 (1973).
13. E. T. Butler and M. J. Chamberlin, *J. Biol. Chem.* **257,** 5772 (1982).
14. E. T. Schenborn and R. C. Mierendorf, Jr, *Nucleic Acids Res.* **13,** 6223 (1985).
15. J. Sambrook, E. F. Fritch, and T. Maniatis, "Molecular Cloning: A Laboratory Manual," 2nd ed. Cold Spring Harbor Lab. Press, Cold Spring Harbor, NY, 1989.
16. J. Thompson and D. Gillepsie, *Anal. Biochem.* **163,** 281 (1987).
17. G. S. Firestein, S. M. Gardner, and W. D. Roeder, *Anal. Biochem.* **167,** 381 (1987).
18. M. G. Pellegrino, M. Lewin, W. A. Meyer, III, R. S. Lanciotti, L. Bhaduri-Hauck, K. Volsky, K. Sakai, T. M. Folks, and D. Gillepsie, *BioTechniques* **5,** 452 (1987).

[23] Antisense DNA/RNA-Based Strategies to Analysis of Signal Transduction via G Proteins

Meiling Shih, Christopher M. Moxham, and
Craig C. Malbon

Introduction

Antisense DNA/RNA technology is a powerful approach with which one can probe questions of function by eliminating a "targeted" protein from the cell. The approach can be adapted either to target a specific member within a large family of proteins by choosing a gene sequence that is unique to the protein of interest or to target all members of a family by choosing a gene sequence conserved among family members. In our laboratory we have employed three different types of antisense DNA/RNA-based approaches to address critical questions surrounding signal transduction: (i) oligodeoxynucleotides antisense to a target mRNA for use in cultured cells; (ii) constitutive expression of RNA antisense to a targeted mRNA for cells stably transfected or retrovirally infected in culture; and (iii) inducible, tissue-specific expression of RNA antisense to a targeted mRNA for cells stably transfected *in vitro* or in transgenic mice *in vivo*. Our studies have addressed the role of G proteins in adipocyte differentiation, protein kinases in receptor desensitization, and specific G proteins in the growth and development of transgenic animals. The discussion of antisense DNA/RNA approaches in this chapter includes biochemical analyses used in tandem to characterize the effects of the antisense DNA/RNA molecules.

Selection of the Target Sequence

Specificity and selectivity are the major considerations for selecting target sequences to perform gene suppression. The first step in choosing the target gene sequence is to align the gene sequence of the target protein with those of any related isoforms or subtypes. Depending on the specific aim of the approach (i.e., to specifically knock out one protein or a whole class of proteins), identify the regions of greatest diversity or highest nucleotide identity among the related sequences. For instance, there is a large degree

Methods in Neurosciences, Volume 25

of sequence diversity in the 5′ noncoding region immediately upstream of the ATG among the G protein α-subunits, whereas the translated regions are highly homologous. Therefore, to achieve specific suppression of either $G_{s\alpha}$ or $G_{i\alpha 2}$, the target sequence is confined to the 39 bp in the 5′ noncoding region of the respective genes. On the other hand, to target all known members of the calcium-dependent protein kinase C (PKC) family, a highly conserved 30-bp region among the α-, β-, γ-, and δ-isoforms of PKC located approximately 130 bp downstream from the ATG translation start codon can be selected.

The second step in selecting the target gene sequence is to minimize hybridization of the antisense with other, nontarget mRNAs. Target sequences under consideration always should be compared to the gene sequences available from the GenBank data base. Other alternative target sequences should be explored if there exists a high degree of nucleotide identity to mRNAs of other, nontarget proteins. The third step is to determine the length of the target sequence. Theoretically as the length of the target sequence is increased (>17 bp) the probability for undesirable interactions becomes greater (1, 2). In practice, however, oligodeoxynucleotides of 20 base pairs or longer have been used with equal success (3–5). For expression of antisense RNA in cultured cells or transgenic animals, the length of the antisense RNA displays a wide range from 39 base pairs in the studies described herein to several kilobases (6–8). Based on the theoretical considerations (1, 2) and our experience of scanning the GenBank data base to evaluate the possible crosshybridization with the mRNAs of nontargeted proteins, minimization of any length to 20–40 bp seems a prudent starting point.

Inhibition of Gene Expression by Oligodeoxynucleotides

Differentiation of 3T3-L1 Mouse Fibroblast Cells

It has been found that 3T3-L1 fibroblasts differentiate into adipocytes upon induction with dexamethasone and methylisobutylxanthine (DEX/MIX) (9). During differentiation, the steady-state level of $G_{s\alpha}$ declines significantly. Immunoblot analysis revealed that treating 3T3-L1 fibroblasts with oligodeoxynucleotides antisense to $G_{s\alpha}$ for 3.5 days reduced the steady-state expression of $G_{s\alpha}$ by more than 90%, whereas, oligodeoxynucleotides sense to $G_{s\alpha}$ was ineffective (5). Accompanying the reduction in $G_{s\alpha}$ expression in the fibroblasts treated with the oligodeoxynucleotides antisense to $G_{s\alpha}$ was the extensive accumulation of lipid in the absence of DEX/MIX, inducers of differentiation in these cells. In addition, cells treated with DEX/MIX

TABLE I The Effects of Antisense Oligodeoxynucleotides to $G_{s\alpha}$ on the Differentiation of 3T3-L1 Mouse Fibroblasts to Adipocytes

Incubation time	Treatment	% Differentiation
3.5 days	None	0 (n = 4)
	DEX/MIX	11.7 ± 1.7 (n = 6)
	DEX/MIX + antisense $G_{s\alpha}$	34.7 ± 2.4 (n = 6)
	DEX/MIX + sense $G_{s\alpha}$	13.5 ± 1.7 (n = 6)
7.0 days	None	01.8 ± 0.7 (n = 5)
	DEX/MIX	90.1 ± 1.1 (n = 5)
	Antisense $G_{s\alpha}$	80.0 ± 2.5 (n = 5)
	Sense $G_{s\alpha}$	02.2 ± 0.9 (n = 5)

and the oligodeoxynucleotides antisense to $G_{s\alpha}$ differentiated at a faster rate compared to cells receiving DEX/MIX alone or treated with the oligodeoxynucleotides sense to $G_{s\alpha}$ (Table I). These observations demonstrate the ability of oligodeoxynucleotides antisense to $G_{s\alpha}$ to accelerate the rate of adipogenic differentiation in 3T3-L1 cells and define a critical role for $G_{s\alpha}$ in adipogenesis.

Incubation of 3T3-L1 Cells with Oligodeoxynucleotides

3T3-L1 fibroblasts harvested from culture plates are resuspended in serum-free Dulbecco's modified Eagle's medium (DMEM) and cultured into Lab-Tek chamber slides (Nunc) at a density of 1,500,000 cells/90 μl/chamber. Oligodeoxynucleotides sense and antisense to the 39 bp immediately upstream and including the ATG start codon of $G_{s\alpha}$ are synthesized and HPLC-purified for use in cell culture (Operon, Inc.). The sequences of the sense and antisense oligodeoxynucleotides are 5′-CGCGCCCCGCCGCCGCCATG-3′ and 5′-CATGGCGGCGGCGGGGCGCG-3′, respectively. The oligodeoxynucleotides are dissolved in DMEM to 300 μ*M* as a stock solution and 12 μl of this reagent is added to each cell chamber to give a final concentration of 30 μ*M*. After a 30-min incubation at 37°C, the cells are supplemented with 12 μl of fetal bovine serum (FBS) and treated with or without 6 μl DEX/MIX (250 n*M* and 0.5 m*M* final concentrations, respectively) to induce cell differentiation. The total incubation volume in each chamber is minimized to 120 μl to reduce the consumption of the oligodeoxynucleotides. The incubation continues for 3.5–7.0 days. Every other day, the medium is replaced with DMEM supplemented with 10% FBS and 30 μ*M* of the appropriate oligodeoxynucleotides. Dexamethasone and methylisobutylxanthine are omitted after Day 2.0. Cells in different treatment groups are harvested at Days 1.0, 3.5, and 7.0 with phosphate-buffered saline supplemented with ethylenediaminetetraacetic acid (PBS/EDTA, 10 m*M* Na_2HPO_4, 10 m*M* NaH_2PO_4, 150 m*M* NaCl, pH 7.2, and 2 m*M* EDTA).

Preparation of Crude Membrane Fractions

Cells are lysed with 20 μl of HME buffer containing protease inhibitors [20 m*M* 4-(2-hydroxyethyl)-1-piperazineethanesulfonic acid (HEPES), pH 7.5 2 m*M* $MgCl_2$, 1 m*M* EDTA, 5 μg/ml aprotinin, 5 μg/ml leupeptin, and 0.2 m*M* phenylmethylsulfonyl fluoride] in a 200-μl glass homogenizer (Kontes Glass Inc.). The lysate is centrifuged at 500*g*. The supernatant is then centrifuged in analytical ultracentrifuge (Beckman Optima TLX) at 450,000*g*.

Microscale Sodium Dodecyl Sulfate–Polyacrylamide Gel Electrophoresis (SDS–PAGE)

The pellet from the high-speed centrifugation step is resuspended in a small volume of Laemmli solution (125 m*M* Tris, pH 6.8, 4% SDS, 20% glycerol, and 0.01% bromophenol blue) and heated at 95°C for 2 min. The membrane proteins are resolved by SDS–PAGE using a PHAST system (Pharmacia, Inc.) microelectrophoresis apparatus. One microgram of membrane protein from each sample (4 μl total volume) is loaded on a six-well sample applicator and the samples are automatically applied to a 10% SDS–polyacrylamide gel. After protein separation, the separated proteins are transferred electrophoretically to a nitrocellulose membrane. After transfer, the piece of nitrocellulose is blocked with PBS containing 10% bovine serum albumin (BSA) for 60 min.

Detection of G Protein α-Subunits by Immunoblotting

The $G_{s\alpha}$ anti-peptide antibody is diluted 1 : 200 in PBS/Tween-20 (0.3%) and incubated with the blot at room temperature for 1 hr on a blot rotator. Unbound primary antibody is removed by washing the blot three times with PBS/Tween-20, 30 min each wash. The blot is then incubated with a second antibody conjugated to calf alkaline phosphatase (1 : 1000 dilution) for 1 hr at room temperature. The blot is washed three times with PBS/Tween-20, 10 min each at room temperature. Immune complexes are made visible by incubating the blot with 5 ml of staining solution (50 m*M* glycine, pH 9.0, 1.0 mg *p*-nitrobluetetrazolium chloride, 8 m*M* $MgCl_2$, 100 μl of 5 mg/ml 5-bromo-4-chloro-3-indoyl phosphate dissolved in dimethyl formamide) for 15–30 min at room temperature.

Desensitization of β_2-Adrenergic Receptors in Chinese Hamster Ovary Cells

Desensitization, a rapid decrease in responsiveness to a continuous stimulation, is a general phenomenon among hormonal receptor signaling systems. Cyclic-AMP (cAMP)-dependent protein kinase (PKA) and the β-adrenergic receptor kinase (βARK) have been implicated in the desensitization of the β_2-adrenergic receptor (10). When oligodeoxynucleotides antisense to either PKA or βARK is added to Chinese hamster ovary (CHO) cells stably expressing β_2-adrenergic receptors (11), the agonist-induced receptor desensitization is attenuated, demonstrating a role for these kinases in the desensitization of β_2-adrenergic receptor signaling (3) (Fig. 1).

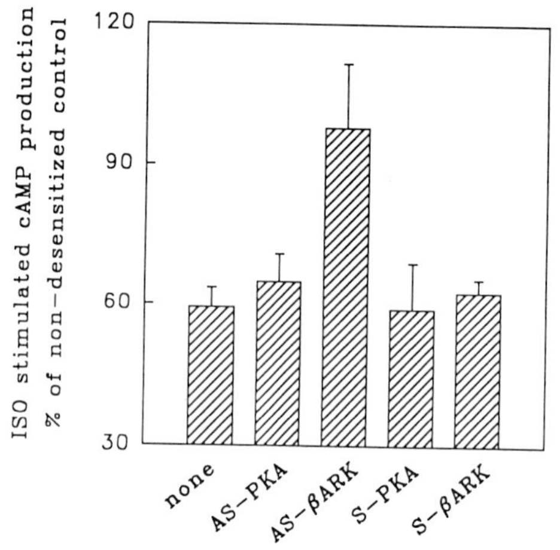

FIG. 1 The effects of sense and antisense oligdeoxynucleotides to PKA and βARK on β_2-adrenergic receptor-mediated cAMP production. Chinese hamster ovary cells stably expressing β_2-adrenergic receptors are grown in the presence or absence of the indicated oligodeoxynucleotide (30 μM) for 2 days prior to the experiments. Cells are then incubated with or without the β-adrenergic agonist isoproterenol (1 μM) for 30 min followed by extensive washing to remove the isoproterenol. The cells are reexposed to medium with or without 1 μM isoproterenol for 15 min. Cyclic-AMP generated in cells challenged twice with agonist is compared to that in nonpretreated cells. The results are expressed as a percentage relative to the non-pretreated control. AS-PKA, antisense oligodeoxynucleotide to cAMP-dependent protein kinase; S-PKA, sense oligodeoxynucleotide to cAMP-dependent protein kinase; AS-βARK, antisense oligodeoxynucleotide to βARK; S-βARK, sense oligodeoxynucleotide to βARK.

Incubation of Chinese Hamster Ovary Cells with Antisense Oligodeoxynucleotides

Chinese hamster ovary cells stably expressing β_2-adrenergic receptors are grown in DMEM containing 10% FBS and 500 μg/ml geneticin (Gibco BRL). Cells are harvested from a 100-mm tissue culture plate with PBS/EDTA and resuspended in serum-free DMEM at a density of 500,000 cells/ml. Oligodeoxynucleotides antisense to the catalytic subunit of the PKA α- and β-isoforms (5′-CTGCTCGCTGCCCTTCTTGG-3′ and 5′-GCGATCGCAGT-GTTCCCCAT-3′, respectively) or βARK (5′-ACCGCCTCCAGGTCCGC-CAT-3′) and the corresponding sense oligodeoxynucleotides are dissolved in DMEM to a final concentration of 300 μM as stock solution. In a 96-well plate, 40 μl of cell suspension and 5 μl of antisense oligodeoxynucleotides are added to each well. After a 30-min incubation at 37°C, the cells are supplemented with 5 μl FBS and the incubation is continued for 48 hr prior to the desensitization assay.

Desensitization Analysis

Cells grown in a 96-well plate are rinsed three times with HEPES-buffered medium (12) (HBM, 10 mM HEPES, pH 7.5, 13.4 mM NaCl, 4.7 mM KCl, 1.2 mM $MgSO_4$, 2.5 mM $NaHCO_3$ and 5 mM, glucose) and incubated with or without the β-adrenergic agonist, isoproterenol (1μM, ISO) at 37°C for 30 min. Cells are then washed three times and treated with the cAMP phosphodiesterase inhibitor, Ro-20-1724 (0.1 mM, Calbiochem Co.) and 0.5 U/ml adenosine deaminase (Sigma Co.) in HBM for 5 min before being rechallenged with 1 μM ISO for 15 min at 37°C. The incubation is terminated by the addition of 2 volumes of 100% ethanol. An aliquot (40 μl) of the ethanolic solution is used for cAMP determination.

Measurement of Cyclic-AMP Accumulation

Cyclic-AMP accumulation is measured according to the procedure described by Brown *et al.* (13) with minor modifications. Briefly, aliquots of the samples are dried under vacuum and reconstituted with 20 μl of KH_2PO_4 buffer (20 mM, pH 6.0). Ten microliters of [^{3}H]cAMP (0.05 μM) is added to each sample as an internal standard to compete with the unlabeled cAMP for binding to a cAMP-binding protein prepared from bovine adrenal cortices. The unbound cAMP is separated by the addition of a solution containing activated charcoal (3.33 mg/ml) and BSA (1.66 mg/ml), followed by centrifugation at 2000g at 4°C. The radioactivity in the supernatant, representing the protein-bound cAMP, is measured by liquid scintillation counting. A standard curve is generated using known amounts of unlabeled cAMP and plotting the recovered radioactivity vs $\log_{10}$ [cAMP]. The amount of cAMP in the samples is determined by extrapolating the recovered radioactivity in the samples from the standard curve.

Vector-Driven Expression of Antisense RNA *in Vitro*

The use of antisense DNA technology for gene suppression is very effective and straightforward for most experiments when performed on a relatively small scale. However, for studies demanding biochemical analyses, it is more practical to introduce vectors containing the antisense RNA sequence of interest into cells by stable transfection or retroviral infection. We have utilized antisense RNA approaches employing vectors designed to produce either abundant, constitutive expression of the antisense RNA or inducible antisense RNA expression.

Expression of Antisense RNA to $G_{i\alpha 2}$ in Retrovirally Infected Mouse F9 Teratocarcinoma Stem Cells

The production of retroviruses harboring a vector capable of stably producing antisense RNA provides a useful means for introduction of antisense RNA into target cells. In our study, the 39-bp sequence antisense to $G_{i\alpha 2}$ was engineered into the retroviral vector pLNCX (Fig. 2). Infection of F9 stem

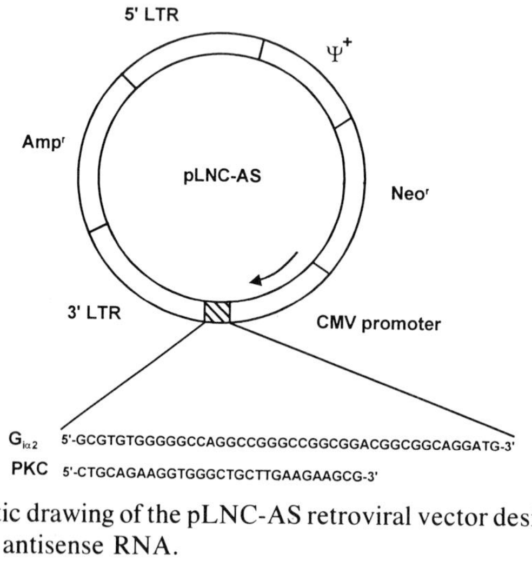

FIG. 2 A schematic drawing of the pLNC-AS retroviral vector designed for constitutive expression of antisense RNA.

cells with retroviruses harboring the antisense RNA construct to $G_{i\alpha 2}$ produced a dramatic decrease in $G_{i\alpha 2}$ expression and attenuated significantly the inhibitory adenylylcyclase response to thrombin (14). In addition the loss of $G_{i\alpha 2}$ expression induced the differentiation of the F9 stem cells to that of primitive endoderm, as determined both morphologically and by the resultant production of tissue plasminogen activator (tPA), a biochemical marker for the differentiated state in these cells (Fig. 3). The differentiation of F9 stem cells to primitive endoderm also can be induced by the morphogen retinoic acid, during which the levels of $G_{i\alpha 2}$ decrease dramatically (15). Thus using antisense RNA technology we have defined a role for $G_{i\alpha 2}$ in the differentiation of F9 stem cells to primitive endoderm and mimicked the effects of retinoic acid on this G protein subunit.

Construction of the pLNC-ASG$_{i\alpha 2}$ Retroviral Vector

The 39 bp of the 5′ noncoding region of $G_{i\alpha 2}$ is engineered into the *Hin*dIII/*Cla*I sites of the pLNCX retroviral vector using standard recombinant DNA techniques (16). The pLNCX vector contains the necessary retroviral packaging sequences (Ψ) and the gene to confer neomycin resistance under the control of the 5′- and 3′-long terminal repeats (LTR) of the mouse Moloney

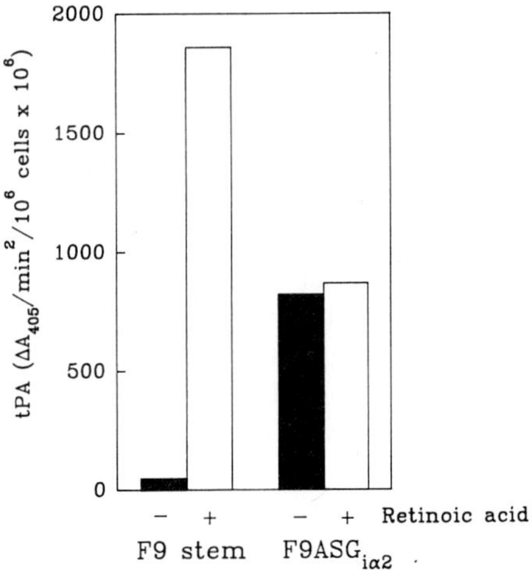

FIG. 3 Suppression of $G_{i\alpha 2}$ expression promotes the differentiation of F9 stem cells. The differentiation state of F9 stem cells infected with the retrovirus harboring the antisense sequence to $G_{i\alpha 2}$ is assessed by the presence of secreted tissue plasminogen activator (tPA) activity into the cell culture media using the amidolytic assay as described in the text.

virus. The expression of the antisense RNA is under the control of the constitutively active cytomegalovirus (CMV) promoter (Fig. 2).

Retroviral Infection of F9 Teratocarcinoma Stem Cells with Retrovirus Harboring the pLNC-ASG$_{i\alpha 2}$ Construct

Recombinant retroviral vectors containing the $G_{i\alpha 2}$ sequence are transfected into Ψ_{GP+E86} packaging cells as previously described (17). Neomycin-resistant colonies are selected for high titer virus production by Northern analysis with a ^{32}P-labeled oligodeoxynucleotide as a probe. Virus from the packaging cells is used to infect F9 stem cells that have been preincubated in hexadimethrine bromide (8 μg/ml, Sigma) for 24 hr (8). Neomycin-resistant transfectants are maintained in DMEM supplemented with 15% FBS and 500 μg/ml G-418 and screened for the expression of $G_{i\alpha 2}$ by immunoblotting with $G_{i\alpha 2}$ anti-peptide antibodies.

Preparation of Crude Membranes

Cells are harvested with PBS/EDTA and pelleted by centrifugation at 1000*g* for 10 min at 4°C. The cell pellet is resuspended in 2 ml of ice-cold HME buffer supplemented with protease inhibitors as described before. The cell suspension is homogenized with 20 strokes of a hand-held Dounce tissue homogenizer, prechilled on ice. The cell homogenate is centrifuged at 750*g* for 10 min at 4°C. The resultant supernatant is centrifuged at 48,000*g* for 20 min at 4°C. The crude membrane pellet is resuspended in 0.5–1.0 ml of ice-cold HME and stored at −70°C until ready for use. The protein concentration is determined by the method of Lowry *et al.* (18).

Sodium Dodecyl Sulfate–Polyacrylamide Gel Electrophoresis and Western Blotting of Crude Membranes

Equal amounts of membrane protein (50–100 μg) are aliquoted into 1.5-ml microcentrifuge tubes and spun for 10 min. The pellet is resuspended in 30 μl of TDS buffer [40 m*M* Tris, pH 6.8, 2m*M* DTT, 2% (w/v) SDS] and placed in a boiling water bath. After 5 min, 10 μl of *N*-ethylmaleimide (25 mg/ml) is added and the samples are incubated at room temperature for 15 min. Prior to electrophoresis, 60 μl of Laemmli solution is added and the samples are boiled for 5 min. The membrane proteins are separated on a 10% SDS–polyacrylamide gel. The electrophoresis of the sample through the stacking gel is performed at a constant current of 20 mA. The current is increased to 30 mA (constant current) as the dye-front enters into the separating gel. Following electrophoresis the separated proteins are transferred to nitrocellulose with a constant current of 0.05 A for 12 hr at 4°C. Immunoblot analysis with polyclonal $G_{i\alpha 2}$ anti-peptide antibodies (1 : 200 dilution) is performed essentially as described in the previous section.

Measurement of Tissue Plasminogen Activator Production

Secretion of tPA from the F9 stem cells and the transfectant clones harboring the pLNC-ASG$_{i\alpha 2}$ construct is measured using the amidolytic assay (19). Cells (5×10^6 cells/dish) are plated in 100-mm dishes containing DMEM supplemented with 15% FBS and maintained at 37°C in the absence or presence of retinoic acid (100 n*M*). At the end of 4 days, the medium is assayed for tPA activity. Following the addition of the plasmin substrate H-D-Val-Leu-Lys-pNA (0.3 m*M*) and Glu- or Lys-plasminogen (0.42 μM), the change in absorbance (ΔA_{405}) at 405 nm is measured at timed intervals using a microplate reader. Tissue plasminogen activator activity is defined as that amount of tPA that results in a reaction rate of $10^{-5}\ \Delta A_{405}\ \text{min}^{-2}$ (change in optical absorbance at 405 nm divided by the square of the time in minutes).

Stable Expression of Antisense RNA to Protein Kinase C in Human Epidermoid Carcinoma Cells (A431)

It has been reported that the treatment of cells with phorbol esters induces the desensitization of β_2-adrenergic receptors (20, 21). In addition, β_2-adrenergic receptors phosphorylated by PKC *in vitro* displayed 60% reduction in their ability to stimulate the GTPase activity of G_s in response to agonist (22), suggesting that PKC contributes to desensitization. To explore these findings further, we expressed an antisense RNA sequence to PKC in human A431 epidermoid carcinoma cells. Cells transfected with the pLNC vector harboring the antisense sequence to PKC displayed a marked reduction in PKC expression (Fig. 4). These PKC-deficient cells are being employed as a model

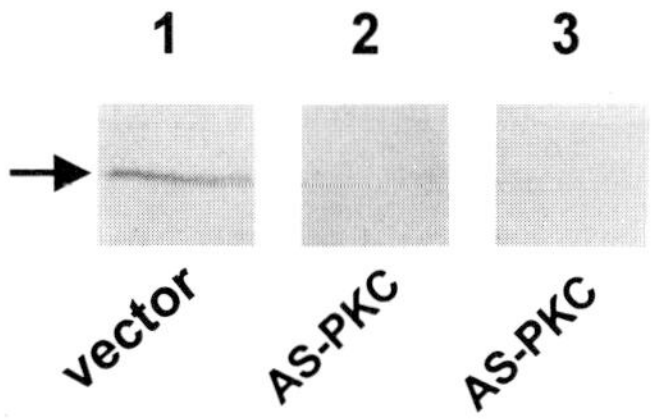

FIG. 4 Immunoblot analysis of protein kinase C (PKC) expression in A431 cells transfected with pLNC-ASPKC. (Lane 1) Cells transfected with the pLNCX vector without the antisense sequence. (Lanes 2 and 3) Two selected clones transfected with the pLNC-ASPKC plasmid. Protein kinase C is probed with antibody specific for α-isoform of PKC.

in which to investigate the modulation of β_2-adrenergic receptor signaling by PKC.

Preparation of the pLNC-ASPKC Construct

A sequence of 29 bases which is 134 bases downstream of the ATG start codon of the human PKC_α gene was chosen as an antisense probe. This sequence is conserved in the α, β, and γ isoforms of PKC and has little homology to other protein kinases. The sense and antisense strands of this sequence are sythesized to include additional bases at both ends to facilitate the subcloning of the oligomer. The two oligodeoxynucleotides are hybridized by heating equimolar amounts of the 5′-phosphorylated oligos to 90°C followed by slow cooling to room temperature. Subcloning of the antisense sequence to PKC into the *Hin*dIII/*Cla*I sites of the pLNCX vector is performed following standard techniques (16). The presence of the insert is confirmed by restriction digestion with *Eae*I. Digestion of the pLNCX vector alone with *Eae*I produces four major fragments with sizes of 3.1, 1.4, 0.5, and 0.4 kb, respectively. Insertion of the antisense sequence introduces one additional *Eae*I site in the 3.1-kb fragment, thereby generating five major fragments, 2.4, 1.4, 0.7, 0.5, and 0.4 kb, respectively. Once the insertion is confirmed, A431 cells are tranfected with the pLNC-ASPKC plasmid using Lipofectin (Gibco BRL) reagent according to the manufacturer's protocol. Positive transfectants are obtained by selection with the neomycin analog G-418 (500 μg/ml).

Identification of Protein Kinase C-deficient Clones

The ability of the pLNC-ASPKC construct to suppress PKC expression in the transfectant clones was determined by immunoblot analysis. Cells collected from a 100-mm plate are transferred to a 1-ml Wheaton glass homogenizer and lysed in 100 μl of ice-cold HME buffer containing protease inhibitors as described in the previous section. Cells are homogenized with 30 strokes and the homogenate is centrifuged at 500*g* for 10 min at 4°C. Fifty micrograms of supernatant protein is subjected to 10% SDS–PAGE and the separated proteins are transferred onto a nitrocellulose membrane. Protein kinase C expression is probed with an antibody (2 μg/ml) raised against a peptide sequence derived from the α-isoform of PKC (Gibco BRL).

Inducible Suppression of $G_{i\alpha2}$ in FTO-2B Cells

We have created an inducible antisense RNA expression systems with applications for both *in vitro* and *in vivo* studies (23, 24). With the pPCK-AS construct, the 39-bp antisense RNA sequence is expressed as a hybrid RNA

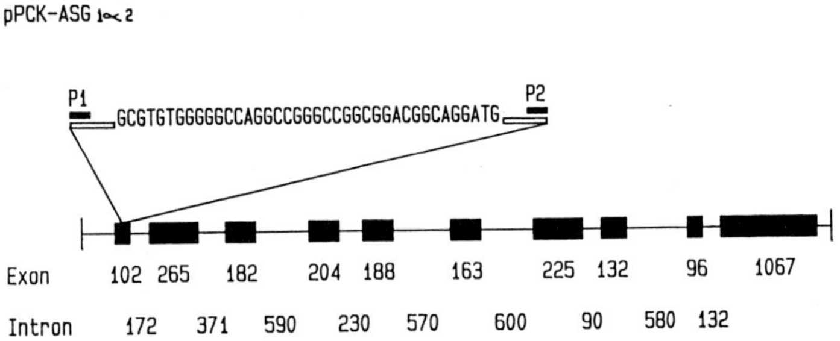

FIG. 5 The pPCK-AS construct designed for inducible and tissue-specific expression *in vitro* and *in vivo*. Reprinted from Moxham *et al.* (23) with permission (copyright AAAS, 1993).

molecule within the larger 2.8-kb mRNA of the phosphoenolpyruvate carboxykinase (PEPCK) gene (Fig. 5). The PEPCK gene was chosen for several reasons (see Table II). Of greatest importance were (i) the lack of gene expression *in utero* (25, 26) which would bypass any deleterious effects of suppressed G protein expression *in utero;* (ii) the tissue-specific expression in liver, kidney, and white adipose tissue (27), tissues with well characterized G protein-linked signaling systems; and (iii) the inducible and suppressible nature of the PEPCK promoter by several different agents which allows for regulated antisense RNA expression (26).

Prior to the production of transgenic mice harboring an antisense RNA construct, it is recommended that the feasibility of the approach be tested

TABLE II Strategy for Suppression of $G_{i\alpha 2}$ *in Vivo* Using Antisense RNA

Consideration	PEPCK gene expression
Developmental restriction	Undetectable expression *in utero* initial appearances after birth
Tissue-specific expression	Predominant expression in liver, kidney, and adipose tissue
Regulated expression	
Inducers	Glucagon, catecholamines (via cAMP)
	Glucocorticoids (liver and kidney)
	Thyroid hormones (synergistic with cAMP)
	Retinoic acid (liver)
	Metabolic acidosis (kidney)
	High-protein diet (via glucagon)
Repressors	Insulin
	Glucocortocoids (adipose)
	Metabolic alkalosis (kidney)
	High-carbohydrate diet (via insulin)

first in cells grown in culture. The FTO-2B rat hepatoma cell line was selected as a screening platform for our pPCK-AS constructs as these cells display cAMP-inducible PEPCK gene expression (28,29) and express $G_{i\alpha2}$. FTO-2B clones transfected with the pPCK-ASG$_{i\alpha2}$ construct displayed normal amounts of $G_{i\alpha2}$ in the absence of cAMP, an inducer of PEPCK gene expression (Fig. 6). $G_{i\alpha2}$ expression declined >85% when these same cells were challenged with the cAMP analog, 8-(4-chlorophenylthio)-cAMP (CPT-cAMP) for 12 days. FTO-2B clones transfected with the vector lacking the antisense sequence to $G_{i\alpha2}$ displayed no change in $G_{i\alpha2}$ expression. In marked contrast to the suppression of $G_{i\alpha2}$, the steady-state levels of $G_{s\alpha}$ and $G_{i\alpha3}$ were not changed in cells expressing the antisense RNA to $G_{i\alpha2}$. These results demonstrate the inducible nature and the specificity of the pPCK-ASG$_{i\alpha2}$ construct as well as the usefulness of the FTO-2B hepatoma cell line as an

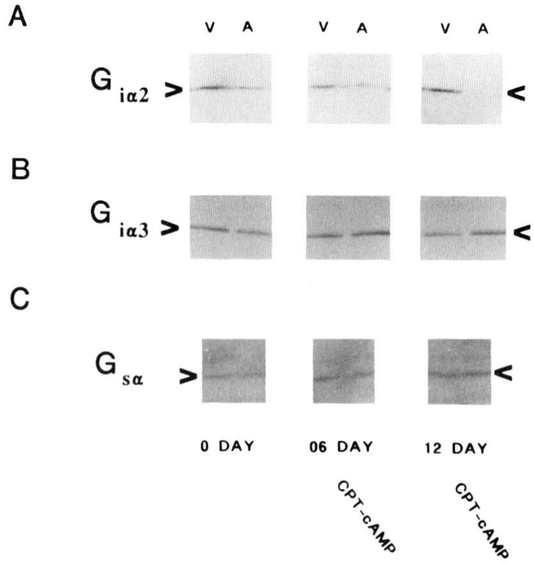

FIG. 6 Cyclic-AMP induces the expression of RNA antisense to $G_{i\alpha2}$ in FTO-2B rat hepatoma cells transfected with pPCK-ASG$_{i\alpha2}$. FTO-2B cells are transfected with either the $G_{i\alpha2}$ antisense RNA construct (A) or vector alone (V) as a control. Neomycin-resistant colonies are selected (G-418, 400 μg/ml) and cultured in the absence or presence of the cAMP analog, CPT-cAMP (25 μM) for either 6 or 12 days. Immunoblot analyses are performed on at least three separate preparations as described and probed with antibodies specific for $G_{i\alpha2}$, $G_{i\alpha3}$, and $G_{s\alpha}$ (A,B,C, respectively). Reprinted from Moxham *et al.* (24) with permission (copyright Wiley-Liss, 1993).

efficient screening system for antisense RNA constructs prior to the production of transgenic mice.

Cell Culture

The cells are maintained in Hams F-12/DMEM (1 : 1) supplemented with 10% FBS at 37°C in an atmosphere of 5% CO_2/95% O_2. Cells are transfected with the vehicle lacking the antisense sequence or the antisense RNA construct using the Lipofectin (Gibco BRL) reagent according to the manufacturer's protocol. Twenty-five micromolar 8-(4-chlorophenylthio)-cAMP (CPT-cAMP) is added for the amount of time indicated and replenished every 24 hr. The CPT-cAMP is prepared in water and sterile-filtered prior to use. On the last day of the induction period, the cells are washed once with ice-cold PBS, harvested with 5 ml of ice-cold PBS/EDTA, and then centrifuged at 1000*g* for 10 min. The preparation of crude membranes from the cells, and the immunoblot analyses with the $G_{i\alpha2}$-, $G_{i\alpha3}$-, or $G_{s\alpha}$-specific antiserum are performed as described for the F9 stem cells above.

Targeted Expression of $G_{i\alpha2}$-Specific Antisense RNA *in Vivo*

As shown in Table II, the target tissues for expression of the $G_{i\alpha2}$ antisense RNA are liver, kidney, and white adipose tissue. Of the three target tissues, adipocytes are easily isolated and possess a well-characterized inhibitory adenylylcyclase response (30, 31). White adipocytes isolated from the pPCK-AS$G_{i\alpha2}$ transgenic mice display a >90% reduction in $G_{i\alpha2}$ expression (23,24). The inhibitory regulation of adenylylcyclase was probed in acutely prepared adipocytes from the transgenic mice using the A1-adenosine receptor agonist (−), R-N^6-phenylisopropyladenosine (R-PIA). In the absence of any inhibitory ligand, cAMP levels were elevated 3.1-fold in the transgenic mice when compared to control mice (Fig. 7A). These data suggest that $G_{i\alpha2}$ exerts a tonic suppression of the basal adenylylcyclase activity *in vivo*. When one compares the inhibitory response to R-PIA in adipocytes from control and transgenic mice, it is apparent that the inhibitory response in the adipocytes from the transgenic mice is severely attenuated after challenge with several different concentrations of R-PIA (Fig. 7B). The residual inhibitory response observed in the adipocytes of the transgenic mice may reflect the residual $G_{i\alpha2}$ or the presence of some other inhibitory G protein such as $G_{i\alpha1}$ or $G_{i\alpha3}$. Not unexpectedly, at higher concentrations of R-PIA which crossover to A2-adenosine receptors, a stimulatory adenylylcyclase response was observed in adipocytes from both transgenic and control mice (Fig. 7B). These results demonstrate that the loss of $G_{i\alpha2}$ is manifest

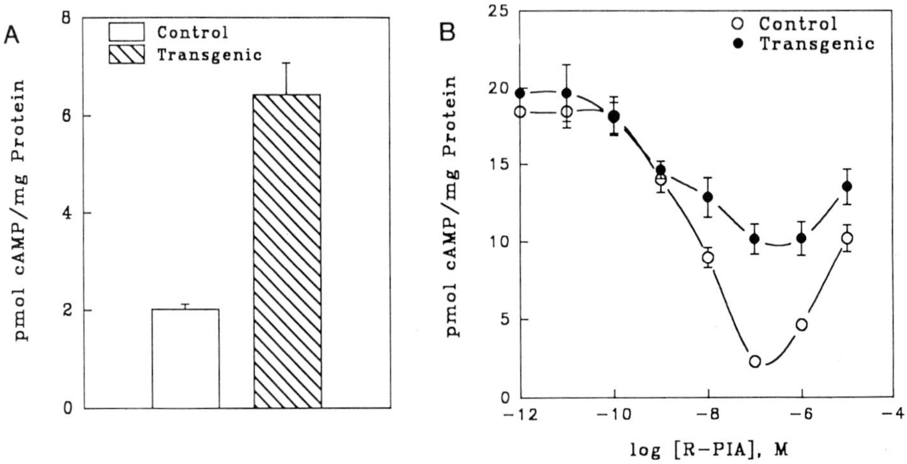

FIG. 7 Inhibitory adenylylcyclase response is attenuated by suppression of $G_{i\alpha 2}$ *in vivo*. The inhibitory adenylylcyclase response is measured using inhibition of cAMP accumulation in adipocytes isolated from mice carrying the pPCK-AS$G_{i\alpha 2}$ transgene and their littermates. Cyclic-AMP accumulation is measured in unstimulated cells (basal, panel A) as well as cells stimulated with 10 μM epinephrine in the absence or presence of increasing concentrations of the inhibitory agonist R-N^6-phenylisopropyladenosine (R-PIA, panel B). The data are expressed as the mean values in pmol of cAMP (± SEM) per mg cellular protein from three independent trials, each performed in triplicate. Reprinted from Moxham *et al.* (24) with permission (copyright Wiley-Liss, 1993).

at the level of the inhibitory adenylylcyclase response, suggesting that *in vivo* $G_{i\alpha 2}$ is the major transducer of the inhibitory adenylylcyclase response in adipose tissue.

White Adipocyte Isolation

White adipocytes are isolated from epididymal and parametrial fat pads by collagenase digestion, as described previously (32). Briefly, 0.5–1.0 g of white adipose tissue is excised from male and female mice, weighed, and added to an equal volume of Krebs–Ringer phosphate (KRP) buffer (120 m*M* NaCl, 4.75 m*M* KCl, 1.2 m*M* Mg_2SO_4, 10 m*M* Na_2HPO_4, 1.2 m*M* $CaCl_2$) containing 3% BSA prewarmed to 37°C. The tissue is digested for 30 min using collagenase (Type I, Worthington, 1 mg/ml) at 37°C in an orbital, shaking water bath. The isolated adipocytes are washed twice with the KRP/BSA buffer and then resuspended to a final concentration of 62.5 mg/ml in the same buffer. The KRP/BSA buffer is supplemented with adenosine deaminase (Sigma) at a concentration of 0.5 U/ml.

Measurement of cAMP Accumulation

Cyclic-AMP accumulation in adipocytes is measured as described previously (33). Briefly, 80 μl of fat cells (~5 mg/tube) are treated with various agents for 6 min at 37°C. The reaction is stopped by the addition of HCl (0.1 *N* final) and boiling for 1 min. The samples are neutralized with NaOH and assayed for cAMP content using a competitive binding assay (13).

Hormonal inhibition of cAMP accumulation is measured in adipocytes stimulated with epinephrine (10 μM) and the indicated concentrations of the inhibitory, A1-adenosine receptor agonist, R-PIA. The results are normalized for total cellular protein added per tube using the protein determination of Lowry *et al.* (18).

Summary

The ability to selectively suppress the expression of specific signaling elements has provided a new strategy with which to probe the complex regulatory networks of signal transduction pathways. From an economical perspective, the use of antisense DNA oligodeoxynucleotides is practical for studies which require relatively small-scale culture of cells, for pilot studies seeking to test the antisense DNA strategy, and for cell systems amenable to single-cell assays (i.e., patch clamping or histochemical analysis). Vector-driven antisense RNA expression, both constitutive and inducible, in cell culture allows for large-scale cell growth capacities enabling biochemical analyses. Expanding the antisense RNA approach to transgenic animals provides the means to generate unique mouse models with which to explore the role of transmembrane signaling elements in complex biological processes *in vivo*.

In our studies, the use of antisense oligodeoxynucleotides for suppression of $G_{s\alpha}$ or for suppression of specific protein kinases provided powerful insights into the roles of these proteins in differentiation and receptor desensitization, respectively. Similarly the role of $G_{i\alpha 2}$ in stem cell differentiation and the role of PKC in receptor desensitization have been addressed in cells stably expressing antisense RNA. Finally, investigation of the role of $G_{i\alpha 2}$ in adipose tissue and liver function as well as its role in whole-body metabolism, growth, and development has been made possible only through the hybrid PEPCK gene construct employed in our laboratory. Using a panel of different antisense DNA/RNA-based approaches, one can explore the roles of signaling elements at several distinctly different levels by selectively suppressing either a single target or a family of targets in cells *in vitro* or in tissues *in vivo*.

Our knowledge of the role of transmembrane signaling elements in disease is growing rapidly. Our success with antisense DNA/RNA-based approaches *in vitro* and *in vivo* highlights the potential applications of this technology for use in gene therapy to treat pathological disorders. The delivery of antisense DNA oligodeoxynucleotides or retroviruses harboring antisense RNA sequences to tissues as well as the ability to express antisense RNA in a narrowly defined and specific set of tissues has great implications not only for our basic understanding of how signal transduction pathways impinge on these complex events but also for the treatment of human disease.

Acknowledgments

We gratefully acknowledge the support provided in part by United States Public Health Service Grants DK25410 and DK30111 from the National Institutes of Health, by American Cancer Society Grant BE-188, and by a National Research Service Award (T32DK07521) as a fellowship (to CMM).

References

1. T. M. Woolf, D. A. Melton, and C. G. B. Jennings, *Proc. Natl. Acad. Sci. U.S.A.* **89,** 7350 (1992).
2. B. P. Monia, J. F. Johonston, D. J. Ecker, M. A. Zounes, W. F. Lima, and S. M. Freier, *J. Biol. Chem.* **267,** 19954 (1992).
3. M. Shih and C. C. Malbon, *Proc. Natl. Acad. Sci. U.S.A.* (1994) in press.
4. C. Kleuss, H. Scherubl, J. Hescheler, G. Schultz, and B. Wittig, *Science* **259,** 832 (1993).
5. H. Y. Wang, D. C. Watkins, and C. C. Malbon, *Nature (London)* **358,** 334 (1992).
6. C. Godson, K. S. Bell, and P. A. Insel, *J. Biol. Chem.* **268,** 11946 (1993).
7. M. Katsuki, M. Sato, M. Kimura, M. Yokoyama, K. Kobayashi, and T. Nomura, *Science* **241,** 593 (1988).
8. M.-C. Pepin, F. Pothier, and N. Barden, *Nature (London)* **355,** 725 (1992).
9. H. Green and O. Kehinde, *Cell (Cambridge, Mass.)* **5,** 19 (1975).
10. W. P. Hausdorff, M. G. Caron, and R. J. Lefkowitz, *FASEB J.* **4,** 2881 (1990).
11. S. T. Geroge, M. Berrios, J. R. Hadcock, H. Y. Wang, and C. C. Malbon, *Biochem. Biophys. Res. Commun.* **150,** 665 (1988).
12. K. Sho, F. Okajima, M. A. Majid, and Y. Kondo, *J. Biol. Chem.* **266,** 12180 (1991).
13. B. L. Brown, D. M. Albano, R. P. Ekins, and A. M. Sgherzi, *Biochem. J.* **121,** 561 (1971).
14. D. C. Watkins, G. L. Johnson, and C. C. Malbon, *Science* **258,** 1373 (1992).
15. P. Galvin-Parton, D. C. Watkins, and C. C. Malbon, *J. Biol. Chem.* **265,** 17771 (1990).
16. J. Sambrook, E. F. Fritsch, and T. Maniatis, Southern hybridization, "Molecular

Cloning: A Laboratory Manual," 2nd ed. Cold Spring Harbor Lab., Cold Spring Harbor, NY, 1989.
17. D. Markowitz, S. Goff, and A. Bank, *Virology* **167,** 400 (1988).
18. O. H. Lowry, N. J. Rosebrough, A. L. Farr, and R. J. Randall, *J. Biol. Chem.* **193,** 265 (1951).
19. P. Andrade-Gordon and S. Strickland, *Biochemistry* **25,** 4033 (1986).
20. J. A. Johnson, T. J. Goka, and R. B. Clarke, *J. Cyclic Nucleotide Protein Phosphorylation Res.* **11**(3), 199 (1986).
21. J. A. Johnson, R. B. Clark, J. Friedman, R. A. F. Dixon, and C. D. Strader, *Mol. Pharmacol.* **38,** 289 (1990).
22. J. Pitcher, M. J. Lohse, J. Codina, M. G. Caron, and R. Lefkowitz, *Biochemistry* **31,** 3193 (1992).
23. C. M. Moxham, Y. Hod, and C. C. Malbon, *Science* **260,** 991 (1993).
24. C. M. Moxham, Y. Hod, and C. C. Malbon, *Dev. Genet.* **14,** 266 (1993).
25. J. P. Garcia Ruiz, R. Ingram, and R. W. Hanson, *Proc. Natl. Acad. Sci. U.S.A.* **75,** 4189 (1978).
26. F. J. Ballard and R. W. Hanson, *Biochem. J.* **104,** 866 (1967).
27. R. W. Hanson and M. A. Mehlman, "Gluconeogenesis: Its Regulation in Mammalian Species" Wiley, New York, 1976.
28. Y. Hod, and R. W. Hanson, *J. Biol. Chem.* **263,** 7747 (1988).
29. J. Liu, E. A. Park, A. L. Gurney, W. J. Roesler, and R. W. Hanson, *J. Biol. Chem.* **266,** 19095 (1991).
30. J. N. Fain and C. C. Malbon, *Mol. Cell Biochem.* **25,** 143 (1979).
31. C. C. Malbon, P. J. Rapiejko, and T. J. Mangano, *J. Biol. Chem.* **260,** 2558 (1985).
32. C. C. Malbon, R. C. Hert, and J. N. Fain, *J. Biol. Chem.* **253,** 3114 (1978).
33. M. P. Czech, C. C. Malbon, K. Kerman, W. Gitomer, and P. F. Pilch, *J. Clin. Invest.* **66,** 574 (1980).

Index